MODERN CURRICULUM PRESS
MATHEMATICS

Teacher's Edition

Level D

MW00875141

Royce Hargrove   **Richard Monnard**

Acknowledgments

Content Writers
Babs Bell Hajdusiewicz
Phyllis Rosner
Laurel Sherman

Contributors
Linda Gojak
William Hunt
Christine Bhargava
Jean Laird
Roger Smalley
Erdine Bajbus
Rita Kuhar
Vicki Palisin
Jeanne White
Kathleen M. Becks
Jean Antonelli
Sandra J. Heldman
Susan McKenney
Nancy Toth
Nancy Ross
Connie Gorius
Denise Smith

Project Director   Dorothy A. Kirk
Editors   Martha Geyen
                       Phyllis Sibbing
Editorial Staff   Sharon M. Marosi
                       Ann Marie Murray
                       Patricia Kozak
                       Ruth Ziccardi
Design   The Remen-Willis
                       Design Group

Cover Art . . . . © 1993 Adam Peiperl

ISBN 0-8136-3119-X (Teacher's Edition)   **ISBN 0-8136-3112-2** (Pupil's Edition)

4   5   6   7   8   9   10   98   97

K

Modern Curriculum Press

Mathematics

A

Modern Curriculum Press

Mathematics

B

Modern Curriculum Press

Mathematics

D

Modern Curriculum Press

Mathematics

E

Modern Curriculum Press

Mathematics

F

Modern Curriculum Press

Mathematics

C

Modern Curriculum Press

Mathematics

Teacher's Guide

C

Modern Curriculum Press

Mathematics

INTRODUCING
MODERN CURRICULUM PRESS MATHEMATICS

A COMPLETE, ECONOMICAL MATH SERIES TEACHING PROBLEM-SOLVING STRATEGIES, CRITICAL-THINKING SKILLS, ESTIMATION, MENTAL-MATH SKILLS, AND ALL BASIC MATH CONCEPTS AND SKILLS!

Modern Curriculum Press Mathematics is an alternative basal program for students in grades K-6. This unique developmental series is perfect for providing the flexibility teachers need for ability grouping. Its design encourages thinking skills, active participation, and mastery of skills within the context of problem-solving situations, abundant practice to master those skills, developed models students actively work with to solve problems, and reinforcement of problem-solving and strategies. Other features like these provide students with solid math instruction.

- Each lesson begins with a developed model that teaches algorithms and concepts in a problem-solving situation.

- Students are required to interact with the model by gathering data needed to solve the problem.

- A developmental sequence introduces and extends skills taught in the basal curriculum—including statistics, logic, and probability.

- An abundant practice of math skills ensures true mastery of mathematics.

- Estimation and mental math skills are stressed in all computational and problem-solving activities.

- Calculator activities introduce students to basic calculator skills and terms.

- Comprehensive **Teacher's Editions** provide abundant additional help for teachers in features like **Correcting Common Errors, Enrichment,** and **Extra Credit,** and the complete **Table of Common Errors.**

Modern Curriculum Press Mathematics is a comprehensive math program that will help students develop a solid mathematics background. This special sampler will show you how:

- **Developed Models** begin each lesson, demonstrate the algorithm and concept in a problem-solving situation, and get students actively involved with the model.

- **Getting Started** provides samples of the concept or skill that is taught and allows the teacher to observe students' understanding.

- **Practice, Apply,** and **Copy and Do** activities develop independent skills where students practice the algorithm and apply what they have learned in the lesson or from a previous lesson. **Excursion** activities extend the math skill and are fun to do.

- **Problem Solving** pages introduce students to the techniques of problem solving using a four-step model. **Apply** activities on these pages allow students to use problem-solving strategies they have learned in everyday situations. The second half of the page focuses on higher-order thinking skills.

- **Chapter Test** pages provide both students and teachers with a checkpoint that tests all the skills taught in the chapter. There are alternative Chapter Tests based on the same objectives at the end of each student book.

- **Cumulative Review** pages maintain skills that have been taught not only in the previous chapter, but all skills taught up to this point. A standardized test format is used beginning at the middle of the second grade text.

- **Calculator** pages teach students the various functions and the basic skills needed to use calculators intelligently.

- **Teacher Edition** pages feature reduced student pages with answers, objectives, suggestions for **Teaching the Lesson, Materials, Correcting Common Errors, Enrichment,** and more.

A DEVELOPED MODEL GETS STUDENTS TO THINK, ACTIVELY PARTICIPATE, AND UNDERSTAND MATH SKILLS!

The major difference between *Modern Curriculum Press Mathematics* and other math programs is the developed model in which students actively work. Every lesson of *Modern Curriculum Press Mathematics* features concept development based on this developed model. Students are required to interact with this model discriminating what data is needed to solve the problem. This process teaches and reinforces their thinking skills and gets them actively involved providing the motivation to read and understand. The four-step teaching strategy of SEE, PLAN, DO, CHECK successfully increases students' understanding and provides a firm foundation for total math master of skills.

- One major objective is the focus of every two-page lesson.
- An algorithm or a model word problem keeps students interested and involved and provides a purpose for learning.

Dividing by 4

Therese is using baskets of flowers to decorate the tables.
~~How many flowers~~

r of flowers that
asket.

o make up.

the baskets.

rs, we divide

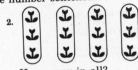

wers into each basket.

nplete the number sentences.

2.

How many in all? _____

How many groups? _____

How many in each group? _____

12 ÷ 4 = _____

nces.

4 = _____   5. 8 ÷ 4 = _____   6. 32 ÷ 4 = _____

Reviewing Addition Facts

Aaron left home early one morning to walk to the library, before he went to school. How many blocks did he walk on his way to school?

Home ●————————● Library

● School

We want to know the number of blocks Aaron walked all together.

We know that he walked _____ blocks from his house to the library.

He walked another _____ blocks from the library to school.
To find the total number of blocks, we add

_____ and _____.

0 1 2 3 4 5 6 7 8 9 10 11 12 13 14 15 16 17 18

7 + 6 = _____
↗ ↑ ↑
addends sum

$\begin{array}{r} 7 \\ + 6 \end{array}$ ← addends
← sum

7 + 6 = 13 is called a **number sentence**.

Aaron walked _____ blocks from his home to school.

Getting Started

Complete the number sentences.

1. 4 + 2 = _____   2. 7 + 9 = _____   3. 8 + 3 = _____

4. 2 + 9 = _____   5. 5 + 6 = _____   6. 8 + 8 = _____

Add.

7. $\begin{array}{r} 8 \\ + 7 \end{array}$ 8. $\begin{array}{r} 4 \\ + 1 \end{array}$ 9. $\begin{array}{r} 9 \\ + 9 \end{array}$ 10. $\begin{array}{r} 5 \\ + 5 \end{array}$ 11. $\begin{array}{r} 3 \\ + 6 \end{array}$ 12. $\begin{array}{r} 9 \\ + 4 \end{array}$

3

- Students interact with the artwork to gather data needed to solve problems. This interaction helps develop higher-order thinking skills.

- Each objective is introduced in a problem-solving setting developing problem-solving thinking skills.

- The four-step teaching method of SEE, PLAN, DO, CHECK guides students easily through the development of each skill.

- Students SEE the "input" sentences and the artwork and use them to help solve the problems. This allows them to be actively involved in their work.

- Students PLAN how they are going to solve problems using their reasoning skills to determine what operations are needed.

- Students use the model to help DO the problem. Each developed model shows students how to do the algorithm.

- To CHECK understanding of the math skill, a concluding sentence reinforces the problem-solving process.

- Important math vocabulary is bold-faced throughout the text and defined in context and in the glossary.

- A check (√) points out important concepts to which students should give special attention.

Place Value through Thousands

The government space agency plans to sell used moon buggies to the highest bidders. What did Charley pay for the one he bought?

We want to understand the cost of Charley's moon buggy.

Charley paid exactly _____.
To understand how much money this is, we will look at the place value of each digit in the price.

✔ The numbers 0, 1, 2, 3, 4, 5, 6, 7, 8 and 9 are called **digits**. The position of the digit decides its place value.

thousands	hundreds	tens	ones
_____	_____	_____	_____

In 7,425, the digit 4 represents hundreds, and the

digit 7 represents _____.
Numbers can be written in **standard** or **expanded form**.

Standard Form   Expanded Form
7,425   7,000 + 400 + 20 + 5

We say Charley paid **seven thousand, four hundred twenty-five dollars**. We write _____.

Getting Started

Write in standard form.

1. five thousand, six hundred fifty-eight _____

2. 3,000 + 50 + 8 _____

Write in words.

3. 6,497

4. 823

5. 9,045

_____

_____

Write the place value of the red digits.

6. 3,948

7. 9,603

8. 7,529

9. $5,370

7

Subtracting Fractions with Unlike

Duncan is feeding the chickens on his uncle's farm. When he started, there were $4\frac{1}{2}$ buckets of chicken feed. How much feed has he used?

We want to know how much chicken feed Duncan has used.

We know that he started with _____ buckets

of feed, and he has _____ buckets left.

To find the amount used, we subtract the amount left from the original amount.

We subtract _____ from _____.

To subtract fractions with unlike denominator follow these steps:

Rename the fractions as equivalent fractions with the least common denominator

$$4\frac{1}{2} = 4\frac{2}{4}$$
$$-1\frac{1}{4} = 1\frac{1}{4}$$

Subtract the fractions.

$$4\frac{1}{2} = 4\frac{2}{4}$$
$$-1\frac{1}{4} = 1\frac{1}{4}$$
$$\frac{1}{4}$$

Duncan has used _____ buckets of feed.

Getting Started

Subtract.

1. $15\frac{5}{8}$
   $- 7\frac{1}{3}$

2. $87\frac{2}{3}$
   $- 39\frac{1}{6}$

3.

Copy and subtract.

5. $\frac{7}{8} - \frac{1}{4} =$ _____

6. $\frac{5}{6} - \frac{1}{2} =$ _____

7. $\frac{9}{10} - \frac{6}{15} =$ _____

127

TEACHER-GUIDED PRACTICE ACTIVITIES CHECK STUDENTS' UNDERSTANDING OF MATH CONCEPTS!

Getting Started activities provide the opportunity for students to try to do what they've just learned and for teachers a chance to check understanding. These activities also allow the teacher to evaluate students' progress in a particular objective before continuing on in the lesson. A complete **Table of Common Errors** can be found in the **Teacher's Editions.** This list helps the teacher diagnose and correct those errors identified by research to be the most common. Lesson plans offer specific suggestions for dealing with each individual error, so the teacher can concentrate on those areas where students need help. Showing the teacher ways to keep errors from happening by alerting to common mistakes, will make teaching math go more smoothly.

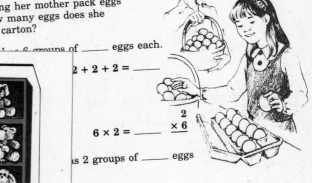

Multiplying, the Factor 2

Sun Li is helping her mother pack eggs in cartons. How many eggs does she pack into each carton?

_____ 6 groups of _____ eggs each.

2 + 2 + 2 = _____

6 × 2 = _____ $\begin{array}{r} 2 \\ \times 6 \\ \hline \end{array}$

_____ 2 groups of _____ eggs

2 × 6 = _____ $\begin{array}{r} 6 \\ \times 2 \\ \hline \end{array}$

_____ into each carton.

_____ltiplication to show how many

2.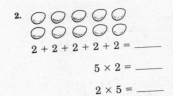

2 + 2 + 2 + 2 + 2 = _____

5 × 2 = _____

2 × 5 = _____

4. 2 × 6 = _____   5. $\begin{array}{r} 4 \\ \times 2 \\ \hline \end{array}$ 6. $\begin{array}{r} 2 \\ \times 2 \\ \hline \end{array}$

Subtracting 2-digit Numbers

Annie collects stuffed animals. She must take 17 of them to school for a display. How many are left at home?

We want to know how many animals she left at home.

Annie has _____ stuffed animals.

She is taking _____ animals to school. To find how many animals she left at home, we subtract _____ from _____.

✔ Subtract the ones first.

Do you need more ones?	Trade 1 ten to get 10 ones.	Subtract the ones.	Subtract the tens.
6 − 7 = ? Yes, you need more ones.	Now there are 2 tens and 16 ones.	16 − 7 = 9	2 − 1 = 1

tens	ones
3	6
−1	7
	?

tens	ones
³3̸	¹⁶6̸
−1	7
	9

tens	ones
²3̸	¹⁶6̸
−1	7
1	9

Annie left _____ stuffed animals at home.

Getting Started

Subtract. Trade if needed.

	tens	ones		tens	ones		tens	ones		tens	ones
1.	9	3	2.	6	2	3.	8	4	4.	8	8
	−5	9		−3	6		−2	9		−1	8

Subtracting 2-digit numbers, with trading

- Samples that the students work allow the teacher to check students' understanding of the skill.

- Students gain both confidence and competence in working these problems.

- If the objective is not fully grasped by the student, the **Table of Common Errors** will help the teacher deal with each individual type of error.

- Students gain a deeper understanding of the basic algorithm introduced in the developed model.

- New skills are reinforced through the sample problems students work right on the spot.

- Teachers observe any typical student errors before continuing additional work in the lesson.

- Teacher-guided practice activities will encourage classroom discussion.

- **Getting Started** activities help the teacher to single out predictable errors quickly.

- All samples found in the **Getting Started** activities prepare students to work the exercises found in the next part of the lesson.

Using Customary Units of Length

Robert, Janis and Jonathan are being measured for band uniforms. What is Robert's height in inches?

We want to rename Robert's height in inches.

We know that he is ____ feet ____ inches tall.

____ches as inches, we multiply ____ the number of inches in a ____ extra inches.

____ and add ____.

12 inches (in.)	= 1 foot (ft)
3 feet	= 1 yard (yd)
36 inches	= 1 yard
5,280 feet	= 1 mile (mi)
1,760 yards	= 1 mile

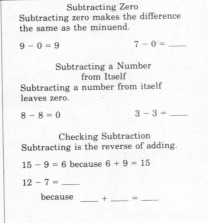

____ll.

____ber like $5\frac{1}{2}$ feet as inches, we

____ inches

____ as larger units, like 48 inches

____.

____o rename larger units as ____ name smaller units as larger ones.

____easurements of length.

$$\begin{array}{r} 3 \text{ yd } 1 \text{ ft} \\ - 1 \text{ yd } 2 \text{ ft} \\ \hline 1 \text{ yd } 2 \text{ ft} \end{array}$$

1 yd = 3 ft
3 ft + 1 ft = 4 ft

____2 in.)

$1\frac{8}{9}$ yd

Add or subtract.

3. $\begin{array}{r} 6 \text{ ft } 9 \text{ in.} \\ - 2 \text{ ft } 11 \text{ in.} \end{array}$

4. $\begin{array}{r} 7 \text{ yd } 2 \text{ ft } 3 \text{ in.} \\ - 5 \text{ yd } 1 \text{ ft } 6 \text{ in.} \end{array}$

217

Addition and Subtraction Properties

Properties are like special tools. They make the job of adding and subtracting much easier.

(Speech bubble, boy): Twelve minus nine is three.

(Speech bubble, girl): That's right because nine plus three is twelve.

Addition

Order Property
We can add in any order.

$5 + 2 = 7$ $2 + 5 = 7$

$3 + 6 + 7 =$ ____ $7 + 3 + 6 =$ ____

Grouping Property
We can change the grouping.
✔ Remember to add the numbers in parentheses first.

$(6 + 3) + 5 = 14$ $6 + (3 + 5) = 14$

$(8 + 2) + 4 =$ ____ $8 + (2 + 4) =$ ____

Zero Property
Adding zero makes the sum the same as the other addend.

$5 + 0 = 5$ $0 + 7 = 7$

$0 + 1 =$ ____ $8 + 0 =$ ____

Subtraction

Subtracting Zero
Subtracting zero makes the difference the same as the minuend.

$9 - 0 = 9$ $7 - 0 =$ ____

Subtracting a Number from Itself
Subtracting a number from itself leaves zero.

$8 - 8 = 0$ $3 - 3 =$ ____

Checking Subtraction
Subtracting is the reverse of adding.

$15 - 9 = 6$ because $6 + 9 = 15$

$12 - 7 =$ ____

because ____ + ____ = ____

✔ **Solving for n** is finding the value for the n in the equation.

Getting Started

Solve for n.

1. $0 + 0 = n$ 2. $0 + 6 = n$

   $n =$ ____ $n =$ ____

Subtract. Check by adding.

3. $\begin{array}{r} 15 \\ - 9 \end{array}$ 4. $\begin{array}{r} 12 \\ - 7 \end{array}$ 5. $\begin{array}{r} 18 \\ - 9 \end{array}$

Add. Check by grouping the addends another way.

6. $\begin{array}{r} 5 \\ 3 \\ + 4 \end{array}$ 7. $\begin{array}{r} 2 \\ 6 \\ + 3 \end{array}$ 8. $\begin{array}{r} 6 \\ 3 \\ + 4 \end{array}$ 9. $(5 + 2) + 6 = n$

                         $n =$ ____   10. $3 + (5 + 4) = n$

                                                    $n =$ ____

3

INDEPENDENT PRACTICE ACTIVITIES PROVIDE PLENTY OF DRILL, PRACTICE, AND EXTENSION IN A VARIETY OF FORMATS!

The purpose of building skills is to ensure that students can use and apply those skills. That goal can only be reached when skills are clearly and systematically taught and then practiced. With *Modern Curriculum Press Mathematics,* the teacher can be as-sured that students will have abundant opportunities to practice their newly-learned math skills. The variety of practice activities allows the teacher to meet the needs of every student. Working independently helps students strengthen new skills, become more confident, and increase their under-standing. Practice helps students learn Some students need more practice than others to help them catch on. *Modern Curriculum Press Mathematics* offers a variety of practice situations so that students stay on target with what they are learning.

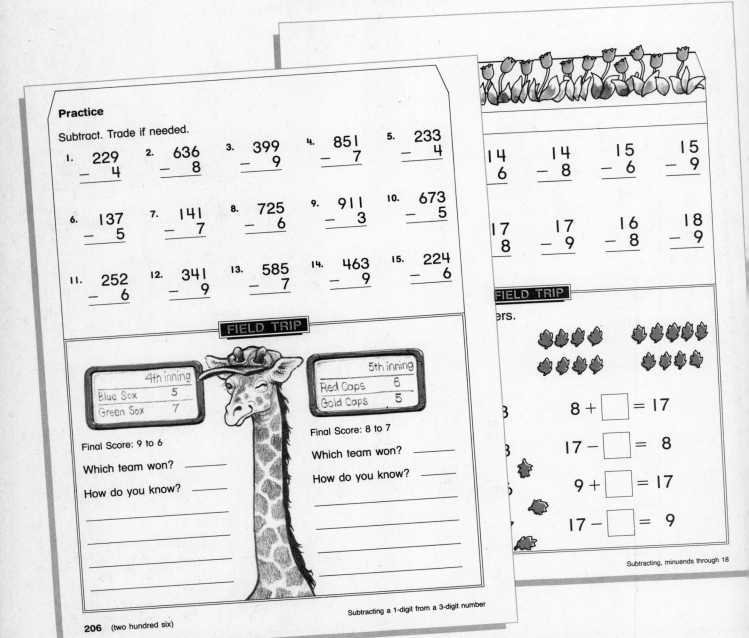

Practice

Subtract. Trade if needed.

1. 229 − 4
2. 636 − 8
3. 399 − 9
4. 851 − 7
5. 233 − 4
6. 137 − 5
7. 141 − 7
8. 725 − 6
9. 911 − 3
10. 673 − 5
11. 252 − 6
12. 341 − 9
13. 585 − 7
14. 463 − 9
15. 224 − 6

FIELD TRIP

4th inning
Blue Sox 5
Green Sox 7

Final Score: 9 to 6

Which team won? _____

How do you know? _____

5th inning
Red Caps 6
Gold Caps 5

Final Score: 8 to 7

Which team won? _____

How do you know? _____

206 (two hundred six)

Subtracting a 1-digit from a 3-digit number

14 − 6
14 − 8
15 − 6
15 − 9
17 − 8
17 − 9
16 − 8
18 − 9

FIELD TRIP

...ers.

8 + ☐ = 17

17 − ☐ = 8

9 + ☐ = 17

17 − ☐ = 9

Subtracting, minuends through 18

T-8

The teacher can begin the process of individual mastery by assigning **Practice** exercises that students can work independently.

■ *Modern Curriculum Press Mathematics* integrates problem solving into the practice activities with **Apply** problems. Some of these problems relate to the algorithm. However, some require previously-learned skills encouraging students to think and maintain skills.

■ Both vertical and horizontal forms of problems are used making students more comfortable with forms found in standardized test formats.

■ An emphasis on practical skills encourages learning by applying math to everyday situations.

■ Independent practice provides more opportunities for application and higher-order thinking skills.

■ The variety of practice activities keeps students motivated and interested in learning.

■ **Copy and Do** exercises check students' ability to assemble an algorithm from an equation and gives them practice in transferring information.

■ **Excursion** activities extend the basic skill work and are fun to do. The teacher can challenge the more capable students with these mind-stretching activities.

■ Giving students ample opportunities to practice and strengthen new skills builds solid skill development and helps the teacher more easily measure the results.

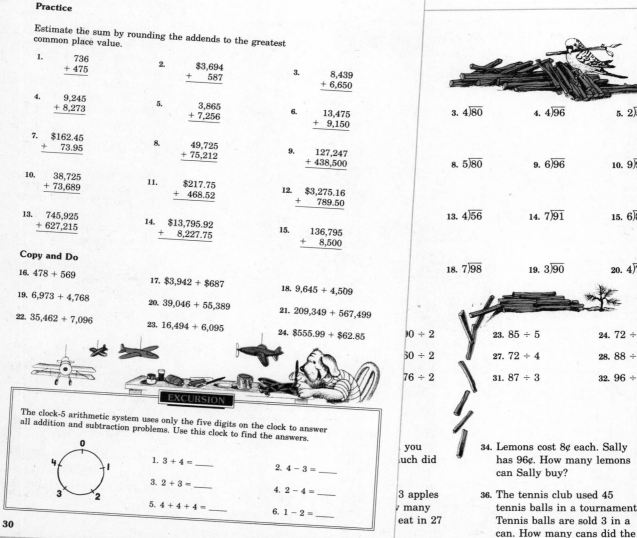

Practice

Estimate the sum by rounding the addends to the greatest common place value.

1.
```
  736
+ 475
```

2.
```
$3,694
+   587
```

3.
```
  8,439
+ 6,650
```

4.
```
  9,245
+ 8,273
```

5.
```
  3,865
+ 7,256
```

6.
```
 13,475
+ 9,150
```

7.
```
$162.45
+  73.95
```

8.
```
 49,725
+ 75,212
```

9.
```
 127,247
+ 438,500
```

10.
```
 38,725
+ 73,689
```

11.
```
$217.75
+ 468.52
```

12.
```
$3,275.16
+   789.50
```

13.
```
 745,925
+ 627,215
```

14.
```
$13,795.92
+  8,227.75
```

15.
```
 136,795
+   8,500
```

Copy and Do

16. 478 + 569

17. $3,942 + $687

18. 9,645 + 4,509

19. 6,973 + 4,768

20. 39,046 + 55,389

21. 209,349 + 567,499

22. 35,462 + 7,096

23. 16,494 + 6,095

24. $555.99 + $62.85

EXCURSION

The clock-5 arithmetic system uses only the five digits on the clock to answer all addition and subtraction problems. Use this clock to find the answers.

1. 3 + 4 = ___

2. 4 – 3 = ___

3. 2 + 3 = ___

4. 2 – 4 = ___

5. 4 + 4 + 4 = ___

6. 1 – 2 = ___

30

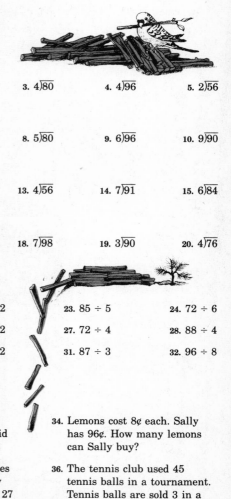

3. 4)80   4. 4)96   5. 2)56

8. 5)80   9. 6)96   10. 9)90

13. 4)56   14. 7)91   15. 6)84

18. 7)98   19. 3)90   20. 4)76

. . . 0 ÷ 2   23. 85 ÷ 5   24. 72 ÷ 6

. . . 60 ÷ 2   27. 72 ÷ 4   28. 88 ÷ 4

. . . 76 ÷ 2   31. 87 ÷ 3   32. 96 ÷ 8

34. Lemons cost 8¢ each. Sally has 96¢. How many lemons can Sally buy?

36. The tennis club used 45 tennis balls in a tournament. Tennis balls are sold 3 in a can. How many cans did the club use?

MATH COMES ALIVE WHEN STUDENTS LEARN TO INTEGRATE COMPUTATION, PROBLEM-SOLVING STRATEGIES, AND REASONING TO MAKE DECISIONS FOR THEMSELVES!

Problem-solving pages present lessons that increase understanding with a four-step teaching strategy: SEE, PLAN, DO, CHECK. *Modern Curriculum Press Mathematics* offers step-by-step instruction in how to understand word problems as well as varied practice in actually using the skills learned. Each lesson focuses on a different problem-solving strategy. These strategies develop students' higher-order thinking skills and help them success-fully solve problems. Step-by-step, students will understand the question, find the information needed, plan a solution, and then check it for accuracy. This develops students' critical-thinking skills and ability to apply what they've learned to solve problems that go beyond basic operations.

■ Word problems utilize high-interest information and focus on everyday situations.

PROBLEM SOLVING

Drawing a Picture

A parking lot has 9 rows of 8 parking spaces each. The fourth and fifth spaces in every third row have trees in them. The outside spaces in every row are reserved for the handicapped or for emergency vehicles. How many regular parking spaces are there in the lot?

★ SEE
We want to know how many spaces are left for regular parking.

There are _____ rows of parking spaces.

There are _____ spaces in each row.

In every third row, _____ spaces are lost to trees.

In every row _____ spaces are used for special vehicles.

★ PLAN
We can draw a picture of the parking lot, crossing out the closed parking spaces. Then we can count the regular spaces left.

★ DO
We count _____ spaces left for regular parking.

★ CHECK
We can check by adding the spaces open in each row.

$4 + 6 + 6 + 4 + 6 + 6 + 4 + 6 + 6 =$ _____

173

---

ach problem.

2. A Super-Duper ball bounces twice its height when it is dropped. Carl dropped a Super-Duper ball from the roof of a 12-foot garage. How high will the ball bounce after 5 bounces?

4. The distance around a rectangle is 10 centimeters. The length of each of the two longer sides is 3 centimeters. What is the length of each of the two shorter sides?

6. What 7 coins together make 50 cents?

- Step by step, students learn to understand the question, find the information they need, plan a method of solution, find an answer, and check it for accuracy.

- Every step of the process is organized so that students truly understand how to arrive at the solution.

- The problem-solving banner alerts students that they are involved in a problem-solving lesson. These focused lessons remind students how to approach problems and how to use skills and specific strategies already learned.

- Learning to integrate computation, problem-solving strategies, and reasoning makes math come alive for students.

- Problems incorporate previously taught computational skills—focusing students' minds on the problem-solving process itself.

- Problem-solving applications appear in every problem-solving lesson. This frequent practice reduces apprehension and builds confidence.

- Practice in applying the strategies gives students a chance to use skills in routine and non-routine problems.

- In every chapter, problem-solving strategies and critical-thinking skills are developed, applied, and reinforced.

- Students choose appropriate strategies to solve problems and are challenged to formulate their own problems and to change the conditions in existing problems.

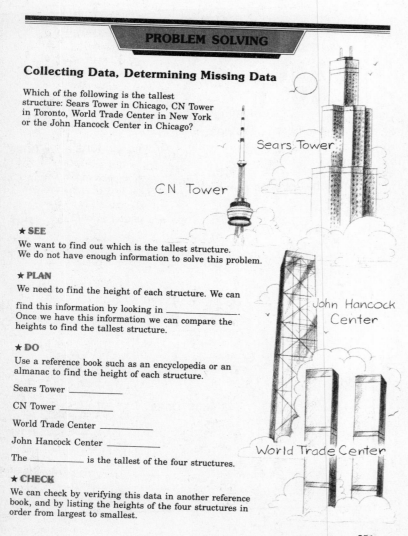

PROBLEM SOLVING

Collecting Data, Determining Missing Data

Which of the following is the tallest structure: Sears Tower in Chicago, CN Tower in Toronto, World Trade Center in New York or the John Hancock Center in Chicago?

Sears Tower

CN Tower

★ SEE

We want to find out which is the tallest structure. We do not have enough information to solve this problem.

★ PLAN

We need to find the height of each structure. We can find this information by looking in _____. Once we have this information we can compare the heights to find the tallest structure.

★ DO

Use a reference book such as an encyclopedia or an almanac to find the height of each structure.

Sears Tower _____

CN Tower _____

World Trade Center _____

John Hancock Center _____

The _____ is the tallest of the four structures.

★ CHECK

We can check by verifying this data in another reference book, and by listing the heights of the four structures in order from largest to smallest.

John Hancock Center

World Trade Center

251

Empire State Building

2. Roll a pair of dice 30 times and record the number of times each sum appears. Perform the experiment a second time. What sum appears most often? What sum appears least often?

4. Toss a coin 50 times and record the number of heads and tails. Which side of the coin appears most often?

6. Record the dates of the coins available in your classroom. How many years difference exist between the newest and oldest coin?

8. An arithmetic game is created by adding the values of certain U.S. currency. Since a portrait of George Washington appears on a $1 bill and a portrait of Abraham Lincoln appears on the $5 bill, we say that George Washington + Abraham Lincoln = $6. Find the value of Thomas Jefferson + Alexander Hamilton + Woodrow Wilson.

CHAPTER TEST PAGES PROVIDE A VEHICLE FOR STUDENT EVALUATION AND FEEDBACK!

Every chapter in *Modern Curriculum Press Mathematics* concludes with a **Chapter Test**. These tests provide the opportunity for students to demonstrate their mastery of recently acquired skills. **Chapter Test** pages enable the teacher to measure all the basic skills students have practiced in the lesson and evaluate their understanding. The focus of these pages is the assessing of mastery of algorithms. An adequate number of sample problems are provided to accomplish this. This important checkpoint helps the teacher to better meet individual student-computational needs.

■ **Chapter Test** pages are carefully correlated to what has been taught throughout the entire series.

■ **Chapter Test** pages assess students' mastery of all the skills taught in the lesson.

■ All directions are written in an easy-to-follow format.

■ Both vertical and horizontal forms of problems are used making students more comfortable with exercises found in standardized tests.

■ In the back of each student book, there is an alternate **Chapter Test** for each chapter based on the same objectives covered in the first test.

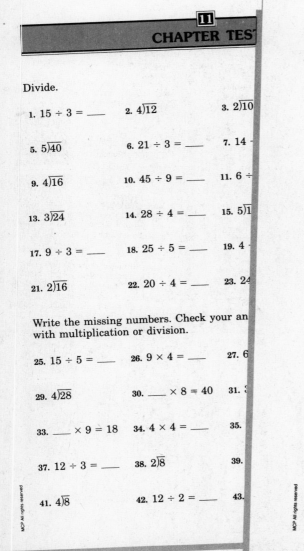

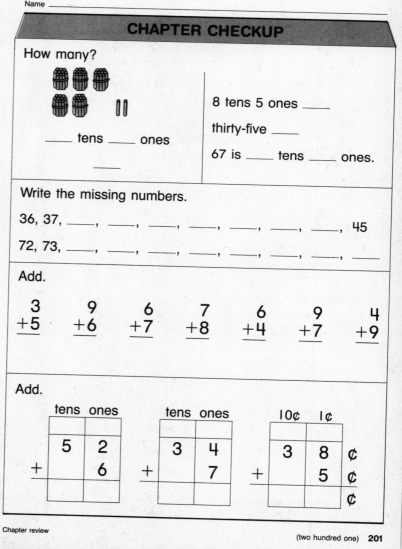

SYSTEMATIC MAINTENANCE IS PROVIDED AT EVERY LEVEL WITH CUMULATIVE REVIEW PAGES!

Every chapter contains a **Cumulative Review** page that provides an on-going refresher course in basic skills. These pages maintain the skills that have been taught in the chapter plus the skills learned in previous chapters.

Cumulative Review pages actually reach back into the text for a total maintenance of skills. **Cumulative Review** pages are progressive instruction because they build on the foundation laid earlier for a thorough and sequential program of review. A standardized test format is used beginning at the middle of the second grade. Students will benefit by gaining experience in dealing with this special test format.

■ A variety of problems done in standardized test format give students a better chance to score well on these tests.

■ Directions are minimal and easy to understand.

■ Design elements on every test are the same found on standardized tests.

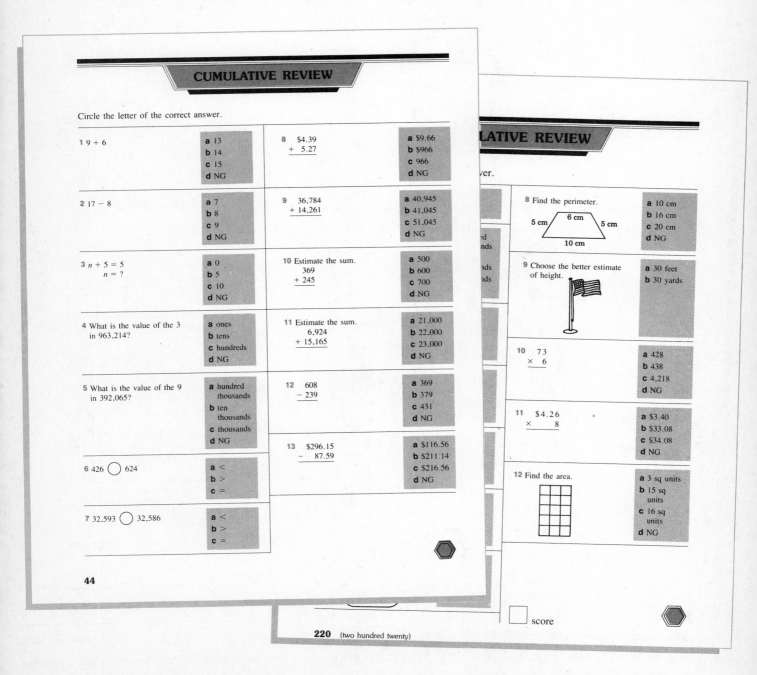

CUMULATIVE REVIEW

Circle the letter of the correct answer.

1 9 + 6
  a 13
  b 14
  c 15
  d NG

2 17 − 8
  a 7
  b 8
  c 9
  d NG

3 $n + 5 = 5$
  $n = ?$
  a 0
  b 5
  c 10
  d NG

4 What is the value of the 3 in 963,214?
  a ones
  b tens
  c hundreds
  d NG

5 What is the value of the 9 in 392,065?
  a hundred thousands
  b ten thousands
  c thousands
  d NG

6 426 ◯ 624
  a <
  b >
  c =

7 32,593 ◯ 32,586
  a <
  b >
  c =

8 $4.39
  + 5.27
  a $9.66
  b $966
  c 966
  d NG

9 36,784
  + 14,261
  a 40,945
  b 41,045
  c 51,045
  d NG

10 Estimate the sum.
  369
  + 245
  a 500
  b 600
  c 700
  d NG

11 Estimate the sum.
  6,924
  + 15,165
  a 21,000
  b 22,000
  c 23,000
  d NG

12 608
  − 239
  a 369
  b 379
  c 431
  d NG

13 $296.15
  − 87.59
  a $116.56
  b $211.14
  c $216.56
  d NG

44

LATIVE REVIEW

8 Find the perimeter.
  5 cm / 6 cm \ 5 cm
  10 cm
  a 10 cm
  b 16 cm
  c 20 cm
  d NG

9 Choose the better estimate of height.
  a 30 feet
  b 30 yards

10 73
  × 6
  a 428
  b 438
  c 4,218
  d NG

11 $4.26
  × 8
  a $3.40
  b $33.08
  c $34.08
  d NG

12 Find the area.
  a 3 sq units
  b 15 sq units
  c 16 sq units
  d NG

☐ score

CALCULATOR LESSONS PROVIDE EXCITING LEARNING ACTIVITIES AND ADD INTEREST AND PRACTICALITY TO MATH!

Calculator lessons are found throughout *Modern Curriculum Press Mathematics*. The activities are used in many ways—to explore num-ber patterns, to do calculations, to check estimations, and to investigate functions. Each **Calculator** lesson is designed to help students learn to use and operate calculators while they re-inforce and improve their mathematical skills.

■ **Calculator** lessons teach students to use simple calculators while reinforc-ing chapter content.

■ **Calculator** lessons introduce students to basic calculator skills and terms.

■ Practical calculator activities promote student involvement as they take an active part in what they are learning.

■ Students learn, practice, and apply critical-thinking skills as they use cal-culators.

Practice

Complete these calculator codes.

1. $85 \div 5 =$ ▭

2. $57 \div 3 =$ ▭

4. $96 \div 6 =$ ▭

6. $90 \div 9 =$ ▭

8. $63 \div 7 \times 8 =$ ▭

10. $75 \div 5 \times 6 =$ ▭

12. $216 - 158 \div 2 =$ ▭

14. Nathan can jog 5 miles in 65 minutes. How long will it take Nathan to jog 8 miles?

16. Bananas are on sale at 6 for 96¢. How much do 8 bananas cost?

2. The sum of 2 numbers is 60. Their difference is 12. What are the numbers?

4. Five times one number is three more than six times another number. The difference between the numbers is 1. What are the numbers?

Calculators, the Division Key

Natalie is packing lunches for a picnic. She needs to buy 5 apples. How much will Natalie pay for the 5 apples?

Apples 3 for 51¢

We want to know the price for 5 apples.

We know that _____ apples cost _____.

To find the cost of 5 apples, we first find the cost of 1 by dividing _____ by _____. Then, we multiply the cost of 1 apple by _____.

This can be done on the calculator in one code.

$. 51 \div 3 \times 5 =$ ▭

Natalie will pay _____ for 5 apples.

Complete these calculator codes.

1. $42 \div 7 =$ ▭

3. $96 \div 4 =$ ▭

5. $36 \div 9 \times 7 =$ ▭

7. $72 \div 6 \times 9 =$ ▭

2. $76 \div 2 =$ ▭

4. $52 \div 4 =$ ▭

6. $84 \div 4 \times 3 =$ ▭

8. $75 \div 5 \times 9 =$ ▭

139

Plan Classroom-Ready Math Lessons In Minutes With Comprehensive Teacher's Editions!

The **Teacher's Editions** of *Modern Curriculum Press Mathematics* are designed and organized with the teacher in mind. The full range of options provides more help than ever before and guarantees efficient use of the teacher's planning time and the most effective results for efforts exerted.

Each **Teacher's Edition** provides an abundance of additional **Enrichment, Correcting Common Errors** and application activities. Plus they contain a complete **Error Pattern Analysis.** The teacher will also find reduced student pages with answers, objectives, suggestions for teaching lessons, materials, **Mental Math** exercises, and more.

■ There's no need for the teacher to struggle with two separate books because student pages are reduced in the **Teacher's Edition.**

■ Clear headings and notes make it easy for the teacher to find what is needed before teaching the lesson.

■ The teacher will be more effective with lesson plans that are always complete in two pages and include everything needed.

■ Student **Objectives** set a clear course for the lesson goal.

Time to the Half-hour
pages 163-164
Objective
To practice telling time to the hour and half-hour

Materials
*Demonstration clock
*Two pencils of different lengths

Mental Math
Which is less?
1. 2 dimes or 5 nickels (2 dimes)
2. 14 pennies or 2 nickels (2 nickels)
3. 5 nickels or 4 dimes (5 nickels)
4. 1 quarter or 2 dimes (2 dimes)
5. 6 nickels or 1 quarter (1 quarter)

Skill Review
Show times to the hour and half hour on the demonstration clock. Have students write the time on the board. Now have a student set the clock to show an hour or half-hour. Have the student ask another student to write the time on the board. Have a student write a time for the hour or half-hour and invite another student to place the hands on the clock to show the times.

Teaching page 163
On the demonstration clock, start at 12:00 and slowly move the minute hand around the clock. Ask students to tell what the hour hand does as the minute hand moves slowly around the clock face. (moves slowly toward the next number) Tell students the minute hand moves around the clock face 60 minutes while the hour hand moves from one number to the next. Tell students there are 60 minutes in 1 hour.

Ask students to tell where the hour hand is on the first clock. (between 5 and 6) Ask where the minute hand is. (on 6) Ask the time. (5:30) Tell students to find 5:30 in the center column and trace the line from the clock to 5:30. Tell students to draw a line from each clock to its time.

Name

Match the clocks

Telling time to the hour and half-hour

Multiplying, the Factor 5
pages 125-126
Objective
To multiply by the factor 5

Materials
none

Mental Math
Ask students to multiply 4 by:
1. the number of ears one person has. $(4 \times 2 = 8)$
2. the number of feet two students have. $(4 \times 4 = 16)$
3. the number of noses in a crowd of 7. $(4 \times 7 = 28)$
4. the number of their toes. $(4 \times 10 = 40)$

Skill Review
Have students make up a multiplication chart. Tell them to write 2, 3, 4 along the top, 2 through 10 along the side. Tell them to fill in the chart by multiplying each top number by each side number.

Teaching the Lesson

Introducing the Problem Have students look at the calculator illustrated while you read the problem. Identify the question and explain that there are several ways they could answer it. Have students read the information sentences, filling in the information required. (5 rows, 5 keys) Read each sentence and tell students to do the indicated operation in their text while one student writes it on the board. Read the solution sentence aloud and have a student give the answer while the others complete that sentence in their texts. (30)

Multiplying, the Factor 5

Each key on a calculator has a special job to do. How many keys are there on the calculator keyboard?

We need to find the total number of keys on the calculator.

We can see there are __5__ rows of keys.

Each row has __5__ keys.

We can add. $5 + 5 + 5 + 5 + 5 =$ __25__

We can also multiply.

$5 \times 5 =$ __25__ $\begin{array}{r} 5 \\ \times\, 5 \\ \hline 25 \end{array}$

There are __25__ keys on the calculator keyboard.

Getting Started
Use both addition and multiplication to show how many are in the picture.

1. $5 + 5 + 5 + 5 =$ __20__
   $4 \times 5 =$ __20__
   $5 \times 4 =$ __20__

Multiply.

2. $\begin{array}{r} 3 \\ \times\, 5 \\ \hline 15 \end{array}$
3. $\begin{array}{r} 8 \\ \times\, 5 \\ \hline 40 \end{array}$
4. $5 \times 6 =$ __30__
5. $9 \times 5 =$ __45__

(one hundred twenty-five) **125**

Developing the Skill Have students start at 5 and count aloud by five's through 50. Ask a volunteer to continue through 100. Explain that this may seem easy because they are used to counting out nickels. Because five is also half of ten, and when counting by fives, every other number will be a multiple of ten. Now write these addition problems on the board and have students work them: $5 + 5 =$, $5 + 5 + 5 =$, $5 + 5 + 5 + 5 =$. (10, 15, 20) Next to each of these problems write the multiplication problem that corresponds. $(5 \times 2, 5 \times 3, 5 \times 4)$ Have volunteers put the rest of the addition and multiplication problems for the factor five on the board, 5×5 through 5×10.

■ A list of **Materials** helps the teacher reduce class preparation time.

■ The **Mental Math** exercise gives the teacher an opportunity to brush-up on skills at the beginning of each day's lesson.

■ **Skill Review** bridges new skills with previously-taught skills for total reinforcement.

■ The **Teaching the Lesson** section is meant to give practical suggestions for introducing the problem and developing the skill. Specific suggestions for an effective presentation of the model are made in **Introducing the Problem.**

■ In **Developing the Skill,** the teacher is given suggestions for presenting and developing the algorithm, skill, and/or concept. Where practical, recommendations are made for the use of manipulatives.

Zeros in Minuend

pages 69-70

Objective

To subtract 4- or 5-digit numbers when minuends have zeros

Materials

*thousands, hundreds, tens, ones jars
*place value materials

Mental Math

Tell students to answer true or false:

1. 776 has 77 tens. (T)
2. 10 hundreds < 1,000. (F)
3. 46 is an odd number. (F)
4. 926 can be rounded to 920. (F)
5. 72 hours = 3 days. (T)
6. $42 \div 6 > 7 \times 1$. (F)
7. 1/3 of 18 = 1/2 of 12. (T)
8. perimeter = L × W. (F)

Skill Review

Write 4 numbers of 3- to 5-digits each on the board. Have students arrange the numbers in order, from the least to the greatest, and then read the numbers as they would appear if written from the greatest to the least. Repeat for more sets of 4 or 5 numbers.

Subtracting, More Minuends with Zeros

The Susan B. Anthony School held its annual fall Read-a-Thon. How many more pages did the fifth grade read than the second-place class?

We want to know how many more pages the fifth grade read than the second-place class.

We know the fifth grade read __5,003__ pages.

The second-place class is the __sixth__ grade.

It read __4,056__ pages.
To find the difference between the number of pages, we subtract __4,056__ from __5,003__.

✔ Remember to trade from one place value at a time.

```
    9 9
  4 10 10 13
  5,003
 −4,056
    947
```

The fifth grade read __947__ more pages than the sixth grade.

Getting Started

Subtract.

1.	2.	3.
3,005 − 1,348 1,657	$40.09 − 9.75 $30.34	3,300 − 1,856 1,444
4. $50.00 − 27.26 $22.74	5. 8,512 − 7,968 544	6. $90.17 − 20.87 $69.30

Copy and subtract.

7. 26,007 − 18,759
7,248

8. 70,026 − 23,576
46,450

9. $900.05 − $267.83
$632.22

69

Teaching the Lesson

Introducing the Problem Have a student read the problem. Ask students what 2 problems are to be solved. (which class read the second-highest number of pages and how many more pages the first-place fifth graders read) Ask students how we can find out which class came in second place. (arrange numbers from the table in order from greatest to least) Ask a student to write the numbers from greatest to least on the board. (5,003, 4,056, 3,795) Have students complete the sentences and work through the model problem with them.

Developing the Skill Write **4,000−2,875** vertically on the board. Ask students if a trade is needed to subtract the ones column. (yes) Tell students that since there are no tens and no hundreds, we must trade 1 thousand for 10 hundreds. Show the 3 thousands and 10 hundreds left. Now tell students we can trade 1 hundred for 10 tens. Show the trade with 9 hundreds and 10 tens left. Tell students we can now trade 1 ten for 10 ones. Show the trade so that 9 tens and 10 ones are left. Tell students we can now subtract each column beginning with the ones column and working to the left. Show students the subtraction to a solution of **1,125.** Remind students to add the subtrahend and the difference to check the work. Repeat for more problems with zeros in the minuend.

69

---

ime Factoring

y whole number greater than 1 can writ·ten as a product of prime mber factors. This is called **prime ctoring.** One way to find me number factors is to make a tor tree. There may be different ys to start a factor tree, but final set of prime factors will ays be the same. Use a factor to find the prime factors of 24. e exponents to write this prime torization.

```
     24            24
    /  \          /  \
   8    3        6    4
  / \           / \  / \
 2   2×2      2×3 2×2
```

2 × 2 2 × 3   3 × 2 × 2 × 2
$2^3 \times 3$ 3 × 2^3

✔ Remember, the exponent tells how many times to use the base number as a factor. $2^3 = 2 \times 2 \times 2$

etting Started

omplete each factor tree.

1.	2.	3.
20 5 × 4 2 × 2	36 4 × 9 2 × 2 3 × 3	50 2 × 25 5 × 5

rite each prime factorization using exponents if possible.

8	5. 35	6. 48	7. 72	8. 400
2^3	5 × 7	$2^4 \times 3$	$2^3 \times 3^2$	$2^4 \times 5^2$

99

---

omposite ctors. Tell e factor· n aloud, he tree tor tree on at a num· of prime d. Have f each

Developing the Skill Point out that when a prime number is used more than once in the prime factoring, it can be expressed with an **exponent.** Remind students that 2^3 is the same as 2 × 2 × 2. Stress that the 3 in 2^3 is an exponent, and that this exponent tells the number of times 2 is used as a factor. Have students complete each of the following factor trees and then write each prime factorization using exponents:

```
      50                28
    /    \            /    \
   2  ×  (25)        7  ×  (4)
(2) × (5) × (5)   (7) × (2) × (2)
   (2 × 5²)          (7 × 2²)
```

99

T-16

FREQUENT ATTENTION IS GIVEN TO CORRECTING COMMON ERRORS, ENRICHMENT AND OPTIONAL EXTRA-CREDIT ACTIVITIES!

These comprehensive **Teacher's Editions** are intended to provide the teacher with a convenient, well-structured approach to teaching mathematics. From motivating introductory exercises to challenging extension activities, *Modern Curriculum Press*

Mathematics **Teacher's Editions** suggest a complete step-by-step plan to insure successful learning. The succinct lesson plans help the teacher provide solid math instruction to students.

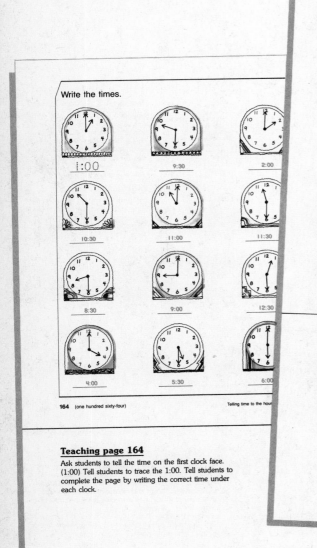

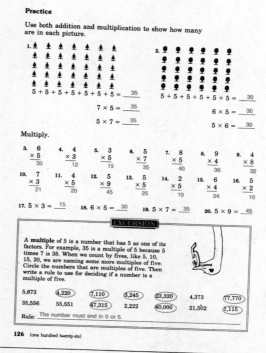

Teaching page 164

Ask students to tell the time on the first clock face. (1:00) Tell students to trace the 1:00. Tell students to complete the page by writing the correct time under each clock.

Practice

Have students do all the problems on the page. Remind the class that they can use addition to figure out any multiplication facts they are not sure of.

Excursion

Have students write the multiples of 5 through 200. Help students to see that any number that ends in 0 or 5 is a multiple of 5. Have students write the rule. Now write several 4- and 5-digit numbers on the board and ask students to circle the numbers that are multiples of 5.

Correcting Common Errors

If students have difficulty learning facts of 5, have them practice with partners. Have them draw a vertical number line from 0 through 50, marking it in intervals of 5. Have one partner write the addition problem to the left of each multiple of five on the number line while the other partner writes the corresponding multiplication problem.

	0	
5	5	(5 × 1)
5 + 5	10	(5 × 2)
5 + 5 + 5	15	(5 × 3)
5 + 5 + 5 + 5	20	(5 × 4)

Enrichment

Ask students how many fives are in 55 if, there are 10 fives in 50. (11) Tell them to complete a multiplication table for fives that goes up to the product 150. Have them use the table to figure the number of nickels in $4.00. (80)

Extra Credit *Logic*

Write the following on the board:

WOW	TOT	POP	BIB
525	969	343	5445

Ask students what all of these have in common. Explain they are palindromes, or words or numbers which are the same whether they are read forward or backward. Also, explain 302 is not a palindrome, but if you reverse the numbers and add, it will make a palindrome:

```
 302
+203
 505
```

Using this method, ask students what palindrome they can make with these numbers: 36; (99) 342; (585) 4,205; (9,229) 3,406 (9,449). Have students list some other numbers which, when reversed and added, will form a palindrome.

- Follow up activities focus on **Correcting Common Errors, Enrichment,** and **Extra Credit** suggestions.

- In the **Correcting Common Errors** feature, a common error pattern is explored and a method of remediation is recommended. Collectively, all the **Correcting Common Errors** features in any chapter constitute a complete set of the common errors likely to be committed by the students when working in that area of mathematics.

- **Enrichment** activities are a direct extension of the skills being taught. Students can do these activities on their own while the teacher works with those students who need more help.

- **Extra Credits** are challenging independent activities to expand the mathematical experiences of the students. The **Extra Credit** section encompasses a wide variety of activities and projects and introduces and extends skills taught in the normal basal curriculum—including statistics, logic, and probability.

Practice

Subtract.

1. 3,004
   − 2,356
   ‾‾‾‾‾
   648

2. 8,002
   − 5,096
   ‾‾‾‾‾
   2,906

3. 3,891
   − 1,750
   ‾‾‾‾‾
   2,141

4. $20.08
   − 15.99
   ‾‾‾‾‾
   $4.09

5. 4,020
   − 1,865
   ‾‾‾‾‾
   2,155

6. $87.00
   − 28.59
   ‾‾‾‾‾
   $58.41

7. 3,007
   − 2,090
   ‾‾‾‾‾
   917

8. $50.06
   − 37.08
   ‾‾‾‾‾
   $12.98

9. 19,006
   − 8,275
   ‾‾‾‾‾
   10,731

10. 20,006
   − 14,758
   ‾‾‾‾‾
   5,248

11. $400.26
   − 236.58
   ‾‾‾‾‾
   $163.68

12. $793.42
   − 253.87
   ‾‾‾‾‾
   $539.55

Copy and Do

13. 4,001 − 2,756 1,245
14. $70.05 − $26.59 $43.46
15. 8,060 − 7,948 112
16. 7,007 − 2,468 4,539
17. 21,316 − 12,479 8,837
18. 14,000 − 8,396 5,604
19. $100.21 − $93.50 $6.71
20. 60,004 − 51,476 8,528
21. 52,006 − 9,037 42,969
22. $800.00 − $275.67 $524.33
23. 34,612 − 29,965 4,647
24. 50,010 − 36,754 13,256

Apply

Use the chart on page 69 to help solve these problems.

25. How many pages did the three classes read all together? 12,854 pages

26. How many more pages did the sixth grade read than the fourth grade? 261 pages

70

Correcting Common Errors

Some students may bring down the numbers that are being subtracted when there are zeros in the minuend.

INCORRECT	CORRECT
3,006 − 1,425 ‾‾‾‾ 2,421	3,006 − 1,425 ‾‾‾‾ 1,581

Have students work in pairs and use play money to model a problem such as $300 − $142, where they see that they must trade 3 hundreds for 2 hundreds, 9 tens, and 10 ones before they can subtract.

Enrichment

Tell students to find out the year in which each member of their family was born, and make a chart to show how old each will be in the year 2000.

3. 7 28
        × ┐
       2 × 2

6. 75
   3 × 25
       5 × 5

10. 64   11. 66
    2⁶   2 × 3 × 11

15. 180   16. 225
    2² × 3² × 5   3² × 5²

7 are examples of twin

43, 71 and 73

irror primes. 13 and 31 are

9 and 97

f its factors except itself.
of the perfect numbers less

Correcting Common Errors

Some students do not write the prime factorization of a number correctly because they cannot identify prime numbers. Have them work with partners to name all the prime numbers from 1 to 50 and write them on an index card. The students can use these cards as a guide for this work.

Enrichment

Provide this alternative method of dividing to find prime factorization of a number. Tell students they must always divide by a prime number.

2 | 36   36 = 2² × 3²
2 | 18
3 | 9
      3

Have students use this method to find the prime factorization of:
120 (2³ × 3 × 5); 250 (5³ × 2); 1,000 (2³ × 5³); 72 (3² × 2³)

Practice

Remind students to begin with the ones column, work to the left and trade from one place value at a time. Have students complete the page independently.

Extra Credit *Biography*

An American inventor, Samuel Morse, struggled for many years before his inventions, the electric telegraph and Morse code were recognized. Morse was born in Massachusetts in 1791, and studied to be an artist. On a trip home from Europe, Morse heard his shipmates discussing the idea of sending electricity over wire. Intrigued, Morse spent the rest of the voyage formulating his ideas about how this could be accomplished. Morse taught at a university in New York City, and used his earnings to continue development of his telegraph. After five years, Morse demonstrated his invention, but found very little support. After years of requests for support, Congress finally granted Morse $30,000 to test his invention. He dramatically strung a telegraph wire from Washington, D.C. to Baltimore, Maryland, and relayed the message, "What hath God wrought" using Morse code. Morse's persistence finally won him wealth and fame. A statue honoring him was unveiled in New York City one year before his death in 1872.

Extra Credit *Applications*

Have a student write the primary United States time zones across the board. Discuss how this pattern continues around the world. Divide students into groups and provide them with globes or flat maps. Have students choose various cities in the United States and elsewhere in the world, and determine what the time would be in those cities when it is 6:00 AM in their home city. Have students make another list of cities without times indicated to exchange with classmates to figure time comparisons.

100

Bibliography

Suggested Mathematics Software for Grades 4–6

The list that follows is only a sampling of the many interesting and effective commercial software programs that are available in the marketplace. These and many of the mathematical programs that are published every year have a high degree of correlation with the objectives, goals and content of **Modern Curriculum Press Mathematics.**

Estimation Quick Solve. Minneapolis, MN: MECC, 1990. (Apple)

This program offers practice in estimation skills in a timed game. Playing against the computer or opponents, students are given enough time to estimate an answer to problems involving operations, whole numbers, fractions, decimals, percents, and measurements.

Exploring Measurement, Time, and Money, Level III. Dayton, NJ: IBM, 1990. (IBM)

Combining a tutorial with drill and practice, this program focuses on rounding, measurements of weight, length, time, money, and temperature.

The Factory: Strategies in Problem Solving. Cappo, Marge and Mike Fish. Scotts Valley, CA: Wings for Learning, 1991.

Students make decisions on how to operate an assembly line that turns out geometric products. They choose the appropriate tools and the most logical order to create a specified end-product.

The Geometric preSupposer: Points and Lines. Education Development Center, Judah L. Schwartz, Michal Yerushalmy. Pleasantville, NY: Sunburst Communications, 1986. (Apple, IBM)

Students learn about basic geometric concepts by drawing shapes and experimenting with lengths, angles, perimeter, area, and circumference.

How the West Was One + Three × Four. Bonnie Seiler. Pleasantville, NY: Sunburst Communications, 1989. (Apple, IBM)

A train and a stagecoach race each other down a number trail. In this problem-solving game, students practice working with the order of operations and the use of parentheses.

Math Blaster Mystery. Torrance, CA: Davidson and Associates, 1991. (Apple, IBM, Mac)

Everyone loves a mystery! In four different activities, students build problem-solving skills as they use whole numbers, fractions, decimals, percents, and pre-algebraic equations.

Math Shop. Jefferson City, MO: Scholastic, 1990. (Apple, IBM, Mac)

In the familiar setting of a mall, students become clerks at stores and practice their math skills involving basic arithmetic operations, fractions, decimals, percents, ratios, linear measurement, coins, and equations with two variables.

New Math Blaster Plus. Torrance, CA: Davidson, 1990. (Apple, IBM, Mac)

Using fast-paced, arcade-like games, this program builds skills in addition, subtraction, multiplication, division, fractions, decimals, and percents. Teachers can print customized tests.

Number Maze Decimals and Fractions. Scotts Valley, CA: Great Wave Software, 1990. (Mac)

Students gain admittance to fascinating mazes by answering questions involving decimals and fractions. Students must add and subtract decimals and fractions and convert from decimals to fractions.

Number Munchers. Minneapolis, MN: MECC, 1986. (Apple, IBM, Mac)

Students control a number-munching monster. If the monster eats the correct answer, the student moves on to the next level. The program drills concepts such as multiples 2–20, factoring of numbers 3–99, prime numbers 1–99, and equality and inequality.

Return of the Dinosaurs. Oklahoma City, OK: American Educational Computer, 1988. (Apple, IBM)

A dinosaur is running through the town and someone must catch it and take it back to its home. In this problem-solving program, students use thinking, observing, and planning skills to guess the best way to catch up with the dinosaur. Along the way, they learn how to use a database and maps.

Stickybear Math 2. Norfolk, CT: Optimum Resource, Inc., 1989. (Apple, IBM)

This program features the adventures of the colorful character Stickybear. Students solve sets of multiplication and division problems of increasing difficulty.

Super Solvers: Outnumbered! Fremont, CA: The Learning Company, 1990. (IBM)

This program, part of a series, immerses students in the treachery of the Master of Mischief who wants to take over the TV station and show boring shows all day. Students use math skills to solve problems and amass enough clues to ruin his plans. This is an exciting, fun-filled program.

Scope and Sequence

	K	1	2	3	4	5	6
READINESS							
Attributes	■						
Shapes	■	■					
Colors	■	■	■				
NUMERATION							
On-to-one correspondence	■						
Understanding numbers	■	■	■				
Writing numbers	■	■					
Counting objects	■	■	■				
Sequencing numbers	■	■	■	■	■		
Numbers before and after	■	■	■	■	■		
Ordering numbers			■	■	■	■	■
Comparing numbers	■	■	■	■	■	■	■
Grouping numbers	■	■	■	■	■		
Ordinal numbers	■	■	■	■			
Number words		■	■	■	■	■	■
Expanded numbers		■	■	■	■	■	■
Place value		■	■	■	■	■	■
Skip-counting		■	■	■	■	■	
Roman numerals			■	■	■		
Rounding numbers				■	■	■	■
Squares and square roots				■			

Scope and Sequence for MODERN CURRICULUM PRESS MATHEMATICS

Scope and Sequence

	K	1	2	3	4	5	6
Primes and composites				■	■	■	■
Multiples					■	■	■
Least common multiples						■	■
Greatest common factors						■	■
Exponents							■
ADDITION							
Addition facts	■	■	■	■	■	■	■
Fact families		■	■	■	■	■	
Missing addends	■	■	■	■	■		
Adding money	■	■	■	■	■	■	■
Column addition		■	■	■	■	■	■
Two-digit addends		■	■	■	■	■	
Multidigit addends			■	■	■	■	■
Addition with trading		■	■	■	■	■	■
Basic properties of addition				■	■	■	■
Estimating sums				■	■	■	
Addition of fractions				■	■	■	■
Addition of mixed numbers				■	■	■	■
Addition of decimals				■	■	■	■
Rule of order				■	■	■	■
Addition of customary measures						■	■

Scope and Sequence for MODERN CURRICULUM PRESS MATHEMATICS

Scope and Sequence

	K	1	2	3	4	5	6
Addition of integers							■
SUBTRACTION							
Subtraction facts	■	■	■	■	■	■	■
Fact families		■	■	■	■	■	
Missing subtrahends		■	■				
Subtracting money	■	■	■	■	■	■	■
Two-digit numbers		■	■	■	■	■	
Multidigit numbers			■	■	■	■	■
Subtraction with trading		■	■	■	■	■	■
Zeros in the minuend				■	■	■	■
Basic properties of subtraction				■	■	■	■
Estimating differences				■	■	■	■
Subtraction of fractions				■	■	■	■
Subtraction of mixed numbers						■	■
Subtraction of decimals				■	■	■	■
Rule of order				■	■	■	■
Subtraction of customary measures						■	■
Subtraction of integers							■
MULTIPLICATION							
Multiplication facts			■	■	■	■	■
Fact families			■	■	■		

Scope and Sequence

	K	1	2	3	4	5	6
Missing factors					■		
Multiplying money			■	■	■	■	■
Multiplication by powers of ten				■	■	■	■
Multidigit factors				■	■	■	■
Multiplication with trading				■	■	■	■
Basic properties of multiplication			■	■	■	■	■
Estimating products				■	■	■	■
Rule of order				■	■	■	■
Multiples					■	■	■
Least common multiples						■	■
Multiplication of fractions						■	■
Factorization						■	■
Multiplication of mixed numbers							■
Multiplication of decimals					■	■	■
Exponents							■
Multiplication of integers							■
DIVISION							
Division facts				■	■	■	■
Fact families				■	■		
Divisibility rules				■		■	■
Two-digit quotients				■	■	■	■

Scope and Sequence for MODERN CURRICULUM PRESS MATHEMATICS

Scope and Sequence

	K	1	2	3	4	5	6
Remainders				■	■	■	■
Multidigit quotients					■	■	■
Zeros in quotients					■	■	■
Division by multiples of ten					■	■	■
Two-digit divisors					■	■	■
Properties of division					■	■	
Averages				■	■	■	■
Greatest common factors						■	■
Division of fractions						■	■
Division of mixed numbers						■	■
Division of decimals						■	■
Division by powers of ten						■	■
MONEY							
Counting pennies	■	■	■	■	■		
Counting nickels	■	■	■	■	■		
Counting dimes	■	■	■	■	■		
Counting quarters		■	■	■	■		
Counting half-dollars				■	■	■	
Counting dollar bills		■	■	■	■		
Writing dollar and cents signs		■	■	■	■	■	■
Matching money with prices	■	■	■				

Scope and Sequence for MODERN CURRICULUM PRESS MATHEMATICS

Scope and Sequence

	K	1	2	3	4	5	6
Determining amount of change	■	■	■				
Determining sufficient amount		■	■				
Determining which coins to use		■	■				
Addition	■	■	■	■	■	■	■
Subtraction	■	■	■	■	■	■	■
Multiplication			■	■	■	■	■
Division					■	■	■
Rounding amounts of money				■	■	■	■
Finding fractions of amounts					■	■	■
Buying from a menu or ad			■	■	■	■	■

FRACTIONS

	K	1	2	3	4	5	6
Understanding equal parts	■	■	■	■			
One half	■	■	■	■			
One fourth	■	■	■	■			
One third	■	■	■	■			
Identifying fractional parts of figures			■	■	■	■	■
Identifying fractional parts of sets			■	■	■	■	■
Finding unit fractions of numbers				■	■	■	
Equivalent fractions				■	■	■	■
Comparing fractions				■	■	■	■
Simplifying fractions					■	■	■

Scope and Sequence

	K	1	2	3	4	5	6
Renaming mixed numbers					■	■	■
Addition of fractions				■	■	■	■
Subtraction of fractions				■	■	■	■
Addition of mixed numbers					■	■	■
Subtraction of mixed numbers						■	■
Multiplication of fractions						■	■
Factorization						■	■
Multiplication of mixed numbers						■	■
Division of fractions						■	■
Division of mixed numbers						■	■
Renaming fractions as decimals							■
Renaming fractions as percents							■
DECIMALS							
Place value				■	■	■	■
Reading decimals				■	■	■	■
Writing decimals				■	■	■	■
Converting fractions to decimals				■	■	■	■
Writing parts of sets as decimals				■	■	■	
Comparing decimals				■	■	■	■
Ordering decimals							■
Addition of decimals				■	■	■	■

Scope and Sequence

	K	1	2	3	4	5	6
Subtraction of decimals				■	■	■	■
Rounding decimals				■		■	■
Multiplication of decimals					■	■	■
Division of decimals						■	■
Renaming decimals as percents							■
GEOMETRY							
Polygons	■	■	■	■	■	■	■
Sides and corners of polygons			■	■	■		
Lines and line segments					■	■	■
Rays and angles					■	■	■
Measuring angles						■	■
Symmetry			■			■	■
Congruency				■	■	■	■
Similar figures					■	■	■
Circles						■	■
MEASUREMENT							
Non-standard units of measure	■	■					
Customary units of measure		■	■	■	■	■	■
Metric units of measure	■	■	■	■	■	■	■
Renaming customary measures					■	■	■
Renaming metric measures					■	■	■

Scope and Sequence for MODERN CURRICULUM PRESS MATHEMATICS

Scope and Sequence

	K	1	2	3	4	5	6
Selecting appropriate units			■	■	■	■	
Estimating measures		■	■	■	■	■	
Perimeter by counting	■	■	■				
Perimeter by formula			■	■	■	■	■
Area of polygons by counting			■	■			
Area of polygons by formula					■	■	■
Volume by counting				■			
Volume by formula					■	■	■
Addition of measures						■	■
Subtraction of measures						■	■
Circumference of circles							■
Area of circles							■
Surface area of space figures							■
Estimating temperatures				■			
Reading temperature scales			■	■			
TIME							
Ordering events	■						
Relative time	■						
Matching values	■	■	■	■	■		
Calendars	■	■	■	■			
Days of the week	■	■	■	■			

Scope and Sequence

	K	1	2	3	4	5	6
Months of the year	■	■	■	■			
Telling time to the hour	■	■	■	■			
Telling time to the half-hour		■	■	■			
Telling time to the five-minutes			■	■	■		
Telling time to the minute			■	■	■		
Understanding AM and PM					■		
Time zones					■		
GRAPHING							
Tables		■	■	■	■	■	■
Bar graphs	■	■	■	■	■	■	■
Picture graphs			■	■	■		■
Line graphs					■	■	■
Circle graphs						■	■
Tree diagrams						■	
Histograms							■
Ordered pairs				■	■	■	■
PROBABILITY							
Understanding probability					■	■	■
Listing outcomes					■	■	■
Means and medians						■	
Circle graphs						■	■

Scope and Sequence

	K	1	2	3	4	5	6
Tree diagrams						■	■
Histograms							■
RATIOS AND PERCENTS							
Understanding ratios					■	■	■
Equal ratios						■	■
Proportions							■
Scale drawings						■	■
Ratios as percents						■	■
Percents as fractions						■	■
Fractions as percents						■	■
Finding the percents of numbers						■	■
INTEGERS							
Understanding integers							■
Addition of integers							■
Subtraction of integers							■
Multiplication of integers							■
Graphing integers on coordinate planes							■
PROBLEM SOLVING							
Creating an algorithm from a word problem		■	■	■	■	■	■
Selecting the correct operation		■	■	■	■	■	■
Using data			■	■	■	■	■

Scope and Sequence

	K	1	2	3	4	5	6
Reading a chart			■	■	■	■	■
Using a four-step plan				■	■	■	■
Drawing a picture				■	■	■	■
Acting it out				■	■	■	■
Making a list				■	■	■	■
Making a tally				■	■	■	
Making a table				■	■	■	■
Making a graph				■	■	■	
Guessing and checking					■	■	■
Looking for a pattern					■	■	■
Making a model					■		
Restating the problem					■	■	■
Selecting notation					■	■	■
Writing an open sentence					■	■	■
Using a formula					■	■	■
Identifying a subgoal						■	■
Working backwards						■	■
Determining missing data						■	■
Collecting data						■	■
Solving a simpler but related problem						■	■
Making a flow chart							■

Scope and Sequence	K	1	2	3	4	5	6
CALCULATORS							
Calculator codes				■	■	■	■
Equal key				■	■	■	■
Operation keys				■	■	■	■
Square root key				■			
Clear key				■	■	■	■
Clear entry key				■	■	■	■
Money				■	■	■	■
Unit prices				■	■	■	
Fractions				■	■		
Percents					■		
Banking				■	■	■	
Inventories					■		
Averages					■		
Rates						■	■
Formulas						■	■
Cross multiplication							■
Functions							■
Binary numbers							■
Repeating decimals							■
Statistics							■

Table of Common Errors

This **Table of Common Errors** is designed to help the teacher understand the thinking patterns and potential errors that students commonly commit in the course of learning the content in *Modern Curriculum Press Mathematics*. Familiarity with this list can help the teacher forestall errant thinking and save much time used in reteaching.

In the **Correcting Common Errors** feature in each lesson in this **Teacher's Guide,** one of the errors that is relevant to that lesson is discussed in detail and a suggestion is given for its remediation. Collectively, in any chapter, all of the common errors for that mathematical topic are discussed and abundant assistance is given to the teacher for rectifying the situations.

Numeration

1. The student mistakes a number for another because it has been carelessly written.

2. When writing numerals for a number that is expressed in words, a student fails to use zeros as placeholders where they are needed.

   Write forty thousand, fifty-two in numerals.

   ▶ **4,052**

3. The student confuses the names of place values.

4. The student confuses the names of periods with those of place values.

   Write 42,652 in words.

   ▶ Four **thousand** two, six hundred fifty-two

5. The student fails to write the money sign and/or decimal point in an answer involving money.

   $$\begin{array}{r} \$5.65 \\ +\ 3.39 \\ \hline \end{array}$$
   ▶ **904**

6. When ordering whole numbers, the student incorrectly compares single digits regardless of place value.

   The student thinks that 203 is less than 45 because 2 is less than 4 and 3 is less than 5.

7. The student rounds down when the last significant digit is 5.

   ▶ $7,546 \approx$ **7,000**

8. The student rounds progressively from digit to digit until the designated place value is reached.

   Round 3,456 to thousands.

   ▶ **3,460 $\approx$ 3,500 $\approx$ 4,000**

9. The student changes the common interval when skip-counting.

   ▶ 2, 4, 6, 8, **11, 14, 17**

10. When evaluating a base number raised to an exponent, the student multiplies the base number by the exponent.

    ▶ $2^4 =$ **2 × 4** $= 8$

Addition and Subtraction

1. The student is unsure of the basic facts of addition and/or subtraction.

2. The student copies the problem incorrectly.

3. The student makes simple addition errors when adding numbers with two or more digits.

$$\begin{array}{r} 64 \\ +28 \\ \hline \blacktriangleright \quad 95 \end{array}$$

4. The student thinks that a number plus zero is zero.

▶ $6 + 0 = 0$

5. The student adds during a subtraction computation, or vice versa.

$$\begin{array}{r} 46 \\ -27 \\ \hline \blacktriangleright \quad 23 \end{array} \qquad \begin{array}{r} 28 \\ +42 \\ \hline 66 \end{array}$$

6. The student computes horizontal equations from left to right regardless of the operations.

▶ $3 + 2 \times 5 = 25$

7. When doing a computation involving several numbers, the student omits a number.

Find the sum of 23, 36, 54, 88, and 75.

$$\begin{array}{r} 23 \\ 36 \\ 88 \\ +75 \\ \hline 222 \end{array}$$

8. The student forgets the partial sum when adding a column of addends.

$$\begin{array}{r} 6 \\ 9 \\ +3 \\ \hline 12 \end{array}$$

9. The student omits the regrouped value.

$$\begin{array}{r} 47 \\ +23 \\ \hline \blacktriangleright \quad 60 \end{array}$$

10. The student fails to rename and places more than one digit in a column in an addition problem.

$$\begin{array}{r} 56 \\ +78 \\ \hline \blacktriangleright \quad 1,214 \end{array}$$

11. In an addition problem, the student writes the tens digit as part of the answer and regroups the ones.

$$\begin{array}{r} 4 \\ 56 \\ +78 \\ \hline \blacktriangleright \quad 161 \end{array}$$

12. The student rounds the answer rather than the components of the problem.

Estimate the sum of 356 and 492.

$$\begin{array}{r} 356 \\ +492 \\ \hline \blacktriangleright \quad 848 \approx 800 \end{array}$$

13. The student does not align the numbers properly when adding or subtracting whole numbers.

$$\begin{array}{r} 43 \\ +\ 29 \\ \hline \blacktriangleright \quad 659 \end{array} \qquad \begin{array}{r} 307 \\ +1\ 2 \\ \hline 409 \end{array}$$

14. The student incorrectly adds or subtracts from left to right.

$$\begin{array}{r} 1 \\ 47 \\ +\ 91 \\ \hline \blacktriangleright \quad 39 \end{array}$$

15. The student confuses addition and subtraction by one with either addition and subtraction of zero or with multiplication by one.

$$6 + 1 = \mathbf{6}$$
▶ $6 - 1 = \mathbf{6}$

16. The student adds or subtracts digits from different columns.

$$
\begin{array}{r}
67 \\
+\ 2 \\
\hline
\mathbf{89}
\end{array}
\qquad
\begin{array}{r}
35 \\
-\ 2 \\
\hline
\mathbf{13}
\end{array}
$$
▶

17. The student makes simple subtraction errors when subtracting numbers with two or more digits.

$$
\begin{array}{r}
5\,1\,4 \\
\cancel{6}\cancel{4} \\
-2\,8 \\
\hline
\mathbf{37}
\end{array}
$$
▶

18. The student thinks that a number minus zero is zero.

▶ $6 - 0 = \mathbf{0}$

19. The student thinks that zero minus another number is zero.

▶ $0 - 6 = \mathbf{0}$

20. When creating fact families, the student applies commutativity to subtraction.

$8 - 5 = 3$ $\mathbf{5 - 8 = 3}$

$8 - 3 = 5$ ▶ $\mathbf{3 - 8 = 5}$

21. The student brings down the digit in the subtrahend when the corresponding minuend digit is a zero.

$$
\begin{array}{r}
20 \\
-13 \\
\hline
\mathbf{13}
\end{array}
$$
▶

22. In multidigit subtraction problems, the student correctly renames the zero in the tens place but does not decrease the digit to the left of the zero.

$$
\begin{array}{r}
9 \\
1\,0\,12 \\
6\ \cancel{0}\ \cancel{2} \\
-4\,3\,7 \\
\hline
2\,6\,5
\end{array}
$$
▶

23. In a multidigit subtraction problem, the student ignores the zero and renames from the digit to the left of the zero.

$$
\begin{array}{r}
5\ 10\,12 \\
\cancel{6}\ \cancel{0}\ \cancel{2} \\
-4\,3\,7 \\
\hline
1\,7\,5
\end{array}
$$
▶

24. In a multidigit subtraction problem, the student correctly trades from the digit to the left of the zero and renames the zero as 10 but fails to reduce the 10 by one when the second regrouping is done.

$$
\begin{array}{r}
5\ \mathbf{10\,12} \\
\cancel{6}\ \cancel{0}\ \cancel{2} \\
-4\,3\,7 \\
\hline
1\,7\,5
\end{array}
$$
▶

25. The student does not trade in a subtraction problem but finds the difference between the smaller digit and the larger regardless of their positions.

$$
\begin{array}{r}
537 \\
-182 \\
\hline
455
\end{array}
$$
▶

26. In a subtraction problem, the student does not decrease the digit to the left after trading.

$$
\begin{array}{r}
12 \\
5\,\cancel{2} \\
-3\,8 \\
\hline
24
\end{array}
$$
▶

Multiplication

1. The student is unsure of the basic multiplication facts.

2. The student mistakes a multiplication sign for an addition sign, or vice versa.

3. The student thinks that one times any number is one.

$$6 \times 1 = 1$$

$$
\begin{array}{r}
36 \\
\times 31 \\
\hline
\blacktriangleright \quad 11 \\
108 \\
\hline
1{,}091
\end{array}
$$

4. The student confuses multiplication by zero with multiplication by one, thinking that any number times zero is that number.

$$6 \times 0 = 6$$

$$
\begin{array}{r}
26 \\
\times 30 \\
\hline
\blacktriangleright \quad 26 \\
78 \\
\hline
806
\end{array}
$$

5. The student ignores the parentheses in a horizontal equation and does not distribute multiplication over addition or subtraction.

▶ $3 \times (5 + 2) = 15 + 2 = 17$

6. The student makes simple multiplication mistakes in multidigit multiplication problems.

$$
\begin{array}{r}
234 \\
\times 12 \\
\hline
\blacktriangleright \quad 498 \\
234 \\
\hline
2{,}838
\end{array}
$$

7. The student is unsure of how many zeros are in the product when multiplying by a multiple of ten.

$$
\begin{array}{r}
756 \\
\times 500 \\
\hline
\blacktriangleright \quad 37{,}800
\end{array}
$$

8. The student makes simple addition errors in multidigit multiplication problems.

$$
\begin{array}{r}
234 \\
\times 12 \\
\hline
468 \\
234 \\
\hline
\blacktriangleright \quad 2{,}838
\end{array}
$$

9. The student misaligns the digits in the partial products in a multiplication problem.

$$
\begin{array}{r}
234 \\
\times 12 \\
\hline
468 \\
\blacktriangleright \quad 234 \\
\hline
702
\end{array}
$$

10. The student writes the tens digit as part of the answer and regroups the ones.

$$
\begin{array}{r}
5 \\
47 \\
\times 5 \\
\hline
\blacktriangleright \quad 253
\end{array}
$$

11. The student does not regroup or fails to add the regrouped value.

$$
\begin{array}{r}
36 \\
\times 7 \\
\hline
\blacktriangleright \quad 212
\end{array}
$$

12. When multiplying whole numbers, the student adds the renamed digit before multiplying.

$$
\begin{array}{r}
4 \\
56 \\
\times 7 \\
\hline
\blacktriangleright \quad 632
\end{array}
$$

13. In a multidigit multiplication problem, the student confuses the regrouped digit for the second partial product with that of the first partial product.

$$
\begin{array}{r}
1 \\
4 \\
36 \\
\times 27 \\
\hline
252 \\
\blacktriangleright \quad 102 \\
\hline
1{,}272
\end{array}
$$

Division

1. The student is unsure of the basic division facts.

2. In a division problem, if either term is one, the student thinks the answer is one.

   ▶ $6 \div 1 = 1$
        $1 \div 6 = 1$

3. The student does not realize that division by zero has no meaning.

   ▶ $6 \div 0 = 6$
        $0 \div 0 = 0$

4. The student places the initial quotient digit over the wrong digit in the dividend.

   ▶   **7,231**
        2)1,462

5. The student ignores initial digits in the dividend that are less than the divisor.

          231
   ▶   2)1,462

6. The student does not align the digits carefully and consequently misses one of the digits in the dividend.

   ▶   **4 7** R1
       3)1,462
        1 2
          22
          21
           1

7. The student fails to record a zero in a quotient.

   ▶   **3 9** R7
      12)3,715
        3 6
          115
          108
           7

8. The first estimated partial quotient is too low so the student subtracts and divides again and places an extra digit in the quotient.

   ▶   **3,106** R7
      23)9,345
        6 9
        24
        23
         145
         138
          7

9. The student fails to subtract the last time or fails to record the remainder as part of the quotient.

         309
      12)3,715
        3 6
         115
   ▶   **108**

10. The student records the remainder as the last digit of the quotient.

    ▶   571
        3)172
         15
         22
         21
          1

11. The student records a remainder that is larger than the divisor.

         155 R**5**
    ▶   4)625

12. The student incorrectly subtracts in a division problem.

         73 R1
       3)230
         21
    ▶   10
          9
          1

13. The student fails to subtract before bringing down the next digit in a division problem.

$$\begin{array}{r} 41 \\ 6\overline{)256} \\ \underline{24} \\ 6 \\ \underline{6} \end{array}$$

14. The student incorrectly multiplies in a division problem.

$$\begin{array}{r} 86\ \text{R2} \\ 3\overline{)230} \\ \underline{21} \\ 20 \\ \underline{18} \\ 2 \end{array}$$

15. The student stops dividing before all the digits in the dividend have been divided.

▶
$$\begin{array}{r} 1{,}257 \\ 6\overline{)75{,}421} \end{array}$$

16. When factoring for primes, the student lists a composite number for a prime number.

▶ $64 = \mathbf{8} \times \mathbf{4} \times \mathbf{2} \times \mathbf{2}$

Measurement

1. The student is confused about how to read fractional measures on a ruler.

▶ **7 inches**

2. The student does not properly align the object to be measured with the point that represents zero on the ruler.

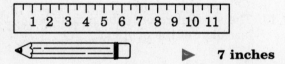

3. The student reads the small hand of a standard clock as minutes and the large hand as hours.

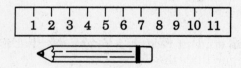

▶ **3:00**

4. The student reads the time on a standard clock incorrectly when the hour hand is between two numbers.

▶ **7:53**

5. The student confuses the meaning of 12:00 AM and 12:00 PM.

6. The student selects an incorrect time zone for a designated city.

7. The student is confused about whether the time is earlier to the east or to the west.

8. When converting from one measure to another, the student uses the incorrect equivalent unit.

▶ *The student thinks that 1 foot = 10 inches, or that 1 meter = 1000 centimeters.*

9. When converting from one measure to another, the student incorrectly divides when he or she should have multiplied, or vice versa.

Change 36 yards to feet.

▶ 36 yards = **12 feet**

10. The student compares measurements expressed in different units.

> 3 ft < 15 in.
> 6 m > 2 km

11. The student fails to simplify measures after adding denominate numbers.

> $\begin{array}{r} 6 \text{ ft} \quad 8 \text{ in.} \\ +4 \text{ ft} \ 10 \text{ in.} \\ \hline 10 \text{ ft } \textbf{18 in.} \end{array}$

12. The student finds the difference between the two numbers instead of regrouping denominate numbers.

> $\begin{array}{r} 6 \text{ ft} \quad 8 \text{ in.} \\ -4 \text{ ft} \ 10 \text{ in.} \\ \hline 2 \text{ ft } \textbf{2 in.} \end{array}$

13. The student regroups an incorrect value when subtracting denominate numbers.

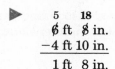

Geometry

1. The student confuses the meanings of line and line segment.

2. The student confuses perpendicular lines with parallel lines.

3. The student confuses the diameter of a circle with its radius.

4. The student does not name the vertex of an angle by the middle of the three letters.

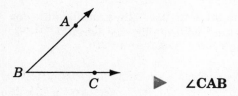

> ∠CAB

5. The student uses the opposite scale on a protractor when measuring an angle.

6. When bisecting a line or angle, the student uses the incorrect point for the compass point and fails to find the midpoint.

7. The student changes the setting of the compass when the construction calls for identical settings.

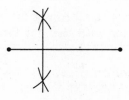

8. The student calculates the measure of a missing angle of a triangle by subtracting only one of the given angles from 180°.

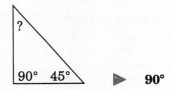

> **90°**

9. The student confuses the names of basic polygons.

10. The student thinks the geometrical term *similar* applies to two figures that are somewhat like each other.

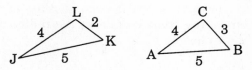

11. The student confuses the meaning of *similar* and *congruent*.

12. The student identifies a line as a line of symmetry, even though it doesn't create two congruent parts.

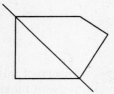

13. The student confuses corresponding parts of congruent figures with different orientations.

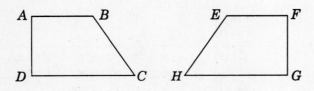

14. The student confuses the formula for finding the perimeter with that for finding the area.

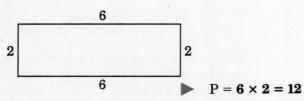

$P = 6 \times 2 = 12$

15. The student omits one or more of the dimensions of a polygon when computing its perimeter.

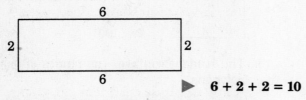

$6 + 2 + 2 = 10$

16. When computing the circumference of a circle, the student multiplies the radius by *pi*.

17. When computing the circumference of a circle, the student squares the radius and multiplies by *pi*.

18. When calculating the area of a parallelogram, the student uses the slant height for one of the measures.

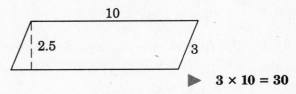

$3 \times 10 = 30$

19. When computing the area of a triangle, the student forgets to divide by 2.

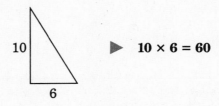

$10 \times 6 = 60$

20. When computing the area of a circle, the student multiplies the diameter by *pi*.

21. When computing the area of a circle, the student squares the diameter and multiplies by *pi*.

22. When computing the area of a circle, the student multiplies the radius by pi and then squares that product.

23. The student confuses the names of basic space figures.

24. The student uses the area formula for finding volume, or vice versa.

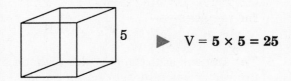

$V = 5 \times 5 = 25$

25. The student confuses the concepts of surface area and volume of space figures.

26. The student finds the surface area of only the visible faces of space figures.

Fractions

1. The student counts the wrong number of parts of a picture when naming equivalent fractions.

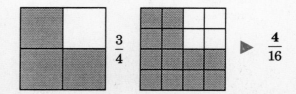

2. The student transposes the numerator and the denominator of a fraction.

3. The student confuses the meanings of greatest common factor and least common multiple.

4. To find equivalent fractions, the student uses addition or subtraction instead of multiplication or division.

▶ $\dfrac{3+4}{4+4} = \dfrac{7}{8}$

$\dfrac{9-8}{12-8} = \dfrac{1}{4}$

5. When raising fractions to a common denominator, the student uses a number that is not a multiple of all the denominators.

▶ $\dfrac{3}{4} = \dfrac{}{8}$

$\dfrac{2}{3} = \dfrac{}{8}$

6. When simplifying a fraction the student does not divide by the greatest common factor.

▶ $\dfrac{16 \div 4}{24 \div 4} = \dfrac{4}{6}$

7. When the denominator is a multiple of the numerator, the student simplifies the fraction by dividing the denominator by the numerator.

▶ $\dfrac{3}{9} = 3$

8. When renaming an improper fraction as a mixed number, the student writes the tens digit in the numerator as the whole number.

▶ $\dfrac{17}{9} = 1\dfrac{7}{9}$

9. When renaming an improper fraction as a mixed number, the student writes only the whole number and ignores the fraction.

▶ $\dfrac{13}{4} = 3$

10. When converting mixed numerals to improper fractions, the student multiplies the whole number by the denominator but fails to add the numerator to this product.

▶ $3\dfrac{1}{2} = \dfrac{6}{2}$

11. When renaming a whole number as a mixed number in a subtraction problem, the student fails to reduce the whole number by one.

▶ $3 = 3\dfrac{5}{5}$

12. When comparing fractions with the same numerator, the student compares only the denominators.

▶ $\dfrac{1}{3} < \dfrac{1}{4}$

13. When using cross multiplication to compare two fractions, the student confuses the products and reverses the sign.

▶ $\begin{array}{cc} \dfrac{2}{3} > \dfrac{3}{4} \\ 9 \quad 8 \end{array}$

14. When renaming a fraction as a decimal, the student divides the denominator by the numerator instead of vice versa.

▶ $\dfrac{3}{4} = 1.33\dfrac{1}{3}$

15. When adding or subtracting fractions, the student adds or subtracts both the numerators and the denominators.

▶ $\dfrac{3}{5} + \dfrac{2}{3} = \dfrac{5}{8}$

16. When adding fractions, the student finds a common denominator but does not rename the numerators.

▶ $\dfrac{2}{3} + \dfrac{1}{5} = \dfrac{2}{15} + \dfrac{1}{15} = \dfrac{3}{15} = \dfrac{1}{5}$

17. When adding mixed numbers, the student fails to recognize an opportunity to simplify an answer.

$$3\frac{2}{3} = 3\frac{4}{6}$$
$$+4\frac{1}{2} = 4\frac{3}{6}$$

▶ $\qquad 7\frac{7}{6}$

18. When adding or subtracting mixed numbers, the student works only with the fractions.

$$5\frac{1}{2} = \frac{3}{6}$$
$$+6\frac{1}{3} = \frac{2}{6}$$

▶ $\qquad \frac{5}{6}$

19. When regrouping a mixed number in a subtraction problem, the student regroups the whole number but simply affixes a one to the existing numerator.

▶ $\quad 5\frac{1}{3} = 5\frac{5}{15} = 4\frac{15}{15}$
$$-2\frac{3}{5} = 2\frac{9}{15} = 2\frac{9}{15}$$

$$\qquad\qquad 2\frac{6}{15} = 2\frac{2}{5}$$

20. The student adds the one regrouped from the whole number to the numerator of the fraction.

▶ $\quad 7\frac{2}{5} = 6\frac{3}{5}$
$$-\quad \frac{3}{5} = \frac{3}{5}$$

$$\qquad\qquad 6$$

21. When finding a fractional part of a number, the student fails to either multiply by the numerator or divide by the denominator.

▶ $\frac{3}{7}$ of $21 = 63$

$\frac{3}{7}$ of $21 = 3$

22. The student confuses multiplication of fractions with addition and finds a common denominator before operating only on the numerators.

▶ $\frac{2}{3} \times \frac{3}{4} = \frac{8}{12} \times \frac{9}{12} = 72$

23. When multiplying two fractions, the student cross multiplies to find the product.

▶ $\frac{2}{3} \times \frac{3}{4} = \frac{8}{9}$

24. The student confuses multiplication of fractions with division and inverts the second factor.

▶ $\frac{2}{3} \times \frac{3}{4} = \frac{2}{3} \times \frac{4}{3} = \frac{8}{9}$

25. When multiplying or dividing mixed numbers, the student operates on the whole numbers and the fractions separately.

▶ $2\frac{3}{5} \times 4\frac{1}{6} = 8\frac{3}{30} = 8\frac{1}{10}$

$6\frac{3}{7} \div 2\frac{3}{5} = \frac{3}{7} \times \frac{5}{3} = 3\frac{15}{21} = 3\frac{5}{7}$

26. The student fails to invert the second fraction in a division problem.

▶ $\frac{3}{4} \div \frac{2}{3} = \frac{3}{4} \times \frac{2}{3} = \frac{1}{2}$

27. When dividing fractions, the student inverts the first fraction instead of the second one.

▶ $\frac{3}{4} \div \frac{2}{9} = \frac{4}{3} \times \frac{2}{9} = \frac{8}{27}$

28. When dividing by a mixed number, the student reciprocates only the fraction.

▶ $2\frac{2}{3} \div 1\frac{1}{2} = 2\frac{2}{3} \times 1\frac{2}{1} = \frac{8}{3} \times \frac{3}{1} = \frac{24}{3} = 8$

29. When the divisor is a mixed number, the student renames it as an improper fraction but fails to invert it before multiplying.

▶ $\frac{3}{4} \div 1\frac{1}{2} = \frac{3}{4} \times \frac{3}{2} = \frac{9}{8} = 1\frac{1}{8}$

Ratios, Proportion, and Percents

1. The student transposes the terms of a ratio.

   The student thinks that a 7 to 5 ratio is the same as a 5 to 7 ratio.

2. The student does not multiply or divide the terms of a ratio by the same number when finding equal ratios.

   ▶ $3:5 = 9:\textbf{25}$

3. When calculating a missing term in two equal ratios, the student multiplies by the other term.

   ▶ $\dfrac{3}{4} = \dfrac{\textbf{12}}{24}$

4. When finding a distance based on a scale, the student uses the actual distance and ignores the scale.

5. When solving a proportion for a missing term, the student confuses the means and the extremes of the proportion.

   $$\textbf{2:3} :: 4:x$$
   ▶ $$\textbf{2} \times \textbf{3} = 4 \times x$$
   $$6 = 4x$$
   $$1\tfrac{1}{2} = x$$

6. The student renames a ratio or fraction as a percent but fails to write the percent sign.

   ▶ $\dfrac{1}{2} = 0.50 = \textbf{50}$

7. When renaming a percent as a decimal, the student simply omits the percent sign.

   ▶ $3.5\% = \textbf{3.5}$

8. When renaming a decimal number with more than two decimal places as a percent, the student moves the point as far right as the last place.

   ▶ $0.275 = \textbf{275}\%$

9. When renaming a fraction as a percent, the student renames the fraction as a decimal and then simply affixes a percent sign to the decimal equivalent.

   ▶ $\dfrac{3}{4} = 0.75 = \textbf{.75}\%$

10. When finding the percent of a number, the student multiplies by the percent without renaming it as a decimal or fraction.

    20% of 45

    ▶ $\textbf{20} \times 45 = 900$

Decimals

1. The student confuses the terms used for place values of decimals with those of whole numbers.

   Find the place value of the underlined digit in 4.6̲39.

    tens

2. When writing decimal numbers, the student does not place the nonzero digits in the correct places.

   Write three and four thousandths in numerals.

   ▶ 3.040

3. The student omits the decimal point in a decimal number.

   Write fourteen hundredths in numerals.

   ▶ **14**

4. When ordering decimals, the student's answer is based on the number of digits.

The student thinks that because 0.2381 has five digits and 0.47 has only three, the first number is larger.

5. The student rounds to the wrong decimal place.

Round 0.6524 to the nearest thousandths.

▶ **0.65**

6. When rounding decimal numbers, the student replaces values beyond the designated place value with zeros.

Round 0.6548 to hundredths.

▶ 0.65**00**

7. The student places the repeating bar over digits which don't repeat in a decimal number.

▶ $0.91 \ldots = 0.\overline{91}$

8. When interpreting a repeating decimal, the student thinks the next digit to the right of the overlined digit is always zero.

9. When adding or subtracting decimals, the student operates on the whole number parts and the decimal parts of the numbers separately.

$$\begin{array}{r} 22.3 \\ +17.9 \\ \hline \end{array}$$
▶ 39.**12**

10. The student confuses multiplication of decimals with addition and aligns the decimal point of the product with the decimal points of the factors.

$$\begin{array}{r} 2.6 \\ \times 3.2 \\ \hline 5\,2 \\ 78 \\ \hline \end{array}$$
▶ 83.2

11. The student counts decimal places in the factors in a multiplication problem from the left instead of from the right.

$$\begin{array}{r} 3.01 \\ \times 1.12 \\ \hline 6\,02 \\ 30\,1 \\ 301 \\ \hline \end{array}$$
▶ **33.**712

12. The student annexes zeros to the right of a nonzero digit in a product.

$$\begin{array}{r} 0.3 \\ \times 0.02 \\ \hline \end{array}$$
▶ 0.6**00**

13. When dividing a number by a decimal number, the student ignores the decimal point in the divisor and places the decimal point in the quotient over that in the dividend.

▶ $0.3\overline{)1.68}$ with quotient 0.56

14. The student fails to use a zero as a place holder in a quotient.

▶ $5\overline{)0.230}$ with quotient $.\,46$

15. The student moves the decimal point to the left when multiplying by a power of 10 or to the right when dividing by a power of 10.

▶ $7.6 \times 100 = \mathbf{0.076}$
   $7.6 \div 100 = \mathbf{760}$

16. The student moves the decimal point in the divisor but fails to move it accordingly in the dividend.

▶ $0.05_\wedge\overline{)6.25}$ with quotient 1.25

17. The student fails to divide to at least one place beyond the place value to which the answer is to be rounded.

Express $\frac{1}{3}$ as a decimal to the nearest hundredths.

▶ $3\overline{)1.00}$ with quotient **0.33**

Integers

1. When comparing integers, the student ignores the signs and compares absolute values.

   ▶ $^-3 > {}^+1$

2. When adding a positive and negative number, the student adds absolute values and uses the sign of the greater.

   ▶ $^-4 + {}^+7 = {}^+\mathbf{11}$

3. When adding a positive and negative number, the student finds the difference between them but always makes the answer either positive or negative.

   ▶ $^+7 + {}^-6 = {}^-\mathbf{1}$
   $^-3 + {}^+2 = {}^+\mathbf{1}$

4. When subtracting integers, the student adds them and uses the sign of the greater absolute value.

   ▶ $^+5 - {}^-7 = {}^-\mathbf{2}$

5. When subtracting integers, the student changes the operation sign to addition but fails to change the sign of the subtrahend.

   ▶ $^-7 - {}^-5 = {}^-7 + {}^-5 = {}^-\mathbf{12}$

6. When multiplying or dividing any negative numbers, the student always makes the answer negative.

   ▶ $^-2 \times {}^-2 \times {}^-2 \times {}^-2 = {}^-\mathbf{16}$
   $^-15 \div {}^-3 = {}^-\mathbf{5}$

7. When graphing integers, the student becomes confused about which quadrant to use for any given pair.

Graphing and Probability

1. The student fails to divide by the number of addends when computing the average.

2. The student confuses the meanings of mean and median.

3. The student reverses the numbers in an ordered pair.

   The student thinks (3,2) is the same as (2,3).

4. The student locates points on a grid by first counting up and then over.

   Graph (2,3) and label it point A.

   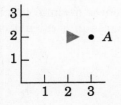

5. The student uses the next larger interval on the scale when reading data on a bar graph.

   ▶ **A is 3.**

6. The student confuses the data from the two categories in creating a double-bar graph.

7. The student reads a point on a line graph by counting forward from the next higher interval.

   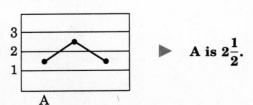

   ▶ **A is $2\frac{1}{2}$.**

8. The student does not refer to a key when interpreting a picture graph.

How many books does Bob own?

$2\frac{1}{2}$

9. The student interprets percents on a circle graph as amounts of the whole rather than as rates.

10. The student confuses the number of possible outcomes with the number of chances not to get the outcome.

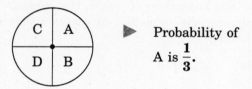

Probability of A is $\frac{1}{3}$.

11. The student may fail to write all possible combinations for each and every choice.

Problem Solving Errors

1. The student uses the wrong operation or operations to solve a problem.

2. The student chooses the operation based on the relative size of the numbers in the problem.

3. The student thinks that all numbers in a word problem must be used to get the solution.

4. The student does not use all the relevant information given in a problem.

5. The student does not read the problem carefully, but selects key words to determine the operation or operations.

6. The student does not answer the question posed in the problem.

7. The student is confused because the problem contains unfamiliar words or situations.

8. The student does not find all the possible solutions because he or she is not systematic when making a list.

9. The student does not collect enough data to establish a pattern.

10. The student misreads a chart or table used in a problem.

11. The student's diagram does not faithfully depict the situation in the problem.

12. The student does not check if the data that is being tested makes sense in the problem.

Calculators

1. The student enters incorrect codes into the calculator.

2. When entering a percent in a calculator, the student forgets to enter the percent sign.

3. When entering a subtraction into a calculator, the student enters the subtrahend before the minuend.

4. When entering a division into a calculator, the student enters the divisor before the dividend.

5. The student fails to enter decimal points at the appropriate places in a calculator code.

Comparing Numbers

pages 1-2

Objective

To compare whole numbers

Materials

*number line 0 through 100

Mental Math

Have students solve:
1. 8×7 (56)
2. $6 \div 2$ (3)
3. 1 dozen + 5 (17)
4. $16 - 11$ (5)
5. 1 foot + 11 inches (23 in.)
6. $6 + 7$ (13)
7. $6 + 7 + 8$ (21)
8. 7¢ less than 8 dimes (73¢)

Skill Review

Have a student write a number between 10 and 99 on the board and tell how many tens and ones are in the number. Have another student write the number that is made by reversing those digits and tell the number of tens and ones. Group students in pairs to continue the activity.

*One or one set per class. Unmarked items for individual students.

---

1 ADDITION AND SUBTRACTION FACTS

Understanding Whole Numbers

Polly is buying dog food for her puppy. Which can will cost Polly the least amount of money?

We want to know which can has the lowest price. We know the prices for the different dog foods are __49¢__, __43¢__ and __45¢__.
To compare the 3 prices, we can find them on a number line.

40 41 42 43 44 45 46 47 48 49 50

✔ The set of **whole numbers** starts with 0 and goes as far as we need it to go. The whole numbers 0, 1, 2, 3, 4, 5, 6, 7, 8 and 9 are called **digits.** On a number line, the number to the right is always greater.

49 is greater than 45 $49 > 45$

✔ A number to the left is always less.

43 is less than 45 $43 < 45$

The dog food that will cost Polly the least costs __43¢__.

Getting Started

Write the missing whole numbers.

1. 26, __27__, 28   2. __90__ comes after 89.   3. 58 is between __57__ and __59__.

Compare these numbers. Write < or > in the circle.

4. 32 ⊘ 38   5. 76 ⊙ 53   6. 27. ⊘ 72

Write the numbers in order from least to greatest.

7. 32, 46, 15   __15__, __32__, __46__   8. 13, 43, 29   __13__, __29__, __43__

1

Teaching the Lesson

Introducing the Problem Have a student read the problem aloud. Tell students to use the picture to tell what they are to find and what facts they already know. (Which of the three cans costs the least, if the prices are 49¢, 43¢ and 45¢?) Read with the students as they complete the sentences to solve the problem. Have students check the number line to see if 43¢ comes before 45¢ and 49¢.

Developing the Skill Display the number line and write **84** and **46** on the board. Ask students to tell which of these **whole numbers** comes first on the number line (46) and why. (4 tens is less than 8 tens.) Repeat for more groups of 2 numbers. Remind students that numbers on the number line get larger to the right and smaller to the left. Write **78** and **72** on the board and ask students to compare the tens. (same) Tell students they must compare the ones when the numbers in the tens columns are the same. Ask students to tell which whole number is greater and why. (78 is greater

because 8 ones is more than 2 ones.) Write **26, 62, 21** on the board and ask students to compare the numbers by placing them in order from least to greatest. (21, 26, 62) Review the **greater than** (>) and **less than** (<) signs and remind students that the point of the sign will always be aimed at the smaller number, with the wide end next to the larger number.

1

Practice

Write the missing whole numbers.

1. 35, 36, __37__, __38__

2. __46__, __47__, 48, 49

3. 63, 62, __61__, __60__

4. 91, __92__, __93__, 94

5. 7, __8__, __9__, 10, __11__

6. __17__, __18__, __19__, 20, 21

7. 82, __81__, 80, __79__, 78

8. __50__ comes after 49.

9. __57__ is between 56 and 58.

10. 77 is between __76__ and __78__.

Compare these numbers. Write < or > in the circle.

11. 39 ⊙> 36

12. 73 ⊙> 17

13. 21 ⊙< 30

14. 17 ⊙> 16

15. 81 ⊙< 89

16. 32 ⊙< 42

17. 63 ⊙> 36

18. 22 ⊙< 33

19. 89 ⊙> 40

20. 25 ⊙< 29

21. 57 ⊙> 51

22. 48 ⊙> 40

23. 48 ⊙< 50

24. 96 ⊙< 99

25. 15 ⊙< 35

Write the numbers in order from least to greatest.

26. 75, 36, 48

__36__, __48__, __75__

27. 35, 87, 29

__29__, __35__, __87__

28. 23, 57, 45

__23__, __45__, __57__

29. 83, 47, 58

__47__, __58__, __83__

30. 22, 57, 39

__22__, __39__, __57__

31. 18, 81, 88

__18__, __81__, __88__

32. 25, 36, 12, 19

__12__, __19__, __25__, __36__

33. 47, 58, 75, 21

__21__, __47__, __58__, __75__

34. 67, 9, 42, 83

__9__, __42__, __67__, __83__

2

Correcting Common Errors

Some students may have trouble writing missing numbers in order. Point to 46 on a number line and ask students to name a number that is:

1 more. (47)

1 less. (45)

between 47 and 49. (48)

Enrichment

Tell students they are to write a beverage menu of 5 different-priced drinks so that no drink costs less than 29¢ and the purchase of any 2 drinks is not more than 85¢.

Practice

Remind students to compare 2-digit numbers by comparing the tens and then the ones. Encourage students to use the number line to check their work. Have students complete the page independently.

Extra Credit   *Numeration*

Challenge students to think of one thing they do that does not involve numbers. On the board list all the activities they think of that do not involve numbers. Encourage students to be critical of each other's suggestions and try to think of a way in which the activity is impossible without numbers. For example, if they suggest that painting or drawing is something they can do without numbers, point out that drawing paper comes in numbered packs, and is sold for a given price. See if they can think of a single thing that can be done without numbers.

Addition Facts

pages 3-4

Objective

To review basic addition facts

Materials

*carpet squares or masking tape
*number line through 20
number lines

Mental Math

Ask students to tell how old each will
be in 1978, if:
1. Eli, born in 1952? (26)
2. Selma, born in 1969? (9)
3. Grant, born in 1892? (86)
4. Corey, born in 1966? (12)
5. Evan, born in 1918? (60)
6. Katina, born in 1924? (54)
7. Cliff, born in 1971? (7)
8. Chim, born in 1953? (25)

Skill Review

Write the number names for 0 through
18 in mixed order on the board. Have
students write the number names and
their corresponding numbers in order
on paper.

Reviewing Addition Facts

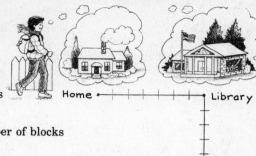

Aaron left home early one
morning to walk to the library,
before he went to school. How
many blocks did he walk on his
way to school?

We want to know the number of blocks
Aaron walked all together.

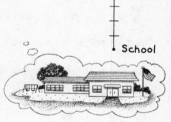

We know that he walked ___7___ blocks
from his house to the library.

He walked another ___6___ blocks from the
library to school.
To find the total number of blocks, we add

___7___ and ___6___.

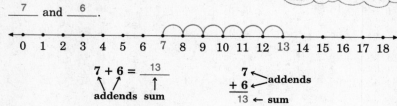

0 1 2 3 4 5 6 7 8 9 10 11 12 13 14 15 16 17 18

$7 + 6 =$ ___13___
         addends sum

$\begin{array}{r} 7 \\ + 6 \\ \hline 13 \end{array}$ ← addends
                    ← sum

$7 + 6 = 13$ is called a **number sentence.**

Aaron walked ___13___ blocks from his home to school.

Getting Started

Complete the number sentences.

1. $4 + 2 =$ ___6___   2. $7 + 9 =$ ___16___   3. $8 + 3 =$ ___11___

4. $2 + 9 =$ ___11___   5. $5 + 6 =$ ___11___   6. $8 + 8 =$ ___16___

Add.

7. $\begin{array}{r} 8 \\ + 7 \\ \hline 15 \end{array}$ 8. $\begin{array}{r} 4 \\ + 1 \\ \hline 5 \end{array}$ 9. $\begin{array}{r} 9 \\ + 9 \\ \hline 18 \end{array}$ 10. $\begin{array}{r} 5 \\ + 5 \\ \hline 10 \end{array}$ 11. $\begin{array}{r} 3 \\ + 6 \\ \hline 9 \end{array}$ 12. $\begin{array}{r} 9 \\ + 4 \\ \hline 13 \end{array}$

3

Teaching the Lesson

Introducing the Problem   Have a student read the
problem aloud to the class. Ask students to tell what they
are to find and what facts are given in the problem and on
the map. (The total distance Aaron went if he walked 7
blocks and then 6 more blocks.) Read with students as they
complete the information and solution sentences.

Developing the Skill   Use carpet squares or tape to
make a number line on the floor. Write **6 + 8 =** on the
board. Have a student stand on 6 and go 8 spaces forward
toward the larger numbers. Ask students to tell the number
where the student stopped. (14) Write on the board:

$6 + 8 = 14$
addends sum

$\begin{array}{r} 6 \\ +8 \\ \hline 14 \end{array}$ ← addends
                    ← sum

Tell students the first **addend** tells the first position, the second **addend** tells how many to go forward and the stopping
place is the **sum.** Write more addition problems in horizontal and vertical forms and have students use the number line
to find the sums. Tell students that an addend plus an addend equals a sum. Write **6 + 8 = 14** on the board and tell
students this is called a **number sentence.**

Practice

Complete the number sentences.

1. $7 + 1 = \underline{8}$ 2. $7 + 7 = \underline{14}$ 3. $2 + 1 = \underline{3}$ 4. $6 + 9 = \underline{15}$

5. $8 + 5 = \underline{13}$ 6. $4 + 4 = \underline{8}$ 7. $1 + 1 = \underline{2}$ 8. $5 + 9 = \underline{14}$

9. $9 + 3 = \underline{12}$ 10. $6 + 8 = \underline{14}$ 11. $8 + 2 = \underline{10}$ 12. $7 + 6 = \underline{13}$

13. $7 + 8 = \underline{15}$ 14. $2 + 4 = \underline{6}$ 15. $8 + 9 = \underline{17}$ 16. $8 + 4 = \underline{12}$

Add.

17. $\begin{array}{r}7\\+4\\\hline 11\end{array}$	18. $\begin{array}{r}5\\+2\\\hline 7\end{array}$	19. $\begin{array}{r}4\\+3\\\hline 7\end{array}$	20. $\begin{array}{r}9\\+5\\\hline 14\end{array}$	21. $\begin{array}{r}2\\+6\\\hline 8\end{array}$	22. $\begin{array}{r}1\\+8\\\hline 9\end{array}$
23. $\begin{array}{r}3\\+8\\\hline 11\end{array}$	24. $\begin{array}{r}5\\+4\\\hline 9\end{array}$	25. $\begin{array}{r}9\\+1\\\hline 10\end{array}$	26. $\begin{array}{r}9\\+8\\\hline 17\end{array}$	27. $\begin{array}{r}2\\+7\\\hline 9\end{array}$	28. $\begin{array}{r}9\\+7\\\hline 16\end{array}$
29. $\begin{array}{r}1\\+3\\\hline 4\end{array}$	30. $\begin{array}{r}5\\+7\\\hline 12\end{array}$	31. $\begin{array}{r}1\\+6\\\hline 7\end{array}$	32. $\begin{array}{r}3\\+9\\\hline 12\end{array}$	33. $\begin{array}{r}3\\+3\\\hline 6\end{array}$	34. $\begin{array}{r}4\\+9\\\hline 13\end{array}$
35. $\begin{array}{r}2\\+8\\\hline 10\end{array}$	36. $\begin{array}{r}1\\+7\\\hline 8\end{array}$	37. $\begin{array}{r}1\\+2\\\hline 3\end{array}$	38. $\begin{array}{r}8\\+6\\\hline 14\end{array}$	39. $\begin{array}{r}1\\+9\\\hline 10\end{array}$	40. $\begin{array}{r}3\\+4\\\hline 7\end{array}$
41. $\begin{array}{r}3\\+2\\\hline 5\end{array}$	42. $\begin{array}{r}6\\+1\\\hline 7\end{array}$	43. $\begin{array}{r}8\\+1\\\hline 9\end{array}$	44. $\begin{array}{r}2\\+2\\\hline 4\end{array}$	45. $\begin{array}{r}4\\+7\\\hline 11\end{array}$	46. $\begin{array}{r}5\\+8\\\hline 13\end{array}$

Apply

Solve these problems.

47. Megan bought a wool scarf for $7 and a pair of mittens for $6. How much did she spend?
$13

48. Earle's club has 9 members. Each member has asked one friend to join the club. How many members will there be?
18 members

4

4

Column Addition

pages 5-6

Objective

To add 1-digit numbers in a column

Materials

*floor number line through 20 counters

Mental Math

Ask students to name:
1. 1/2 of 10 (5)
2. 6 − 4 + 10 (12)
3. 6 × (3 + 6) (54)
4. 7 ÷ 1 + 12 (19)
5. 6 quarters + 8 dimes ($2.30)
6. 45 min before 10:35 (9:50)
7. The value of the 6 in 168 (tens)
8. 3/8 + 5/8 (8/8 or 1)

Skill Review

Write on the board:

10	13	15	16
(9 + 1)	(9 + 4)	(9 + 6)	(9 + 7)
(8 + 2)	(8 + 5)	(8 + 7)	(8 + 8)
(7 + 3)	(7 + 6)	(7 + 8)	(7 + 9)
(6 + 4)	(6 + 7)	(6 + 9)	
(5 + 5)	(5 + 8)		
(4 + 6)	(4 + 9)		
(3 + 7)			
(2 + 8)			
(1 + 9)			

Have students write all the basic addition facts under each corresponding sum.

Column Addition

The first United States astronauts orbited the earth in 1962. How many orbits did these Americans complete in that year?

Date	Astronaut	Orbits
February 20	John Glenn	3
May 24	Scott Carpenter	3
October 3	Wally Schirra	6

We want to find the total number of orbits all the astronauts made in 1962.

We know that Glenn orbited __3__ times; Carpenter, __3__ times; and Schirra, __6__ times.

To find this **total** or **sum,** we add __3__, __3__ and __6__.

We can add only two numbers at a time.

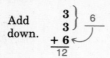

Add down.
$$\begin{array}{r} 3 \\ 3 \\ +6 \\ \hline 12 \end{array} \Big\} 6$$

$$\begin{array}{r} 3 \\ 3 \\ +6 \\ \hline 12 \end{array} \Big\} 9$$ Add up to check.

The American astronauts completed __12__ orbits in 1962.

Getting Started

Add and check.

1. $\begin{array}{r}1\\3\\+5\\\hline 9\end{array}$	2. $\begin{array}{r}2\\6\\+3\\\hline 11\end{array}$	3. $\begin{array}{r}3\\4\\+2\\\hline 9\end{array}$	4. $\begin{array}{r}6\\3\\+1\\\hline 10\end{array}$	5. $\begin{array}{r}7\\1\\+7\\\hline 15\end{array}$
6. $\begin{array}{r}6\\3\\2\\+4\\\hline 15\end{array}$	7. $\begin{array}{r}3\\2\\4\\+9\\\hline 18\end{array}$	8. $\begin{array}{r}5\\4\\5\\+3\\\hline 17\end{array}$	9. $\begin{array}{r}7\\2\\3\\5\\+1\\\hline 18\end{array}$	10. $\begin{array}{r}8\\1\\6\\2\\+2\\\hline 19\end{array}$

5

Teaching the Lesson

Introducing the Problem Direct the students' attention to the chart as you tell them that on February 20, John Glenn made 3 orbits around the earth. Ask students to tell about Scott Carpenter's adventure in space. (He made 3 orbits on May 24.) Ask students what else the chart tells us. (Wally Shirra made 6 orbits on October 3.) Have students read the problem to the left of the chart and tell what it asks them to do. (Tell how many times the Americans orbited the earth in 1962.) Have students fill in the information and solve the problem in the model. Have them complete the solution sentence.

Developing the Skill Write **3 + 2 + 5** vertically on the board. Have students place counters in 3 groups of 3, 2 and 5. Tell students to combine the first 2 groups and tell how many. (5) Have students combine the 2 remaining groups and tell how many in all. (10) Tell students to again place their counters in 3 groups of 3, 2 and 5, and combine the last 2 groups and tell how many. (7) Have students combine the 2 remaining groups. (10) Ask students if they have the same sum when they add forward and backward. (yes) Talk through the problem on the board as you add down and then check by adding up the column. Repeat for more problems of 3 addends, being careful that the sum of any 2 addends is 9 or less and no addends are 0.

Practice

Add and check.

1. 6
    3
    + 6
    ——
    15

2. 2
    3
    + 4
    ——
    9

3. 2
    7
    + 2
    ——
    11

4. 8
    1
    + 4
    ——
    13

5. 6
    3
    + 2
    ——
    11

6. 4
    5
    + 3
    ——
    12

7. 1
    6
    + 3
    ——
    10

8. 5
    2
    + 5
    ——
    12

9. 5
    2
    + 1
    ——
    8

10. 7
    2
    + 6
    ——
    15

11. 2
    5
    + 3
    ——
    10

12. 6
    2
    + 3
    ——
    11

13. 4
    2
    + 7
    ——
    13

14. 3
    1
    + 6
    ——
    10

15. 9
    1
    + 5
    ——
    15

16. 8
    1
    5
    1
    + 4
    ——
    19

17. 1
    6
    3
    + 7
    ——
    17

18. 5
    3
    4
    + 1
    ——
    13

19. 5
    1
    6
    + 2
    ——
    14

20. 6
    1
    3
    5
    + 3
    ——
    18

EXCURSION

Complete the boxes by adding each number at the top to each number on the left. Look for patterns.

+	5	3	7
8	13	11	15
18	23	21	25
28	33	31	35
38	43	41	45
48	53	51	55
58	63	61	65

+	2	6	4
9	11	15	13
19	21	25	23
29	31	35	33
39	41	45	43
49	51	55	53
59	61	65	63

+	9	7	5
7	16	14	12
27	36	34	32
47	56	54	52
67	76	74	72
87	96	94	92
97	106	104	102

6

6

Addition Properties

pages 7-8

Objective

To learn the order, grouping and zero properties of addition

Materials

*floor number line through 20

Mental Math

Ask students what number tells:
1. shoes in 7 pairs. (14)
2. minutes in 5 1/2 hours. (330)
3. 1/4 of 20. (5)
4. tens in 10,786. (8)
5. inches in 4 feet. (48)
6. eggs in 3 1/4 dozen. (39)
7. cups in a pint. (2)

Skill Review

Have students give all the basic addition facts with sums greater than 13 but less than 16. (5 + 9, 9 + 5, 8 + 6, 6 + 8, 7 + 7, 7 + 8, 8 + 7, 9 + 6, 6 + 9) Write the facts on the board as they are given.

Understanding Addition Properties

Understanding the basic properties of addition can help you find sums more easily.

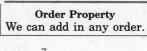

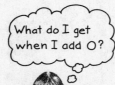

> **Order Property**
> We can add in any order.

$3 + 4 = \underline{}7$ $4 + 3 = \underline{}7$

> **Grouping Property**
> We can change the grouping. Remember to add the numbers in the parentheses first.

$(5 + 3) + 6 = ?$ $5 + (3 + 6) = ?$

$\underline{}8 + 6 = \underline{}14$ $5 + \underline{}9 = \underline{}14$

> **Zero Property**
> Adding zero does not affect the answer.

$6 + 0 = \underline{}6$ $0 + 3 = \underline{}3$

Getting Started

Complete the number sentences.

1. $5 + 0 = \underline{}5$
2. $(6 + 3) + 2 = \underline{}11$
3. $0 + 9 = \underline{}9$

4. $4 + (0 + 6) = \underline{}10$
5. $(2 + 7) + 0 = \underline{}9$
6. $5 + (3 + 5) = \underline{}13$

Add and check.

7.	8.	9.	10.	11.
6	3	9	1	7
2	9	3	4	5
+ 4	+ 4	0	5	0
12	16	+ 2	3	3
		14	+ 4	+ 2
			17	17

7

Teaching the Lesson

Introducing the Problem Have a student read aloud the thoughts of the children pictured. Read to the students the statement beside the picture. Now have students read the property names and examples as they complete the number sentences with you.

Developing the Skill Tell students they are going to test the three **addition properties.** Have a student write **4 + 2** on the board. Tell students the **order property** tells us we will get the same sum if we add forward or backward. Have the student find the sum both ways and tell if the sums are the same. (The sums are both 6.) Have other students test the order property by adding 2 numbers from 1 through 9. Write **4 + 5 + 2 + 3** on the board and tell students the **grouping property** tells us we will get the same sum if we add 2 and then 3 and then 5 and then 4, as when we add 4 + 5 and then 2 and then 3. Have a student work the problem aloud and write the sum. (14) Have another stu-

dent work the problem differently. Now write the problem vertically and have a student work the problem aloud. Write **3 + (4 + 2)** on the board and remind students that they are to always add the numbers in parentheses first. Have students similarly test the **zero property.**

Practice

Complete the number sentences.

1. 7 + 0 = __7__
2. (4 + 2) + 7 = __13__
3. 0 + 8 = __8__
4. (8 + 0) + 2 = __10__
5. 5 + (8 + 1) = __14__
6. (7 + 2) + 3 = __12__
7. (0 + 6) + 9 = __15__
8. (6 + 0) + 9 = __15__
9. 4 + (6 + 3) = __13__
10. (2 + 5) + 8 = __15__
11. 8 + (5 + 2) = __15__
12. 3 + (4 + 5) = __12__
13. 6 + (2 + 0) = __8__
14. 4 + (3 + 6) = __13__
15. (5 + 0) + 5 = __10__

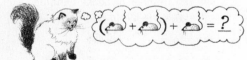

Add and check.

16.	17.	18.	19.	20.
5	2	0	1	6
3	7	8	7	0
+ 2	+ 0	+ 6	+ 2	+ 8
10	9	14	10	14

21.	22.	23.	24.	25.
4	3	8	7	9
0	8	1	1	0
+ 7	+ 0	+ 5	+ 8	+ 4
11	11	14	16	13

26.	27.	28.	29.	30.
0	3	5	2	8
5	6	9	2	0
+ 0	+ 1	+ 1	+ 5	+ 6
5	10	15	9	14

31.	32.	33.	34.	35.
9	1	4	6	7
0	4	5	8	2
6	3	3	0	5
+ 4	+ 0	+ 2	+ 5	+ 2
19	8	14	19	16

36.	37.	38.	39.	40.
3	2	5	1	4
0	4	4	0	4
8	3	2	5	3
2	4	1	8	3
+ 4	+ 2	+ 6	+ 2	+ 3
17	15	18	16	17

8

Subtraction Facts

pages 9-10

Objective
To review basic subtraction facts

Materials
*addition fact cards
*floor number line through 20

Mental Math
Dictate the following problems:
1. 10,000 + 400 (10,400)
2. 60 − 8 (52)
3. 100 + 40 − 60 (80)
4. 7 times 8 plus 3 (59)
5. 10 ÷ 2 (5)
6. 6 + (8 × 2) (22)
7. 3 dozen − 2 dozen (12)

Skill Review
Give each student an addition fact card. Have students group themselves in fact families so that all facts having the same sum are together. Then have students arrange themselves in fact family order, within each group.

Reviewing Subtraction Facts

Rinaldo's goal for this year is to read 12 books. So far, he has read 5 books. How many books must Rinaldo read to reach his goal?

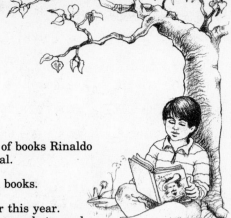

We want to know the number of books Rinaldo must still read to reach his goal.

Rinaldo's goal is to read __12__ books.

He has read __5__ books so far this year.
To find the number of books he needs to read,

we subtract __5__ from __12__.

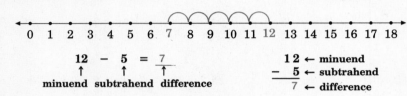

$$12 - 5 = 7$$

minuend subtrahend difference

$$
\begin{array}{r}
1\,2 \leftarrow \text{minuend} \\
-5 \leftarrow \text{subtrahend} \\
\hline
7 \leftarrow \text{difference}
\end{array}
$$

Rinaldo wants to read __7__ more books this year.

Getting Started

Complete the number sentences.

1. 8 − 7 = __1__   2. 11 − 3 = __8__   3. 15 − 9 = __6__   4. 12 − 5 = __7__

5. 10 − 8 = __2__   6. 8 − 6 = __2__   7. 18 − 9 = __9__   8. 13 − 4 = __9__

Subtract.

9.	10.	11.	12.	13.	14.
11 − 4 7	5 − 3 2	8 − 3 5	14 − 8 6	6 − 3 3	10 − 4 6

9

Teaching the Lesson

Introducing the Problem   Have a student read the problem aloud. Ask students what they need to find. (how many more books Rinaldo needs to read) Ask students what they already know. (He wants to read 12 books and has read 5 so far.) Have students read the sentences aloud as they solve the problem.

Developing the Skill   Write **8 − 4 =** on the board. Have a student stand on 8 on the number line and go 4 spaces backward, toward the smaller numbers. Ask another to tell where the student stopped. (4) Write on the board:

minuend

8   −   4 = 4 ← difference   8 minuend
          ↖   −4 subtrahend
       subtrahend   4 difference

Explain that 8 − 4 = 4 is a number sentence for subtraction. Tell students the first number is the **minuend** and tells where to start on the number line, the second is the **subtrahend** and tells how far to walk backwards and the stopping place is the **difference.** Give several examples using the horizontal and vertical notation. Have students act out the problems.

Practice

Complete the number sentences.

1. $9 - 2 = \underline{7}$ 2. $6 - 2 = \underline{4}$ 3. $4 - 1 = \underline{3}$ 4. $7 - 3 = \underline{4}$

5. $15 - 8 = \underline{7}$ 6. $4 - 3 = \underline{1}$ 7. $11 - 8 = \underline{3}$ 8. $13 - 5 = \underline{8}$

9. $2 - 1 = \underline{1}$ 10. $10 - 3 = \underline{7}$ 11. $5 - 1 = \underline{4}$ 12. $6 - 5 = \underline{1}$

13. $9 - 8 = \underline{1}$ 14. $13 - 9 = \underline{4}$ 15. $14 - 5 = \underline{9}$ 16. $8 - 1 = \underline{7}$

17. $15 - 6 = \underline{9}$ 18. $9 - 9 = \underline{0}$

Subtract.

19. $10 \\ \underline{-\ 1} \\ 9$	20. $9 \\ \underline{-\ 6} \\ 3$	21. $12 \\ \underline{-\ 8} \\ 4$	22. $10 \\ \underline{-\ 7} \\ 3$	23. $9 \\ \underline{-\ 4} \\ 5$	24. $3 \\ \underline{-\ 1} \\ 2$	
25. $12 \\ \underline{-\ 4} \\ 8$	26. $11 \\ \underline{-\ 2} \\ 9$	27. $4 \\ \underline{-\ 2} \\ 2$	28. $14 \\ \underline{-\ 9} \\ 5$	29. $14 \\ \underline{-\ 6} \\ 8$	30. $11 \\ \underline{-\ 6} \\ 5$	
31. $17 \\ \underline{-\ 8} \\ 9$	32. $9 \\ \underline{-\ 1} \\ 8$	33. $10 \\ \underline{-\ 2} \\ 8$	34. $16 \\ \underline{-\ 7} \\ 9$	35. $7 \\ \underline{-\ 4} \\ 3$	36. $13 \\ \underline{-\ 8} \\ 5$	
37. $16 \\ \underline{-\ 9} \\ 7$	38. $10 \\ \underline{-\ 6} \\ 4$	39. $9 \\ \underline{-\ 7} \\ 2$	40. $15 \\ \underline{-\ 6} \\ 9$	41. $12 \\ \underline{-\ 7} \\ 5$	42. $11 \\ \underline{-\ 9} \\ 2$	
43. $10 \\ \underline{-\ 9} \\ 1$	44. $7 \\ \underline{-\ 6} \\ 1$	45. $12 \\ \underline{-\ 3} \\ 9$	46. $11 \\ \underline{-\ 7} \\ 4$	47. $10 \\ \underline{-\ 5} \\ 5$	48. $14 \\ \underline{-\ 7} \\ 7$	

Apply

Solve these problems.

49. Butch made 9 sandwiches. His brothers ate 7 of them for lunch. How many sandwiches does Butch have left to eat?
2 sandwiches

50. Suzanne bought toothpaste for $2, and a toothbrush for $1. How much change will she receive from a $10 bill?
$7

10

10

Subtraction Properties

pages 11-12

Objectives

To subtract using zero property
To recognize that addition and subtraction check each other

Materials

*floor numbers line through 20
counters

Mental Math

Have students name the numbers that are 20 more and 20 less than:

1. 87. (107, 67)
2. 2 tens 1 one. (41, 1)
3. 365. (385, 345)
4. 1,000. (1,020, 980)
5. 705. (725, 685)
6. 194. (214; 174)
7. 8 tens 6 ones. (106, 66)

Skill Review

Write **7○2 = 5** on the board. Ask students what sign would complete this number sentence. (−) Continue for more addition and subtraction sentences in both horizontal and vertical forms. Do not use zero in the problems.

Understanding Subtraction Properties

Understanding the properties of subtraction can help you find **differences** more easily.

What do I get when I subtract 0?

Zero Properties	
Subtracting zero does not affect the answer.	Subtracting a number from itself leaves zero.

$$8 - 0 = \underline{8} \qquad 7 - 7 = \underline{0}$$

$$3 - 0 = \underline{3} \qquad 0 - 0 = \underline{0}$$

Checking Subtraction
Addition and subtraction check each other.

$$8 - 5 = \underline{3} \qquad 8 - 3 = \underline{5}$$

$$3 + 5 = \underline{8} \qquad 5 + 3 = \underline{8}$$

What's the best way to check subtraction?

These four number sentences are called a **fact family**.

✔ Remember that the order of the numbers in a subtraction sentence is always important.

$$3 - 2 \text{ is } not \text{ the same as } 2 - 3.$$

Getting Started

Complete the number sentences.

1. $7 - 7 = \underline{0}$ 2. $5 - 0 = \underline{5}$ 3. $4 - 0 = \underline{4}$ 4. $6 - 6 = \underline{0}$

5. $8 - 8 = \underline{0}$ 6. $9 - 0 = \underline{9}$ 7. $5 - 5 = \underline{0}$ 8. $3 - 0 = \underline{3}$

Subtract and check.

9.	10.	11.	12.	13.	14.
$\begin{array}{r}14\\-\ 6\\\hline 8\end{array}$	$\begin{array}{r}9\\-\ 2\\\hline 7\end{array}$	$\begin{array}{r}6\\-\ 0\\\hline 6\end{array}$	$\begin{array}{r}17\\-\ 8\\\hline 9\end{array}$	$\begin{array}{r}12\\-\ 7\\\hline 5\end{array}$	$\begin{array}{r}10\\-\ 6\\\hline 4\end{array}$

11

Teaching the Lesson

Introducing the Problem Have a student read aloud the thoughts of the children pictured, and the statement beside the picture. Have students complete the number sentences.

Developing the Skill Have a student stand on 9 on the number line and then go zero spaces backward. Write **9 − 0 =** on the board and ask students to tell the difference, when zero is subtracted from 9. (9) Have the student move 9 spaces backward and tell the difference. (0) Write **9 − 9 = 0** on the board. Have students practice with more subtraction sentences. Now write the numbers **5, 3** and **2** on the board. Have students write addition or subtraction sentences using these 3 numbers. (5 − 3 = 2, 2 + 3 = 5, 5 − 2 = 3, 3 + 2 = 5). Have other students write these 4 sentences in vertical form. Tell students that since these 4 sentences are the only ones that can be made with 2, 3, and 5, they are called a **fact family**. Write **5 − 3 = 2** on

the board and ask students which of the other sentences could be used to check this difference. (2 + 3 = 5 or 3 + 2 = 5) Experiment with subtraction sentences from other fact families.

Practice

Complete the number sentences.

1. $7 - 0 = \underline{7}$ 2. $0 - 0 = \underline{0}$ 3. $10 - 3 = \underline{7}$ 4. $6 - 6 = \underline{0}$

5. $8 - 3 = \underline{5}$ 6. $2 - 1 = \underline{1}$ 7. $9 - 0 = \underline{9}$ 8. $3 - 0 = \underline{3}$

9. $4 - 4 = \underline{0}$ 10. $11 - 3 = \underline{8}$ 11. $7 - 4 = \underline{3}$ 12. $1 - 1 = \underline{0}$

Subtract and check.

13. $\begin{array}{r}11\\-\ 6\\\hline 5\end{array}$	14. $\begin{array}{r}6\\-1\\\hline 5\end{array}$	15. $\begin{array}{r}14\\-\ 7\\\hline 7\end{array}$	16. $\begin{array}{r}9\\-5\\\hline 4\end{array}$	17. $\begin{array}{r}4\\-1\\\hline 3\end{array}$	18. $\begin{array}{r}12\\-\ 4\\\hline 8\end{array}$
19. $\begin{array}{r}11\\-\ 7\\\hline 4\end{array}$	20. $\begin{array}{r}10\\-\ 7\\\hline 3\end{array}$	21. $\begin{array}{r}8\\-5\\\hline 3\end{array}$	22. $\begin{array}{r}16\\-\ 8\\\hline 8\end{array}$	23. $\begin{array}{r}13\\-\ 8\\\hline 5\end{array}$	24. $\begin{array}{r}10\\-\ 4\\\hline 6\end{array}$
25. $\begin{array}{r}15\\-\ 9\\\hline 6\end{array}$	26. $\begin{array}{r}5\\-0\\\hline 5\end{array}$	27. $\begin{array}{r}13\\-\ 4\\\hline 9\end{array}$	28. $\begin{array}{r}11\\-\ 8\\\hline 3\end{array}$	29. $\begin{array}{r}6\\-3\\\hline 3\end{array}$	30. $\begin{array}{r}14\\-\ 9\\\hline 5\end{array}$
31. $\begin{array}{r}9\\-2\\\hline 7\end{array}$	32. $\begin{array}{r}13\\-\ 6\\\hline 7\end{array}$	33. $\begin{array}{r}12\\-\ 7\\\hline 5\end{array}$	34. $\begin{array}{r}16\\-\ 9\\\hline 7\end{array}$	35. $\begin{array}{r}4\\-0\\\hline 4\end{array}$	36. $\begin{array}{r}8\\-2\\\hline 6\end{array}$
37. $\begin{array}{r}15\\-\ 8\\\hline 7\end{array}$	38. $\begin{array}{r}1\\-0\\\hline 1\end{array}$	39. $\begin{array}{r}11\\-\ 4\\\hline 7\end{array}$	40. $\begin{array}{r}18\\-\ 9\\\hline 9\end{array}$	41. $\begin{array}{r}12\\-\ 8\\\hline 4\end{array}$	42. $\begin{array}{r}9\\-6\\\hline 3\end{array}$
43. $\begin{array}{r}3\\-3\\\hline 0\end{array}$	44. $\begin{array}{r}10\\-\ 6\\\hline 4\end{array}$	45. $\begin{array}{r}8\\-8\\\hline 0\end{array}$	46. $\begin{array}{r}10\\-\ 9\\\hline 1\end{array}$	47. $\begin{array}{r}13\\-\ 7\\\hline 6\end{array}$	48. $\begin{array}{r}17\\-\ 9\\\hline 8\end{array}$
49. $\begin{array}{r}9\\-4\\\hline 5\end{array}$	50. $\begin{array}{r}9\\-9\\\hline 0\end{array}$	51. $\begin{array}{r}10\\-\ 2\\\hline 8\end{array}$	52. $\begin{array}{r}8\\-0\\\hline 8\end{array}$	53. $\begin{array}{r}14\\-\ 5\\\hline 9\end{array}$	54. $\begin{array}{r}12\\-\ 9\\\hline 3\end{array}$
55. $\begin{array}{r}7\\-5\\\hline 2\end{array}$	56. $\begin{array}{r}17\\-\ 8\\\hline 9\end{array}$	57. $\begin{array}{r}9\\-8\\\hline 1\end{array}$	58. $\begin{array}{r}7\\-7\\\hline 0\end{array}$	59. $\begin{array}{r}6\\-0\\\hline 6\end{array}$	60. $\begin{array}{r}13\\-\ 9\\\hline 4\end{array}$
61. $\begin{array}{r}14\\-\ 8\\\hline 6\end{array}$	62. $\begin{array}{r}5\\-5\\\hline 0\end{array}$	63. $\begin{array}{r}2\\-0\\\hline 2\end{array}$	64. $\begin{array}{r}15\\-\ 6\\\hline 9\end{array}$	65. $\begin{array}{r}2\\-2\\\hline 0\end{array}$	66. $\begin{array}{r}12\\-\ 5\\\hline 7\end{array}$

12

Correcting Common Errors

Some students may be confused by the two uses of zero in subtraction; e.g., as a number being subtracted and as an answer in a subtraction problem. To correct, give pairs of students sets of exercises similar to those shown below to help them see how zero can be a number to be subtracted and a number that is an answer.

$\begin{array}{r}5\\-0\\\hline 5\end{array}$	$\begin{array}{r}5\\-1\\\hline 4\end{array}$	$\begin{array}{r}5\\-2\\\hline 3\end{array}$	$\begin{array}{r}5\\-3\\\hline 2\end{array}$	$\begin{array}{r}5\\-4\\\hline 1\end{array}$	$\begin{array}{r}5\\-5\\\hline 0\end{array}$

Enrichment

Tell students to write all the addition facts that have an addend or sum of 9. Then write a subtraction sentence to check each fact.

Practice

Tell students to complete all the problems on the page. Remind students to check subtraction by adding in reverse order.

Extra Credit *Applications*

Have students make math skill clusters. Give each student or pair of students a diagram that shows a center circle as a hub, and six more circles on the ends of spokes pointing out from the center.

In the center circle of each cluster, write a math skill such as making change, bookkeeping, measurement, telling time, etc. Ask students to fill the other six circles around it with pictures of occupations that require that skill. They may cut pictures from magazines or draw them. Have students label each picture. The finished projects can be displayed on a bulletin board and used to discuss what importance math skills might have in their future occupations.

Missing Addends

pages 13-14

Objective

To find the missing addend

Materials

*floor number line through 20

Mental Math

Have students count by 2's, 3's, 5's,
10's or 25's to tell the total:
1. 7 quarters ($1.75)
2. 15 rows of 10 seats (150)
3. socks in 13 pairs (26)
4. 9 boxes of 25 pencils (225)
5. 8 groups of 3 girls (24)
6. plugs in 16 outlets (32)
7. feet in 9 yards (27)

Skill Review

Write __ + __ =10 on the board.
Have students complete the sentence
with any addition fact for 10. Ask students to write other facts for 10 which
could make the sentence true. Write
__ − __ =7 and continue the activity. Repeat for other addition and subtraction facts.

Finding Missing Addends

Yan-Wah belongs to the
photography club and she is
saving money to buy a camera.
So far, she has saved $9. How
much more does Yan-Wah need?

We want to find the amount that
Yan-Wah still has to save.

We know that the camera costs __$17__.

Yan-Wah has saved __$9__ so far.
We can write this problem as an addition
sentence, and let the letter n stand for the
missing addend.

$9 + n = 17

To solve this problem, we think of a related
subtraction sentence from the same fact family.

$17 − $9 = n$

Since $17 − $9 = __$8__, $n = __$8__.

Yan-Wah still needs __$8__ to buy the camera.

Getting Started

Write related subtraction sentences.

1. $5 + n = 12$
$12 − 5 = n$

2. $n + 7 = 16$
$16 − 7 = n$

3. $6 + n = 14$
$14 − 6 = n$

4. $n + 7 = 13$
$13 − 7 = n$

Use subtraction sentences to help find the missing addends.

5. $6 + n = 10$
$n = \underline{4}$

6. $n + 3 = 7$
$n = \underline{4}$

7. $n + 4 = 12$
$n = \underline{8}$

8. $7 + n = 15$
$n = \underline{8}$

9.
$$\begin{array}{r} 8 \\ + n \\ \hline 11 \end{array}$$
$n = \underline{3}$

10.
$$\begin{array}{r} n \\ + 9 \\ \hline 16 \end{array}$$
$n = \underline{7}$

11.
$$\begin{array}{r} 8 \\ + n \\ \hline 15 \end{array}$$
$n = \underline{7}$

12.
$$\begin{array}{r} n \\ + 9 \\ \hline 18 \end{array}$$
$n = \underline{9}$

13

Teaching the Lesson

Introducing the Problem Have a student read the
problem aloud. Have students study the picture to tell what
they are to find and what they already know. (How much
more than $9 is needed to buy a $17 camera.) Read aloud
with the students as they complete the sentences to solve
the problem.

Developing the Skill Have one student stand on 5 on
the number line and another student on 14. Tell students
we want to know how many spaces the first student will
have to move to reach the second student. Write **5 +
n =14** on the board and tell students the **n** stands for the
missing number. Have the first student walk ahead to 14, as
others count the steps. (9) Ask students what number n
stands for. (9) Write **n =9** under 5 + n =14. Now have the
student at 14 walk backward 5 spaces. Ask students the
number for n. (9) Tell students the student on 14 found the

missing addend by subtracting. Write **n =9** under 14 − 5 =
n. Write 5 x's on the board, and draw a circle around
them. Tell students this circle should have 12 x's in it and
they must find how many more x's need to be added. Write
5 + n =12 and **n + 5 = 12** on the board. Ask students to
give a subtraction problem which will help us find n. (12 −
5 = n) Continue for more missing addend problems.

Write related subtraction sentences.

1. $7 + n = 10$
$10 - 7 = n$

2. $4 + n = 12$
$12 - 4 = n$

3. $n + 9 = 15$
$15 - 9 = n$

4. $n + 6 = 14$
$14 - 6 = n$

5. $9 + n = 18$
$18 - 9 = n$

6. $n + 6 = 13$
$13 - 6 = n$

7. $3 + n = 3$
$3 - 3 = n$

8. $n + 8 = 10$
$10 - 8 = n$

9. $6 + n = 14$
$14 - 6 = n$

10. $8 + n = 14$
$14 - 8 = n$

11. $9 + n = 16$
$16 - 9 = n$

12. $8 + n = 12$
$12 - 8 = n$

Use subtraction sentences to help find the missing addends.

13. $3 + n = 5$
$n = \underline{2}$

14. $7 + n = 8$
$n = \underline{1}$

15. $n + 6 = 12$
$n = \underline{6}$

16. $4 + n = 10$
$n = \underline{6}$

17. $7 + n = 14$
$n = \underline{7}$

18. $n + 9 = 16$
$n = \underline{7}$

19. $n + 8 = 16$
$n = \underline{8}$

20. $9 + n = 14$
$n = \underline{5}$

21. $n + 6 = 13$
$n = \underline{7}$

22. $n + 8 = 12$
$n = \underline{4}$

23. $7 + n = 16$
$n = \underline{9}$

24. $8 + n = 13$
$n = \underline{5}$

25. $\begin{array}{r} 6 \\ + n \\ \hline 11 \end{array}$
$n = \underline{5}$

26. $\begin{array}{r} n \\ + 9 \\ \hline 14 \end{array}$
$n = \underline{5}$

27. $\begin{array}{r} 4 \\ + n \\ \hline 13 \end{array}$
$n = \underline{9}$

28. $\begin{array}{r} 8 \\ + n \\ \hline 15 \end{array}$
$n = \underline{7}$

29. $\begin{array}{r} n \\ + 2 \\ \hline 5 \end{array}$
$n = \underline{3}$

30. $\begin{array}{r} 8 \\ + n \\ \hline 9 \end{array}$
$n = \underline{1}$

31. $\begin{array}{r} 3 \\ + n \\ \hline 7 \end{array}$
$n = \underline{4}$

32. $\begin{array}{r} n \\ + 5 \\ \hline 8 \end{array}$
$n = \underline{3}$

33. $\begin{array}{r} 7 \\ + n \\ \hline 12 \end{array}$
$n = \underline{5}$

34. $\begin{array}{r} n \\ + 3 \\ \hline 3 \end{array}$
$n = \underline{0}$

35. $\begin{array}{r} 9 \\ + n \\ \hline 15 \end{array}$
$n = \underline{6}$

36. $\begin{array}{r} 8 \\ + n \\ \hline 16 \end{array}$
$n = \underline{8}$

14

Correcting Common Errors

If students have difficulty writing a related subtraction sentence to find a missing addend, have them practice writing fact families. Have students work with partners. Give each pair of students an addition fact and have them write the other facts in the fact family. For example:

$$7 + 8 = 15 \qquad 15 - 8 = 7$$
$$8 + 7 = 15 \qquad 15 - 7 = 8$$

Enrichment

Tell students they must write an addition problem and a related subtraction sentence where n plus 126 equals 3 more than 200, and then solve for n.

Practice

Remind students that a missing addend can be found by identifying a related subtraction sentence. Have students complete the page independently.

Extra Credit *Measurement*

Introduce students to non-standard units of body measure, especially the **cubit,** the distance from elbow to end of middle finger with arm bent; and the **foot,** the length of the human foot. Have them measure assigned items in the classroom using these non-standard units, and record their results. Have students suggest reasons in favor of using standard units of measure rather than non-standard units.

Problem Solving
Use a Plan

pages 15-16

Objective

To use a plan to solve problems

Materials

Mental Math

Ask students how to get to 186 if you start at:
1. 100 (add 86)
2. 198 (subtract 12)
3. 2 × 90 (add 6)
4. 1/2 of 300 (add 36)
5. 1,000 (subtract 814)
6. 93 (multiply by 2)
7. 500 ÷ 2 (subtract 64)
8. 372 (divide by 2)

Using a Four-step Plan

Of the 1,576 children enrolled in Roosevelt School, 757 are girls. Of the 76 students absent today, 27 are girls. How many boys attend Roosevelt? How many boys are present?

★ SEE

We need to find:
    the number of boys who attend Roosevelt School.
    the number of boys who are absent.
    the number of boys who are present.
We know:
    the total number of students is __1,576__ .

    the number of girls is __757__ .

    there are __76__ students who are absent.

    there are __27__ girls who are absent.

★ PLAN

To find out how many boys attend Roosevelt

School, we subtract __757__ from __1,576__ .
To find out how many boys are absent, we subtract

__27__ from __76__ . To find the number of boys who are present, we find the difference between the total number of boys, and the number of boys who are absent.

★ DO

$$
\begin{array}{r} 1{,}576 \\ -\ 757 \\ \hline 819 \text{ boys} \end{array}
\qquad
\begin{array}{r} 76 \\ -27 \\ \hline 49 \text{ boys absent} \end{array}
\qquad
\begin{array}{r} 819 \\ -\ 49 \\ \hline 770 \text{ boys present} \end{array}
$$

__819__ boys attend Roosevelt School. __770__ boys are present today.

★ CHECK

$$
\begin{array}{r} 757 \\ +\ \boxed{819} \\ \hline 1{,}576 \text{ students enrolled} \end{array}
\qquad
\begin{array}{r} 49 \\ +27 \\ \hline \boxed{76} \text{ students absent} \end{array}
\qquad
\begin{array}{r} \boxed{770} \\ +49 \\ \hline \boxed{819} \text{ boys} \end{array}
$$

15

Teaching the Lesson

Have students think about what they would do if they cut or scraped their skin. Tell students that they would think to wash the cut and apply some ointment and a bandage. Tell students they would do what they had planned and then check to see if the cut felt better and if the bandage was the right size for the wound. Remind students that they would solve a problem of a cut or scrape in this order, as you write on the board: **SEE, PLAN, DO, CHECK.** Tell students that this same plan can be used to solve many problems in daily life.

Have a student read the problem aloud. Remind students that the SEE stage is done first as they read and complete the sentences. Tell students that we know what the problem is and we know some facts about it, so we can plan what to do. Remind students that just as there were several steps to care for their wound, there are often several steps in working any problem. Have students read through the PLAN stage with you and complete the sentences. Tell students to

work the problem, write the answers and then check to see if the solution makes sense. Remind students that just as a sling would be a silly solution to caring for a small cut, sometimes a solution to a problem does not make sense and we need to go back to the SEE and PLAN stages and start again. Tell students to complete the problems to check their solution.

15

Apply

Solve these problems. Remember to use the four-step plan.

1. Marcia wants to buy a camera for $49.99. She has saved $31.27 and will earn $5.00 babysitting on Saturday. How much more will she need to buy the camera?
$13.72

2. There are 157 fourth graders at Watson School. At lunch, 43 students go home to eat, 27 bring their lunches and the rest buy their lunches. How many students buy their lunches?
87 students

3. The book Frank is reading has 135 pages. He has finished reading 81 pages. If Frank is to finish the book in six days, how many pages must he read each day?
9 pages

4. Karen wants to bring a birthday treat for her class. There are 24 students and one teacher in the class. She bought 4 boxes of granola bars with 5 granola bars in each. How many more granola bars does she need?
5 granola bars

5. Read Exercise 3 again. What if Frank is to finish the book in nine days instead of six? Now how many pages must he read each day?
6

6. Donuts cost $3.50 a dozen at Aunt Molly's Bakery and 30¢ each at Uncle Don's Diner. At whose place would it cost less to buy 3 dozen donuts?
Aunt Molly's

7. There are 157 fourth graders at Watson School. Forty-three of them walk to school. Some of the others ride bicycles and others ride on school buses. What do you need to know to tell how many fourth-grade students ride in school buses?
The number who ride bicycles

8. You have $10.00 to spend. It costs 50¢ to take the bus downtown. You buy a pencil-and-pen set for $2.98, a notebook for $3.98, and paper for $1.99. Can you buy a ruler for 49¢ and still have enough money left to take the bus home?
No

16

Extra Credit *Numeration*

Tell students they are looking for a magic sum. Have the students write each of the following numbers on separate cards: 2, 4, 6, 8, 10, 12, 14, 16, 18, 22, 24. Tell students to arrange the cards in addition pairs, so that each pair will have the same magic sum. Have them record their pairs and the magic sum. (All pairs total 26. The pairs should be: 2-24, 4-22, 6-20, 8-18, 10-16, 12-14.)

Solution Notes

1. Have students find the total of Marcia's savings by adding. This amount is then subtracted from the cost of the camera.
2. First, students need to find the total number of students who either go home to eat or bring their lunches. They then subtract this from the fourth grade enrollment.
3. First, have students find the number of pages Frank has left to read. To find the average number of pages Frank will need to read each day, we divide the number of pages by the number of days.
4. Several subgoals exist in this problem. Remind students to include the teacher in the number of people getting a treat. Students must also find the number of bars Karen has by multiplying the number of boxes by the number of bars in each. The difference between these two numbers is the number of granola bars Karen still needs.

Higher-Order Thinking Skills

5. [Analysis] Students should recognize the same basic fact is used in both problems: the original problem (54 pages ÷ 6 days = 9 pages a day) and the "what if" problem (54 pages ÷ 9 days = 6 pages a day).
6. [Evaluation] Students might find the unit cost or the cost of one dozen to compare. Others might find the cost of 3 dozen in each store.
8. [Evaluation] Encourage students to use mental computation rounding $2.98, $3.98, and $1.99 to whole dollars, or about $9. This leaves about $1 left, 50¢ of which has already been spent on the bus downtown.

100 Basic Facts

pages 17-18

Objective

To practice addition and subtraction facts

Materials

*addition and subtraction fact cards

Mental Math

Dictate the following:
1. 14 + 1/4 + 2/4 (14 3/4)
2. 7 × 7 + 2 (51)
3. 100 × $5 ($500)
4. 40 ÷ 5 (8)
5. 16 − 0 (16)
6. 3 parts of 4 (3/4)
7. average of 4 and 6 (5)
8. sides on a pentagon (5)

Skill Review

Write on the board:

> Thirty days hath September,
> April, June and November.
> All the rest have 31
> Except February alone,
> Which has four and twenty-four
> Till leap year
> Gives it one day more.

Have students read this Mother Goose rhyme with you. Have students write the months of the year in order, with each month's number of days beside it.

Practicing Addition Facts

2 +1 = 3	7 +2 = 9	7 +4 = 11	1 +3 = 4	1 +8 = 9	2 +3 = 5	3 +2 = 5	7 +9 = 16	1 +4 = 5	4 +4 = 8
3 +0 = 3	5 +5 = 10	8 +6 = 14	6 +0 = 6	2 +9 = 11	0 +6 = 6	3 +1 = 4	6 +3 = 9	6 +5 = 11	7 +6 = 13
9 +1 = 10	7 +0 = 7	5 +2 = 7	2 +2 = 4	1 +2 = 3	2 +6 = 8	9 +4 = 13	5 +3 = 8	7 +3 = 10	6 +8 = 14
0 +9 = 9	9 +9 = 18	3 +3 = 6	1 +6 = 7	9 +0 = 9	0 +2 = 2	7 +7 = 14	0 +3 = 3	4 +7 = 11	0 +5 = 5
7 +8 = 15	5 +6 = 11	1 +1 = 2	2 +4 = 6	5 +9 = 14	2 +5 = 7	5 +4 = 9	1 +0 = 1	7 +1 = 8	3 +5 = 8
8 +8 = 16	9 +2 = 11	5 +7 = 12	1 +7 = 8	5 +8 = 13	0 +4 = 4	6 +9 = 15	6 +7 = 13	4 +9 = 13	7 +5 = 12
9 +6 = 15	5 +1 = 6	4 +3 = 7	6 +1 = 7	8 +1 = 9	2 +0 = 2	8 +2 = 10	3 +8 = 11	6 +2 = 8	3 +7 = 10
6 +4 = 10	4 +0 = 4	9 +5 = 14	0 +7 = 7	3 +4 = 7	4 +1 = 5	8 +9 = 17	1 +8 = 9	0 +1 = 1	9 +3 = 12
8 +7 = 15	1 +5 = 6	8 +0 = 8	3 +9 = 12	8 +4 = 12	4 +2 = 6	3 +6 = 9	8 +3 = 11	5 +0 = 5	2 +8 = 10
1 +9 = 10	0 +8 = 8	6 +6 = 12	0 +0 = 0	2 +7 = 9	4 +5 = 9	9 +7 = 16	9 +8 = 17	4 +6 = 10	8 +5 = 13

17

Teaching the Lesson

Introducing the Problem Ask students to give examples to show how addition and subtraction facts are used constantly in our daily lives. Tell students that because we use these facts so often, it is very important that we learn to say them as quickly as we say our names, addresses, phone numbers, etc. when asked.

Developing the Skill Remind students that addition and subtraction facts are related so if they know that 4 + 2 = 6, they also know 2 + 4 = 6, 6 − 2 = 4 and 6 − 4 = 2. Write the more difficult facts having addends of 6, 7, 8 or 9 and have students write the related facts on the board.

Practicing Subtraction Facts

6 −5 **1**	13 −9 **4**	9 −6 **3**	12 −8 **4**	11 −2 **9**	9 −1 **8**	2 −2 **0**	10 −6 **4**	7 −6 **1**	13 −6 **7**
16 −9 **7**	15 −7 **8**	7 −5 **2**	14 −9 **5**	17 −8 **9**	12 −6 **6**	8 −5 **3**	12 −7 **5**	5 −4 **1**	10 −9 **1**
11 −9 **2**	9 −9 **0**	12 −9 **3**	16 −8 **8**	9 −5 **4**	6 −1 **5**	8 −8 **0**	16 −7 **9**	9 −7 **2**	5 −2 **3**
6 −4 **2**	11 −7 **4**	4 −2 **2**	11 −6 **5**	14 −7 **7**	15 −6 **9**	7 −4 **3**	2 −0 **2**	14 −6 **8**	17 −9 **8**
6 −6 **0**	13 −8 **5**	10 −5 **5**	3 −0 **3**	11 −5 **6**	9 −3 **6**	0 −0 **0**	7 −2 **5**	13 −7 **6**	12 −3 **9**
3 −1 **2**	1 −0 **1**	12 −4 **8**	9 −4 **5**	10 −7 **3**	7 −7 **0**	8 −1 **7**	14 −5 **9**	6 −0 **6**	9 −8 **1**
15 −9 **6**	8 −7 **1**	10 −8 **2**	18 −9 **9**	8 −0 **8**	13 −4 **9**	7 −0 **7**	3 −2 **1**	12 −5 **7**	14 −8 **6**
8 −4 **4**	11 −3 **8**	8 −6 **2**	4 −0 **4**	7 −3 **4**	10 −4 **6**	11 −4 **7**	5 −3 **2**	1 −1 **0**	8 −3 **5**
8 −2 **6**	6 −3 **3**	5 −5 **0**	4 −1 **3**	9 −0 **9**	3 −3 **0**	11 −8 **3**	2 −1 **1**	10 −1 **9**	4 −4 **0**
9 −2 **7**	7 −1 **6**	6 −2 **4**	5 −0 **5**	15 −8 **7**	4 −3 **1**	13 −5 **8**	10 −3 **7**	5 −1 **4**	10 −2 **8**

18

Correcting Common Errors

If students continue to have difficulty mastering their basic facts, have them practice with their fact cards. Have students work in pairs. Give each pair the fact cards, mixed up, for five fact families. Have them form the fact families by placing the related fact cards in groups.

Enrichment

Have students write the fact families which have fewer than 4 addition and subtraction facts.

Practice

Pages 17 and 18 can be used as timed tests and repeated often throughout the year to assure retention. Students may make a chart to record their times and use the chart later in the year to construct a graph of their progress.

Chapter Test

page 19

Item	Objective
1-4	Compare and order numbers less than 100 (See pages 1-2)
5-11	Compute basic addition facts (See pages 3-4)
12-16	Add three, four or five 1-digit addends (See pages 5-6)
17-22	Understand grouping and zero properties of addition (See pages 7-8)
23-30	Compute basic subtraction facts (See pages 9-10)
31-34	Find missing addend (See pages 13-14)

Compare these numbers.

1. 15 $\bigcirc$ 19   2. 26 $\bigcirc$ 62   3. 36 $\bigcirc$ 39   4. 78 $\bigcirc$ 70

Add.

5. $4 + 4 = \underline{8}$ 6. $7 + 3 = \underline{10}$ 7. $8 + 7 = \underline{15}$ 8. $5 + 9 = \underline{14}$

9. $\begin{array}{r} 5 \\ + 6 \\ \hline 11 \end{array}$ 10. $\begin{array}{r} 9 \\ + 9 \\ \hline 18 \end{array}$ 11. $\begin{array}{r} 8 \\ + 3 \\ \hline 11 \end{array}$ 12. $\begin{array}{r} 4 \\ 6 \\ + 2 \\ \hline 12 \end{array}$

13. $\begin{array}{r} 3 \\ 1 \\ + 4 \\ \hline 8 \end{array}$ 14. $\begin{array}{r} 5 \\ 2 \\ 6 \\ + 1 \\ \hline 14 \end{array}$ 15. $\begin{array}{r} 8 \\ 1 \\ 5 \\ + 3 \\ \hline 17 \end{array}$ 16. $\begin{array}{r} 4 \\ 5 \\ 3 \\ 3 \\ 6 \\ + 1 \\ \hline 22 \end{array}$

17. $(5 + 2) + 3 = \underline{10}$ 18. $2 + (8 + 0) = \underline{10}$ 19. $(7 + 5) + 2 = \underline{14}$

20. $3 + (6 + 3) = \underline{12}$ 21. $(4 + 0) + 9 = \underline{13}$ 22. $7 + (5 + 2) = \underline{14}$

Subtract.

23. $9 - 6 = \underline{3}$ 24. $14 - 7 = \underline{7}$ 25. $11 - 8 = \underline{3}$ 26. $15 - 6 = \underline{9}$

27. $\begin{array}{r} 8 \\ - 3 \\ \hline 5 \end{array}$ 28. $\begin{array}{r} 16 \\ - 8 \\ \hline 8 \end{array}$ 29. $\begin{array}{r} 12 \\ - 9 \\ \hline 3 \end{array}$ 30. $\begin{array}{r} 7 \\ - 7 \\ \hline 0 \end{array}$

Write the missing addends.

31. $5 + n = 8$

$n = \underline{3}$

32. $n + 7 = 13$

$n = \underline{6}$

33. $\begin{array}{r} 6 \\ + n \\ \hline 13 \end{array}$

$n = \underline{7}$

34. $\begin{array}{r} 11 \\ + n \\ \hline 19 \end{array}$

$n = \underline{8}$

19

CUMULATIVE REVIEW

Circle the letter of the correct answer.

1 23 ◯ 36
  (a) <
  b >

2 46 ◯ 41
  a <
  (b) >

3 3 + 4
  a 5
  b 6
  (c) 7
  d NG

4 5 + 6
  (a) 11
  b 12
  c 13
  d NG

5 9 + 6
  a 12
  b 14
  c 16
  (d) NG

6 4 + 8
  a 11
  (b) 12
  c 13
  d NG

7 4 3 + 5
  a 11
  b 13
  (c) 12
  d NG

8 8 + (3 + 5)
  a 8
  (b) 16
  c 17
  d NG

9 7 − 3
  a 6
  b 8
  c 10
  (d) NG

10 15 − 7
  a 6
  b 7
  (c) 8
  d NG

11 14 − 6
  (a) 8
  b 7
  c 6
  d NG

12 8 − 8
  (a) 0
  b 4
  c 8
  d NG

13 7 + n = 12, n = ?
  a 7
  b 19
  (c) 5
  d NG

14 n + 2 = 8, n = ?
  (a) 6
  b 8
  c 10
  d NG

☐ score

20

Cumulative Review

page 20

Item	Objective
1-2	Compare and order numbers less than 100 (See pages 1-2)
3-6	Compute basic addition facts (See pages 3-4)
7	Add three 1-digit numbers (See pages 5-6)
8	Understand grouping property of addition (See pages 7-8)
9-12	Compute basic subtraction facts (See pages 9-10)
13-15	Find missing addend (See pages 13-14)

Alternate Cumulative Review
Circle the letter of the correct answer.

1 63 ◯ 36
  (a) >
  b <
  c =

2 91 ◯ 95
  a >
  (b) <
  c =

3 5 + 3 =
  a 7
  (b) 8
  c 9
  d NG

4 9 + 3 =
  a 11
  b 13
  c 15
  (d) NG

5 8 + 6
  a 15
  (b) 14
  c 13
  d NG

6 5 + 7
  (a) 12
  b 13
  c 14
  d NG

7 3 6 + 2
  a 8
  b 9
  (c) 11
  d NG

8 (9 + 1) + 4 =
  a 10
  b 13
  (c) 14
  d NG

9 6 − 4 =
  (a) 2
  b 3
  c 4
  d NG

10 18 − 9 =
  a 10
  (b) 9
  c 8
  d NG

11 13 − 6
  a 8
  b 9
  c 10
  (d) NG

12 7 − 7
  (a) 0
  b 1
  c 2
  d NG

13 6 + n = 14, n = ?
  a 6
  (b) 8
  c 20
  d NG

14 n + 6 = 12, n = ?
  (a) 6
  b 12
  c 18
  d NG

20

Hundreds, Tens and Ones

pages 21-22

Objective
To use place value to read and write 2- and 3-digit numbers

Materials
*3 dice
hundred-squares, ten-strips, single counters

Mental Math
Ask students how to find:
1. area (length × width)
2. feet in 12 yards (12 × 3)
3. pints in 20 cups (20 ÷ 2)
4. perimeter (sum of sides)
5. volume (length × width × height)
6. quotient (dividend ÷ divisor)
7. sum (addend + addend)

Skill Review
Write **fifty-seven** on the board. Ask a student to write the number. (57) Repeat for forty-six. (46) Now have a student write a number through 99 on the board and choose a friend to write the number word. Continue the activity until every student has participated.

Understanding Hundreds, Tens and Ones

Linda bought flower stickers to add to her collection. How many did she buy?

Linda bought ___2___ pages of 100 stickers each.

She bought ___4___ strips of 10 stickers each.

She also bought ___3___ individual stickers.
We say she bought **two hundred forty-three**

stickers, and write it as ___243___.

Study the number in this place value chart.

In the number, **243**,

hundreds	tens	ones
2	4	3

— the digit ___3___ is in the **ones** place.

— the digit ___4___ is in the **tens** place.

— the digit ___2___ is in the **hundreds** place.

Linda bought ___243___ stickers.

Getting Started

Write the numbers.

1. two hundred seventy-seven ___277___   2. nine hundred twenty-six ___926___

Write the place value of the red digits.

3. 2**7**6 ___tens___   4. **3**85 ___hundreds___   5. 62**0** ___ones___

Write the missing words.

6. 132 one ___hundred___ thirty-two   7. 753 ___seven___ hundred fifty-three

21

Teaching the Lesson

Introducing the Problem   Have a student read the problem aloud and tell what is to be done. (tell how many stickers Linda bought) Ask students what information is given in the problem and the picture. (Linda bought two 100-stamp sheets plus four 10-stamp sheets plus 3 single stamps.) Discuss with students how 243 fits into the place value chart in the model. Then have students complete the solution sentence.

Developing the Skill   Write **35** on the board. Have students lay out 3 ten-strips and 5 singles. Draw on the board:

hundreds	tens	ones
	3	5

Continue the activity for other 2-digit numbers and have students write each number in the chart. Now write **167** on the board and repeat the activity using hundred-squares, ten-strips and singles counters. Remind students that 1 hundred equals 10 tens or 100 ones. Ask students how many ones equal 200. (200) Repeat for other equivalents.

Practice

Write the numbers.

1.  _640_

2. 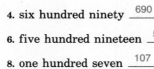 _806_

3. three hundred fifty-two _352_

4. six hundred ninety _690_

5. two hundred seventy-eight _278_

6. five hundred nineteen _519_

7. four hundred ninety-nine _499_

8. one hundred seven _107_

9. eight hundred sixty-five _865_

10. seven hundred eleven _711_

Write the place value of the red digits.

11. 18**6** _ones_

12. 2**6** _ones_

13. **1**59 _hundreds_

14. 6**3**7 _tens_

15. **5**75 _hundreds_

16. 1**5**0 _tens_

17. 32**0** _ones_

18. 2**1**5 _tens_

19. **2**59 _hundreds_

20. **7**07 _hundreds_

21. **6**0 _tens_

22. 9**9** _ones_

23. 5**6**1 _tens_

24. 80**0** _ones_

25. **8**25 _hundreds_

Write the missing words.

26. 333 three hundred _thirty_ -three

27. 923 nine _hundred_ twenty-three

28. 600 _six_ hundred

29. 201 _two_ hundred _one_

30. 575 five _hundred_ seventy- _five_

31. 815 eight hundred _fifteen_

22

Correcting Common Errors

Some students may name the place value of digits incorrectly. Have them place each three-digit number, such as 294, on a place-value chart and give the place-value names for all 3 digits.

hundreds	tens	ones
2	9	4

2 hundreds
9 tens
4 ones

Enrichment

Write 3 single-digit numbers on the board. Ask students to write all the 3-digit numbers possible. Then have them arrange them in order from smallest to largest.

Practice

Remind students to place a zero in a 3-digit number when there are no tens or ones. Have students complete the page independently.

Mixed Practice

1. 3 + 6 + 4 + 1 (14)
2. 15 − 8 (7)
3. (2 + 8) + 7 (17)
4. 9 + 7 (16)
5. 14 − 6 (8)
6. 5 + 4 + 2 + 3 + 4 (18)
7. 8 + (3 + 2) (13)
8. 12 − 7 (5)
9. 7 + 6 + 4 + 9 (26)
10. 17 − 8 (9)

Extra Credit *Applications*

Provide the students with a list of five major cities in the world. Using an atlas, have them find the distance from their hometown to each of these cities. Have them list the cities in order from the farthest away, to the closest. Then divide students into groups and assign each group a city. Tell each group to find out as much numerical information as they can about their city, and report it to the class. Students should see that they could report on everything from monetary units to annual rainfall and population.

Dollars and Cents

pages 23-24

Objective

To read and write amounts less than $10

Materials

*figure with 6 equal sides
play or real money
hundred-squares, ten-strips, single
counters

Mental Math

Dictate the following numbers for students to round to the nearest ten:

1. 87 (90)
2. 218 (220)
3. 61 (60)
4. 612 (610)
5. 439 (440)
6. 1,317 (1,320)
7. 999 (1,000)

Skill Review

Trace a 6-sided figure on the board. Label each side **5 inches.** Have students count by 5's to find the perimeter. (30 inches) Change the length of each side to **10 centimeters** and have students count by 10's to find the perimeter. (60 centimeters) Repeat for **25 miles** to count by 25's, **100 yards** to count by 100's, etc.

Understanding Dollars and Cents

Artiss is the treasurer for the drama club. He is sorting the coins he has collected for membership dues. He puts the pennies into penny tubes. How much money, in pennies, has Artiss collected?

There are __100__ pennies in a dollar.

Artiss has __300__ pennies in the penny tubes.

He has __46__ more pennies that are not in a tube. We say Artiss has saved **three dollars and**

forty-six cents, and write it as __$3.46__.
To understand the total amount of money that Artiss has, study this place value chart.

In the amount, **$3.46,**

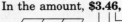

the digit __6__ is in the **pennies** place.

the digit __4__ is in the **dimes** place.

the digit __3__ is in the **dollars** place.

dollars	dimes	pennies
3	4	6

✔ Remember to include the dollar sign and decimal point in writing amounts of money.

Artiss has collected __$3.46__ in pennies.

Getting Started

Write the amounts.

1.  $3.43

2. 8 dollars and 30 pennies __$8.30__

3. 85 cents __$0.85__

4. 4 dollars __$4.00__

23

Teaching the Lesson

Introducing the Problem Have a student read the problem and tell what question is to be answered. (the total amount of money Arliss collected in pennies) Ask students what information is given in the problem and the picture. (Artiss has 3 groups of 100 pennies and 46 more pennies.) Discuss with students how $3.46 fits into the place value chart in the model. Then have students complete the solution sentence.

Developing the Skill Write **327¢** on the board. Ask a student to read the amount. (three hundred twenty-seven cents) Ask students to name the place value of each number. (3 hundreds, 2 tens, 7 ones) Draw on the board:

hundreds	tens	ones		dollars	dimes	pennies
3	2	7		(3)	(2)	(7)

Ask students how many pennies equal 7 ones. (7) Ask how many dimes equal 2 tens. (2) Repeat for 3 hundreds and write each in the money chart. Write **327¢ = $3.27** on the board and have students read with you the dollar amount. (three dollars and twenty-seven cents) Repeat for other 3-digit amounts. Write **nine cents, $.09, seventeen cents** and **$.17** on the board and have students read each amount. Repeat for more amounts under $1.00.

Practice

Write the amounts.

1. _____ $4.58

2. 3 dollars and 15 cents $3.15

3. 2 dollars and 4 dimes $2.40

4. 5 dimes and 6 cents $0.56

5. 7 dollars and 27 cents $7.27

6. 4 dollars, 2 dimes, and 8 pennies $4.28

7. 6 dollars and 2 dimes $6.20

8. 9 dollars and 99 cents $9.99

9. 1 dollar and 5 cents $1.05

10. 5 dollars and 89 cents $5.89

11. 2 dollars, 7 dimes and 0 pennies $2.70

EXCURSION

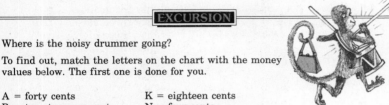

Where is the noisy drummer going?

To find out, match the letters on the chart with the money values below. The first one is done for you.

A = forty cents
B = twenty-seven cents
E = five cents
H = zero
K = eighteen cents
N = four cents
O = ninety-three cents
T = fifty cents

T	O	T	H	E	B	A	N	K
50¢	93/100	half dollar	$0.00	5¢	2 dimes 7 pennies	$0.40	$0.04	18¢

T	O	T	H	E	B	A	N	K
50 pennies	$0.93	50/100	00/100	$0.05	2 dimes 1 nickel 2 pennies	40¢	4 pennies	$0.18

T	O	T	H	E	B	A	N	K
$0.50	93¢	5 dimes	0¢	05/100	27¢	40/100	04/100	1 dime 1 nickel 3 pennies

B	A	N	K	B	A	N	K
$0.27	4 dimes	4¢	18/100	27/100	8 nickels	$0.04	18¢

24

Counting Money

pages 25-26

Objective

To identify and count money through $20

Materials

play or real money

Mental Math

Have students complete each comparison

A nickel is to 5 as a:
1. dime is to ____. (10)
2. dollar is to ____. (100)
3. quarter is to ____. (25)
4. penny is to ____. (1)
5. $5 bill is to ____. (500)
6. $10 bill is to ____. (1,000)
7. half-dollar is to ____. (50)
8. $2 bill is to ____. (200)

Skill Review

Write **10, 20,** and **30** on the board. Ask students to read the numbers aloud and decide how they are counting. (by 10's) Write **35, 40,** and **45** and ask students how they are counting now. (by 5's) Continue the activity for counting by 20's. Repeat for more practice with counting by 1's, 5's, 10's, 20's and 100's in any order. Then have students give each group of 3 numbers and tell how they are counting.

Counting Money

Help Onida count the money she has collected from the customers on her paper route.

We count:

We say:

twenty
forty
sixty

seventy

seventy-five
eighty

eighty-one
eighty-two
eighty-three

eighty-three fifty
eighty-three seventy-five
eighty-three eighty-five
eighty-three ninety
eighty-three ninety-one
eighty-three ninety-two
eighty-three ninety-three

✔ Remember to include the dollar signs and decimal points.

Onida has collected ____$83.93____.

Getting Started

Write the total amounts.

1. ____ $50.95

2. ____ $85.71

25

Teaching the Lesson

Introducing the Problem Have students tell about the picture. (A girl is counting money.) Have a student read the problem aloud. Ask students what they are to find. (total amount of money Onida has collected) Ask students to identify the bills and coins pictured. Have students read with you to count the money. Tell students to complete the solution sentence giving the amount of money Onida has collected.

Developing the Skill Have students examine a $20 bill and tell its value. (20 dollars) Continue to have students identify the following bills and coins and tell the value of each: $10 bill, $5 bill, $1 bill, half-dollar, quarter, dime, nickel and penny. Have students lay out nickels as they count by 5's through 100. Repeat with dimes for counting by 10's. Now have students count aloud by 20's and then by 25's through 200. Ask students to lay out 2 quarters, 3 dimes and 3 nickels and count with you: 25, 50, 60, 70, 80, 85, 90, 95¢. Have students name other amounts, lay out the money and then write the amounts on the board.

Practice

Write the total amounts.

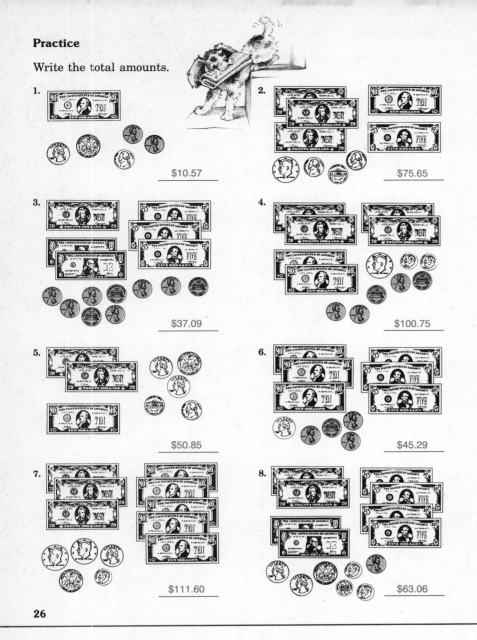

1. $10.57

2. $75.65

3. $37.09

4. $100.75

5. $50.85

6. $45.29

7. $111.60

8. $63.06

26

Thousands

pages 27-28

Objective
To use place value to read and write 4-digit numbers

Materials
hundred-squares, ten-strips, single counters

Mental Math
Have students name the number that tells:
1. freezing Fahrenheit temperature (32°)
2. digits in a phone number (7 or 10)
3. years before 2000
4. decimal for 4/10 (.4)
5. hours in 1/12 of a day (2)
6. freezing Celsius temperature (0°)
7. years in 5 decades (50)

Skill Review
Write on the board:

5 hundreds   4 tens   6 ones
2 hundreds   8 tens   9 ones

Have a student add the 2 numbers, write the sum and tell the value of each digit. (835, 8 hundreds 3 tens 5 ones) Now have a student subtract the 2 numbers, write the difference and tell the value of each digit. (257, 2 hundreds 5 tens 7 ones) Continue for more problems using numbers that have sums and differences less than 1,000.

Understanding Thousands

Roberto is taking inventory of the stickers that the bookstore has to sell. How many stickers are in stock?

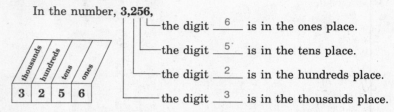

Roberto counts ___3___ envelopes of 1,000 stickers each.

He has ___2___ sheets of 100 stickers each.

Roberto also has ___5___ strips of 10 stickers each, and ___6___ individual stickers.
We say that Roberto counts **three thousand, two hundred fifty-six** stickers. We write the number as ___3,256___.

To understand the total number of stickers, study this place value chart.

In the number, **3,256,**

thousands	hundreds	tens	ones
3	2	5	6

- the digit ___6___ is in the ones place.
- the digit ___5___ is in the tens place.
- the digit ___2___ is in the hundreds place.
- the digit ___3___ is in the thousands place.

Roberto counts ___3,256___ stickers in the bookstore.

Getting Started

Write the number.

1. five thousand, two hundred twenty-three ___5,223___

Write the place value of the red digits.

2. 3,2**7**5 ___tens___   3. **1**,086 ___thousands___   4. 4,39**0** ___ones___   5. 6,**0**05 ___hundreds___

Write the missing words.

6. 1,900 one ___thousand___, nine hundred   7. 6,040 six thousand, ___forty___

27

Teaching the Lesson

Introducing the Problem Have a student read the problem aloud and tell what is to be found. (the number of stickers in stock) Have students tell what facts are given in the picture, and then have them complete the information sentences. (Roberto has 3 groups of 1,000, 2 groups of 100, 5 groups of 10 and 6 single stickers.) Discuss with students how 3,256 fits into the place value chart in the model. Then have students complete the solution sentence.

Developing the Skill Hold up 1 hundred-square and ask a student to write 100 on a hundreds, tens, ones chart on the board. Place a second hundred-square on top of the first and ask a student to change the 100 to 200. Continue adding a 1 hundred-square at a time through 900 as students change the numbers on the chart. Hold up a ten-strip and ask how many ones there are. (10) Remind students that when we have 10 ones we can trade for 1 ten and when we have 10 tens we can trade for 1 hundred. Ask students what we do when we have 10 hundreds. (trade for 1 thousand) Add a thousands column on the board chart and change the 900 to 1,000. Have students record dictated 4-digit numbers on 4-digit place value charts. Ask students to give the value of each digit.

Practice

Write the numbers.

1. three thousand, six hundred fifty-seven _____3,657_____

2. one thousand, ninety-four _____1,094_____

3. eight thousand, two hundred eighty-three _____8,283_____

4. four thousand, eight hundred eleven _____4,811_____

5. two thousand one _____2,001_____

Write the place value of the red digits.

6. 5,075 _____hundreds_____   7. 7,421 _____tens_____   8. 2,555 _____tens_____

9. 4,020 _____ones_____   10. 9,500 _____tens_____   11. 6,226 _____thousands_____

Write the missing words.

12. 7,946 _____seven_____ thousand, nine _____hundred_____ forty-six

13. 9,216 nine thousand, _____two_____ hundred _____sixteen_____

EXCURSION

Checks can be used in place of cash to buy things.

> October 1, 1988
> Payable to *Jiffy Auto Sales* $8,799.00
> *Eight thousand, seven hundred ninety-nine and 00/100* Dollars
> *John E. Cashmore*

Write these dollar amounts in words, as you would write them on a check.

1. $3,075.00 _____three thousand, seventy-five and 00/100_____ Dollars

2. $9,103.75 _____nine thousand, one hundred three and 75/100_____ Dollars

3. $1,240.16 _____one thousand, two hundred forty and 16/100_____ Dollars

4. $5,815.01 _____five thousand, eight hundred fifteen and 01/100_____ Dollars

28

28

Comparing Numbers

Objective

To compare numbers with up to 4 digits

Materials

Mental Math

Dictate the following problems:
1. 1/8 + 4/8 (5/8)
2. 40 × 9 (360)
3. 5,000 − 100 (4,900)
4. 63 ÷ 7 (9)
5. 2.1 + 6.1 (8.2)
6. 1/6 of 24 (4)
7. 1/5 of (20 + 5) (5)
8. 18 ÷ 9 (2)

Skill Review

Write **79** ◯ **26** on the board and have a student write the greater than or less than sign in the circle. (>) Continue writing more pairs of 2-digit numbers on the board. Have students write the sign to compare the numbers.

Comparing Numbers

The Parana River is in South America. The Mississippi River is in North America. Which river is longer?

Principal Rivers of the World	
River	Length
Amazon	4,000 miles
Mississippi	2,348 miles
Nile	4,145 miles
Parana	2,485 miles

We want to know if the Parana or the Mississippi River is longer.
We know that the Parana River is __2,485__ miles long.
The length of the Mississippi River is __2,348__ miles.
To find the longer river, we compare their lengths.

	thousands	hundreds
Parana	**2,485**	**2,485**
Mississippi	**2,348**	**2,348**
	2 = 2	4 > 3

We say 2,485 is greater than 2,348.

We write __2,485__ > __2,348__.

The __Parana__ River is longer than the __Mississippi__ River.

Getting Started

Compare these numbers. Write < or > in the circle.

1. 583 ⊖ 581 2. 325 ⊖ 383 3. 216 ⊖ 410
4. 7,919 ⊖ 8,215 5. 8,096 ⊖ 8,039 6. 5,416 ⊖ 5,420

Write the numbers in order from least to greatest.

7. 3,715, 4,210, 1,650   8. 3,748, 3,750, 3,746   9. 8,050, 8,196, 8,100

   __1,650__, __3,715__, __4,210__   __3,746__, __3,748__, __3,750__   __8,050__, __8,100__, __8,196__

10. Write the names of the principal rivers in order, from shortest to longest.

   __Mississippi__   __Parana__   __Amazon__   __Nile__

29

Teaching the Lesson

Introducing the Problem   Have a student read the problem and tell what is to be found. (the longer of 2 rivers, the Mississippi or the Parana) Ask students to tell what information is given in the table. (The Amazon River is 4,000 mi, the Mississippi is 2,348 mi, the Nile is 4,145 mi and the Parana is 2,485 mi.) Ask students if all the information is needed to solve the problem. (no) Ask what information they will use. (The Mississippi River is 2,348 mi and the Parana is 2,485 mi.) Work through the problem in the model comparing the lengths of the rivers. Then have students complete the solution sentence.

Developing the Skill   Draw a number line showing the numbers from 260 through 270. Have students find 263 and 267 on the number line. Ask which number is greater. (267) Write **267 > 263** on the board. Tell students that using place value to compare larger numbers is usually quicker than using a number line. Write **1,763** ◯ **1,784** on the board. Tell students that both numbers have the same number of thousands and hundreds but 1,784 has more tens so we know that 1,784 is greater than 1,763. Write < in the problem. Refer students to the chart of rivers on page 29 and have students compare the river lengths for more practice.

Practice

Compare these numbers. Write < or > in the circle.

1. 468 ⊚> 463
2. 297 ⊚< 300
3. 897 ⊚> 879
4. 3,246 ⊚< 3,252
5. 6,485 ⊚> 6,481
6. 8,296 ⊚> 8,290
7. 5,084 ⊚< 5,163
8. 7,006 ⊚> 7,002
9. 4,561 ⊚< 4,651
10. 8,000 ⊚< 9,000
11. 1,250 ⊚> 850
12. 2,080 ⊚> 2,079
13. 6,361 ⊚< 6,362
14. 8,207 ⊚< 8,702
15. 3,500 ⊚> 3,499

Write the numbers in order from least to greatest.

16. 468, 686, 560
    __468__, __560__, __686__

17. 212, 235, 210
    __210__, __212__, __235__

18. 376, 372, 378
    __372__, __376__, __378__

19. 3,210, 3,240, 3,260
    __3,210__, __3,240__, __3,260__

20. 8,512, 7,416, 7,800
    __7,416__, __7,800__, __8,512__

21. 5,286, 5,179, 5,280
    __5,179__, __5,280__, __5,286__

22. 7,427, 7,430, 7,425
    __7,425__, __7,427__, __7,430__

23. 9,000, 8,967, 9,967
    __8,967__, __9,000__, __9,967__

24. 4,241, 4,421, 4,124
    __4,124__, __4,241__, __4,421__

Apply

Solve these problems.

25. The Ohio River is 975 miles long. The Red River is 1,270 miles long. Which river is longer?
    The Red River

26. Great Bear Lake is 1,463 feet deep. Lake Superior is 1,333 feet deep. Which lake is deeper?
    Great Bear Lake

27. The Verrazano-Narrows Bridge is 4,260 feet long. The Golden Gate Bridge is 4,200 feet long. Which bridge is longer?
    The Verrazano-Narrows Bridge

28. The Lincoln Tunnel is 8,216 feet long. The Holland Tunnel is 8,557 feet long. Which tunnel is longer?
    The Holland Tunnel

30

Rounding Numbers

pages 31-32

Objective

To round to the nearest ten, hundred or dollar

Materials

numbers 20 through 30 on cards
newspaper grocery ads

Mental Math

Have students name the year that came first:

1. 1776 or 1976. (1776)
2. 1986 or 1968. (1968)
3. 1072 or 865. (865)
4. 1615 or 1516. (1516)
5. 1326 or 1218. (1218)
6. 1812 or 1912. (1812)
7. 1971 or 1969. (1969)
8. 1421 or 1399. (1399)

Skill Review

Write **26 + 48 + 137** vertically on the board. Have a student find the sum. (211) Have another student write the problem and the sum as money amounts. ($.26 + .48 + 1.37 = $2.11) Repeat for more problems.

Rounding Numbers

Mr. Sanchez is buying his wife a birthday present along with paper to wrap it. Estimate the cost of each item by rounding.

We want to estimate the cost of each item by rounding.
We know the cost of wrapping paper is __39¢__.

The earrings cost __$62__ and the gold bracelet costs __$135__.

> To round a number to a particular place value, locate the digit to be rounded.

If the digit to the right is 0, 1, 2, 3 or 4, the digit we are rounding stays the same. All the digits to the right are replaced by zeros.

$62 is rounded to __$60__.

✔ Sometimes it makes more sense to round a number to another place value. We round $135 to $140, rather than $100, because $140 is a closer estimate to the cost of the bracelet.

If the digit to the right is 5, 6, 7, 8 or 9, the digit we are rounding is raised one. All digits to the right are replaced by zeros.

39¢ is rounded to __40¢__.

✔ When rounding money to the nearest dollar, the rounded number is written without a decimal point, and without zeros in the dimes and pennies places. For example, $8.45 is rounded to $8.

Mr. Sanchez will have to pay about __40¢__ for the wrapping paper, about __$60__ for the earrings and about __$140__ for the bracelet.

Getting Started

Round to the nearest

1. ten. 38 __40__   2. hundred. 316 __300__   3. dollar. $3.81 __$4__

31

Teaching the Lesson

Introducing the Problem Have a student read the problem and tell what is to be done. (estimate each item's cost) Ask students to use the picture to tell what facts are given and then fill in the information sentences. (wrapping paper costs 39¢, earrings $62 and the bracelet $135) Work through the rounding process in the model with the students. Then have them complete the solution sentences.

Developing the Skill Tell students that we often need to know a sum quickly but do not need to know the exact amount. We can then **round** the number to get an **estimate.** Tell students that an example might be if you had $2 and wanted to quickly know if you could afford to buy 3 items costing 79¢, 59¢ and 83¢. Help students name other times when rounding numbers could be helpful. (referring to the speed travelled, pages in a book, etc.) Draw a number line from 10 to 20 on the board. Write **16** on the board. Ask students if 16 is closer to 10 or to 20. (20) Show stu-

dents that 16 is rounded to 20 rather than to 10 because it is closer to 20. Tell students that we always round a 5 up, so 15 is rounded to 20. Repeat the procedure for rounding 162 to the nearest ten. Now draw a number line from 100 through 200 and show students how 162 is rounded to the nearest hundred. Give more examples of rounding 2- and 3-digit numbers.

31

Practice

Round to the nearest ten.

1. 26 __30__   2. 67 __70__   3. 59¢ __60¢__   4. 35 __40__

5. 47¢ __50¢__   6. 81 __80__   7. 87 __90__   8. 78¢ __80¢__

9. 23 __20__   10. 17 __20__   11. 25¢ __30¢__   12. 38 __40__

13. 72 __70__   14. 91¢ __90¢__   15. 42 __40__   16. 76 __80__

Round to the nearest hundred.

17. 351 __400__   18. 286 __300__   19. 142 __100__   20. $350 __$400__

21. $187 __$200__   22. 427 __400__   23. 583 __600__   24. 791 __800__

25. 820 __800__   26. $275 __$300__   27. 321 __300__   28. 468 __500__

Round to the nearest dollar.

29. $8.51 __$9__   30. $6.35 __$6__   31. $2.50 __$3__   32. $3.16 __$3__

33. $7.89 __$8__   34. $5.29 __$5__   35. $4.10 __$4__   36. $7.38 __$7__

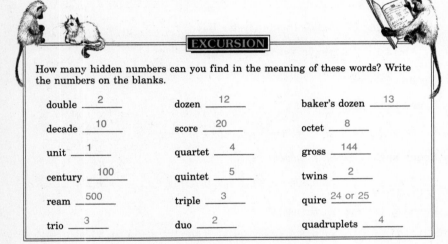

EXCURSION

How many hidden numbers can you find in the meaning of these words? Write the numbers on the blanks.

double __2__	dozen __12__	baker's dozen __13__
decade __10__	score __20__	octet __8__
unit __1__	quartet __4__	gross __144__
century __100__	quintet __5__	twins __2__
ream __500__	triple __3__	quire __24 or 25__
trio __3__	duo __2__	quadruplets __4__

32

Practice

Remind students that when rounding money to the nearest dollar, we use only the dollar sign and the number of dollars. Have students complete the problems.

Excursion

Tell students to write all the answers they are sure of before consulting a dictionary. Extend the activity by having students research more words that have a numerical connotations.

Extra Credit *Logic*

Tell students to draw a box that is divided into 9 squares. Using the numbers 1 through 9, have students put one number in each square arranging them in such a way that when adding the numbers in any row, vertical, horizontal, or diagonal, the answer will be the same.

(4   9   2

3   5   7

8   1   6)

Tell them they have created a magic square.

Rounding Numbers

pages 33-34

Objective

To round to nearest hundred or thousand

Materials

newspapers

Mental Math

Ask if the following questions are true or false:

1. $7 \times 8 = 56$ (T)
2. $10 \times 0 = 10$ (F)
3. $45 \div 3 = 15$ (T)
4. $52 + 15 + 21$ is about 90 (T)
5. 119 can be rounded to 100 (T)
6. 56 can be rounded to 50 (F)
7. $2,246 < 2,217$ (F)
8. $2/10 - 2/10 = 0$ (T)

Skill Review

Write **3 thousands + 9 hundreds + 7 tens + 4 ones** on the board. Have students write the number that is 160 less. (3,814) Repeat using similar kinds of problems.

Rounding Greater Numbers

Theodore Roosevelt made his home in Oyster Bay, New York. His cousin Franklin D. Roosevelt made his home in Hyde Park, New York. About how many people live in each of these towns?

Small Towns in New York

Town	Population
Geneseo	6,746
Herkimer	8,383
Hyde Park	2,550
Larchmont	6,308
Oyster Bay	6,497

We want to estimate the population of Oyster Bay and Hyde Park.
We know the population of Oyster Bay is

__6,497__ and that of Hyde Park is __2,550__ .
We can estimate these populations to thousands place, to hundreds or to tens.

	Population	Rounded to thousands	Rounded to hundreds	Rounded to tens
Oyster Bay	6,497	6,000	6,500	6,500
Hyde Park	9,801	10,000	9,800	9,800

About __6,500__ people live in Oyster Bay.

About __2,600__ people live in Hyde Park.

Getting Started

Round to the nearest ten.

1. 5,438 __5,440__   2. 7,015 __7,020__   3. 7,212 __7,210__   4. 3,896 __3,900__

Round to the nearest hundred.

5. 3,248 __3,200__   6. 5,598 __5,600__   7. 6,750 __6,800__   8. 8,290 __8,300__

Round to the nearest thousand.

9. 6,585 __7,000__   10. 3,296 __3,000__   11. 4,500 __5,000__   12. 8,196 __8,000__

33

Teaching the Lesson

Introducing the Problem Have a student read the problem and tell what is to be found. (about how many people live in Oyster Bay and Hyde Park, New York) Have students reread the problem and use the table to tell what information is needed to solve the problem. (the populations of Oyster Bay and Hyde Park) Have students fill in the information sentences. Ask students what they will need to do. (round the population figures) Work through each rounding process with students and then have them complete the solution sentences.

Developing the Skill Write **2,619** on the board and ask students to round the number to the nearest ten (2,620) and the nearest hundred. (2,600) Draw a number line on the board for 2,000 through 3,000 in intervals of 100. Ask students if 2,619 is closer to 2,000 or to 3,000. (3,000) Repeat more numbers. Discuss with students that newspaper headlines and stories often report large numbers in rounded figures. Ask students why an article might report that 7,000 rather than 7,416 people attended a rock concert. (The exact number may not be known; the estimate is sufficient to show that there was a large audience; etc.) Discuss each headline or article and have students tell the range of numbers each rounded figure includes. For example, 7,000 could be any number from 6,500 through 7,499.

Practice

Round to the nearest ten.

1. 3,276 _3,280_
2. 4,391 _4,390_
3. 7,549 _7,550_
4. 4,212 _4,210_

5. 6,831 _6,830_
6. 4,287 _4,290_
7. 6,375 _6,380_
8. 7,126 _7,130_

9. 5,650 _5,650_
10. 8,296 _8,300_
11. 9,632 _9,630_
12. 5,864 _5,860_

Round to the nearest hundred.

13. 2,975 _3,000_
14. 3,710 _3,700_
15. 6,823 _6,800_
16. 7,879 _7,900_

17. 4,176 _4,200_
18. 5,275 _5,300_
19. 8,048 _8,000_
20. 3,119 _3,100_

21. 8,225 _8,200_
22. 6,867 _6,900_
23. 5,500 _5,500_
24. 3,271 _3,300_

Round to the nearest thousand.

25. 2,659 _3,000_
26. 8,960 _9,000_
27. 7,016 _7,000_
28. 6,400 _6,000_

29. 4,867 _5,000_
30. 3,450 _3,000_
31. 6,650 _7,000_
32. 2,905 _3,000_

33. 7,221 _7,000_
34. 8,949 _9,000_
35. 3,738 _4,000_
36. 5,615 _6,000_

Small Towns in New York	
Town	Population
Bayport	9,282
Brockport	9,776
Delmar	8,423
Great Neck	9,168
Herricks	8,123

	Round each population to the nearest ten.	Round to the nearest hundred.
37. Bayport	9,280	9,300
38. Brockport	9,780	9,800
39. Delmar	8,420	8,400
40. Great Neck	9,170	9,200
41. Herricks	8,120	8,100

34

Correcting Common Errors

Some students may round to the wrong place when they are working with large numbers. Have students first draw an arrow above the place to which they are rounding. Then have them look at the digit to the right to determine whether the digit under the arrow should stay the same or be increased by 1. Then have them replace all the digits to the right of the arrow with zeros.

Enrichment

Have students use newspapers, magazines or books to locate 4-digit numbers which have been rounded to the nearest hundred or thousand. Then using the context, ask them to tell why each was so rounded.

Practice

Have students complete the problems on the page. At the bottom of the page, tell students they are to find the population of each town in the table and then round each to the nearest ten and hundred.

Extra Credit   *Applications*

Have students call and order a set of bus schedules from your local transit system. Discuss and demonstrate how to follow the schedules. Invite a representative of the transit system to come and talk to students about how schedules are made. Ask the students to bring in a variety of other schedules such as sports season schedules, television guides, school calendars etc. Make a display of them on a bulletin board. Finally, have students make a schedule of their school day for the whole week.

Hundred Thousands

pages 35-36

Objective

To read and write numbers in ten thousands and hundred thousands

Materials

*number cards 0 through 9

Mental Math

Have students name the number that is:

1. the reverse of 1,632. (2,361)
2. 2 times 6 + 9. (30)
3. 14 less than 16 + 14. (16)
4. 200 more than 1/4 of 8. (202)
5. less than 415 but more than 413. (414)
6. tenth if counting by 100's. (1,000)
7. .1 of 100. (10)

Skill Review

Dictate numbers having 1, 2, 3, or 4 digits for students to write on the board. Have students name the value of each digit. Now have students write the number representing 2 thousands 4 hundreds 6 tens 8 ones and have a student write the number. Repeat similar problems.

Understanding Ten Thousands and Hundred Thousands

Normal Heart Rates	
Beats	Time
72	minute
4,320	hour
103,680	day

A human heart normally beats 72 times every minute. How many times does a normal heart beat in one day?

We know that the normal heart will beat

__103,680__ times in one day. To understand this number, study this place value chart.

```
hundred thousands | ten thousands | thousands | hundreds | tens | ones
        1         |       0       |     3     |    6     |  8   |  0
```

thousands period , ones period

A comma separates the periods.

We say a normal heart beats **one hundred three thousand, six hundred eighty** times a day.

We write the number as __103,680__.

Getting Started

Write the place value of the red digits.

1. 3**4**1,219   2. 120,**2**10   3. **1**56,035
   ten thousands   hundreds   hundred thousands

Write the number.

4. two hundred forty-six thousand, five hundred fifteen __246,515__

Write the missing word.

5. 470,050 four hundred seventy __thousand__, fifty

35

Teaching the Lesson

Introducing the Problem   Have a student read the problem and tell what question is to be answered. (how many times a normal heart beats per day) Tell students to study the table, reread the problem and determine what fact is needed to complete the answer. (The heart beats 103,680 times per day.) Help students study the place value chart until they understand how 103,680 fits in. Point out that ones, tens and hundreds together comprise the *ones period.* Thousands, ten thousands and hundred thousands comprise the *thousands period.* Have students complete the last sentence with the correct number.

Developing the Skill   Write **10,928** on the board and have students read it with you as you write **ten thousand, nine hundred twenty-eight.** Ask students where the comma is written in both the number and its number name. (after the thousands digit and the word thousand) Go through the place value of each digit introducing the ten thousands place. Write **42,081** and ask students the value of the 4. (ten thousands) Write **436,962** and ask students the value of the 3. (ten thousands) Tell students the 4 is in hundred thousands place. Write **four hundred thirty-six thousand, nine hundred sixty-two** on the board and read it together. Be sure students do not use the word **and** to separate the periods. Repeat the procedure for more examples of 6-digit numbers.

Practice

Write the place value of the red digits.

1. 63,215
   thousands

2. 139,073
   hundreds

3. 21,875
   ten thousands

4. 370,150
   hundred thousands

5. 929,175
   ten thousands

6. 836,207
   tens

7. 180,000
   thousands

8. 57,329
   ones

Write the numbers.

9. four hundred twenty-seven thousand, three hundred twelve   427,312

10. two hundred nine thousand, one hundred fifty-six   209,156

11. eighty-seven thousand, nine hundred nine   87,909

12. ten thousand, two hundred eighty-three   10,283

Write the missing words.

13. 64,419 sixty-four thousand , four hundred nineteen

14. 406,060 four hundred six thousand, sixty

15. 515,291 five hundred fifteen thousand, two hundred ninety-one

Apply

Write the numbers.

16. The highest mountain in the world is Mt. Everest. It is twenty-nine thousand, twenty-eight feet high.
    29,028 feet

17. Captain Joe B. Jordan flew a jet plane to the altitude of three hundred fourteen thousand, seven hundred fifty feet.
    314,750 feet

18. Earth is about two hundred thirty-eight thousand, eight hundred sixty miles from the moon. 238,860 miles

19. The earth travels in orbit at about sixty-six thousand, six hundred miles per hour.
    66,600 miles per hour

36

Millions

pages 37-38

Objective

To read and write numbers in millions

Materials

cards with numbers 0 through 9
2 comma cards

Mental Math

Have students name a number that is:

1. evenly divisible by 2. (2, 4, etc.)
2. 1,000 more than 100. (1,100)
3. odd and less than 25. (1, 3, 5, . . . 23)
4. 27 pairs. (54)
5. 1,000 less than 100,000. (99,000)
6. 47 more than 20,100. (20,147)
7. 64 when multiplied by itself. (8)

Skill Review

Write **100,000** on the board. Ask students to tell the next number if counting by 10,000's. (110,000) Have a student write **110,000** under 100,000. Continue counting and writing the numbers through 200,000. Now have students count and write the numbers by 20,000's through 300,000. Continue to count and write by 50,000's through 500,000 and then by 100,000's through 900,000.

Understanding Millions

Chicago is the home of the world's busiest airport. Over 20 million passengers take off from O'Hare Airport each year. How many travellers have departed from O'Hare according to the meter?

PASSENGER DEPARTURES
2 5 6 3 6 4 8 3

We know the meter shows ____25,636,483____ passengers have departed from O'Hare Airport. To understand this number, study this place value chart.

hundred millions	ten millions	millions	hundred thousands	ten thousands	thousands	hundreds	tens	ones
2	5	6	3	6	4	8	3

millions period , thousands period , ones period

Commas separate the periods.

We say that **twenty-five million, six hundred thirty-six thousand, four hundred eighty-three** travellers have departed from O'Hare Airport this year. We write this number as ____25,636,483____.

Getting Started

Write the numbers.

1. two hundred twelve million, four hundred sixty-five thousand,

   twenty-nine ___212,465,029___

Write the place value of the red digits.

2. 23,465,183

   _____millions_____

3. 706,341,212

   _____ten thousands_____

4. 5,962,159

   _____hundred thousands_____

37

Teaching the Lesson

Introducing the Problem Have a student read the problem and tell what is to be found. (exact number of travellers who have flown out of O'Hare Airport in one year) Ask students if the exact number is given in the problem or picture. (yes) Tell students to write the number in the information sentence. Help students study the place value chart until they understand how 21,356,483 fits in. Read the number word with students and then have them complete the last sentence with the correct number. Have students complete the sentences as they read aloud with you.

Developing the Skill Have students count by 10,000's and write the numbers from 900,000 through 990,000. Tell students one million is 1,000 thousands and is written 1,000,000. Write **1,000,000** under 990,000. Ask students if they can guess how long it would take to spend $1,000,000 if they spent $10 each day. (100,000 days or almost 274 years) Draw a place value grid for 9-digit numbers. Write **273,646,901** on the grid. Help students tell the place value of each digit. (2 hundred millions 7 ten millions 3 millions 6 hundred thousands 4 ten thousands 6 thousands 9 hundreds 0 tens 1 one) Write **two hundred seventy-three million, six hundred forty-six thousand, nine hundred one** on the board. Repeat the activity for more 8- and 9-digit numbers. Tell students a comma separates the millions period from the thousands period, and the thousands period from the hundreds period.

Practice

Write the numbers.

1. three hundred seventy-five million, two hundred five thousand,

   sixty-seven ___375,205,067___

2. four hundred ten million, five hundred sixteen thousand,

   four hundred twenty ___410,516,420___

3. thirty-eight million, sixty-three thousand,

   eight hundred forty-nine ___38,063,849___

Write the place value of the red digits.

4. 6,241,573   5. 14,903,124   6. 136,248,796

   _____millions_____   __hundred thousands__   ___ten millions___

7. 828,297,159   8. 16,274,302   9. 460,760,858

   __hundred millions__   _____hundreds_____   ___ten thousands___

10. 3,926,575   11. 47,239,105   12. 906,400,000

    ___ten thousands___   __hundred thousands__   _____millions_____

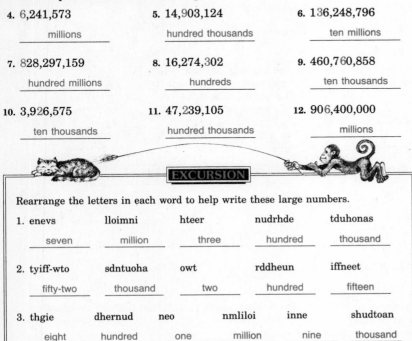

EXCURSION

Rearrange the letters in each word to help write these large numbers.

1. enevs   lloimni   hteer   nudrhde   tduhonas

   seven   _million_   _three_   _hundred_   _thousand_

2. tyiff-wto   sdntuoha   owt   rddheun   iffneet

   fifty-two _thousand_   _two_   _hundred_   _fifteen_

3. thgie   dhernud   neo   nmliloi   inne   shudtoan

   eight   _hundred_   _one_   _million_   _nine_   _thousand_

38

Practice

Remind students to use commas to separate the periods in numbers of 1,000 or more. Have students complete the page independently.

Excursion

Have students rearrange the letters into numerical words. Extend the activity by having them write the numbers as numerals and by writing some scrambled number words of their own.

Correcting Common Errors

Students may omit zeros as place-holders or write too many zeros when they are writing numbers in the millions. Have them write the numbers on a place-value chart using period names—millions, thousands, ones—to guide them. Every place on the chart to the right of the left-most digit must also have a digit.

Enrichment

Have students write 4 numbers each having 9 digits. They should then round each to the nearest million and thousand.

Extra Credit   *Measurement*

Conduct a classroom "Meter-Olympics" spread over several class periods. Students choose a variety of athletic events whose results can be measured in meters. Some examples might be: standing broad jump, softball throw and running long jump. In the school yard or gymnasium, have students take turns trying each event. Use a meter stick to record the results. Keep a chart of the number of meters each person jumped or threw. Continue this activity for several days and keep a daily record. Students may wish to compare their first efforts with later jumps and note their individual improvement.

Problem Solving Guess and Check

pages 39-40

Objective

To solve problems using the guess and check strategy

Materials

small pieces of paper

Mental Math

Tell students to write an equation, using 3 numbers, and any operation, to arrive at:

1. 42 (50 − 10 + 2, etc.)
2. 100 (50 + 25 + 25, etc.)
3. 1 (1/3 + 1/3 + 1/3, etc.)
4. 4.1 (10 − 3.9 − 2, etc.)
5. 8 (7.4 + 2.6 − 2, etc.)
6. 75 (50 + 100 − 75, etc.)
7. $5.00 ($20 − 10 − 5, etc.)
8. 3,000 (1,500 + 1,000 + 500, etc.)

Guessing and Checking

Holly's age is 4 times Phillip's age. Eric's age is twice Phillip's age. The sum of all of their ages is 21. How old is each child?

★ SEE

We need to find each child's age.

We know:

Holly is __4__ times older than Phillip.

Eric is __2__ times older than Phillip.

The sum of all three ages is __21__.

★ PLAN

We can make an **educated guess** of each child's age using the given information. Since the facts seem to show that Phillip is the youngest, we should start with him. To know when we have reached the solution, it is important to check after each guess. We can also use guesses to help us make a better guess on the next try.

★ DO

	Phillip's Age	Holly's Age	Eric's Age	Sum of Ages	
Guess 1	5	20	10	35	(too high)
Guess 2	2	8	4	14	(too low)
Guess 3	3	12	6	21	

Phillip is __3__ years old. Holly is __12__ and Eric is __6__.

★ CHECK

We can check further by making sure our solution fits the problem.
Phillip's age: 3
Holly's age: 12 (4 × 3 = 12)
Eric's age: 6 (2 × 3 = 6)   __3__ + __12__ + __6__ = __21__

39

Teaching the Lesson

Tell students that in life we often use a strategy called **Guess and Check.** Discuss the term **educated guess** and help students understand that an educated guess is a guess made after considering all known facts, whereas a wild guess is not based on facts. Ask students if they would be giving an educated guess or a wild guess if asked to simply name the number another person has in mind. (wild) Ask students what kind of guess they would give if the person said the number was between 1 and 100. (educated guess based on the limits of 1 to 100) Tell students that in the strategy of Guess and Check, it is important to record all guesses to prevent repetition and therefore save time. Help students name other times when they use the Guess and Check strategy and then discuss how guesses and checks are made in each.

Have a student read the problem and complete the **See** portion. Read and discuss the **Plan** section with students. Work through the guess chart until students arrive at the solution. Have them fill in the sentences with the correct answers. They should then check their answers at the bottom of the page. Discuss with students how a table or list provides an organized way to keep track of their guesses.

Apply

Guess and check to help solve these problems.

1. Put each of the digits from 1 to 9 in a circle so each side of the triangle adds up to exactly 20. *Answers may vary.*

2. Put the digits 1 through 7 in the circles so that any 3 circles connected by a line have the same sum. *Answers may vary.*

3. Use the digits 1 through 7 only once to make an addition problem that will have a sum of 100. Find 2 ways to solve this.
Answers may vary.

```
  24        26
  16        14
  53        57
+  7       +  3
-----      -----
 100       100
```

4. Complete this multiplication table.

×	5	3	7	9
8	40	24	56	72
4	20	12	28	36
2	10	6	14	18
5	25	15	35	45

5. If the sum of two numbers is 8 and the product is 0, what are the two numbers?
0 and 8

6. How can a 6-digit number with 5 in the ten-thousands place be greater than a number with 9 in the ten-thousands place?
Answers will vary.

7. One of the digits in a seven-digit number is the digit 6. If the digit in the next place to the left has a value 10 times the value of the digit 6, what digit is to the left?
6

8. Write a guess-and-check problem about four numbers for a classmate to solve. Use information about the numbers shown below.
    20 + 40 + 60 + 80 = 200
Answers will vary.

40

Extra Credit *Numeration*

Have students devise a list of fifteen addition and subtraction problems in which they have left out the operational sign. 62 (+) 53 = 115, 32 (−) 17 = 15, etc. Have them exchange lists with a partner and complete each other's problems. Have students correct the problems they devised and record the student's score.

Solution Notes

Tell students that in problems 1, 2, 3, 5 and 6, where the solution is dependent on the arrangement of numbers, it is helpful to write the numbers they will use on small pieces of paper which they can move around to try different arrangements.

1. Other solutions include:

```
       1                  5
     3   8              4   8
   7       6          2       6
  9 4     2         5 9     3   7 1
```

2. Other solutions include:

```
  5   6   1         6   5   1
      4                 4
  7   2   3         7   3   2
```

3. Other solutions include:

```
   67      12      27
   25      73      54
    1       5      13
    3       4       6
    4       6     ----
  ----    ----    100
  100     100
```

4. If students have difficulty getting started, remind them that 72 is the product of 8 and another number and that 12 is the product of 4 and another number. Tell students to think of the factors of 25 and 35 to complete the missing factors.

Higher-Order Thinking Skills

5. [Analysis] Students should recognize that a product of 0 means that one factor, at least, is 0. Thus, if the sum is 8, the other factor must be 8.

6. [Synthesis] A possible answer is that the first number has a larger digit in the hundred-thousands place than does the second number.

8. [Synthesis] A possible problem is that the sum of 5 numbers is 200. Each number is 20 more than the preceding number.

Chapter Test

page 41

Item	Objective
1, 3, 5	Identify place value of digit in number less than 10,000 (See pages 27-28)
2	Identify place value of digit in number less than 1,000,000 (See pages 35-36)
4, 6, 7, 8	Identify place value in 7-, 8- and 9-digit numbers (See pages 37-38)
9-13	Read and write numbers less than 1 billion (See pages 21-22, 27-28, 35-38)
14-16	Compare and order numbers less than 10,000 (See pages 29-30)
17-19	Round numbers to nearest 10 (See pages 31-34)
20-25	Round numbers to nearest 100 or dollar (See pages 31-34)

Write the place value of the red digits.

1. 6,508   2. 13,279   3. 8,111   4. 149,000,432
   hundreds   thousands   tens   hundred millions

5. 3,406   6. 126,031,210   7. 2,479,036   8. 75,493,000
   ones   ten millions   millions   hundred thousands

Write the numbers.

9. three hundred sixty-five   10. four thousand, two hundred twenty-four
   365   4,224

11. three million, three hundred thousand, sixty-two
    3,300,062

Write the missing words.

12. 56,483 fifty-six __thousand__, four __hundred__ eighty-three

13. 216,050,000 two __hundred__ sixteen __million__, fifty __thousand__

Compare these numbers.

14. 6,247 ⊙> 4,396   15. 7,743 ⊙< 7,748   16. 5,029 ⊙< 5,039

Round to the nearest ten.

17. 1,463 __1,460__   18. 3,555 __3,560__   19. 4,096 __4,100__

Round to the nearest hundred or dollar.

20. 398 __400__   21. 456 __500__   22. $7.48 __$7__

23. $8.21 __$8__   24. 1,612 __1,600__   25. 550 __600__

41

Circle the letter of the correct answer.

1 53 ◯ 35
   a <
   b >

2 4 + 8
   a 4
   b 10
   c 12
   d NG

3   8
  + 7
   a 14
   b 15
   c 16
   d NG

4   3
   2
  + 6
   a 5
   b 8
   c 13
   d NG

5 7 + (6 + 3)
   a 2
   b 9
   c 16
   d NG

6 14 − 8
   a 6
   b 7
   c 8
   d NG

7   12
  − 9
   a 3
   b 4
   c 5
   d NG

8   9
  − 0
   a 0
   b 9
   c 10
   d NG

9 $n + 6 = 13$
   $n = ?$
   a 5
   b 6
   c 7
   d NG

10 What is the value of the 5 in 4,256?
   a ones
   b tens
   c hundreds
   d NG

11 3,279 ◯ 3,289
   a <
   b >

12 Round 896 to the nearest hundred
   a 800
   b 900
   c NG

13 Round 6,750 to the nearest thousand.
   a 6,000
   b 6,800
   c 7,000
   d NG

14 What is the value of the 9 in 29,067?
   a tens
   b hundreds
   c thousands
   d NG

☐ score

42

Cumulative Review

page 42

Item	Objective
1	Compare and order numbers less than 100 (See pages 1-2)
2-3	Compute basic addition facts (See pages 3-6)
4	Add three 1-digit numbers (See pages 5-6)
5	Understand grouping property of addition (See pages 7-8)
6-8	Compute basic subtraction facts (See pages 9-10)
9	Find missing addend (See pages 13-14)
10	Identify place value of digit in a number less than 10,000 (See pages 27-28)
11	Compare and order numbers less than 10,000 (See pages 29-30)
12	Round numbers to nearest 100 (See pages 31-34)
13	Round numbers to nearest 1,000 (See pages 33-34)
14	Identify place value of digit in a number less than 1,000,000 (See pages 35-36)

Alternate Cumulative Review

Circle the letter of the correct answer.

1 67 ◯ 76
   a >
   b <
   c =

2 3 + 9 =
   a 9
   b 11
   c 13
   d NG

3   9
  + 6
   a 13
   b 14
   c 15
   d NG

4   5
   4
  + 3
   a 12
   b 9
   c 7
   d NG

5 3 + (7 + 2) =
   a 9
   b 6
   c 12
   d NG

6 18 − 7 =
   a 11
   b 10
   c 9
   d NG

7   14
  − 8
   a 5
   b 7
   c 9
   d NG

8   7
  − 7
   a 14
   b 1
   c 0
   d NG

9 $n + 7 = 16$
   $n = ?$
   a 8
   b 9
   c 23
   d NG

10 What is the value of the 3 in 3,621?
   a thousands
   b hundreds
   c tens
   d NG

11 6,927 ◯ 6,279
   a >
   b <
   c =

12 Round 723 to the nearest hundred.
   a 600
   b 700
   c 800
   d NG

13 Round 4,869 to the nearest thousand.
   a 4,900
   b 4,000
   c 5,000
   d NG

14 What is the value of the 2 in 37,256?
   a tens
   b hundreds
   c thousands
   d NG

2-Digit Addition

pages 43-44

Objective

To add 2-digit numbers with trading

Materials

*jar marked hundreds
*jar marked tens
*jar marked ones
place value materials

Mental Math

Ask students to name a related fact you would use to check:

1. $18 \div 2.$ $(9 \times 2, 2 \times 9)$
2. $4 + 9.$ $(13 - 4, 13 - 9)$
3. $8 \times 8.$ $(64 \div 8)$
4. $15 - 8.$ $(8 + 7, 7 + 8)$
5. $54 \div 6.$ $(9 \times 6, 6 \times 9)$
6. $16 \div 4.$ (4×4)
7. $7 \times 8.$ $(56 \div 7, 56 \div 8)$
8. $9 + 7.$ $(16 - 7, 16 - 9)$

Skill Review

Remind students that $4 + 4 = 8$ is called a double. Ask students to name the sum of that double plus 1. $(4 + 5 = 9)$ Continue until all doubles, doubles plus 1 and their sums are given.

---

3 ADDITION OF WHOLE NUMBERS

Adding 2-digit Numbers

Carla eats fruit for breakfast each morning. This morning, she ate a peach and half a grapefruit. How many calories does Carla get in her fruit?

We want to know the total number of calories that Carla gets from the fruit.

We know a peach provides __36__ calories

and half a grapefruit __47__ calories.

To find the total, we add __36__ and __47__.

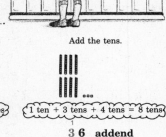

Add the ones.	Make a trade.	Add the tens.
6 ones + 7 ones = 13 ones	13 ones = 1 ten + 3 ones	1 ten + 3 tens + 4 tens = 8 tens

$$\begin{array}{r} 36 \\ +47 \\ \hline \end{array}$$

$$\begin{array}{r} 36 \\ +47 \\ \hline 3 \end{array}$$

$$\begin{array}{rl} 3\,6 & \text{addend} \\ +4\,7 & \text{addend} \\ \hline 8\,3 & \text{sum} \end{array}$$

Carla gets __83__ calories from the fruit she eats.

Getting Started

Add.

1. $\begin{array}{r} 43 \\ +18 \\ \hline 61 \end{array}$

2. $\begin{array}{r} 79 \\ +35 \\ \hline 114 \end{array}$

3. $\begin{array}{r} 26 \\ +53 \\ \hline 79 \end{array}$

4. $\begin{array}{r} 86 \\ +58 \\ \hline 144 \end{array}$

Copy and add.

5. $17 + 59$
   76

6. $76 + 8$
   84

7. $61 + 48$
   109

8. $87 + 67$
   154

43

Teaching the Lesson

Introducing the Problem Have a student tell about the picture. (A girl is looking at fruit and a chart of the calories in fruits.) Have a student read the problem aloud and tell what is to be solved. (how many calories Carla had if she ate a peach and half a grapefruit) Ask students what information they need to use from the picture. (A peach has 36 calories and half of a grapefruit has 47 calories.) Have students read and complete the sentences to solve the problem. Guide students through the 3-step algorithm, explaining each trade depicted.

Developing the Skill Write **29 + 16** on the board. Ask students to tell how many tens and ones in 29. (2, 9) Put 2 ten-strips in the tens jar and 9 singles in the ones jar. Ask students how many ones all together when we add 6 singles to the ones jar. (15) Dump the singles out and trade 10 for 1 ten. Ask students how many ones are left. (5) Put the 5 singles in the ones jar and a ten-strip in the tens jar. Ask how many tens-strips in all. (4) Ask students to tell the sum of 29 and 16. (45) Repeat for more sums of two 2-digit numbers and use the hundreds jar similarly when sums are more than 99.

Practice

Add.

1. 38 + 26 64	**2.** 49 + 8 57	**3.** 36 + 81 117	**4.** 9 + 86 95	**5.** 78 + 57 135
6. 76 + 43 119	**7.** 45 + 55 100	**8.** 26 + 9 35	**9.** 59 + 63 122	**10.** 71 + 65 136
11. 6 + 58 64	**12.** 67 + 96 163	**13.** 80 + 17 97	**14.** 46 + 74 120	**15.** 98 + 96 194

Copy and Do

16. 32 + 19 51	**17.** 75 + 48 123	**18.** 82 + 57 139	**19.** 25 + 96 121
20. 43 + 58 101	**21.** 22 + 67 89	**22.** 9 + 86 95	**23.** 82 + 53 135
24. 74 + 8 82	**25.** 82 + 48 130	**26.** 93 + 19 112	**27.** 57 + 70 127

Apply

Solve these problems.

28. Robert poured a cup of skim milk over his bowl of oatmeal. The oatmeal contained 68 calories and the skim milk had 85. How many calories did Robert eat?
153 calories

29. Juanita ate a scrambled egg and a piece of toast. The egg contained 96 calories and the toast had 79. How many calories did Juanita eat?
175 calories

44

Correcting Common Errors

Students may add incorrectly because they add each column separately failing to regroup.

INCORRECT	CORRECT
	1
47	47
+ 68	+ 68
1015	115

Correct by having students use place-value materials to model the problem.

Enrichment

Tell students to find the number of cans of fruit punch you would need to buy to serve the 27 students in your class and 28 students across the hall if each student will have 3 ounces and one can of fruit punch contains 55 ounces.

Practice

Have students complete the page independently.

Extra Credit *Logic*

Provide students with the following observations collected after a class field trip to the park. Ask them to use these facts to figure out what color the Tanager was and which bird the class liked best.
The Blue Jay was blue and the Goldfinch, yellow.
The hawk was not the red one.
The Oriole and the brown bird were in the same tree as the Blue Jay and the red bird.
They all liked the orange bird best.
(Red; Oriole)

3-Digit Addition

pages 45-46

Objective

To add 3-digit numbers with trading

Materials

*jar marked hundreds
*jar marked tens
*jar marked ones
*jar marked thousands
place value materials

Mental Math

Have students name the pattern:

1. 7, 10, 13, 16, 19 (+3)
2. 92, 88, 84, 80, 76 (−4)
3. 4, 3, 5, 4, 6, 5, 7 (−1, +2)
4. 5, 10, 20, 40, 80 (×2)
5. 36, 27, 18, 9 (−9)
6. 1/8, 2/8, 3/8, 4/8 (+1/8)

Skill Review

Have students identify facts whose sums would require trading. (any fact with sum of 10 or more) Have students name addition facts whose sums would not require trading. (any fact with sum of 9 or less) Now ask students to tell if they would trade if the fact 8 + 7 were in the tens column. (yes) Ask what trade would be made. (10 tens for 1 hundred) Continue for other facts.

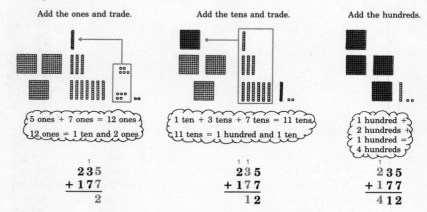

Adding 3-digit Numbers

Jerry is the left tackle for Saratoga High School. Bill is the left guard. Jerry and Bill work well when they block together. What is their total weight?

We want to know the total weight of the two players.

Jerry weighs __235__ pounds and Bill weighs __177__.

To find their total weight, we add __235__ and __177__.

Add the ones and trade.

5 ones + 7 ones = 12 ones
12 ones = 1 ten and 2 ones

$$\begin{array}{r} 1 \\ 235 \\ +177 \\ \hline 2 \end{array}$$

Add the tens and trade.

1 ten + 3 tens + 7 tens = 11 tens
11 tens = 1 hundred and 1 ten

$$\begin{array}{r} 11 \\ 235 \\ +177 \\ \hline 12 \end{array}$$

Add the hundreds.

1 hundred +
2 hundreds +
1 hundred =
4 hundreds

$$\begin{array}{r} 1 \\ 235 \\ +177 \\ \hline 412 \end{array}$$

The total weight for both boys is __412__ pounds.

Getting Started

Add.

1. $\begin{array}{r} 138 \\ +147 \\ \hline 285 \end{array}$
2. $\begin{array}{r} 384 \\ +135 \\ \hline 519 \end{array}$
3. $\begin{array}{r} 473 \\ +798 \\ \hline 1,271 \end{array}$
4. $\begin{array}{r} 157 \\ +60 \\ \hline 217 \end{array}$

Copy and add.

5. 683 + 294
   977
6. 64 + 743
   807
7. 811 + 496
   1,307
8. 426 + 788
   1,214

45

Teaching the Lesson

Introducing the Problem Have students tell what the picture shows. (Jerry is number 77 and weighs 235 pounds, Bill is number 66 and weighs 177 pounds.) Have a student read the problem aloud and tell what they are to find. (find the total weight of the boys) Discuss with students that in football there is a guard and a tackle on each side of the center on the offensive line. Have students complete the sentences and guide students through the 3-step model problems as they find the boys' total weight.

Developing the Skill Write **164 + 749** vertically on the board. Talk through the problem as you add and make the trades. Write **606 + 587** vertically on the board and talk through the problem as you work it. Remind students that a trade is not needed as you add and record the tens. Ask students if trading is necessary in the hundreds column. (yes) Record the trade for 1 thousand and complete the problem. Remind students we use a comma in numbers with 4 or more digits. Have a student place a comma in the answer and read the sum. (1,193; one thousand, one hundred ninety-three) Repeat for more problems of two 3-digit addends.

Practice

Add.

1. 256 + 129 385	2. 627 + 438 1,065	3. 581 + 276 857	4. 875 + 89 964	5. 67 + 483 550
6. 373 + 876 1,249	7. 541 + 212 753	8. 347 + 96 443	9. 709 + 584 1,293	10. 295 + 99 394
11. 560 + 348 908	12. 883 + 910 1,793	13. 396 + 264 660	14. 79 + 389 468	15. 543 + 377 920

Copy and Do

16. 386 + 409
795
17. 718 + 534
1,252
18. 475 + 221
696
19. 256 + 908
1,164
20. 415 + 344
759
21. 690 + 283
973
22. 596 + 19
615
23. 753 + 188
941
24. 93 + 815
908
25. 126 + 568
694
26. 836 + 684
1,520
27. 314 + 375
689

Apply

Solve these problems.

28. The Amelia Earhart School library contains 472 fiction books. The librarian is ordering 255 more. How many fiction books will the library have?
727 fiction books

29. The boys at Clara Barton School collected 267 cans of food for the Thanksgiving food drive. The girls brought in 278 cans. Which group collected the greater number of cans?
The girls

46

46

4-Digit Addition

pages 47-48

Objective

To add 4-digit numbers with trading

Materials

Mental Math

Tell students to name the number that tells:

1. 1/4 inches in 1 foot (48)
2. area of square if 1 side is 4 (16)
3. half of a 98-page book (49)
4. milliliters in 1 Liter (1,000)
5. dimes in $4 (40)
6. ounces in 2 1/2 pounds (40)
7. pencils in 4 1/4 dozen (51)

Skill Review

Write **60 + 90** on the board. Tell students the sum of these numbers requires a tens trade for 1 hundred. Show the trade as you work the problem. Ask a student to write and solve a problem of two 3-digit addends whose sum would require a hundreds trade for 1 thousand. (600 + 800 = 1,400) Continue to have students write and solve similar problems which do or do not require trading.

Adding 4-digit Numbers

The Missouri-Mississippi River system is the third longest river in the world. The Missouri starts at Three Forks, Montana, and joins the lower Mississippi River just north of St. Louis. It empties into the Gulf of Mexico near New Orleans. How long is the Missouri-Mississippi River system?

We want to know the total length of the Missouri-Mississippi River system.

We know the Missouri River from Three Forks to St. Louis measures __3,726__ kilometers.

The Mississippi River from St. Louis to New Orleans measures __2,295__ kilometers.

To find the system's total length, we add __3,726__ and __2,295__.

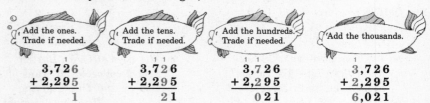

Add the ones. Trade if needed.	Add the tens. Trade if needed.	Add the hundreds. Trade if needed.	Add the thousands.
$\begin{array}{r}\overset{1}{3,72}6\\+2,295\\\hline 1\end{array}$	$\begin{array}{r}\overset{1\ 1}{3,7}26\\+2,295\\\hline 21\end{array}$	$\begin{array}{r}\overset{1\ 1}{3,7}26\\+2,295\\\hline 021\end{array}$	$\begin{array}{r}\overset{1}{3},726\\+2,295\\\hline 6,021\end{array}$

The Missouri-Mississippi River system is __6,021__ kilometers long.

Getting Started

Add.

1.
$$\begin{array}{r}7,836\\+1,193\\\hline 9,029\end{array}$$

2.
$$\begin{array}{r}4,827\\+9,162\\\hline 13,989\end{array}$$

3.
$$\begin{array}{r}6,256\\+3,498\\\hline 9,754\end{array}$$

Copy and add.

4. 4,275 + 907
5,182

5. 8,246 + 3,766
12,012

6. 659 + 5,941
6,600

47

Teaching the Lesson

Introducing the Problem Discuss the map with students. Ask students what river is shown. (Missouri-Mississippi) Discuss the importance of this river to the early settlers of the U.S. Use a map of the Central States to show how a river and its tributaries contribute to the development of cities. Tell students the Missouri River was discovered first and thus it is called the tributary. Have a student read the problem aloud and tell what is to be found. (the length of the Missouri-Mississippi River) Ask students what data is given in the map. (The Missouri River is 3,726 kilometers and the Mississippi is 2,295 kilometers.) Ask students how they can find the total length of the river system. (add 3,726 km and 2,295 km) Talk through the addition process and have students complete the solution sentence.

Developing the Skill Write **9,063 + 957** vertically on the board. Remind students we begin with the ones and work to the left. Have a student record the work on the board as other students talk through each step. Remind students to place the comma between the thousands and hundreds in the sum. Repeat for other problems having sums of 5 or less digits.

Practice

Add.

1. $\begin{array}{r} 4,263 \\ + 3,514 \\ \hline 7,777 \end{array}$

2. $\begin{array}{r} 3,482 \\ + 6,279 \\ \hline 9,761 \end{array}$

3. $\begin{array}{r} 6,257 \\ + 5,767 \\ \hline 12,024 \end{array}$

4. $\begin{array}{r} 8,546 \\ + 1,829 \\ \hline 10,375 \end{array}$

5. $\begin{array}{r} 4,395 \\ + 7,826 \\ \hline 12,221 \end{array}$

6. $\begin{array}{r} 2,054 \\ + 9,975 \\ \hline 12,029 \end{array}$

7. $\begin{array}{r} 8,370 \\ + 7,851 \\ \hline 16,221 \end{array}$

8. $\begin{array}{r} 6,975 \\ + 6,648 \\ \hline 13,623 \end{array}$

Copy and Do

9. 5,483 + 807
6,290

10. 4,563 + 8,789
13,352

11. 7,625 + 9,498
17,123

12. 3,748 + 6,943
10,691

13. 673 + 4,287
4,960

14. 8,496 + 7,827
16,323

15. 5,356 + 7,409
12,765

16. 4,358 + 97
4,455

17. 2,195 + 3,284
5,479

Apply

Solve these problems.

18. The enrollment of Colgate University in one year included 1,423 men and 1,127 women. What was the total number of students?
2,550 students

19. It is 1,411 miles from Pittsburgh to Denver, and 1,174 miles from Denver to Los Angeles. How far is it from Pittsburgh to Los Angeles, through Denver?
2,585 miles

Pittsburgh

Denver

Los Angeles

EXCURSION

Fill in the blanks on this magic square so that the sum of each row down, across and diagonally equals 65. You may use the numbers 1 through 25 only once.

17	24	1	8	15
23	5	7	14	16
4	6	13	20	22
10	12	19	21	3
11	18	25	2	9

48

Correcting Common Errors

Some students may be regrouping incorrectly because they are adding from left to right.

INCORRECT   CORRECT

$\begin{array}{r} {\scriptstyle 1\,1\,1} \\ 7,652 \\ + 9,463 \\ \hline 6,126 \end{array}$ $\begin{array}{r} {\scriptstyle 1\ \ 1} \\ 7,652 \\ + 9,463 \\ \hline 17,115 \end{array}$

Have students use counters and a place-value chart to show the two addends. Then have them join the counters in each place, from right to left, regrouping as they go along.

Enrichment

Tell students that 464, 383 and 4,664 are palindromes. Tell them to write and solve an addition problem of 3 different palindromic addends.

Practice

Have students complete the page independently.

Excursion

Tell students to first determine which numbers from 1 through 25 are unused.
Ask the students to find the sum of all the numbers in the completed square. Since they know the sum of each row, a shortcut would be to multiply 5 × 65 = 325.

Extra Credit   *Measurement*

Tell students that at one time people used parts of their bodies to take measurements and that our system of measurement comes from these ancient people. Ask students to research and report on the origins of these words: cubit, span, hand, inch, yard, fathom, foot. Ask them to tell if they think these types of measurements were accurate. Why or why not?

Estimating Sums

pages 49-50

Objective

To estimate sums of 4-digit numbers by rounding

Materials

Mental Math

Tell students to find the cost of one if items are sold:

1. 10 for $1 (10¢)
2. 3 for 99¢ (33¢)
3. 5 for 75¢ (15¢)
4. 4 for $10 ($2.50)
5. 6 for $1.50 (25¢)
6. 2 for $1.98 (99¢)
7. 20 for $100 ($5)
8. 4 for $6 ($1.50)

Skill Review

Dictate numbers of 9 or fewer digits. Have students write each number on the board and then round it to the nearest ten, hundred and thousand.

Estimating Sums

Sometimes we don't need an exact answer to a problem. We only need to estimate the answer. For example, approximately how many gallons of gasoline are stored in these two tanks?

We want an estimate of the total number of gallons in both tanks.
Tank number 1 contains __1,934__ gallons of gasoline.
Tank number 2 contains __3,587__ gallons.
To estimate the sum, we can round each addend to the nearest thousand, and add these rounded numbers.

1,934 is rounded to __2,000__.

3,587 is rounded to __4,000__.

There are about __6,000__ gallons of gasoline stored in the two tanks.

Getting Started

Estimate the sum after rounding each addend to the nearest ten.

1.	56 + 87 150	2.	34 + 415 450	3.	163 + 589 750	4.	3,019 + 2,936 5,960

Estimate the sum after rounding each addend to the nearest hundred.

5.	368 + 526 900	6.	1,946 + 758 2,700	7.	968 + 412 1,400	8.	3,750 + 1,895 5,700

Estimate the sum after rounding each addend to the nearest thousand.

9.	5,260 + 1,680 7,000	10.	8,576 + 2,750 12,000	11.	6,749 + 7,284 14,000	12.	8,500 + 3,850 13,000

49

Teaching the Lesson

Introducing the Problem   Have a student describe the picture. (There are 1,934 gallons of gas in one tank and 3,587 gallons in the other.) Have a student read the problem aloud and tell what is to be found. (an estimate of the total gallons in both tanks) Have students read aloud with you as they complete the sentences to solve the problem.

Developing the Skill   Tell students that we use rounding to estimate an answer in certain situations. Remind students of the example in a previous lesson of the number of people attending a rock concert. Tell students we estimated the attendance to be 7,000. If we wanted to know the total fans at that concert and another concert, we would round the number of fans at the second concert to the nearest thousand also, and add the 2 numbers. Write **4,950** and **6,201** on the board. Ask students to round the 2 numbers to the nearest thousand, and add to find an estimate of fans. (5,000 + 6,000 = 11,000) Now have students round the numbers to the nearest hundred and add them. (5,000 + 6,200 = 11,200) Repeat for rounding to the nearest ten. (4,950 + 6,200 = 11,150)

Estimate the sum after rounding each addend to the nearest ten.

1. 23 + 58 80	2. 59 + 16 80	3. 275 + 88 370	4. 546 + 682 1,230	5. 7,367 + 4,431 11,800

Estimate the sum after rounding each addend to the nearest hundred.

6. 393 + 167 600	7. 485 + 836 1,300	8. 1,941 + 287 2,200	9. 1,755 + 4,628 6,400	10. 2,808 + 1,576 4,400

Estimate the sum after rounding each addend to the nearest thousand.

11. 3,965 + 2,646 7,000	12. 8,397 + 6,845 15,000	13. 5,545 + 7,851 14,000	14. 4,096 + 9,850 14,000

Copy and Do

15. 73 + 86
159

16. 93 + 18
111

17. 27 + 76
103

18. 41 + 88
129

19. 375 + 916
1,291

20. 876 + 652
1,528

21. 248 + 796
1,044

22. 850 + 525
1,375

23. 5,964 + 8,572
14,536

24. 8,863 + 1,650
10,513

25. 9,452 + 3,925
13,377

26. 229 + 4,076
4,305

Apply

Solve these problems.

27. One storage tank can hold 4,246 gallons of gasoline. Another tank can hold 3,575 gallons. About how many gallons of gasoline can both tanks hold?
About 8,000 gallons

28. It is 1,717 miles from Albany, New York to Dallas, Texas. Dallas is another 2,151 miles from Seattle, Washington. About how far is it from Albany to Seattle, if you drive through Dallas?
About 4,000 miles

Correcting Common Errors

Sometimes students will find the exact sum first, which they then round to give their estimate. Have students discuss instances where estimates are used as the only answer required, encouraging them to see that, once numbers are rounded, it is much easier to use mental computation to find the answer—the estimate—than to add the original addends.

Enrichment

Tell students to write an addition problem whose estimated sum is the smallest possible 6-digit number, and whose exact sum is 12 less than its estimated sum.

Practice

Tell students they are to work each problem in the first 3 rows by rounding each addend and then finding the sum of the rounded numbers. Tell students to check their work in the Copy and Do problems by estimating. Have students complete the page independently.

Mixed Practice

1. 492 + 127 (619)
2. 13 − 8 (5)
3. 437 + 2,725 (3,162)
4. 7 + 4 + 8 + 7 + 2 (28)
5. 11 − 5 (6)
6. 6,208 + 1,395 (7,603)
7. 89 + 2,176 (2,265)
8. 18 − 9 (9)
9. 4,382 + 7,438 (11,820)
10. (8 + 6) + 3 (17)

Extra Credit   *Geometry*

Show students a pair of gloves. Explain that many things have a **handedness** just like the gloves. Have one student identify the right-handed glove and another the left-handed one. Tell students to find several objects that have a handedness. Explain that this property can be called bilateral symmetry. Examples might include: shoes, boots, and mittens, etc. Many toys and dolls are bilaterly symmetrical. The human body has a right-hand and left-hand side. Many houses are symmetric about a line down the middle. Have students explain why bilateral symmetry is sometimes called "mirror symmetry."

5-Digit Addition

pages 51-52

Objective

To add 5-digit numbers with trading

Materials

Mental Math

Tell students to complete each comparison. 10 is to 100 as:

1. 1 is to __ (10)
2. 100 is to __ (1,000)
3. __ is to 200 (20)
4. 4,000 is to __ (40,000)
5. __ is to 10 (1)
6. 60,000 is to __ (600,000)
7. __ is to 1,000,000 (100,000)
8. 50 is to __ (500)

Skill Review

Write several 4- through 9-digit numbers on the board. Have students tell the place value of each digit and then read each number. Have students arrange the numbers in order from least to greatest.

Adding 5-digit Numbers

The Dallas Cowboys and the Washington Redskins play each other twice each year. What was the total attendance for this year's games?

We want to know the total number of fans who attended both games.

We know that __78,475__ fans attended the game in Dallas.

The Washington game drew __76,386__ fans.

To find the sum, we add __78,475__ and __76,386__.

> **ADMIT ONE**
> To add two 5-digit numbers, begin by adding the ones. Then add each column to the left. Trade if needed.

$$
\begin{array}{r} {}^{1} \\ 78,47\!5 \\ +\,76,38\!6 \\ \hline 1 \end{array}
\quad
\begin{array}{r} {}^{1\ 1} \\ 78,4\!7\!5 \\ +\,76,3\!8\!6 \\ \hline 61 \end{array}
\quad
\begin{array}{r} {}^{1} \\ 78,4\!75 \\ +\,76,3\!86 \\ \hline 861 \end{array}
\quad
\begin{array}{r} 7\,8,475 \\ +\,7\,6,386 \\ \hline 4,861 \end{array}
\quad
\begin{array}{r} {}^{1} \\ 78,475 \\ +\,76,386 \\ \hline 154,861 \end{array}
$$

The total attendance for both games was __154,861__.

Getting Started

Add.

1.　17,836
　+ 1,295
　　19,131

2.　　4,827
　+ 19,254
　　24,081

3.　16,875
　+ 58,596
　　75,471

4.　39,454
　+ 12,926
　　52,380

5.　49,025
　+ 38,484
　　87,509

6.　22,456
　+ 67,381
　　89,837

Copy and add.

7. 16,214 + 8,295
　24,509

8. 2,897 + 56,850
　59,747

9. 37,488 + 55,746
　93,234

51

Teaching the Lesson

Introducing the Problem   Have students tell about the picture. (78,475 fans were at Dallas Stadium and 76,386 fans were at Washington Stadium.) Discuss events that often draw large crowds. Tell students that soccer is the most watched sport in the world, and often draws over 100,000 fans a game in other countries. Have a student read the problem and tell what is to be found. (total attendance at a game of Dallas Cowboys vs. Washington Redskins) Ask students it they are given the necessary data. (yes) Have students complete the sentences. Guide students through each step of the 5-digit addition to solve the problem.

Developing the Skill   Write **2,061 + 7,853** vertically on the board. Have a student talk through the problem as another student records the work. Remind students to work the ones column first, and work to the left. Now write **42,061 + 47,853** and tell students we have added 40,000 to each of the numbers. Again have a student talk through the problem as another student records the work. Provide more problems of 5-digit addends for students to talk through and solve.

51

Practice

Add.

1.	16,354 + 2,865 19,219	2.	4,967 + 15,096 20,063	3.	17,858 + 8,392 26,250	4.	11,456 + 56,729 68,185
5.	96,254 + 39,687 135,941	6.	9,560 + 83,785 93,345	7.	26,586 + 76,253 102,839	8.	18,609 + 25,996 44,605
9.	35,694 + 47,828 83,522	10.	83,467 + 25,704 109,171	11.	75,496 + 86,894 162,390	12.	48,182 + 69,778 117,960

Copy and Do

13. 56,965 + 38,758
95,723

14. 41,853 + 26,348
68,201

15. 77,486 + 67,398
144,884

16. 47,836 + 9,548
57,384

17. 64,495 + 40,867
105,362

18. 87,187 + 56,838
144,025

19. 14,390 + 78,956
93,346

20. 60,086 + 88,956
149,042

21. 47,539 + 26,211
73,750

EXCURSION

Write the missing numbers in these problems.

1.
```
   4 6 , 5 0 7
+  1 2 , 9 2 4
   5 9 , 4 3 1
```

2.
```
   2 3 , 8 7 4
+  2 5 , 4 2 9
   4 9 , 3 0 3
```

3.
```
   3 8 , 0 5 3
+  2 6 , 1 9 7
   6 4 , 2 5 0
```

4.
```
   1 7 , 0 4 3
+  5 4 , 2 1 8
   7 1 , 2 6 1
```

52

Column Addition

pages 53-54

Objective

To add multi-digit numbers in a column

Materials

*deck of cards

Mental Math

Ask students to name the number:

1. change 4 cups to pints (2)
2. add 200 to 6 × 4 (224)
3. 9 days after October 24 (November 2)
4. 35 years before 1984 (1949)
5. 4° below freezing F (28°)
6. 2 hours after 2:40 (4:40)
7. 3 decades (30 years)
8. shape of a ball (sphere)

Skill Review

Have a student write **2 + 6 + 8 + 1** vertically on the board and find the sum. (17) Have another student write **7¢ + 6¢ + 2¢** on the board and find the sum. (15¢) Repeat for other column addition problems of 1-digit numbers with and without ¢ signs.

Column Addition

The Metro All-Stars celebrated their tournament win by going out for pizza. They ordered one large Supreme, one family-sized Special and one large Vegetable. How much was their bill?

Pizza			
	Supreme	Special	Vegetable
Family	$12.95	$11.59	$10.29
Large	$9.95	$8.59	$7.29
Small	$6.95	$5.59	$4.29

We want to know the total bill.

We know that a large Supreme pizza costs ___$9.95___.

A family-sized Special costs ___$11.59___.

A large Vegetable pizza costs ___$7.29___.

To find the sum of all the prices, we add

___$9.95___, ___$11.59___ and ___$7.29___.

✔ When adding numbers in a column, add two digits at a time, and then add that sum to the next digit.

✔ When adding amounts of money, remember to line up the decimal points. Include the dollar sign and decimal point in the sum.

$$
\begin{array}{r} \overset{2}{\$\ 9.9}5 \\ 11.59 \\ +\ \ 7.29 \\ \hline 3 \end{array}
\quad
\begin{array}{l} 5 + 9 = 14 \\ \\ 14 + 9 = 23 \end{array}
\qquad
\begin{array}{r} \$\ 9.95 \\ 11.59 \\ +\ \ 7.29 \\ \hline .83 \end{array}
\qquad
\begin{array}{r} \overset{1\ 1}{\$\ 9.9}5 \\ 11.59 \\ +\ \ 7.29 \\ \hline 8.83 \end{array}
\qquad
\begin{array}{r} \overset{1}{\$\ 9.9}5 \\ 11.59 \\ +\ \ 7.29 \\ \hline \$28.83 \end{array}
$$

The total pizza bill was ___$28.83___.

Getting Started

Add.

1.
$$\begin{array}{r} 43 \\ 96 \\ +\ 83 \\ \hline 222 \end{array}$$

2.
$$\begin{array}{r} \$\ 37.46 \\ 158.49 \\ +\ \ 65.53 \\ \hline \$261.48 \end{array}$$

Copy and add.

3. 756 + 4,096 + 8,749
   13,601

4. $52.46 + $96.75 + $43.38
   $192.59

53

Teaching the Lesson

Introducing the Problem   Have a student read the problem and tell what is to be solved. (total pizza bill) Ask students which pizzas were ordered. (large Supreme, family-sized Special and large Vegetable) Ask students how we find the cost of each pizza. (across Large row and down from Supreme to $9.95; across Family row and down from Special to $11.59; across Large row and down from Vegetable to $7.29) Give added practice using the table by asking the cost of other sizes and types of pizza. Have students read and complete the sentences. Guide students through the column addition steps in the model, to find the total pizza bill.

Developing the Skill   Write **64,356 + 2,491 + 984** vertically on the board. Remind students that when adding 6, 1 and 4 we find the sum of 2 numbers and then add the third number to that sum. Talk through the problem as you call attention to the trades. (67,831) Now write **$643.56 + 24.91 + 9.84** vertically on the board and show students how to line up the decimal points and then add. ($678.31) Remind students that we can check the sum by adding up the column. Have students check the problem with you. Repeat for more problems of 3 or 4 addends, with 3 to 5 digits each.

53

Practice

Add.

1.	86 94 + 27 ____ 207	**2.**	215 48 + 135 ____ 398	**3.**	$2.47 6.29 + 5.15 _____ $13.91	**4.**	$8.37 0.96 + 4.15 _____ $13.48
5.	168 457 + 816 ____ 1,441	**6.**	$4.86 2.95 + 6.18 _____ $13.99	**7.**	738 915 + 2,316 _____ 3,969	**8.**	$25.10 6.15 + 8.85 _____ $40.10
9.	1,643 851 + 1,175 _____ 3,669	**10.**	$126.21 27.39 + 436.51 _____ $590.11	**11.**	$ 16.43 537.82 4.57 + 89.75 _____ $648.57	**12.**	37,265 15,586 39,509 + 4,228 _____ 96,588

Copy and Do

13. 48 + 87
135

14. 64 + 168 + 452
684

15. 869 + 47 + 1,750
2,666

16. 984 + 5,465 + 3,750
10,199

17. $41.56 + $1.97 + $562.27
$605.80

18. $39.75 + $18.56 + $27.38
$85.69

19. $50.37 + $392.47 + $88.50
$531.34

20. $65.46 + $39.75 + $132.59 + $3.40
$241.20

21. 13,750 + 27,967 + 37,875
79,592

22. 27,296 + 8,275 + 11,750 + 2,068
49,389

Apply

Use the pizza order board on page 53 to find
the total cost of these orders.

23. 1 family Vegetable
1 large Vegetable
1 small Vegetable
$21.87

24. 1 family Supreme
1 family Special
1 large Special
1 small Vegetable
$37.42

54

Problem Solving
Make a Table

pages 55-56

Objective

To make a table or list to solve a problem

Materials

3 numbered marbles in bag
4 numbered marbles in bag
nickels, dimes, pennies

Mental Math

Have students tell the greatest number using the digits:

1. 7, 9, 6, 2 (9,762)
2. 0, 6, 4, 5 (6,540)
3. 6, 9, 1, 0, 2 (96,210)
4. 4, 8, 8, 1 (8,841)
5. 3, 1, 6, 4, 1 (64,311)
6. 2, 5, 4, 9, 0 (95,420)
7. 1, 2, 3, 4, 5 (54,321)
8. 8, 8, 9, 9 (9,988)

Making a Systematic Listing or Table

Peanuts are packaged in 1-pound and 2-pound cans. List all the different ways you could buy 10 pounds of peanuts.

★ SEE

We need to find:
  all possible arrangements for purchasing 10 pounds of peanuts.
We know:
  peanuts come in __1__-pound and __2__-pound cans.

  A total of __10__ pounds are being purchased.

★ PLAN

We can make an organized list of all the possibilities.

★ DO

Number of 1-lb Cans	10	9	8	7	6	5	4	3	2	1	0
Number of 2-lb Cans	0	0	1	1	2	2	3	3	4	4	5
Total Weight	10	9	10	9	10	9	10	9	10	9	10
Possible Solution	Yes	No	Yes	No	Yes	No	Yes	No	Yes	No	Yes

★ CHECK

The total weight of the 1-pound cans plus the total weight of the 2-pound cans should equal 10 pounds.

$10 + \underline{0} = \underline{10}$ $8 + \underline{2} = \underline{10}$ $6 + \underline{4} = \underline{10}$

$4 + \underline{6} = \underline{10}$ $2 + \underline{8} = \underline{10}$ $0 + \underline{10} = \underline{10}$

55

Teaching the Lesson

Ask students if they've ever made a list of things to do. Tell students they probably look at such a list and decide which thing they will do first, second, etc. Tell students this process of deciding what to do, in what order, is called organizing or planning a **systematic** way to do the tasks. Tell students that planning things in systematic order helps to organize their thinking for them to complete tasks or solve problems.

Have a student read the peanut problem. Tell students they will see how to systematically organize their thoughts to solve this problem. Have a student read the SEE step as all students complete the sentences. Continue similarly through the next 3 steps. Now ask students how the headings would be changed if the peanuts were packaged in 2- and 4-lb bags and we wanted 18 lbs of peanuts. (1-lb cans would become 2-lb bags, 2-lb cans would be 4-lb bags.) Ask how the headings would read if we wanted 40¢ worth of 2¢, 5¢ and 10¢ stamps. (2¢ stamps, 5¢ stamps, 10¢ stamps, total cost)

Apply

Make a systematic listing or table to help solve these problems.

1. A snail crawls five inches up a wall the first minute and then slides back two inches during the second minute. It moves five inches forward the third minute and slides back two inches the fourth minute. If the snail continues in this manner, how far will the snail have traveled in five minutes?
11 inches

2. The three numbered marbles shown below are placed in a bag.

(3) (5) (1)

Without looking at the numbers pick two marbles from the bag. Your score is the sum of the numbers on the two marbles. What are all possible scores?
4, 6, and 8

3. The four numbered marbles shown below are placed in a bag.

(1) (2) (4) (8)

Without looking at the numbers, draw two marbles from the bag and find the sum of their numbers. What scores are possible?
3, 5, 9, 6, 10, and 12
See Solution Notes.

4. Two cuckoo clocks are hanging on the wall. The cuckoo from the first clock comes out every 6 minutes, while the cuckoo from the second clock comes out every 8 minutes. If both cuckoos come out at noon, when will they come out together again?
12:24 PM

5. Marcy is making a list to solve a problem. She has written 21, 32, 43, and 54. Tell what number you think she will write next and explain why.
65

6. Read Exercise 1 again. If the snail has traveled 1 foot up the wall, how many minutes has it been crawling? (Remember: 1 foot = 12 inches.)
8 minutes

7. Read Exercise 2 again. What if the digits on the marbles were 4, 6, and 2. How would this affect the scores you listed?
Each digit would be increased by 1.

8. If you increase each addend in an addition problem by 100, by how much would you increase the sum?
See Solution Notes.

56

Extra Credit *Applications*

Display a large wall map of the United States. Have students research the air distance between major U. S. cities you have selected. Have students mark the air routes with string and attach cards marked with the mileage between cities. Help students discuss the meanings of such terms as "round trip", "by way of" and "direct flight". Now ask students to create problems involving the air distances shown on the map, such as: How far is it from Los Angeles to Minneapolis by way of Denver? How far is a round trip from Cleveland to Miami?

Solution Notes

1. Students may need to act out the problem on a vertical number line before making this table:

   Minute: 1 2 3 4 5
   Distance: 5 3 8 6 11

2. Have students make a table and then check it with the bag of marbles:

1st:	3	3	5
2nd:	5	1	1
Sum:	8	4	6

3.
1st:	1	1	1	2	2	4
2nd:	2	4	8	4	8	8
Sum:	3	5	9	6	10	12

4.
1st	2nd
12:00	12:00
12:06	12:08
12:12	12:16
12:18	12:24
12:24	

   Ask students which cuckoo came out more times. (1st) Have students tell the time when they will again come out together. (12:48) Note: This problem and the next one are real-life uses of common multiples.

Higher-Order Thinking Skills

5. [Analysis] Each digit in its respective place is increased by 1 to get the next number.

6. [Synthesis] This requires a continuation of the table started for Exercise 1, except now the students are interested in the number of minutes rather than the number of inches.

7. [Analysis] Each digit has been increased by 1. Since each score is the result of drawing two marbles, each score would be increased by 2.

8. [Synthesis] The sum would be increased by the number of addends times 100.

Calculator Codes

pages 57-58

Objective

To use a calculator to add numbers

Materials

*large drawing of simple calculator
calculators

Mental Math

Have students name the numbers that
round to:

1. 30 (25, . . . 34)
2. 100 (95, . . . 104; 50, . . . 149)
3. 460 (455, . . . 464)
4. 700 (695, . . . 704; 650, . . . 749)
5. 990 (985, . . . 994)
6. 4,040 (4,035, . . . 4,044)
7. 810 (805, . . . 814)

Skill Review

Write **47,059 + 786** on the board.
Ask students how to estimate the sum
to the nearest thousand. (round each
number to nearest thousand and add)
Ask a student to write and solve the
problem. (47,000 + 1,000 = 48,000)
Repeat for more problems with addends of 5 digits or less. Have students
find estimated sums to the nearest ten,
hundred or thousand.

Calculator Codes

A calculator has **number keys, operation
keys** and **special keys** on its **keyboard.**
The calculator also has an **on/off key.** When
you turn your calculator on, a zero should
show on the screen. When you press a number
key or operation key, an **entry** is made in
the calculator and shows on the screen.
When you press the $\boxed{C}$ key, all entries are
cleared from the calculator. When you press
the $\boxed{CE}$ key, the last entry is cancelled, but
the previous entries are remembered.

Number Keys	Operation Keys	Special Keys
		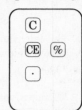

A **calculator code** gives the order for pressing
the keys on your calculator. Try these codes and
write the results on the screen.

127 $\boxed{+}$ 326 $\boxed{=}$ ⟨ 453 ⟩

5162 $\boxed{+}$ 463 $\boxed{+}$ 2000 $\boxed{=}$ ⟨ 7625 ⟩

18036 $\boxed{+}$ 6259 $\boxed{+}$ 2475 $\boxed{+}$ 72405 $\boxed{=}$ ⟨ 99175 ⟩

✔ When you enter an amount of money into the calculator,
the zeros to the far right of the decimal point do not have to
be entered. The calculator will not print them, or a dollar
sign, in the answer.

$9.20 will be entered as and appear as ⟨ 9.2 ⟩.

Enter this code and write the answer on the screen:

65 $\boxed{\cdot}$ 45 $\boxed{+}$ 72 $\boxed{\cdot}$ 05 $\boxed{+}$ 6 $\boxed{\cdot}$ 4 $\boxed{=}$ ⟨ 143.9 ⟩

57

Teaching the Lesson

Introducing the Problem Refer to the large drawing of
a calculator as you read the paragraph to students. Have
students find each number key on the drawing. Remind students that all numbers are made by using combinations of
these 10 numbers. Have students find each operation key
on the drawing. Ask students how many operations there
are. (5) Have students find each special key on the drawing
as you repeat its use. Tell students the decimal point is used
when working with amounts of money. Tell students the ⅙
key will not be used at this time.

Developing the Skill Have students find each **number
key, operation key** and **special key** on their calculator
keyboard. Read about the calculator code to the students.
Have students press the **on/off key,** work the 3 problems
and write the sums. Remind students to press the $\boxed{C}$ to
clear the screen before each new problem. Write more calculator codes on the board for extra practice. Have students
estimate each sum by rounding to the nearest hundred or
thousand. Now continue reading with the students to learn
about entering money amounts into a calculator. Have students enter the code at the bottom and write the answer.
Write more codes with money amounts on the board for
continued practice.

Practice

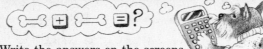

Complete these codes. Write the answers on the screens.

1. 5 [·] 16 [+] 9 [·] 84 = (15)   2. 7691 [+] 12465 [=] (20156)

3. 36615 [+] 9475 [=] (46090)   4. 6458 [+] 605 [+] 11927 [=] (18990)

Use a calculator to find the sum.
Use estimation to be sure the answer seems correct.

5.	247	6.	1,643	7.	42,621	8.	16,941
	629		851		27,436		40,432
	+ 515		+ 1,175		16,948		895
	1,391		3,669		+ 11,753		+ 6,748
					98,758		65,016

Copy and Do

9. $3.57 + $2.96 + $8.38
$14.91

10. $7.95 + $24.63 + $87.51
$120.09

11. 8,207 + 28,416 + 3,796
40,419

12. 6,240 + 14,256 + 3,901
24,397

Apply

Use a calculator to solve this problem.

13. The three tallest mountains in the United
States are Mt. McKinley, St. Elias and Mt.
Foraker. These mountains are all in Alaska.
If you climbed all 3 mountains, how many
feet of mountain have you scaled?
55,728 feet

Mountain	Height
Mt. McKinley	20,320 ft
St. Elias	18,008 ft
Mt. Foraker	17,400 ft

EXCURSION

What does Sally do at the seashore?
To figure out this mystery, complete each code on the calculator. Then turn the
calculator upside down and print each word that shows on the screen.

1. 113 [+] 96 [+] 136 [=]   S H E

2. 44772 [+] 12963 [=]   S E L L S

3. 175596 [+] 39501 [+] 362248 [=]   S H E L L S .

58

58

Chapter Test

page 59

Item	Objective
1-4	Add two 2-digit numbers (See pages 43-44)
5-8	Add two 3-digit numbers (See pages 45-46)
9-16	Add two 4-digit numbers (See pages 47-48)
17-24	Estimate sums up to 4 digits (See pages 49-50)
25-28	Add more than two numbers (See pages 53-54)

Add.

1.
```
  63
+ 31
  94
```

2.
```
  52
+ 26
  78
```

3.
```
  84
+ 29
 113
```

4.
```
  75
+ 86
 161
```

5.
```
  225
+ 134
  359
```

6.
```
  604
+ 138
  742
```

7.
```
  541
+ 183
  724
```

8.
```
  958
+ 476
 1,434
```

9.
```
  2,426
+ 1,273
  3,699
```

10.
```
  4,625
+ 2,419
  7,044
```

11.
```
  7,219
+ 5,825
 13,044
```

12.
```
  4,926
+ 8,286
 13,212
```

13.
```
  4,952
+ 6,098
 11,050
```

14.
```
  7,846
+ 9,752
 17,598
```

15.
```
  9,476
+ 3,697
 13,173
```

16.
```
  6,543
+ 9,758
 16,301
```

Estimate the sum after rounding to the nearest hundred or dollar.

17.
```
  459
+ 675
1,200
```

18.
```
  352
+ 687
1,100
```

19.
```
  5,290
+ 7,500
12,800
```

20.
```
  $18.96
+  37.25
   $56
```

Estimate the sum after rounding to the nearest thousand.

21.
```
  5,296
+ 3,475
 8,000
```

22.
```
  1,256
+ 7,185
 8,000
```

23.
```
  6,295
+ 8,921
15,000
```

24.
```
  8,377
+ 2,934
11,000
```

Add.

25.
```
  892
  437
+ 186
1,515
```

26.
```
  7,275
  4,687
+ 2,651
 14,613
```

27.
```
  11,751
   2,627
+ 37,583
  51,961
```

28.
```
  $29.47
    7.55
+  58.43
  $95.45
```

Circle the letter of the correct answer.

1 5 + 9
a 13
(b) 14
c 15
d NG

2 4
+ 8
(a) 12
b 13
c 14
d NG

3 4
3
+ 8
a 7
b 11
(c) 15
d NG

4 16 − 9
a 8
(b) 7
c 6
d NG

5 11
− 8
a 4
b 5
c 6
(d) NG

6 7 + n = 9
n = ?
(a) 2
b 5
c 16
d NG

7 What is the value of the 3 in 2,436?
(a) tens
b hundreds
c thousands
d NG

8 2,358 ◯ 2,538
(a) <
b >

9 Round 750 to the nearest hundred.
a 600
b 700
(c) NG

10 Round 5,285 to the nearest thousand.
(a) 5,000
b 6,000
c NG

11 What is the value of the 5 in 357,286?
(a) ten thousands
b thousands
c hundreds
d NG

12 54 + 87
a 131
(b) 141
c 1,311
d NG

13 257 + 189
a 336
b 346
(c) 446
d NG

14 3,279
+ 5,726
(a) 9,005
b 9,015
c 9,105
d NG

◻ score

60

Item	Objective
1-2	Compute basic addition facts (See pages 1-2)
3	Add three 1-digit numbers (See pages 5-6)
4-5	Compute basic subtraction facts (See pages 9-10)
6	Find missing addend (See pages 13-14)
7	Identify place value of digit in a number less than 10,000 (See pages 27-28)
8	Compare and order numbers less than 10,000 (See pages 29-39)
9	Round numbers to nearest 100 (See pages 31-34)
10	Round numbers to nearest 1,000 (See pages 33-34)
11	Identify place value of digit in a number less than 1,000,000 (See pages 35-36)
12	Add two 2-digit numbers (See pages 43-44)
13	Add two 3-digit numbers (See pages 45-46)
14	Add two 4-digit numbers (See pages 47-48)

Alternate Cumulative Review

Circle the letter of the correct answer.

1 6 + 8 =
a 16
b 15
(c) 14
d NG

2 7
+ 9
(a) 16
b 17
c 18
d NG

3 5
6
+ 3
a 9
b 11
(c) 14
d NG

4 18 − 9 =
a 10
(b) 9
c 8
d NG

5 14
− 8
a 9
b 8
c 7
(d) NG

6 9 + n = 15
n = ?
a 24
(b) 6
c 9
d NG

7 What is the value of the 7 in 6,723?
a tens
(b) hundreds
c thousands
d NG

8 4,697 ◯ 4,679
(a) >
b <
c =

9 Round 345 to the nearest hundred.
a 200
(b) 300
c 400
d NG

10 Round 6,586 to the nearest thousand.
a 6,000
b 6,500
(c) 7,000
d NG

11 What is the value of the 2 in 267,954?
a ten thousands
(b) hundred thousands
c millions
d NG

12 67 + 35 =
a 912
(b) 102
c 92
d NG

13 366 + 257 =
a 513
b 613
(c) 623
d NG

14 4,368
+ 4,635
a 8,993
(b) 9,003
c 9,113
d NG

2-Digit Subtraction

pages 61-62

Objective

To subtract 2-digit numbers with trading

Materials

*hundreds, ten and ones jars
addition and subtraction fact cards
place value materials

Mental Math

Tell students to complete the statement with >, <, or =:

1. 82 ◯ 9 × 9 (>)
2. 6 + 14 ◯ 1/2 of 40 (=)
3. 12 pints ◯ 6 1/2 quarts (<)
4. 400 − 6 ◯ 90 × 2 (>)
5. 5 feet ◯ 2 yards (<)
6. 18 ÷ 3 ◯ 36 ÷ 6 (=)
7. days in November ◯ days in December (<)

Skill Review

Have students work in pairs. One student shows an addition fact and the second student gives the sum, a related subtraction fact and its difference. Have students change roles after 20 facts. It is valuable practice to give the 100 addition and subtraction facts tests often.

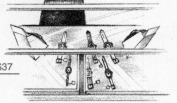

Subtracting 2-digit Numbers

Abby, who is on the swim team, has saved $50 to buy a waterproof watch. How much will she have left after she buys the watch?

We want to know how much Abby will have left after her purchase. We know that Abby has saved __$50__.

The watch costs __$37__. To find the amount left, we subtract __$37__

from __$50__.

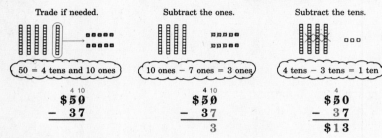

Trade if needed.	Subtract the ones.	Subtract the tens.
50 = 4 tens and 10 ones	10 ones − 7 ones = 3 ones	4 tens − 3 tens = 1 ten

$$\begin{array}{r} {\scriptstyle 4\ 10}\\ \$50 \\ -\ 37 \\ \hline \end{array} \qquad \begin{array}{r} {\scriptstyle 4\ 10}\\ \$\not5\not0 \\ -\ 37 \\ \hline 3 \end{array} \qquad \begin{array}{r} {\scriptstyle 4}\\ \$\not5\not0 \\ -\ 37 \\ \hline \$13 \end{array}$$

Abby will have __$13__ left.

Getting Started

Subtract.

1. $\begin{array}{r} 79 \\ -16 \\ \hline 63 \end{array}$ 2. $\begin{array}{r} 25 \\ -\ 9 \\ \hline 16 \end{array}$ 3. $\begin{array}{r} 97 \\ -28 \\ \hline 69 \end{array}$ 4. $\begin{array}{r} 52 \\ -14 \\ \hline 38 \end{array}$

Copy and subtract.

5. 80 − 14   6. 57 − 20   7. 63 − 61   8. 84 − 27
   66   37   2   57
                                                   61

Teaching the Lesson

Introducing the Problem Have students tell about the picture. (A girl looks at a $37 watch.) Have a student read the problem aloud and tell what is being asked (how much money Abby will have left if she buys the watch) Ask students what information is given. (Abby has $50 and the watch costs $37.) Have a student read and complete the sentences to solve the problem.

Developing the Skill Write **35−19** on the board. Put 3 ten-strips in the tens jar and 5 singles in the ones jar. Ask how we can take away 9 singles. (trade 1 ten-strip for 10 singles) Make the trade and ask how many singles now. (15) Have students continue the subtraction process with the manipulatives to find the difference of 1 ten and 6 singles or 16. Check the work by adding. Repeat for more problems of 2-digit minuends and subtrahends.

Practice

Subtract.

1. 43 −13 = 30	2. 72 −34 = 38	3. 58 −36 = 22	4. 84 −29 = 55	5. 29 − 5 = 24
6. 74 −25 = 49	7. 60 −28 = 32	8. 37 −18 = 19	9. 82 −26 = 56	10. 94 −67 = 27
11. 39 −17 = 22	12. 76 − 9 = 67	13. 41 −38 = 3	14. 57 −14 = 43	15. 63 −45 = 18

Copy and Do

16. 19 − 12
7
17. 43 − 16
27
18. 70 − 15
55
19. 47 − 17
30

20. 37 − 20
17
21. 61 − 43
18
22. 58 − 17
41
23. 30 − 19
11

24. 72 − 36
36
25. 15 − 12
3
26. 82 − 37
45
27. 96 − 58
38

Apply

Solve these problems.

28. Marge spent $24 on a new blouse. Alice spent $38 on a new outfit. How much more did Alice spend for clothes?
$14

29. Chen earned $53 in one month. He spent $38 on birthday gifts for 3 friends. How much did Chen have left?
$15

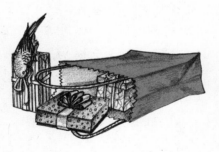

62

Correcting Common Errors

Students may not trade but subtract the lesser digit from the greater digit in a subtraction problem.

INCORRECT   CORRECT
                        713
83   8̶3̶
− 45   − 45
42   38

Correct by having students work in pairs and use place-value materials to model the problems.

Enrichment

Have students solve this problem: If you pay 6¢ tax for every dollar you spend, how much more tax would you pay if you bought a $47 skateboard than if you bought one that cost $39?

Practice

Have students complete the page independently.

Extra Credit *Logic*

Have students draw a grid containing 4 rows of 4 squares each. Next, have them make 4 red markers and four black markers.

Tell students they are to place markers on the squares so that no two of the same color are in a line either horizontally, vertically, or diagonally. Have the first person to find the correct placement draw his grid on the board.

```
  R R
B     R
R     B
  B R
```

3-Digit Subtraction

pages 63-64

Objective

To subtract 3-digit numbers with trading

Materials

*dollar bills, dimes, pennies

Mental Math

Find the total cost of stamps:

1. twelve 5¢ (60¢)
2. three 22¢ (66¢)
3. twenty 10¢ ($2)
4. fifteen 3¢ (45¢)
5. eleven 10¢ ($1.10)
6. eight 20¢ ($1.60)
7. one hundred 10¢ ($10.00)
8. thirty 5¢ ($1.50)

Skill Review

Draw a 5-digit place value chart on the board. Have one student write a 5-digit number on the chart for another student to read and tell the value of each digit. Repeat for more numbers.

Subtracting 3-digit Numbers

The Lopez family drove from Phoenix to the Grand Canyon for a vacation. Their California relatives, the Ruiz family, drove from San Francisco to Yosemite National Park. How much farther did the Lopez family drive than the Ruiz family?

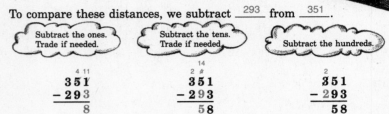

Driving Distances to Parks	
Denver—Pikes Peak	134 kilometers
Los Angeles—Death Valley	416 kilometers
Phoenix—Grand Canyon	351 kilometers
Portland—Crater Lake	394 kilometers
San Francisco—Yosemite	293 kilometers

We want to compare the distances to find how much farther the Lopez family drove.

We know that Phoenix is __351__ kilometers from the Grand Canyon.

San Francisco is __293__ kilometers from Yosemite.

To compare these distances, we subtract __293__ from __351__.

Subtract the ones. Trade if needed.

$$\begin{array}{r} {\scriptstyle 4\ 11} \\ 35\!\!\!/1 \\ -293 \\ \hline 8 \end{array}$$

Subtract the tens. Trade if needed.

$$\begin{array}{r} {\scriptstyle 14} \\ {\scriptstyle 2\ \cancel{4}} \\ 351 \\ -293 \\ \hline 58 \end{array}$$

Subtract the hundreds.

$$\begin{array}{r} {\scriptstyle 2} \\ \cancel{3}51 \\ -293 \\ \hline 58 \end{array}$$

The Lopez family drove __58__ kilometers farther than the Ruiz family.

Getting Started

Subtract.

1.
$$\begin{array}{r} 836 \\ -275 \\ \hline 561 \end{array}$$

2.
$$\begin{array}{r} \$4.57 \\ -1.39 \\ \hline \$3.18 \end{array}$$

3.
$$\begin{array}{r} 983 \\ -259 \\ \hline 724 \end{array}$$

4.
$$\begin{array}{r} 821 \\ -637 \\ \hline 184 \end{array}$$

5.
$$\begin{array}{r} 654 \\ -406 \\ \hline 248 \end{array}$$

6.
$$\begin{array}{r} \$7.93 \\ -1.87 \\ \hline \$6.06 \end{array}$$

7.
$$\begin{array}{r} 519 \\ -483 \\ \hline 36 \end{array}$$

8.
$$\begin{array}{r} \$9.42 \\ -0.96 \\ \hline \$8.46 \end{array}$$

Copy and subtract.

9. 341 − 170   171
10. 523 − 487   36
11. $8.51 − $3.75 $4.76
12. $7.12 − $6.88 $0.24

63

Teaching the Lesson

Introducing the Problem
Have a student read the problem and tell what is to be solved. (how much farther the Lopez family drove than the Ruiz family) Ask students what information is needed from the problem and the table. (Lopez family drove 351 kilometers from Phoenix to the Grand Canyon and the Ruiz family drove 293 kilometers from San Francisco to Yosemite.) Ask students if they need the rest of the information in the table to solve the problem. (no) Ask if any students have visited any of the places named. Have students complete the sentences to solve the problem. Guide them through the 3-step subtraction in the model, and have them complete the solution sentence.

Developing the Skill
Write **946 − 182** vertically on the board. Ask if trading is needed in the ones column. (no) Ask if trading is needed in the tens column. (yes) Ask why. (8 tens is greater than 4 tens.) Have a student work the problem and talk through each step. Check the problem by adding. Repeat for more problems of 3-digit numbers.

Practice

Subtract.

1. 645
  − 437
  208

2. 592
  − 358
  234

3. 728
  − 283
  445

4. $4.16
  − 1.53
  $2.63

5. 721
  − 258
  463

6. 926
  − 305
  621

7. 823
  − 647
  176

8. 615
  − 339
  276

9. 752
  − 283
  469

10. $4.35
  − 0.97
  $3.38

11. 632
  − 108
  524

12. 723
  − 659
  64

Copy and Do

13. 275 − 88
  187

14. 732 − 486
  246

15. $6.26 − $1.54
  $4.72

16. 924 − 808
  116

17. 437 − 158
  279

18. 821 − 480
  341

19. 515 − 326
  189

20. 740 − 196
  544

21. 527 − 83·
  444

22. $4.73 − $1.29
  $3.44

23. 752 − 481
  271

24. 916 − 548
  368

25. 626 − 206
  420

26. 816 − 539
  277

27. 420 − 396
  24

28. $9.38 − $8.42
  $0.96

Apply

Use the chart on page 63 to solve these problems.

29. The Peterson family is driving from San Francisco to Yosemite. They stop for lunch after driving 157 kilometers. How much farther do the Petersons have to drive?
  136 kilometers

30. How far is it from Portland to Crater Lake and back again?
  788 kilometers

64

Correcting Common Errors

Students may trade but forget to decrease the digit to the left after trading.

INCORRECT   CORRECT
      12   2 1 2
   43̷2̷   43̷2̷
  − 3 1 5   − 3 1 5
   1 2 7   1 1 7

Have students first determine if trading is necessary and then have them rewrite the portion of the problem that has been traded; e.g., 3 tens 2 ones = 2 tens 12 ones.

Enrichment

Have students use a map to locate each park listed in the chart on page 63. Tell them to find the population of the city near each park, to the nearest thousand.

Practice

Remind students to use dollar signs and decimal points in the answers to money problems. Have students complete the page independently.

Mixed Practice

1. 4 + (9 + 2) (15)
2. 725 − 432 (293)
3. 157 + 2,651 + 65 (2,873)
4. 9 + 5 + 4 + 8 + 8 (34)
5. $8.43 − 5.67 ($2.76)
6. 748 − 309 (439)
7. 6,427 + 3,722 (10,149)
8. 80 − 23 (57)
9. $27.95 + 14.35 ($42.30)
10. 967 − 432 (535)

Extra Credit   *Applications*

Review how to make flow charts. Ask students to design a chart for tasks that they do often, such as going through the lunch line, playing a record, multiplying a two digit number, etc. Tell students to choose a specific task from a career they are familiar with and design a flow chart for that task. For example, they could chart a short order cook making a cheeseburger. Students may need to observe or interview a person in that career before they can make the chart.

Zeros in Minuend

pages 65-66

Objective

To subtract 3-digit numbers with zeros in the minuend

Materials

ten-strips and single counters

Mental Math

Have students find the total cost of:

1. 5 @ 21¢ ($1.05)
2. 2 @ $1.14 ($2.28)
3. 15 @ $4 ($60)
4. 110 @ 2¢ ($2.20)
5. 40 @ $3 ($120)
6. 6 @ $1.50 ($9)
7. 18 @ $1 per dozen ($1.50)
8. 3 @ $2.50 ($7.50)

Skill Review

Write **$4.26** on the board. Have students tell how many hundreds, tens and ones are in the number. (4, 2, 6) Have students tell how many dollars, dimes and pennies are in the number. (4, 2, 6) Repeat for more amounts through $9.99.

Subtracting, Minuends with Zeros

The Mayflower landed at Plymouth Rock in 1620. There were 87 Pilgrims among the passengers. How many passengers were not Pilgrims?

We want to know how many of the passengers were not Pilgrims.

We know there were ___101___ people aboard the Mayflower when it landed at Plymouth Rock.

Of all the people on the ship, only ___87___ were Pilgrims.

To find the number of non-Pilgrims we subtract

___87___ from ___101___.

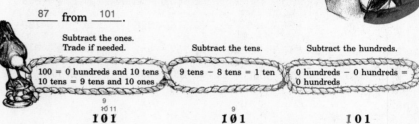

Subtract the ones. Trade if needed.
100 = 0 hundreds and 10 tens
10 tens = 9 tens and 10 ones

Subtract the tens.
9 tens − 8 tens = 1 ten

Subtract the hundreds.
0 hundreds − 0 hundreds = 0 hundreds

$$\begin{array}{r} \overset{9}{}\overset{10\ 11}{101} \\ -\ \ 87 \\ \hline 4 \end{array} \qquad \begin{array}{r} 1\overset{9}{0}1 \\ -\ 87 \\ \hline 14 \end{array} \qquad \begin{array}{r} 101 \\ -\ 87 \\ \hline 14 \end{array}$$

There were ___14___ passengers on the Mayflower who were not Pilgrims.

Getting Started

Subtract.

1.	2.	3.	4.
503	706	521	$4.06
− 231	− 88	− 126	− 1.85
272	618	395	$2.21

Copy and subtract.

5. $3.05 − $2.95
$0.10

6. 603 − 245
358

7. 905 − 728
177

8. $7.04 − $5.28
$1.76

65

Teaching the Lesson

Introducing the Problem Have students describe the picture. (The captain of the Mayflower sees that there are 101 passengers on board.) Discuss the landing of the Mayflower with the students. Have a student read the problem and tell what is to be found. (number of passengers who were not Pilgrims) Ask students what information is needed to solve the problem. (total passengers and number of Pilgrims) Have students complete the sentences and work through each step of the subtraction problem in the model, with them. Have students complete the solution sentence.

Developing the Skill Write **403 − 69** vertically on the board. Tell students that since 9 ones cannot be taken from 3 ones, we need to trade 1 ten for 10 ones. Ask students if there are any tens. (no) Tell students we then trade 1 of the 4 hundreds for 10 tens. Make the trade showing 3 hundreds left. Now ask students if we can trade 1 ten for 10 ones. (yes) Make the trade showing 9 tens left. Continue the subtraction to find the difference. (354) Remind students to check work by adding the subtrahend and the difference and that this must equal the minuend of 403. Repeat for more problems requiring trading in the tens or hundreds column or both.

Practice

Subtract.

1.	604 − 136 468	2.	502 − 185 317
3.	988 − 358 630	4.	$6.05 − 4.83 $1.22

1. 604
   − 136
   468

2. 502
   − 185
   317

3. 988
   − 358
   630

4. $6.05
   − 4.83
   $1.22

5. 408
   − 391
   17

6. $6.00
   − 4.75
   $1.25

7. 201
   − 128
   73

8. 708
   − 220
   488

9. $3.05
   − 2.95
   $0.10

10. 613
   − 245
   368

11. 905
   − 728
   177

12. $7.04
   − 5.28
   $1.76

Copy and Do

13. 307 − 184
    123

14. 509 − 436
    73

15. 802 − 656
    146

16. $9.05 − $6.47
    $2.58

17. 408 − 227
    181

18. 506 − 309
    197

19. 328 − 253
    75

20. $4.07 − $2.28
    $1.79

21. 107 − 98
    9

22. $8.00 − $5.53
    $2.47

23. 604 − 586
    18

24. 705 − 489
    216

Apply

Solve these problems.

25. There are 206 bones in the human body. Of these, 29 bones are in the head. How many bones are there in the rest of the body?
    177 bones

26. A bunch of daisies costs $3.79. If you give the clerk a $5 bill, how much change will you receive?
    $1.21

EXCURSION

Build two more magic word squares like the first one. Use the same words on the horizontal as you do on the vertical. Answers will vary.

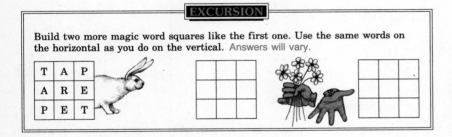

T	A	P
A	R	E
P	E	T

66

Subtracting Large Numbers

pages 67-68

Objective

To subtract multi-digit numbers

Materials

place value materials

Mental Math

Ask students to name the number that is the:

1. smallest 3-digit (100)
2. largest 4-digit (9,999)
3. smallest 5-digit (10,000)
4. largest 3-digit (999)
5. smallest 6-digit (100,000)
6. smallest 9-digit (1 million)
7. largest 8-digit (99,999,999)
8. largest 6-digit (999,999)

Skill Review

Write several 4-, 5- and 6-digit numbers on the board. Have students rename numbers by telling the place value word for each digit.

Subtracting Multi-digit Numbers

Craters of the Moon National Monument, in Idaho, was dedicated in 1924. Dinosaur National Monument, in Colorado and Utah, is an older and larger park. It was dedicated in 1915. How much larger is Dinosaur than Craters of the Moon?

National Monuments		
Name	**Opened**	**Size**
Craters of the Moon	1924	53,545 acres
Dinosaur	1915	211,272 acres
Joshua Tree	1936	559,960 acres
Sunset Crater	1930	3,040 acres
Walnut Canyon	1915	2,249 acres

We want to know how much larger Dinosaur is than Craters of the Moon. We know that Craters of the Moon is ___53,545___ acres large.

Dinosaur National Monument measures ___211,272___ acres.

To find the difference in size between the two parks, we subtract ___53,545___ from ___211,272___.

To subtract multi-digit numbers, begin by subtracting the ones. Trade if needed. Continue subtracting each column to the left.

✔ Remember to trade from one place value at a time. Sometimes you will need to trade from a trade.

```
      10 10
  1  0  0  12 6 12
  2  1  1 , 2  7  2
 −    5  3 , 5  4  5
 ─────────────────
      1  5  7 , 7  2  7
```

Dinosaur is ___157,727___ acres larger than Craters of the Moon.

Getting Started

Subtract.

1.
```
   8,475
 − 2,243
 ───────
   6,232
```

2.
```
  $123.27
 −  86.43
 ────────
   $36.84
```

3.
```
  176,043
 − 29,228
 ────────
  146,815
```

Copy and subtract.

4. 48,758 − 21,475
   27,283

5. 192,175 − 39,857
   152,318

6. $436.25 − $348.99
   $87.26

67

Teaching the Lesson

Introducing the Problem Have a student read the problem and tell what is to be solved. (find out how much larger the Dinosaur National Monument is than Craters of the Moon) Ask students what information is needed from the problem and the table. (sizes of the 2 parks) Have students describe their visits to any of the parks listed. Read with students as they complete the sentences and solve the algorithm in the model. Point out the trading rule.

Developing the Skill Write **111 − 62** vertically on the board and have a student solve the problem, showing the trades. (49) Now write **1,110 − 620** vertically on the board and tell students we have added thousands place to the minuend, and hundreds place to the subtrahend. Show students the trades as you work the problem. (490) Help students see that one trade is made at a time and that they may need to trade from a trade. Check the problem by adding. Repeat for more problems with minuends through 6 digits until students are comfortable working the trades.

Practice

Subtract.

1. 6,456 − 2,329 4,127	**2.** 7,502 − 3,296 4,206	**3.** $57.43 − 18.59 $38.84	**4.** 18,275 − 3,596 14,679
5. 4,253 − 1,327 2,926	**6.** 6,851 − 5,926 925	**7.** $91.15 − 63.75 $27.40	**8.** 8,273 − 4,829 3,444
9. 127,243 − 16,158 111,085	**10.** 40,871 − 29,137 11,734	**11.** 281,275 − 29,469 251,806	**12.** $173.84 − 86.96 $86.88

Copy and Do

13. 7,243 − 1,165
6,078

14. $24.75 − $13.72
$11.03

15. 8,527 − 695
7,832

16. 36,741 − 5,196
31,545

17. 20,275 − 6,889
13,386

18. $543.78 − $275.81
$267.97

19. 296,258 − 26,579
269,679

20. $877.73 − $639.86
$237.87

21. 47,868 − 21,975
25,893

22. 79,246 − 37,865
41,381

23. 430,173 − 14,688
415,485

24. 93,182 − 81,996
11,186

Apply

Use the chart on page 67 to help solve these problems.

25. How many acres larger is Joshua Tree National Monument than Dinosaur?
348,688 acres

26. Is Joshua Tree National Monument larger than the other four listed national parks combined?
Yes

68

Zeros in Minuend

pages 69-70

Objective

To subtract 4- or 5-digit numbers when minuends have zeros

Materials

*thousands, hundreds, tens, ones jars
*place value materials

Mental Math

Tell students to answer true or false:

1. 776 has 77 tens. (T)
2. 10 hundreds < 1,000. (F)
3. 46 is an odd number. (F)
4. 926 can be rounded to 920. (F)
5. 72 hours = 3 days. (T)
6. $42 \div 6 > 7 \times 1$. (F)
7. 1/3 of 18 = 1/2 of 12. (T)
8. perimeter = L × W. (F)

Skill Review

Write 4 numbers of 3- to 5-digits each on the board. Have students arrange the numbers in order, from the least to the greatest, and then read the numbers as they would appear if written from the greatest to the least. Repeat for more sets of 4 or 5 numbers.

Subtracting, More Minuends with Zeros

The Susan B. Anthony School held its annual fall Read-a-Thon. How many more pages did the fifth grade read than the second-place class?

We want to know how many more pages the fifth grade read than the second-place class.

We know the fifth grade read __5,003__ pages.

The second-place class is the __sixth__ grade.

It read __4,056__ pages.
To find the difference between the number

of pages, we subtract __4,056__ from __5,003__.

✔ Remember to trade from one place value at a time.

$$\begin{array}{r} \overset{9\ \ 9}{4\ \cancel{10}\cancel{10}13} \\ \mathbf{5,003} \\ -\ \mathbf{4,056} \\ \hline 9\ 4\ 7 \end{array}$$

The fifth grade read __947__ more pages than the sixth grade.

Getting Started

Subtract.

1. $\begin{array}{r} 3,005 \\ -\ 1,348 \\ \hline 1,657 \end{array}$

2. $\begin{array}{r} \$40.09 \\ -\ 9.75 \\ \hline \$30.34 \end{array}$

3. $\begin{array}{r} 3,300 \\ -\ 1,856 \\ \hline 1,444 \end{array}$

4. $\begin{array}{r} \$50.00 \\ -\ 27.26 \\ \hline \$22.74 \end{array}$

5. $\begin{array}{r} 8,512 \\ -\ 7,968 \\ \hline 544 \end{array}$

6. $\begin{array}{r} \$90.17 \\ -\ 20.87 \\ \hline \$69.30 \end{array}$

Copy and subtract.

7. 26,007 − 18,759
   7,248

8. 70,026 − 23,576
   46,450

9. \$900.05 − \$267.83
   \$632.22

69

Teaching the Lesson

Introducing the Problem Have a student read the problem. Ask students what 2 problems are to be solved. (which class read the second-highest number of pages and how many more pages the first-place fifth graders read) Ask students how we can find out which class came in second place. (arrange numbers from the table in order from greatest to least) Ask a student to write the numbers from greatest to least on the board. (5,003, 4,056, 3,795) Have students complete the sentences and work through the model problem with them.

Developing the Skill Write **4,000−2,875** vertically on the board. Ask students if a trade is needed to subtract the ones column. (yes) Tell students that since there are no tens and no hundreds, we must trade 1 thousand for 10 hundreds. Show the 3 thousands and 10 hundreds left. Now tell students we can trade 1 hundred for 10 tens. Show the trade with 9 hundreds and 10 tens left. Tell students we can now trade 1 ten for 10 ones. Show the trade so that 9 tens and 10 ones are left. Tell students we can now subtract each column beginning with the ones column and working to the left. Show students the subtraction to a solution of **1,125.** Remind students to add the subtrahend and the difference to check the work. Repeat for more problems with zeros in the minuend.

Practice

Subtract.

1.	3,004 − 2,356 648	**2.**	8,002 − 5,096 2,906	**3.**	3,891 − 1,750 2,141	**4.**	$20.08 − 15.99 $4.09
5.	4,020 − 1,865 2,155	**6.**	$87.00 − 28.59 $58.41	**7.**	3,007 − 2,090 917	**8.**	$50.06 − 37.08 $12.98
9.	19,006 − 8,275 10,731	**10.**	20,006 − 14,758 5,248	**11.**	$400.26 − 236.58 $163.68	**12.**	$793.42 − 253.87 $539.55

Copy and Do

13. 4,001 − 2,756
1,245

14. $70.05 − $26.59
$43.46

15. 8,060 − 7,948
112

16. 7,007 − 2,468
4,539

17. 21,316 − 12,479
8,837

18. 14,000 − 8,396
5,604

19. $100.21 − $93.50
$6.71

20. 60,004 − 51,476
8,528

21. 52,006 − 9,037
42,969

22. $800.00 − $275.67
$524.33

23. 34,612 − 29,965
4,647

24. 50,010 − 36,754
13,256

Apply

Use the chart on page 69 to help solve these problems.

25. How many pages did the three classes read all together?
12,854 pages

26. How many more pages did the sixth grade read than the fourth grade?
261 pages

Correcting Common Errors

Some students may bring down the numbers that are being subtracted when there are zeros in the minuend.

INCORRECT	CORRECT
	9
	2 10 10
3,006	3,006
− 1,425	− 1,425
2,421	1,581

Have students work in pairs and use play money to model a problem such as $300 − $142, where they see that they must trade 3 hundreds for 2 hundreds, 9 tens, and 10 ones before they can subtract.

Enrichment

Tell students to find out the year in which each member of their family was born, and make a chart to show how old each will be in the year 2000.

Practice

Remind students to begin with the ones column, work to the left and trade from one place value at a time. Have students complete the page independently.

Extra Credit   *Biography*

An American inventor, Samuel Morse, struggled for many years before his inventions, the electric telegraph and Morse code were recognized. Morse was born in Massachusetts in 1791, and studied to be an artist. On a trip home from Europe, Morse heard his shipmates discussing the idea of sending electricity over wire. Intrigued, Morse spent the rest of the voyage formulating his ideas about how this could be accomplished. Morse taught at a university in New York City, and used his earnings to continue development of his telegraph. After five years, Morse demonstrated his invention, but found very little support. After years of requests for support, Congress finally granted Morse $30,000 to test his invention. He dramatically strung a telegraph wire from Washington, D.C. to Baltimore, Maryland, and relayed the message, "What hath God wrought" using Morse code. Morse's persistence finally won him wealth and fame. A statue honoring him was unveiled in New York City one year before his death in 1872.

imating
fferences
ges 71-72

Objective

To estimate the difference by rounding

Materials

Mental Math

Have students find:

1. volume of $4 \times 4 \times 4$ cub (64)
2. minutes in 2 1/5 hours (132)
3. corners in 3 pentagons (15)
4. 26 days after May 8 (June 3)
5. 4 tens more than 1,084 (1,124)
6. dimes to equal $6.70 (67)
7. years in 1/2 century (50)

Skill Review

Remind students that we use estimation often in our everyday lives. Ask students to tell the present time to the nearest hour and then to the nearest half-hour and quarter-hour. Ask students to tell the number of pages in this book to the nearest ten and nearest hundred. Ask the number of students in the class to the nearest ten. Ask why we would not estimate the number of students in the class to the nearest hundred. (It would be zero and that is not meaningful.)

Estimating Differences

Dominic earned $46.75 as a baker's helper. He wants to buy his own stereo. About how much more money does Dominic need to earn?

We want to estimate how much money Dominic still needs to earn. We know the stereo costs ____$89.29____.

So far, Dominic has earned ____$46.75____. To estimate what he still needs, we round both amounts of money to the nearest dollar, and find the difference between the two.

$89.29 ⟶ **$89**
$46.75 ⟶ **− 47**
⟶ 42

Dominic still needs about ____$42____ to buy the stereo.

Getting Started

Estimate each difference.

Round to the nearest hundred.

1.	2.	3.	4.
845	906	726	586
− 236	− 483	− 413	− 275
600	400	300	300

Round to the nearest thousand.

5.	5.	7.	8.
4,796	6,500	4,975	8,279
− 1,926	− 2,375	− 1,610	− 3,758
3,000	5,000	3,000	4,000

Round to the nearest dollar.

9.	10.	11.	12.
$29.35	$76.21	$94.39	$62.83
− 12.50	− 48.76	− 81.56	− 46.15
$16	$27	$12	$17

71

Teaching the Lesson

Introducing the Problem Have a student describe the picture. (A boy is looking at a camera and stereo in a store window.) Have a student read the problem and tell what is to be solved. (about how much more money Dominic needs to buy the stereo) Ask students what information is needed from the problem and picture. (Dominic has $46.75 and the stereo costs $89.29.) Ask students if we need the other information that is given. (no) Ask if we need to know the exact amount of money Dominic needs yet. (no) Have the students read and complete the sentences. Guide them through the estimation in the model, and have them complete the solution sentence.

Developing the Skill Write **2,762−1,128** vertically on the board. Tell students these numbers could represent the number of days 2 people have lived. Tell students we'd like to know about how many more days the older person has lived so we round each number and then subtract. Tell students we can round each number to the nearest ten or hundred or thousand. Have students round each number to the nearest ten. (2,760, 1,130) Have a student subtract the numbers. (2,760 − 1,130 = 1,630) Have a student round each number to the nearest hundred and subtract. (2,800 − 1,100 = 1,700) Have students check their work by finding the exact difference. Repeat for more problems.

Practice

Estimate each difference.
Round to the nearest hundred.

1. 579 − 346 300	2. 481 − 276 200	3. 825 − 279 500	4. 921 − 735 200
5. 791 − 347 500	6. 426 − 139 300	7. 776 − 438 400	8. 861 − 475 400

Round to the nearest thousand.

9. 7,351 − 2,686 4,000	10. 9,251 − 4,460 5,000	11. 6,865 − 4,956 2,000	12. 8,753 − 1,829 7,000
13. 4,629 − 1,510 3,000.	14. 8,475 − 6,832 1,000	15. 4,500 − 3,723 1,000	16. 8,216 − 5,006 3,000

Round to the nearest dollar.

17. $43.27 − 21.95 $21	18. $71.38 − 18.46 $53	19. $92.50 − 46.89 $46	20. $52.89 − 46.75 $6
21. $38.27 − 14.58 $23	22. $86.89 − 71.76 $15	23. $97.17 − 42.45 $55	24. $66.12 − 51.87 $14

Apply

Solve. Use the ad on page 71 to help solve these problems.

25. About how much money would you have left if you bought the camera with a $100 bill?
$3

26. About how much do the stereo and camera cost together?
$186

72

Checking Subtraction

Objective

To use addition to check subtraction

Materials

*addition and subtraction face cards
place value materials

Mental Math

Dictate the following:

1. $726 - (10 \div 2)$. (721)
2. $5,000 \times 3$. (15,000)
3. $(84 \times 2) + 100$. (268)
4. $18 - 6 + (4 \times 3)$. (24)
5. $(45 \div 9) + 2 1/2$. (7 1/2)
6. $333 - 300 - 30 - 3$. (0)
7. $2/10$ of 100. (20)
8. $171 + 29 \div 4$. (25)

Skill Review

Have students give a related addition fact for each subtraction fact shown. Reverse and have students give a related subtraction fact for each addition fact shown.

Checking Subtraction

Start with a number like 628. Subtract any number from it. Then add that same number to your answer. What do you find? Try this using several different subtrahends. Does the sum always equal the minuend?

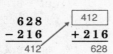

We start with a minuend of __628__.
Then, for example, we subtract the subtrahend 216. We want to see what happens when we add 216 to the difference between 628 and 216.

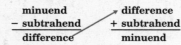

$$\begin{array}{r} 628 \\ -216 \\ \hline 412 \end{array} \qquad \begin{array}{r} 412 \\ +216 \\ \hline 628 \end{array}$$

When we add the subtrahend to the difference,

we must always get the __minuend__.

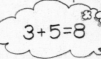

✔ We say that we can check subtraction by addition.

$$\begin{array}{ll} \text{minuend} & \text{difference} \\ -\text{ subtrahend} & +\text{ subtrahend} \\ \hline \text{difference} & \text{minuend} \end{array}$$

Did your sum always equal the minuend? __yes__

Getting Started

Subtract and check.

1. $\begin{array}{r} 639 \\ -156 \\ \hline 483 \end{array}$	2. $\begin{array}{r} 427 \\ -359 \\ \hline 68 \end{array}$	3. $\begin{array}{r} 708 \\ -519 \\ \hline 189 \end{array}$
4. $\begin{array}{r} 4,265 \\ -2,458 \\ \hline 1,807 \end{array}$	5. $\begin{array}{r} 8,056 \\ -3,683 \\ \hline 4,373 \end{array}$	6. $\begin{array}{r} 9,125 \\ -8,136 \\ \hline 989 \end{array}$

Copy, subtract and check.

7. $14,006 - 9,378$
4,628

8. $\$36.86 - \$12.97$
$23.89

9. $37,850 - 19,873$
17,977

73

Teaching the Lesson

Introducing the Problem Have a student describe the picture. (a boy thinking $8 - 3 = 5$ and a girl thinking $3 + 5 = 8$) Ask students if the boy's thought has anything to do with the girl's thought. (They are related facts and the girl's fact can be used to check the boy's fact.) Have a student read the problem and tell what is to be answered. (if the answer plus the subtrahend equals the minuend) Have students complete the sentences and work through the subtraction and check with them. Have students test the solution statement with more subtraction problems.

Developing the Skill Write **476−294** vertically on the board. Have a student work the problem. (182) Have another student write and solve $182 + 294$ to see if the sum is 476. (yes) Give more subtraction problems for students to work at the board and then check by adding. If students make errors in subtracting, allow them to catch the error when the sum of the addition problem does not equal the minuend. Have students go back and re-work the subtraction problem until their work is correct as proven by adding. If students make no errors, incorrectly work a subtraction problem on the board and check by adding, to show students the value of checking.

Practice

Subtract and check.

1.
```
   896
 - 257
   639
```

2.
```
  4,081
 -3,795
    286
```

3.
```
  5,279
 -  875
  4,404
```

4.
```
  8,465
 -7,689
    776
```

5.
```
 11,256
 - 7,587
  3,669
```

6.
```
 $85.40
 - 19.73
 $65.67
```

7.
```
 60,053
 -36,285
 23,768
```

8.
```
 $423.50
 -176.48
 $247.02
```

9.
```
 42,000
 -18,732
 23,268
```

Copy and Do

10. 9,675 − 2,487
7,188

11. 5,001 − 3,974
1,027

12. 27,681 − 8,854
18,827

13. 63,451 − 45,787
17,664

14. $86.53 − $52.45
$34.08

15. 82,741 − 76,857
5,884

Apply

Solve these problems.

16. Queen Victoria ruled Great Britain for 64 years. She was born in 1819 and died in 1901. How old was she when she died?
82 years old

17. The United States admitted 398,613 immigrants in 1976, and 596,600 in 1981. How many more immigrants entered America in 1981 than in 1976?
197,987 immigrants

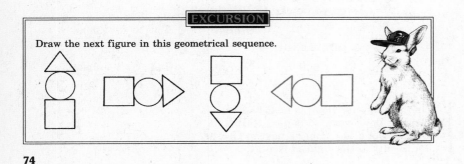

EXCURSION

Draw the next figure in this geometrical sequence.

74

Correcting Common Errors

Some students may have difficulty seeing the relationship between the subtrahend and minuend for checking. Have students work in pairs with place-value materials to model a subtraction problem. One student uses counters to model the minuend; the other takes away the subtrahend and tells what is left. Then they decide what counters must be added to get back to the minuend where they started, and they write the addition problem to show this.

Enrichment

Tell students to follow these steps: record the odometer readings of 4 different cars, estimate how many thousand miles each has travelled, and estimate in thousands how much farther the most-travelled car has been driven than each of the others.

Practice

Remind students to show their addition problems to check each subtraction problem. Tell students to complete the page independently.

Excursion

Have students study the geometrical progression. Advise them to look at each element in the figure and notice what movement is made from one step to another.

Extra Credit   *Numeration*

Assign each letter of the alphabet a number. Give the students problems such as these to solve using the numbers you have chosen. Tell them the answers will match a letter in the alphabet, and spell out the name of a state.

```
  6      30      53      3
 +3     -15     -30     -2
  9      15      23      1
  I       O       W      A
```

When they have solved the first problem, tell students to devise their own state code problems, using addition, subtraction, multiplication or division. Have them exchange with a partner and solve each other's code.

Problem Solving
Act It Out

pages 75-76

Objective

To act out a situation to solve a problem

Materials

56 small objects
20 pennies
deck of cards
4 sheets of paper in 4 colors
scissors
3- and 5-liter containers
8 different objects
17 toothpicks
2 large and 2 small paper clips or similar objects

Mental Math

Ask students how many:

1. boxes of 9 to total 72. (8)
2. seats in each of 6 rows to total 66. (11)
3. 1-in cookies in 8- × 8-in pan. (64)
4. discs in 4 boxes of 10 each. (40)
5. 6 oz cups in 54 oz. (9)
6. decades in 1 1/2 centuries. (15)
7. 1/2 inches on a 12-in ruler. (24)

Acting It Out

The five boys sitting on the bench are Bob, Doug, George, John and Ted. They are not sitting in that order. Neither Bob nor Doug is next to John. Neither Doug nor Bob is next to Ted. Neither John nor Doug is next to George. Ted is just to the right of George. Tell the order in which the boys are sitting.

★ SEE
We need to find:
    the order of the 5 boys in the picture.

★ PLAN
We can act this problem out by following the clues using 5 students and 5 empty chairs.

★ DO
1. We know Ted is just to the right of George.

?	?	George	Ted	?

2. We know Ted is not next to Doug or Bob, which leaves only John to sit there.

?	?	George	Ted	John

3. Of the two boys who are left, Doug is not next to George, so we can sit Bob there.

?	Bob	George	Ted	John

4. Doug is the only one left.

Doug	Bob	George	Ted	John

★ CHECK
Neither Bob nor Doug is next to John.
Neither Doug nor Bob is next to Ted.
Neither John nor Doug is next to George.
Ted is just to the right of George.

75

Teaching the Lesson

Tell students that it is often helpful to act out a situation to find a solution. Tell students they often use acting to solve problems, such as when they try on shoes to see how they look and feel before buying, or when they move furniture in their rooms to see how each arrangement looks. Ask students to give more examples of when they act out a situation. Ask students how we can keep track of where we have been, where we are and where we are going when we act out a situation. (write it down) Tell students that keeping a record of our work helps to keep us organized. Tell students that they can write steps in the form of a list, a table, a picture or a diagram.

Have students read the problem aloud. Tell students there is much information given and they need to organize this information into meaningful parts. Read the SEE step aloud with the students as they complete the sentences. Ask what clue tells us the order of 2 of the boys. (Ted is just to the right of George.) Tell students this clue gives us a good

place to start. Read the PLAN stage and then have students act out the problem. Have them complete the DO sentences. Tell students that in this problem they may find it more helpful to solve the problem and then record the work.

75

Apply

Act out the problems to help you solve them.

1. There are 8 tables in the school cafeteria. If 56 students are eating lunch, how many students are seated at each table?
7 students

3. The 8 members of the chess club are planning a chess tournament. Each member will play one game against each other member. The player who wins the most games will be the champion. How many chess games will be played all together?
28 games

5. Mario Mouse can run at a rate of 2 feet a second. Miranda Mouse can run at a rate of 100 feet a minute. If both run at these rates for 5 minutes, who will run the greater distance?
Mario Mouse

2. Start with 20 pennies. Make stacks of 7, 2, 6 and 5 pennies. A move is taking one or more pennies from a stack and placing them on another stack. How many moves are needed to make the four stacks the same height?
2 moves

4. Choose 4 pieces of colored paper and cut 4 small circles from each. Arrange them in the squares below so that there is a circle in each square, and there are no circles of like color in the same row, column, or in either of the main diagonals.

B	R	G	Y
Y	G	R	B
R	B	Y	G
G	Y	B	R

B = blue
Y = yellow
R = red
G = green

6. Eight mice are equally spaced around a circle of cheese. If every third mouse takes a bite, will all 8 mice get some cheese? Prove your answer.
Yes

76

Solution Notes

1. Students can use manipulatives to act out this problem of division or repeated subtractions.
2. Students may need to begin by using repeated subtractions to group the 20 objects into 4 stacks of 5 each.
3. Students may need to act out this problem several times to be certain that 2 cards will match.
4. Students may find other solutions than these:

R G Y B   B R G Y
B Y G R   Y G R B
G R B Y   R B Y G
Y B R G   G Y B R

Higher-Order Thinking Skills

5. [Evaluation] Students should recognize that the length of time is not important, only the rates. Mario runs 120 feet a minute, or Miranda runs 1 2/3 feet per second.
6. [Analysis] One way for the student to prove the answer is to draw a circle with 8 points and join every third point with segments without lifting the pencil from the paper. The student will "visit" each mouse, or point, and return to the starting point after drawing 8 segments.

Extra Credit Numeration

Have students keep track of examples of Roman numerals they see over the course of a week. Ask students to list the Roman numerals, where they were used and their Hindu-Arabic equivalents. Display these lists and have students compare their findings. Ask them to count how many different places Roman numerals were found. Ask students to list places where they think it would be impractical or confusing to use Roman numerals.

Calculating Money

pages 77-78

Objective

To use a calculator to add and subtract money

Materials

calculators

Mental Math

Ask students how many will fit:

1. 9-in. plates on 54-in. shelf. (6)
2. people in 9 rows of 50 seats. (450)
3. 3- × 5-in. cards in 5-ft space. (20 or 12)
4. 1-in. squares on 6- × 8-in. grid. (48)
5. cars in 2 lots of 52 spaces each. (104)
6. dimes in one $5 roll. (50)
7. yardsticks in 72 inches. (2)
8. nickels in one $2 roll. (40)

Skill Review

Have students work various single-digit column addition problems such as $2 + 6 + 3 + 0 + 4 =$.

Calculators, Data from an Ad

Cecily has $25 to spend on her mother's birthday present. She has decided to buy the camera and the wallet. How much money will Cecily have left?

We want to find how much money Cecily will have left after buying the present. We know the cost of the wallet is $8.97

and the camera costs $15.39.

Cecily has $25 to spend. To find the amount Cecily will have left, add $8.97

to $15.39, and subtract that sum from $25. Enter these codes into your calculator and record the answers on the screens.

Cost of the Camera	Cost of the Wallet	Total Cost
15 $\cdot$ 39 $+$	8 $\cdot$ 97 $=$	24.36

Amount Given	Cost of Items	Change
25 $-$	24.36 $=$	0.64

Cecily will have $0.64 left.

✔ Remember, the calculator will not print zeros to the far right of the decimal point. You do not have to enter these zeros when you enter an amount of money.

13 means $13

6.1 means $6.10

Enter these codes. Write the differences on the screens.

1. 16 $-$ 14 $\cdot$ 25 $=$ 1.75
2. 240 $\cdot$ 15 $-$ 168 $\cdot$ 20 $=$ 71.95
3. 7500 $-$ 5000 $=$ 2500

77

Teaching the Lesson

Introducing the Problem Have students describe the picture. (A girl is reading a newspaper ad that shows a camera, frames, a wallet and pen for sale.) Have a student read the problem and tell what is to be solved. (how much money Cecily will have left from $25 if she buys a camera and wallet) Ask students how they will find out how much money Cecily has left. (add the 2 prices and subtract from $25) Ask students what information is needed to solve the problem. (the cost of the camera and wallet) Have students complete the sentences and work the problems on their calculators.

Developing the Skill Remind students that zeros to the far right of the decimal point are not entered. Write **$6.00 + $12.10** vertically on the board. Have a student write the calculator code for this problem. (6. + 12.1 =) Have students clear their calculator screens and enter the code to find the sum. ($18.10) Have a student write the sum as it appears on the screen. (18.1) Have students write more codes as you dictate 2 amounts of money to be added or subtracted. Remind students to add to check subtraction work. Give ample practice for numbers with zeros to the left and right of the decimal point. Have students work the problems at the bottom of the page. Tell students to check each problem by adding.

Practice

Complete these codes. Write the differences on the screens.

1. 9 [·] 83 [−] 8 [·] 15 [=] (1.68)

2. 324 [·] 60 [+] 14 [·] 68 [−] 43 [·] 76 [=] (295.52)

3. 167 [·] 46 [−] 112 [·] 48 [−] 54 [·] 85 [=] (0.13)

4. 11 [·] 15 [+] 7 [·] 30 [−] 15 [=] (3.45)

Use a calculator to find these differences.
Use estimation to make sure the answers seem correct.

5. $9.86
  − 4.98
  ───────
    $4.88

6. $41.38
  − 16.59
  ───────
    $24.79

7. $397.58
  − 169.59
  ───────
    $227.99

8. $426.37
  − 85.67
  ───────
    $340.70

9. $13.79 − $8.50
$5.29

10. $36.84 − $9.95
$26.89

11. $18,475 − $15,989
$2,486

12. $860.05 − $276.48
$583.57

13. $175 + $68.37 − $115.37
$128

14. $439.20 + $179.56 − $218.73
$400.03

Apply

Use the ad to solve these problems.

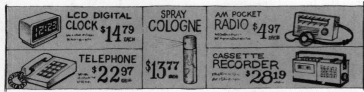

15. Is the cost of the spray cologne and the clock more or less than $28.50?
more

16. How much more is the digital clock than the radio?
$9.82

17. Which three items added together cost exactly $51.53?
clock, spray cologne, telephone

18. Which costs more, the telephone and the spray cologne, or the cassette recorder and the clock?
The cassette recorder and the clock

78

Practice

Remind students they must always clear the calculator screen before working a new problem. Remind students to estimate their answers by rounding each number and finding the sum or difference. Talk through the first word problem with the students to be sure they understand what is to be done. Have students complete the page independently.

Extra Credit *Numeration*

Have students devise their own system of numeration, drawing original symbols to represent numerical values 0 through 50. Ask students to write and answer questions concerning themselves or the class using their system of numeration; for example: How many students are in the class? How old are they? Invite students to compose arithmetic problems using their systems and then solve them.

Chapter Test

page 79

Item	Objective
1-4	Subtract two 2-digit numbers, check subtraction with addition (See pages 61-62, 73-74)
5-8	Subtract two 3-digit numbers, check subtraction with addition (See pages 63-64, 73-74)
9-12	Subtract when middle digit in minuend is zero, check subtraction with addition (See pages 65-66, 73-74)
13-20	Subtract two 4- or 5-digit numbers, check subtraction with addition (See pages 67-68, 73-74)
21-28	Estimate differences of 3- and 4-digit subtractions (See pages 71-72)

Subtract and check.

1. $\begin{array}{r} 68 \\ -34 \\ \hline 34 \end{array}$ 2. $\begin{array}{r} 96 \\ -21 \\ \hline 75 \end{array}$ 3. $\begin{array}{r} 61 \\ -37 \\ \hline 24 \end{array}$ 4. $\begin{array}{r} 85 \\ -19 \\ \hline 66 \end{array}$

5. $\begin{array}{r} 375 \\ -123 \\ \hline 252 \end{array}$ 6. $\begin{array}{r} 485 \\ -149 \\ \hline 336 \end{array}$ 7. $\begin{array}{r} 718 \\ -253 \\ \hline 465 \end{array}$ 8. $\begin{array}{r} 826 \\ -198 \\ \hline 628 \end{array}$

9. $\begin{array}{r} 503 \\ -322 \\ \hline 181 \end{array}$ 10. $\begin{array}{r} 601 \\ -427 \\ \hline 174 \end{array}$ 11. $\begin{array}{r} \$9.03 \\ -5.27 \\ \hline \$3.76 \end{array}$ 12. $\begin{array}{r} \$4.00 \\ -3.72 \\ \hline \$0.28 \end{array}$

13. $\begin{array}{r} 7,294 \\ -3,485 \\ \hline 3,809 \end{array}$ 14. $\begin{array}{r} \$85.37 \\ -29.18 \\ \hline \$56.19 \end{array}$ 15. $\begin{array}{r} 16,375 \\ -9,887 \\ \hline 6,488 \end{array}$ 16. $\begin{array}{r} \$483.29 \\ -136.84 \\ \hline \$346.45 \end{array}$

17. $\begin{array}{r} 4,006 \\ -2,988 \\ \hline 1,018 \end{array}$ 18. $\begin{array}{r} \$30.00 \\ -11.76 \\ \hline \$18.24 \end{array}$ 19. $\begin{array}{r} 70,004 \\ -27,268 \\ \hline 42,736 \end{array}$ 20. $\begin{array}{r} \$300.25 \\ -129.37 \\ \hline \$170.88 \end{array}$

Estimate each difference.

Round to the nearest thousand.

21. $\begin{array}{r} 8,275 \\ -3,927 \\ \hline 4,000 \end{array}$ 22. $\begin{array}{r} 8,069 \\ -4,275 \\ \hline 4,000 \end{array}$ 23. $\begin{array}{r} 7,523 \\ -2,130 \\ \hline 6,000 \end{array}$ 24. $\begin{array}{r} 6,135 \\ -1,852 \\ \hline 4,000 \end{array}$

Round to the nearest hundred.

25. $\begin{array}{r} 796 \\ -249 \\ \hline 600 \end{array}$ 26. $\begin{array}{r} 816 \\ -257 \\ \hline 500 \end{array}$ 27. $\begin{array}{r} 923 \\ -375 \\ \hline 500 \end{array}$ 28. $\begin{array}{r} 695 \\ -96 \\ \hline 600 \end{array}$

Circle the letter of the correct answer.

1 6 + 7
   a 1
   b 11
   c 14
   (d) NG

2 5
  3
 + 6
   a 8
   b 9
   (c) 14
   d NG

3 12 − 4
   a 4
   (b) 8
   c 16
   d NG

4 What is the value of the 0 in 3,076?
   a ones
   b tens
   (c) hundreds
   d NG

5 5,265 ◯ 4,265
   a <
   (b) >

6 Round 873 to the nearest hundred.
   a 700
   b 800
   (c) NG

7 Round 6,750 to the nearest thousand.
   a 6,000
   (b) 7,000
   c NG

8 What is the value of the 8 in 823,075?
   (a) hundred thousands
   b ten thousands
   c thousands
   d NG

9 62 + 49
   a 27
   (b) 111
   c 1,011
   d NG

10 456
  + 324
   a 770
   (b) 780
   c 880
   d NG

11 4,327
  + 1,495
   a 5,722
   b 5,812
   c 6,822
   (d) NG

12 41,615
  + 29,256
   (a) 70,871
   b 70,881
   c 71,871
   d NG

13 626
  − 359
   (a) 267
   b 333
   c 367
   d NG

☐ score

80

Cumulative Review

page 80

Item	Objective
1	Compute basic addition facts (See pages 3-4)
2	Add three 1-digit numbers (See pages 5-6)
3	Compute basic subtraction facts (See pages 9-10)
4	Identify place value of digit in number less than 10,000 (See pages 27-28)
5	Compare, order numbers less than 10,000 (See pages 29-30)
6	Round numbers to nearest 100 (See pages 31-34)
7	Round numbers to nearest 1,000 (See pages 33-34)
8	Identify place value of digit in a number less than 1,000,000 (See pages 35-36)
9	Add two 2-digit numbers (See pages 43-44)
10	Add two 3-digit numbers (See pages 45-46)
11	Add two 4-digit numbers (See pages 47-48)
12	Add two 5-digit numbers (See pages 51-52)
13	Subtract two 3-digit numbers (See pages 63-64)

Alternate Cumulative Review

Circle the letter of the correct answer.

1 8 + 7 =
   a 13
   b 5
   c 17
   (d) NG

2 6
  4
 + 7
   a 10
   b 11
   (c) 17
   d NG

3 14 − 6 =
   a 7
   b 9
   c 11
   (d) NG

4 What is the value of the 9 in 6,497?
   a ones
   (b) tens
   c hundreds
   d NG

5 9,732 ◯ 6,541
   (a) >
   b <
   c =

6 Round 643 to the nearest hundred.
   a 500
   (b) 600
   c 700
   d NG

7 Round 8,965 to the nearest thousand.
   a 7,000
   b 8,000
   (c) 9,000
   d NG

8 What is the value of the 4 in 649,823?
   (a) ten thousands
   b thousands
   c tens
   d NG

9 53 + 39 =
   a 82
   b 812
   (c) 92
   d NG

10 624
  + 217
   a 831
   (b) 841
   c 941
   d NG

11 7,616
  + 2,295
   a 9,801
   b 10,911
   (c) 9,911
   d NG

12 62,427
  + 18,244
   a 81,661
   (b) 80,671
   c 81,771
   d NG

80

Multiplication

pages 81-82

Objective

To understand multiplication

Materials

counters

Mental Math

Tell students to estimate to the nearest ten:

1. $451-149 ($300)
2. 18 years + 61 years (80)
3. 1/2 of 164 (80 years)
4. 111 + 80 (190)
5. $16.16 + .45 ($16.70)
6. 58 ÷ 2 (30)
7. 10:42 AM (10:40)
8. 72 − 18 (50)

Skill Review

Write **2 + 2 + 2** on the board. Ask students the sum. (6) Write **3 + 3 + 3 + 3** and ask students the sum. (12) Continue to present problems of the same number added to itself several times and ask the sums. Alternate the problems so that students add horizontally and in columns.

MULTIPLICATION FACTS

Understanding Multiplication

Charlie packs groceries after school at the Food Mart. How many soup cans can he fit into a box?

We want to know how many cans fit into the box.
We know there are ___4___ rows with ___3___ cans in each row.

We can add the number of cans in each row.

$$3$$
$$3$$
$$3$$
$$+ 3$$
$$\overline{12}$$

We can multiply the number of rows by the number of cans in each.

4 threes = ___12___

$$4 \times 3 = \underline{12}$$
factor × factor = product

We can add the number of cans in each column.

$$4 + 4 + 4 = \underline{12}$$

We can multiply the number of columns by the number of cans in each.

3 fours = ___12___

$$3 \times 4 = \underline{12}$$
factor × factor = product

Charlie can fit ___12___ cans into one box.

Getting Started

Use both addition and multiplication to show how many are in each picture.

1.
$$2 + 2 + 2 + 2 = \underline{8}$$
$$4 \times 2 = \underline{8}$$
$$2 \times 4 = \underline{8}$$

2.
$$5$$
$$5$$
$$+ 5$$
$$\overline{15}$$
$$3 \times 5 = \underline{15}$$
$$5 \times 3 = \underline{15}$$

3.
$$4 + 4 + 4 + 4 = \underline{16}$$
$$4 \times 4 = \underline{16}$$

81

Teaching the Lesson

Introducing the Problem Have a student read the problem aloud and tell what is to be solved. (the number of soup cans Charlie can pack into the box) Tell students they can count all the cans to solve this simple problem but that sometimes it takes too long to count every item. Work through the two models with students to make sure they understand the relationship between addition and multiplication, and that multiplication is the quicker method of finding the total. Have students complete the solution sentence.

Developing the Skill Draw 4 rows of 5 x's on the board. Have students count the x's one-by-one to find the total. (20) Tell students we could also add 5 + 5 + 5 + 5 or 4 + 4 + 4 + 4 + 4 to get a total of 20. Tell students that this adding method also takes a lot of time. Ask students how many rows of 5 x's there are. (4) Write **4 fives = 20** and **4 × 5 = 20** on the board. Tell students the 4 and 5 are called **factors** and 20 is called the **product.** Tell students we can also write 5 fours = 20 and 5 × 4 = 20 since we could say there are 5 rows of 4 in each row. Write **2 × 6 = 12** on the board and have students tell the factors (2 and 6) and the product. (12) Have a student draw 2 rows of 6 x's to verify the product of 12. Have a student write 6 × 2 = 12 and verify by showing the 6 rows of 2. Have students name the factors and products for more multiplication facts.

Practice

Use both addition and multiplication to show how many are in each picture.

1.
$$3 + 3 + 3 + 3 + 3 + 3 = \underline{18}$$
$$6 \times 3 = \underline{18}$$
$$3 \times 6 = \underline{18}$$

2.
$$5$$
$$+\ 5$$
$$\overline{10}$$
$$2 \times 5 = \underline{10}$$
$$5 \times 2 = \underline{10}$$

3.
$$1 + 1 + 1 + 1 + 1 + 1 + 1 = \underline{7}$$
$$7 \times 1 = \underline{7}$$
$$1 \times 7 = \underline{7}$$

4.
$$6$$
$$6$$
$$6$$
$$+\ 6$$
$$\overline{24}$$
$$4 \times 6 = \underline{24}$$
$$6 \times 4 = \underline{24}$$

5.
$$9$$
$$+\ 9$$
$$\overline{18}$$
$$2 \times 9 = \underline{18}$$
$$9 \times 2 = \underline{18}$$

6.
$$4 + 4 + 4 + 4 + 4 + 4 + 4 = \underline{28}$$
$$7 \times 4 = \underline{28}$$
$$4 \times 7 = \underline{28}$$

7.
$$7 + 7 + 7 = \underline{21}$$
$$3 \times 7 = \underline{21}$$
$$7 \times 3 = \underline{21}$$

8.
$$5$$
$$5$$
$$5$$
$$5$$
$$+\ 5$$
$$\overline{25}$$
$$5 \times 5 = \underline{25}$$

82

Correcting Common Errors

Some students may not make the connection between multiplication and repeated addition. Have them work with counters. For example, to find 4 threes, or 4×3, have them lay out 4 sets of 3 counters and skip count to find the total: 3, 6, 9, 12.

Enrichment

Have students draw a seating arrangement for a long narrow room to seat 72 people at a meeting. Half of the people should be on each side of an aisle.

Practice

Have students complete the page independently.

Extra Credit *Logic*

Dictate the following problem for students to solve: A man found he could read 80 pages of a book in 80 minutes, when he wore his red shirt. But when he wore his blue shirt, it took him an hour and 20 minutes to read the same number of pages. How could this be true? (The shirt, of course, has nothing to do with it. Eighty minutes and an hour and 20 minutes are the same length of time.)

82

Factors 2 and 3

pages 83-84

Objective

To multiply by 2 or 3

Materials

counters
1/4-in grid paper

Mental Math

Tell students to halve each number and subtract 9:

1. 120 (51)
2. 200 (91)
3. 1,050 (516)
4. 52 (17)
5. 300 (141)
6. 98 (40)
7. 60 (21)
8. 76 (29)

Skill Review

Write on the board:
5 + 5 + 5 = 15
$\quad$**3 × 5 = 15**
$\quad$**5 × 3 = 15**
Now write **3 + 3 + 3 + 3 = 12** and have students write the 2 multiplication facts for 4 threes. (4 × 3 = 12, 3 × 4 = 12) Repeat for other addition sentences. Vary the activity by writing more multiplication facts and have students write the corresponding addition problems.

Multiplying, the Factors 2 and 3

Rose helped her uncle pick cherries. She saved some of the fruit for her lunch. How many cherries did Rose have for lunch?

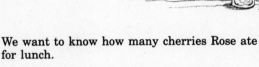

We want to know how many cherries Rose ate for lunch.

We know there are ___4___ bunches of ___3___ cherries each.

To find out how many cherries she ate, we multiply ___4___ by ___3___.

$$4 \times 3 = \underline{12} \qquad \begin{array}{r} 3 \\ \times\, 4 \\ \hline 12 \end{array}$$

We say: **four times three equals twelve.**

Rose saved ___12___ cherries for lunch.

Getting Started

Complete the table.

1. **The Facts of 2**

1	2	3	4	5	6	7	8	9
2	4	6	8	10	12	14	16	18

Multiply.

2. $5 \times 3 = \underline{15}$

3. $4 \times 2 = \underline{8}$

4. $\begin{array}{r} 2 \\ \times\, 9 \\ \hline 18 \end{array}$

5. $\begin{array}{r} 3 \\ \times\, 7 \\ \hline 21 \end{array}$

83

Teaching the Lesson

Introducing the Problem Have a student read the problem aloud and tell what is to be solved. (how many cherries Rose has on her plate) Ask students to count each cherry to tell the total. (12) Work through the model with students showing them how multiplication can be used to quickly solve the problem. Then have students complete the solution sentence.

Developing the Skill Have students lay out 1 group of 2 counters and tell how many in all. (2) Write **1 + 1 = 2, 1 two = 2, 1 times 2 equals 2** and **1 × 2 = 2** across the board. Have students read each with you. Have students tell the total number of counters. (2) Tell students to lay out another group of 2 counters. Have students write 2 + 2 = 4, 2 twos = 4, 2 times 2 equals 4 and 2 × 2 = 4 under the facts for 1 group of 2. Have students read each with you and then count by 2's to tell the total number of 2's. (2, 4) Continue to develop the facts for 2 through 9 × 2. Repeat the procedure for the facts for 3. Remind students that in 2 × 2 = 4, each 2 is a factor and 4 is the product. When all facts for 2 and 3 are developed, have students read some of the facts and tell the factors and the product of each.

83

Practice

Complete the table.

1. The Facts of 3

1	2	3	4	5	6	7	8	9
3	6	9	12	15	18	21	24	27

Multiply.

2. $2 \times 3 =$ ___6___   3. $3 \times 4 =$ ___12___   4. $3 \times 5 =$ ___15___   5. $2 \times 6 =$ ___12___

6. $2 \times 2 =$ ___4___   7. $3 \times 9 =$ ___27___   8. $2 \times 8 =$ ___16___   9. $2 \times 7 =$ ___14___

10. $3 \times 8 =$ ___24___   11. $3 \times 2 =$ ___6___   12. $2 \times 5 =$ ___10___   13. $3 \times 3 =$ ___9___

14. $2 \times 9 =$ ___18___   15. $3 \times 7 =$ ___21___   16. $3 \times 6 =$ ___18___   17. $2 \times 4 =$ ___8___

18. $\begin{array}{r} 2 \\ \times 8 \\ \hline 16 \end{array}$ 19. $\begin{array}{r} 3 \\ \times 9 \\ \hline 27 \end{array}$ 20. $\begin{array}{r} 2 \\ \times 4 \\ \hline 8 \end{array}$ 21. $\begin{array}{r} 2 \\ \times 5 \\ \hline 10 \end{array}$ 22. $\begin{array}{r} 3 \\ \times 6 \\ \hline 18 \end{array}$ 23. $\begin{array}{r} 3 \\ \times 7 \\ \hline 21 \end{array}$

24. $\begin{array}{r} 3 \\ \times 8 \\ \hline 24 \end{array}$ 25. $\begin{array}{r} 2 \\ \times 9 \\ \hline 18 \end{array}$ 26. $\begin{array}{r} 2 \\ \times 5 \\ \hline 10 \end{array}$ 27. $\begin{array}{r} 3 \\ \times 5 \\ \hline 15 \end{array}$ 28. $\begin{array}{r} 3 \\ \times 4 \\ \hline 12 \end{array}$ 29. $\begin{array}{r} 2 \\ \times 7 \\ \hline 14 \end{array}$

Apply

Solve these problems.

30. Fred has outgrown 4 pairs of tennis shoes in one year. How many shoes has he outgrown?
8 shoes

31. There are 9 vases, each containing 3 daisies. How many daisies are there altogether?
27 daisies

32. There are 5 study tables in the library. Each table has 2 chairs. How many chairs are there?
10 chairs

33. Bill has 9 model cars to build. Don has 3 cars. How many more cars does Bill have?
6 cars

34. Betty has 2 records. Each record has 8 songs. How many songs can Betty listen to if she plays both records?
16 songs

35. Ilonda ate 3 apples on Tuesday, 5 apples on Wednesday and 2 apples on Friday. How many apples did she eat altogether?
10 apples

84

Correcting Common Errors

Some students may have trouble learning their facts of 3. Have them use grid paper. In the first row, have them color 3 squares and write $1 \times 3 = 3$. In the next row, have them color 3 squares with one color and the adjacent 3 squares with another color and write $2 \times 3 = 6$. In the third row, have them show 3 sets of 3, alternating colors, and write $3 \times 3 = 9$. Have them continue in this manner to show all the facts of 3 through $9 \times 3 = 27$.

Enrichment

Have students make multiplication fact cards for facts of 2's and 3's through 15×2, 2×15, 15×3 and 3×15.

Practice

Have students complete the table and then work the problems. Tell students they must decide which operation to use in each of the word problems.

Extra Credit   *Numeration*

Have students who previously devised their own numeration system, continue the project. Working in groups, tell them to imagine their own country where their numeration system would be used. Have them describe the country's size, people, flag, monetary system, etc. and prepare a poster detailing this information. Have each group present their poster in a report to the class.

Factors 4 and 5

pages 85-86

Objective

To multiply by 4 or 5

Materials

*multiplication fact cards for 2's and 3's

Mental Math

Ask if the following statements are true or false:

1. 1912 was 40 years before 1952 (T)
2. 20 nickels = $1.50 (F)
3. 1/4 of $1.00 = 25¢ (T)
4. 5 cups > 1 pint (T)
5. n = 2 if n + 3 + 8 = 15 (F)
6. 1 year = 52 weeks (T)
7. 230 = 22 tens and 10 ones (T)
8. 500 < 1/2 of 1,000 (F)

Skill Review

Show the 3 × 4 fact card and have students tell a related addition problem to find the total. (4 + 4 + 4 = 12) Repeat for other facts.

Multiplying, the Factors 4 and 5

Kerry's mother is making pickles. How many quart canning jars did Kerry buy for her?

We want to know how many jars Kerry bought.

We know Kerry bought ___5___ boxes

of ___4___ quart jars each.

To find out how many jars there are in the boxes, we think of 5 sets

of 4 each. We multiply ___5___ by ___4___.

4	8	12	16	20

$1 \times 4 = $ _4_ $2 \times 4 = $ _8_ $3 \times 4 = $ _12_ $4 \times 4 = $ _16_ $5 \times 4 = $ _20_

Kerry bought ___20___ jars for his mother.

Getting Started

Complete the table.

1. The Facts of 4

1	2	3	4	5	6	7	8	9
4	8	12	16	20	24	28	32	36

Multiply.

2. $6 \times 4 = $ _24_   3. $9 \times 5 = $ _45_

4. $\begin{array}{r} 5 \\ \times 2 \\ \hline 10 \end{array}$

5. $\begin{array}{r} 4 \\ \times 7 \\ \hline 28 \end{array}$

85

Teaching the Lesson

Introducing the Problem Have a student read the problem aloud and tell what is to be solved. (how many jars Kerry is buying) Ask students what information is given. (the number of boxes and the number of jars in each) Work through the model with students having them supply answers for each fact. Then have them complete the solution sentence.

Developing the Skill Draw a number line across the board. Mark intervals to accommodate the numbers from 0 through 32. Have a student write the numbers 1 through 4 along the line and tell how many 4's there are. (1) Write **1 × 4 = 8** on the board. Have another student write the numbers from 5 through 8 along the line and tell how many 4's in all. (2) Have the student write 2 × 4 = 8 on the board. Have students continue to develop the multiples of 4 along the line and write the facts. Have students count by 4's as you draw arcs from 0 to 4 to 8 to 12, etc. Now point to the products of 4 at random as students tell the 2 factors which equal each product. Repeat this process with the factor 5.

Practice

Complete the table.

1.

The Facts of 5

1	2	3	4	5	6	7	8	9
5	10	15	20	25	30	35	40	45

Multiply.

2. $2 \times 4 = \underline{8}$ 3. $4 \times 5 = \underline{20}$ 4. $5 \times 5 = \underline{25}$ 5. $7 \times 4 = \underline{28}$

6. $5 \times 4 = \underline{20}$ 7. $6 \times 5 = \underline{30}$ 8. $8 \times 5 = \underline{40}$ 9. $3 \times 4 = \underline{12}$

10. $4 \times 4 = \underline{16}$ 11. $2 \times 5 = \underline{10}$ 12. $3 \times 5 = \underline{15}$ 13. $6 \times 4 = \underline{24}$

14. $9 \times 4 = \underline{36}$ 15. $7 \times 5 = \underline{35}$ 16. $9 \times 5 = \underline{45}$ 17. $8 \times 4 = \underline{32}$

18. $\begin{array}{r}5\\ \times 5\\ \hline 25\end{array}$	19. $\begin{array}{r}5\\ \times 8\\ \hline 40\end{array}$	20. $\begin{array}{r}2\\ \times 2\\ \hline 4\end{array}$	21. $\begin{array}{r}4\\ \times 7\\ \hline 28\end{array}$	22. $\begin{array}{r}3\\ \times 4\\ \hline 12\end{array}$	23. $\begin{array}{r}2\\ \times 9\\ \hline 18\end{array}$
24. $\begin{array}{r}4\\ \times 9\\ \hline 36\end{array}$	25. $\begin{array}{r}5\\ \times 4\\ \hline 20\end{array}$	26. $\begin{array}{r}3\\ \times 6\\ \hline 18\end{array}$	27. $\begin{array}{r}5\\ \times 7\\ \hline 35\end{array}$	28. $\begin{array}{r}5\\ \times 2\\ \hline 10\end{array}$	29. $\begin{array}{r}4\\ \times 6\\ \hline 24\end{array}$
30. $\begin{array}{r}5\\ \times 9\\ \hline 45\end{array}$	31. $\begin{array}{r}3\\ \times 3\\ \hline 9\end{array}$	32. $\begin{array}{r}4\\ \times 8\\ \hline 32\end{array}$	33. $\begin{array}{r}3\\ \times 7\\ \hline 21\end{array}$	34. $\begin{array}{r}2\\ \times 3\\ \hline 6\end{array}$	35. $\begin{array}{r}4\\ \times 5\\ \hline 20\end{array}$

Apply

Solve these problems.

36. Each package holds 5 sticks of gum. How many sticks of gum are in 8 packages?
40 sticks

37. Pamela has 8 seedlings. She bought 5 more at the florist. How many seedlings does Pamela have?
13 seedlings

38. Louis had $5. He spent $4 at the movies. How much does Louis have left?
$1

39. Robert and Mike each bought 4 cans of apple juice. How many cans of juice did they buy together?
8 cans

86

Practice

Have students complete the table and then work the problems on the page. Tell students they must decide what operation they will need to use to solve each word problem.

Mixed Practice

1. $7,040 - 3,236$ (3,804)
2. $169.32 + 38.78$ ($208.10)
3. 3×8 (24)
4. $378 + 265 + 674$ (1,317)
5. $6,483 - 4,252$ (2,231)
6. $570.00 - 376.28$ ($193.72)
7. 5×8 (40)
8. $7 + 3 + 5 + 8 + 3 + 2$ (28)
9. $17.56 + 21.74 + 31.58$ ($70.88)
10. 9×3 (27)

Extra Credit   *Measurement*

Write **millimeter, centimeter, kilometer,** on the board. Ask students to copy the words and underline the root word. Ask students to find out what the origin of each prefix is, including both its language and meaning.

Multiplication Properties

Objective

To understand the order, zero and one properties of multiplication

Materials

Mental Math

Ask if the following would be measured in inches, feet or yards:

1. a room (feet)
2. notebook (inches)
3. fabric (yards)
4. TV screen (inches)
5. scissors (inches)
6. football field (yards)
7. screwdriver (inches)
8. parking space (feet)

Skill Review

Review with students that any number plus zero equals that number and any number minus zero equals that number. Write sample addition and subtraction problems with zeros as addends or subtrahends on the board. Have students supply the answers. Remind students that zero always means nothing.

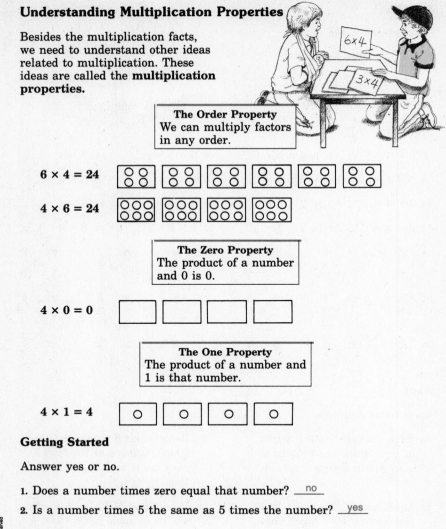

Understanding Multiplication Properties

Besides the multiplication facts, we need to understand other ideas related to multiplication. These ideas are called the **multiplication properties.**

The Order Property
We can multiply factors in any order.

$6 \times 4 = 24$

$4 \times 6 = 24$

The Zero Property
The product of a number and 0 is 0.

$4 \times 0 = 0$

The One Property
The product of a number and 1 is that number.

$4 \times 1 = 4$

Getting Started

Answer yes or no.

1. Does a number times zero equal that number? __no__
2. Is a number times 5 the same as 5 times the number? __yes__

Multiply.

3. $1 \times 7 =$ __7__   4. $7 \times 1 =$ __7__   5. $0 \times 8 =$ __0__   6. $8 \times 0 =$ __0__

87

Teaching the Lesson

Introducing the Problem Read the first paragraph on the page to the students. Then read each property one by one and discuss the examples with the class.

Developing the Skill Have a student draw 3 circles of 5 x's each on the board. Have students tell the total. (15) Have another student draw 5 circles of 3 x's each. Ask the total. (15) Ask students if 3×5 has the same product as 5×3. (yes) Tell students that 2 numbers can be multiplied in any order and the product is the same. Now have 2 students stand with nothing in their hands. Ask students how many books each student is holding. (none) Ask how many books the 2 students are holding in all. (none) Write **2 × 0 = 0** on the board. Have 1 more student join the 2 students and ask how many students are holding no books. (3) Ask the product of 3×0. (0) Tell students that 3 times nothing or zero is always zero. Now give each student 1 book to hold. Ask how many books each student has. (1)

Ask how many students in all. (3) Ask the product of 3×1. (3) Add 2 more students with 1 book each. Ask what 5 times 1 equals. (5) Tell students that any number times 1 equals that number.

Practice

Answer yes or no.

1. If you multiply a number by zero, will the product be zero? __yes__

2. Is 6 times a number the same as the number times 6? __yes__

3. If you multiply 1 by 0, is the answer 1? __no__

4. If you multiply a number by 1, is the product always the same as the other factor? __yes__

5. If you multiply 1 by 1, is the product 2? __no__

Multiply.

6. $3 \times 3 =$ __9__   7. $8 \times 2 =$ __16__   8. $3 \times 1 =$ __3__   9. $9 \times 0 =$ __0__

10. $6 \times 5 =$ __30__   11. $5 \times 6 =$ __30__   12. $0 \times 2 =$ __0__   13. $4 \times 4 =$ __16__

14. $8 \times 3 =$ __24__   15. $3 \times 8 =$ __24__   16. $1 \times 7 =$ __7__   17. $7 \times 1 =$ __7__

18. $\begin{array}{r} 5 \\ \times 3 \\ \hline 15 \end{array}$ 19. $\begin{array}{r} 3 \\ \times 4 \\ \hline 12 \end{array}$ 20. $\begin{array}{r} 9 \\ \times 5 \\ \hline 45 \end{array}$ 21. $\begin{array}{r} 1 \\ \times 7 \\ \hline 7 \end{array}$ 22. $\begin{array}{r} 0 \\ \times 8 \\ \hline 0 \end{array}$ 23. $\begin{array}{r} 2 \\ \times 5 \\ \hline 10 \end{array}$

24. $\begin{array}{r} 3 \\ \times 7 \\ \hline 21 \end{array}$ 25. $\begin{array}{r} 1 \\ \times 9 \\ \hline 9 \end{array}$ 26. $\begin{array}{r} 8 \\ 5 \\ \hline 40 \end{array}$ 27. $\begin{array}{r} 0 \\ \times 4 \\ \hline 0 \end{array}$ 28. $\begin{array}{r} 4 \\ \times 6 \\ \hline 24 \end{array}$ 29. $\begin{array}{r} 1 \\ \times 2 \\ \hline 2 \end{array}$

30. $\begin{array}{r} 0 \\ \times 0 \\ \hline 0 \end{array}$ 31. $\begin{array}{r} 4 \\ \times 5 \\ \hline 20 \end{array}$ 32. $\begin{array}{r} 7 \\ \times 4 \\ \hline 28 \end{array}$ 33. $\begin{array}{r} 2 \\ \times 4 \\ \hline 8 \end{array}$ 34. $\begin{array}{r} 1 \\ \times 5 \\ \hline 5 \end{array}$ 35. $\begin{array}{r} 8 \\ \times 5 \\ \hline 40 \end{array}$

36. $\begin{array}{r} 3 \\ \times 2 \\ \hline 6 \end{array}$ 37. $\begin{array}{r} 5 \\ \times 9 \\ \hline 45 \end{array}$ 38. $\begin{array}{r} 2 \\ \times 9 \\ \hline 18 \end{array}$ 39. $\begin{array}{r} 3 \\ \times 6 \\ \hline 18 \end{array}$ 40. $\begin{array}{r} 5 \\ \times 6 \\ \hline 30 \end{array}$ 41. $\begin{array}{r} 9 \\ \times 4 \\ \hline 36 \end{array}$

88

Correcting Common Errors

Some students may confuse multiplication by 1 with multiplication by 0 and think that 0 times any number is that number. Have students draw 5 stick figures on paper. Then tell them to draw zero tennis balls for each figure and tell how many balls in all they drew and have them write the number sentence $0 \times 5 = 0$ to represent this.

Enrichment

Have students draw a picture to illustrate this looks-can-be-deceiving situation: The coach had 9 cans, each of which should hold 3 tennis balls. The cans looked new so the coach assumed he had 27 new balls. But he found he had only 13 tennis balls because 3 cans were empty and only 2 cans were full.

Practice

Tell students to answer each of the questions and then work the problems on the page.

Extra Credit   *Measurement*

Using newspapers and magazines, working in groups or individually, have the students find as many examples of units of weight and measure as they can. Emphasize that they must find examples of both metric and customary measures. Each example should be cut out and pasted on a chart. Have them compare their findings and identify the examples that are the least and greatest for each unit of measure.

Rule of Order

pages 89-90

Objective

To understand the rule of order

Materials

*counters
*cards numbered 0 through 9

Mental Math

Have students solve the following:

1. 6 tens plus 308 (368)
2. value of n if $10 + 7 - n = 5$ (12)
3. 10 tens minus 47 (53)
4. 9 threes + 3 (30)
5. 5 minus 1.5 (3.5)
6. 1/5 of 50 (10)
7. $48 \div 6 + 12$ (20)
8. $1 + 9 - (6 \div 2)$ (7)

Skill Review

Have students tell the product of 4×6 as you write the fact on the board. (24) Ask students to tell the product of 6×4. (24) Continue reviewing the facts for 2's through 5's.

Understanding Rule of Order

Working an operation in a mathematical sentence to find the value of n is called **solving for n**. Solve for n in the sentence on the chalkboard.

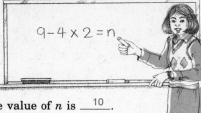

If we subtract and then multiply, the value of n is ___10___.

If we multiply and then subtract, the value of n is ___1___.

Both answers can't be correct. We must follow **rule of order** to know how to solve this sentence correctly.

✔ First, work all multiplications left to right.

✔ Then, work all additions and subtractions left to right.

$$9 - 4 \times 2 = n$$
$$9 - \underline{8} = n$$
$$\underline{1} = n$$

✔ In rule of order, operations within parentheses should be worked before multiplications.

$$(9 - 2) \times 4 = n$$
$$\underline{7} \times 4 = n$$
$$\underline{28} = n$$

Rule of order has been followed in these three mathematical sentences. Solve for n.

$$3 \times 4 - 3 = n$$
$$\underline{12} - 3 = n$$
$$\underline{9} = n$$

$$(3 \times 4) - 3 = n$$
$$\underline{12} - 3 = n$$
$$\underline{9} = n$$

$$3 \times (4 - 3) = n$$
$$3 \times \underline{1} = n$$
$$\underline{3} = n$$

The correct answer for the sentence on the chalkboard is ___1___.

Getting Started

Solve for n. Follow rule of order.

1. $5 + (3 \times 4) = n$ $\underline{17} = n$

2. $3 + 4 \times 5 = n$ $\underline{23} = n$

3. $(6 - 0) \times 4 = n$ $\underline{24} = n$

89

Teaching the Lesson

Introducing the Problem Write on the board: **Slow Cattle Crossing** Discuss the confusion one could have when seeing this sign. Help students see that it could mean that slow cattle are crossing or that one should go slowly because cattle are crossing. Ask students to suggest ways to punctuate the words to help the reader understand its meaning. Have students read the problem in the picture. Tell students that math problems can also be confusing unless we know whether to subtract first or to multiply first. Read the problem beside the picture to the students. Work through the rules of order with students. Then have them complete the solution sentence.

Developing the Skill Tell students that parentheses and the rule of order are used in math to help us understand how to work a problem. Write **4 + 2 × 6 =** on the board. Tell students the **rule of order** tells us to multiply from left to right first so we have $4 + 12 = 16$. Write **4 + (2 × 6) =** on the board. Ask students if the parentheses changed this problem. (No, the rule of order tells us to multiply first anyway.) Now write **(4 + 2) × 6 = 36** on the board and tell students the parentheses have changed this problem because the rule of order also tells us to work the problem within parentheses first. Present more problems with and without parentheses.

89

Practice

Solve for n. Follow rule of order.

1. $(2 \times 5) + 3 = n$

   $\underline{13} = n$

2. $8 + (3 \times 4) = n$

   $\underline{20} = n$

3. $(7 \times 3) - 9 = n$

   $\underline{12} = n$

4. $4 \times 3 + 7 = n$

   $\underline{19} = n$

5. $(7 \times 4) + 15 = n$

   $\underline{43} = n$

6. $(6 \times 5) - 18 = n$

   $\underline{12} = n$

7. $(24 - 16) \times 4 = n$

   $\underline{32} = n$

8. $(2 \times 3) \times 4 = n$

   $\underline{24} = n$

9. $(5 - 3) \times 7 = n$

   $\underline{14} = n$

10. $8 - (6 \times 1) = n$

    $\underline{2} = n$

11. $5 + 3 \times 6 = n$

    $\underline{23} = n$

12. $9 \times (8 - 8) = n$

    $\underline{0} = n$

13. $(7 - 6) \times 4 = n$

    $\underline{4} = n$

14. $(7 \times 5) + 26 = n$

    $\underline{61} = n$

15. $56 - (4 \times 8) = n$

    $\underline{24} = n$

16. $46 + (9 \times 0) = n$

    $\underline{46} = n$

17. $(3 \times 9) + 46 = n$

    $\underline{73} = n$

18. $(5 \times 2) + 30 = n$

    $\underline{40} = n$

19. $7 \times 2 - 8 = n$

    $\underline{6} = n$

20. $72 + (8 \times 4) = n$

    $\underline{104} = n$

21. $(95 - 95) \times 4 = n$

    $\underline{0} = n$

EXCURSION

How many numbers can you make using four 4's?

$(4 \times 4) \times (4 \times 4) = 256$
$(4 \times 4) + (4 \times 4) = 32$
$4 \times (4 + 4) \times 4 = 128$
$(4 + 4) \times (4 + 4) = 64$
$4 + (4 \times 4) + 4 = 24$
$(4 \times 4) - (4 \times 4) = 0$
$(4 \times 4) - (4 + 4) = 8$
$(4 + 4 + 4) \times 4 = 48$
Give credit for any other reasonable answer.

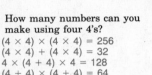

$(4 + 4 + 4) \times 4 = 48$

$()+4 4 4 4 0$

90

Practice

Remind students to look for parentheses and work that part of the problem first. Tell students if there are no parentheses, they must multiply left to right and then do additions or subtractions from left to right. Have students complete the problems independently.

Excursion

Through the use of addition, subtraction, multiplication and parentheses students are to make different numbers using four 4's. Tell students to place parentheses around the operation to be solved first in each problem.

Extra Credit   *Numeration*

Have a student bring in a box of toothpicks. Tell them to solve the following toothpick puzzles by carefully following directions.

1. Take away two toothpicks to make this solution true:
   VII − I = I
   (Remove the V, the II − I = I)
2. You are given 9 toothpicks. Can you make ten out of them? No, you can't break them! (The toothpicks spell the word: TEN)

Factors 6 and 7

pages 91-92

Objective

To multiply by 6 or 7

Materials

*counters
*sheets of paper

Mental Math

Ask students how many:

1. 4's in 32 − 8 (6)
2. 5's in 40 + 5 (9)
3. 3's in 16 − 13 (1)
4. 4's in 125 − 25 (25)
5. 6's in 24 ÷ 4 (1)
6. halves in 2 wholes (4)
7. quarts in 4 gallons (16)
8. legs on 3 octopi (24)

Skill Review

Write **(3 × 2) + (4 × 2) =** on the board. Have students work the problem to tell what 7 × 2 equals. (14) Continue to present more fact-plus-fact problems to have students find the total of 3 fours, 4 sixes, 5 sevens and 3 sixes.

Multiplying, the Factors 6 and 7

It is exactly 8 weeks from New Year's Day to Opal's birthday. How many days does Opal have to wait to celebrate her birthday?

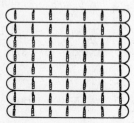

We want to find the number of days before Opal's birthday.

We know that Opal's birthday is __8__ weeks from New Year's Day.

There are __7__ days in one week.

$7 \times 8 =$ __56__

$8 \times 7 =$ __56__

Opal's birthday is __56__ days away.

Getting Started

Complete the table.

1. The Facts of 6

0	1	2	3	4	5	6	7	8	9
0	6	12	18	24	30	36	42	48	54

Multiply.

2. $3 \times 6 =$ __18__   3. $5 \times 6 =$ __30__   4. $4 \times 6 =$ __24__   5. $2 \times 6 =$ __12__

6. $\begin{array}{r} 6 \\ \times 9 \\ \hline 54 \end{array}$ 7. $\begin{array}{r} 7 \\ \times 6 \\ \hline 42 \end{array}$ 8. $\begin{array}{r} 6 \\ \times 6 \\ \hline 36 \end{array}$ 9. $\begin{array}{r} 6 \\ \times 8 \\ \hline 48 \end{array}$

91

Teaching the Lesson

Introducing the Problem Have students tell about the picture. (A girl is looking at January 1st on a calendar.) Have a student read the problem aloud to find the significance of January 1. (Opal's birthday is 8 weeks from January 1.) Ask students what holiday is on January 1. (New Year's Day) Ask what problem is to be solved. (how many days until Opal's birthday) Ask students what information is known. (January 1 is 8 weeks from Opal's birthday.) Work through the model with students and then have them complete the solution sentence.

Developing the Skill Write on the board:

$1 \times 6 = 6$	$2 \times 6 = 12$	$3 \times 6 = 18$	$4 \times 6 = 24$
$(6 \times 1 = 6)$	$(6 \times 2 = 12)$	$(6 \times 3 = 18)$	$(6 \times 4 = 24)$

Have students write the related multiplication fact for each. Ask students what $(5 \times 6) + (1 \times 6)$ would equal. (36) Continue to develop the 6's in this way through $6 \times 9 = 54$. Repeat the activity for the facts for 7 through $7 \times 9 = 63$.

Practice

Complete the table.

1.

The Facts of 7

0	1	2	3	4	5	6	7	8	9
0	7	14	21	28	35	42	49	56	63

Multiply.

2. $5 \times 6 = \underline{30}$ 3. $1 \times 7 = \underline{7}$ 4. $2 \times 6 = \underline{12}$ 5. $7 \times 7 = \underline{49}$

6. $3 \times 7 = \underline{21}$ 7. $9 \times 7 = \underline{63}$ 8. $0 \times 6 = \underline{0}$ 9. $4 \times 6 = \underline{24}$

10. $6 \times 7 = \underline{42}$ 11. $7 \times 6 = \underline{42}$ 12. $4 \times 7 = \underline{28}$ 13. $1 \times 6 = \underline{6}$

14. $3 \times 6 = \underline{18}$ 15. $9 \times 6 = \underline{54}$ 16. $8 \times 7 = \underline{56}$ 17. $0 \times 7 = \underline{0}$

18. $\begin{array}{r} 6 \\ \times 6 \\ \hline 36 \end{array}$ 19. $\begin{array}{r} 7 \\ \times 5 \\ \hline 35 \end{array}$ 20. $\begin{array}{r} 7 \\ \times 2 \\ \hline 14 \end{array}$ 21. $\begin{array}{r} 6 \\ \times 8 \\ \hline 48 \end{array}$ 22. $\begin{array}{r} 6 \\ \times 1 \\ \hline 6 \end{array}$ 23. $\begin{array}{r} 7 \\ \times 4 \\ \hline 28 \end{array}$

24. $\begin{array}{r} 7 \\ \times 5 \\ \hline 35 \end{array}$ 25. $\begin{array}{r} 6 \\ \times 0 \\ \hline 0 \end{array}$ 26. $\begin{array}{r} 6 \\ \times 4 \\ \hline 24 \end{array}$ 27. $\begin{array}{r} 7 \\ \times 0 \\ \hline 0 \end{array}$ 28. $\begin{array}{r} 6 \\ \times 5 \\ \hline 30 \end{array}$ 29. $\begin{array}{r} 6 \\ \times 6 \\ \hline 36 \end{array}$

30. $\begin{array}{r} 7 \\ \times 1 \\ \hline 7 \end{array}$ 31. $\begin{array}{r} 7 \\ \times 7 \\ \hline 49 \end{array}$ 32. $\begin{array}{r} 7 \\ \times 9 \\ \hline 63 \end{array}$ 33. $\begin{array}{r} 6 \\ \times 2 \\ \hline 12 \end{array}$ 34. $\begin{array}{r} 7 \\ \times 3 \\ \hline 21 \end{array}$ 35. $\begin{array}{r} 6 \\ \times 7 \\ \hline 42 \end{array}$

Apply

Solve these problems.

36. I have 9 key rings. Each key ring holds 6 keys. How many keys do I have?
54 keys

37. There are 15 apples in a bag. 7 apples are rotten. How many apples are not rotten?
8 apples

38. One ticket for the ring-toss game costs 5¢. How much do 7 tickets cost?
35¢

39. Brenda waters 6 of her plants each day, in rotation. By the end of the week she has watered all her plants. How many plants does Brenda have?
42 plants

92

92

Factors 8 and 9

pages 93-94

Objective

To multiply by 8 or 9

Materials

Mental Math

Have students give the century of:

1. 1216. (13th)
2. 1977. (20th)
3. 80 years ago. (20th)
4. 20 years prior to 1702. (17th)
5. 2 centuries ago. (18th)
6. 50 years after 1776. (19th)
7. 100 years into the future. (21st)

Skill Review

Write **7 + 2 × 7** on the board. Have students put parentheses in this problem to show which operation is to be done first. (7 + (2 × 7)) Have a student work the problem. (21) Continue with more problems having multiplication and addition or subtraction.

Multiplying, the Factors 8 and 9

	1	2	3	4	5	6	7	8	9	Total
Home	0	0	3	1	0	0	1	1	0	6
Visitor	1	2	0	1	0	0	0	0	1	5

The Crosby County baseball team plays 8 games each summer. If all the games are complete, how many innings does the team play?

We want to know the number of innings Crosby County plays over the summer.

We know that Crosby County plays __8__ games.

A regular baseball game lasts for __9__ innings.

To find the total number of innings

we multiply __8__ by __9__.

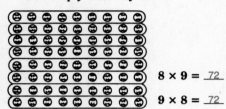

$8 \times 9 =$ __72__

$9 \times 8 =$ __72__

The Crosby County baseball team plays __72__ innings.

Getting Started

Complete the table.

1.   **The Facts of 8**

0	1	2	3	4	5	6	7	8	9
0	8	16	24	32	40	48	56	64	72

Multiply.

2. $3 \times 8 =$ __24__   3. $7 \times 8 =$ __56__   4. $0 \times 8 =$ __0__   5. $5 \times 8 =$ __40__

6. 8
   ×9
   ———
   72

7. 8
   ×8
   ———
   64

8. 8
   ×7
   ———
   56

9. 6
   ×8
   ———
   48

93

Teaching the Lesson

Introducing the Problem Have a student read the problem aloud and tell what is to be solved. (how many innings the Crosby County baseball team plays each summer) Ask students what facts are given in the problem. (The team plays 8 complete games.) Ask students what facts are needed yet. (how many innings are in a game) Ask students if this information is given. (yes) Ask students if we need the scores of each team to answer this problem. (no) Work through the model with students and then have them complete the solution sentence.

Developing the Skill Write on the board:

$1 \times 8 = 8$ $2 \times 8 = 16$ $3 \times 8 = 24$ $4 \times 8 = 32$
$(8 \times 1 = 8)$ $(8 \times 2 = 16)$ $(8 \times 3 = 24)$ $(8 \times 4 = 32)$

Have students write the related multiplication fact for each. Remind students that $(4 \times 8) + (1 \times 8) = 5 \times 8$ or 40. Help students develop the remaining facts for 8 through $8 \times 9 = 72$. Write each fact on the board and have students write the related fact for each. Repeat the activity to develop the facts of 9 through $9 \times 9 = 81$.

Practice

Complete the table.

1. The Facts of 9

0	1	2	3	4	5	6	7	8	9
0	9	18	27	36	45	54	63	72	81

Multiply.

2. $0 \times 9 = \underline{0}$ 3. $8 \times 9 = \underline{72}$ 4. $1 \times 8 = \underline{8}$ 5. $3 \times 8 = \underline{24}$

6. $7 \times 8 = \underline{56}$ 7. $4 \times 9 = \underline{36}$ 8. $9 \times 8 = \underline{72}$ 9. $6 \times 9 = \underline{54}$

10. $4 \times 8 = \underline{32}$ 11. $0 \times 8 = \underline{0}$ 12. $7 \times 9 = \underline{63}$ 13. $3 \times 9 = \underline{27}$

14. $\begin{array}{r} 9 \\ \times 5 \\ \hline 45 \end{array}$ 15. $\begin{array}{r} 8 \\ \times 6 \\ \hline 48 \end{array}$ 16. $\begin{array}{r} 9 \\ \times 9 \\ \hline 81 \end{array}$ 17. $\begin{array}{r} 8 \\ \times 2 \\ \hline 16 \end{array}$ 18. $\begin{array}{r} 8 \\ \times 9 \\ \hline 72 \end{array}$ 19. $\begin{array}{r} 8 \\ \times 8 \\ \hline 64 \end{array}$

20. $\begin{array}{r} 8 \\ \times 0 \\ \hline 0 \end{array}$ 21. $\begin{array}{r} 8 \\ \times 5 \\ \hline 40 \end{array}$ 22. $\begin{array}{r} 9 \\ \times 6 \\ \hline 54 \end{array}$ 23. $\begin{array}{r} 9 \\ \times 1 \\ \hline 9 \end{array}$ 24. $\begin{array}{r} 8 \\ \times 2 \\ \hline 16 \end{array}$ 25. $\begin{array}{r} 9 \\ \times 7 \\ \hline 63 \end{array}$

26. $\begin{array}{r} 8 \\ \times 4 \\ \hline 32 \end{array}$ 27. $\begin{array}{r} 8 \\ \times 7 \\ \hline 56 \end{array}$ 28. $\begin{array}{r} 9 \\ \times 4 \\ \hline 36 \end{array}$ 29. $\begin{array}{r} 9 \\ \times 0 \\ \hline 0 \end{array}$ 30. $\begin{array}{r} 8 \\ \times 1 \\ \hline 8 \end{array}$ 31. $\begin{array}{r} 8 \\ \times 3 \\ \hline 24 \end{array}$

Apply

Solve these problems.

32. Walter practiced his drums for 9 hours each week. How many hours did he practice in 6 weeks?
54 hours

33. Mary bought 8 vases. Each vase cost $8. How much did the vases cost Mary?
$64

34. It costs $8 for a ticket to see Bill's favorite musical group. He has saved $5. How much more does he need for a ticket?
$3

35. Paper plates for a picnic are packed in packages of 8 each. How many plates are there in 7 packages?
56 plates

94

Missing Factors

pages 95-96

Objective

To find missing factors

Materials

counters

Mental Math

Have students name the number that equals:

1. 9 sevens minus 2 tens (43)
2. 3 hundreds plus 7 eights (356)
3. 4 boxes of 7 plus 6 ones (34)
4. 1/2 of 8 eights (32)
5. 1/4 of 2 hours (30 minutes)
6. 6 rows of 12 each (72)
7. 8 nickels plus 1 dime (50¢)
8. $2 \times 6 + 8$ (20)

Skill Review

Write $(7 \times 4) + (2 \times 4) =$ on the board. Have students work the problem to tell what 9×4 equals. (36) Present more fact-plus-fact problems for students to find the total of 9 sixes, 8 sevens, 7 sixes and 9 eights.

Finding Missing Factors

Mike and his sister caught 63 fish on Saturday. Mike will clean the fish and put 7 of them into each freezer bag. Help Mike decide how many freezer bags he will need.

We want to know the number of bags Mike will need to freeze all the fish.

We know he and his sister caught __63__ fish.

He is putting __7__ fish in each freezer bag.

To find the total number of bags needed, we can write a multiplication sentence.

We have n stand for the number of bags.

$$7 \times n = 63$$

We think: 7 times what number equals 63?

$$7 \times \underline{9} = 63$$

$$n = \underline{9}$$

Mike needs __9__ freezer bags.

Getting Started

Solve for n.

1. $n \times 3 = 15$

   $n = \underline{5}$

2. $4 \times n = 28$

   $n = \underline{7}$

3. $9 \times n = 0$

   $n = \underline{0}$

4. $n \times 7 = 42$

   $n = \underline{6}$

5. $8 \times n = 56$

   $n = \underline{7}$

6. $n \times 2 = 16$

   $n = \underline{8}$

95

Teaching the Lesson

Introducing the Problem Have a student read the problem aloud and tell what is to be solved. (how many bags are needed) Ask students what facts are needed to solve the problem. (the total number of fish and number of fish to go into each bag) Ask if information is given. (yes) Work through the model with students and then have them complete the solution sentence.

Developing the Skill Write $8 \times 6 =$ on the board. Have a student write the product to complete the fact. (48) Cover the 6 with a paper or your hand and ask students what is missing in the fact. (6) Write $8 \times n = 48$ on the board and ask students what n stands for. (the missing number or 6) Tell students we can ask ourselves what number times 8 equals 48 or 8 times what equals 48. Write $n = 6$ under the problem. Now cover the 8 in $8 \times 6 = 48$ and ask students what number is missing in the fact. (8) Write $n \times 6 = 48$ and $n = 8$ on the board. Write $7 \times n = 63$ on the board and ask students what times 7 equals 63. (9) Have a student write n = 9 on the board. Repeat for more problems to solve for n.

Practice

Solve for n.

1. $7 \times n = 56$

    $n = \underline{8}$

2. $5 \times n = 30$

    $n = \underline{6}$

3. $n \times 4 = 8$

    $n = \underline{2}$

4. $5 \times n = 45$

    $n = \underline{9}$

5. $n \times 8 = 64$

    $n = \underline{8}$

6. $n \times 7 = 35$

    $n = \underline{5}$

7. $8 \times n = 40$

    $n = \underline{5}$

8. $9 \times n = 54$

    $n = \underline{6}$

9. $7 \times n = 42$

    $n = \underline{6}$

10. $n \times 4 = 36$

    $n = \underline{9}$

11. $4 \times n = 36$

    $n = \underline{9}$

12. $4 \times n = 0$

    $n = \underline{0}$

13. $n \times 8 = 48$

    $n = \underline{6}$

14. $4 \times n = 12$

    $n = \underline{3}$

15. $n \times 1 = 8$

    $n = \underline{8}$

16. $9 \times n = 45$

    $n = \underline{5}$

17. $8 \times n = 16$

    $n = \underline{2}$

18. $n \times 4 = 32$

    $n = \underline{8}$

19. $8 \times n = 56$

    $n = \underline{7}$

20. $9 \times n = 27$

    $n = \underline{3}$

21. $n \times 5 = 0$

    $n = \underline{0}$

EXCURSION

Make the following sentences true by filling in the circle with either an addition symbol or a subtraction symbol.

Example: $8 \; \boxed{+} \; 3 \; \boxed{-} \; 1 = 10$

1. $9 \; \boxed{-} \; 6 \; \boxed{+} \; 2 = 5$

2. $15 \; \boxed{-} \; 7 \; \boxed{-} \; 2 = 6$

3. $8 \; \boxed{+} \; 9 \; \boxed{-} \; 4 = 13$

4. $6 \; \boxed{+} \; 0 = 0 \; \boxed{+} \; 6$

5. $12 \; \boxed{-} \; 7 = 3 \; \boxed{+} \; 2$

6. $14 \; \boxed{-} \; 9 = 11 \; \boxed{-} \; 7 \; \boxed{+} \; 1$

96

Correcting Common Errors

If students have difficulty finding missing factors, have them use counters to model problems. For example, to solve $3 \times n = 21$, they separate 21 counters into 3 equal groups and count to find that there are 7 in each group. Hence, $3 \times 7 = 21$.

Enrichment

Tell students to total their veterinarian bill to cover a visit for their 3 pets for the following: all their distemper shots were $54, combined rabies shots cost $27 and 2 flea baths totaled $32. What was the cost of the distemper and rabies shots for each pet?

Practice

Have students complete the problems on the page by solving for n.

Excursion

In these problems, it is important that operations be done from left to right. No parentheses should be inserted.

Mixed Practice

1. $8 + 5 \times 9$ (53)
2. $20{,}175 - 8{,}254$ (11,921)
3. $\underline{\quad} \times 4 = 32$ (8)
4. $\$65.93 - .78$ ($65.15)
5. $9 + 9 + 9 + 9$ (36)
6. 9×7 (63)
7. $362 + 1{,}758 + 27$ (2,147)
8. $\underline{\quad} \times 6 = 54$ (9)
9. $3{,}688 - 1{,}496$ (2,192)
10. $(98 - 97) \times 5$ (5)

Extra Credit   *Numeration*

Have students create simple division word problems involving missing factors for division sentences. For example:

There are 6 boxes of golf balls. If there are 24 balls in all, how many balls are there in each box? ($24 \div 6 = 4$)

Tell students to illustrate their problems by drawing sets of objects to represent the division sentence. Have students exchange problems to work.

96

Problem Solving, Find a Pattern

pages 97-98

Objective

To find a pattern to solve a problem

Materials

Mental Math

Ask students how many flat surfaces are on:

1. 2 cubes. (12)
2. 3 rectangular prisms. (18)
3. 4 unopened pop cans. (8)
4. 6 coffee cups. (6)
5. a kitchen drawer. (10)
6. a triangular prism. (3)
7. a box of cereal. (6)
8. two 8-pane windows. (32)

Looking for a Pattern

This arrangement of numbers is called Pascal's triangle. What are the numbers in the next 3 rows?

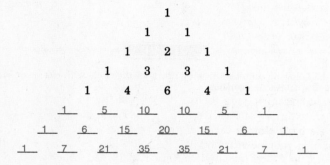

★ **SEE**

We need to find the patterns in Pascal's triangle.
We know:
   the end number in each row is __1__;
   the other numbers in the row are the sum of the two closest numbers in the row

above; and each row has __1__ more number in it than the row above it.

★ **PLAN**

We predict the numbers there are in the next 3 rows by extending the patterns.

★ **DO**

```
                1
            1       1
        1       2       1
     1      3       3       1
   1     4      6      4      1
  1    5    10     10     5     1
 1   6    15    20    15    6    1
1   7   21   35    35   21   7   1
```

★ **CHECK**

We check to see if the numbers continue the patterns.

97

Teaching the Lesson

Have a student read the problem aloud. Tell students that Pascal's triangle is used in higher mathematics. Read the steps with the students as they complete the problem. Guide students through the DO step. Tell students that the method they used for determining each new row of Pascal's triangle is the most common but is not the only method. Allow interested students to discover their own patterns for continuing the triangle. Note: If some students are further intrigued, information should be available in any 2nd year algebra text under the topics binomial expansion or probability, combinations or permutations.

Apply

Look for a pattern.

1. How many numbers are needed for the fiftieth row of Pascal's triangle?

50

2. Complete the last picture.

3. If two 6's are multiplied together the product is 36. If twelve 6's are multiplied together, what is the number in the ones place?

6

4. What number is missing?
1, 2, 4, 7, 11, ~~16~~, 22, 29, 37

5. The first three rows of Pascal's triangle use six numbers.

$$1$$
$$1 \quad 1$$
$$1 \quad 2 \quad 1$$

The first four rows use 10 numbers. How many numbers are needed for the first 10 rows? How can you prove that your answer is correct?

55

6. Think of each row in Pascal's triangle as one number. You can get the number in row two by multiplying 11 × 1. You can get row three by multiplying 11 × 11. Explain how to use 11 as a factor to get more rows.

See Solution Notes.

7. Make five rows of another triangle by multiplying each digit in five rows of Pascal's triangle by 3. How is the pattern in your new triangle like that in Pascal's triangle? How is it different?

See Solution Notes.

8. The sequence of numbers shown below is known as the Fibonacci sequence.
1, 1, 2, 3, 5, 8, 13, . . .
Tell what the next number in the sequence is and explain how you found it.

21

Extra Credit *Measurement*

Have students measure the length and width of their math textbooks in inches, and find the area. Have them repeat the activity for each of their other textbooks. Have students list their texts in order, from smallest in area to largest. Have students circle the length of their largest textbook. Have one student calculate the number of inches in a mile and write this on the board. Now have each student calculate how many of their largest text, laid end to end, it would take to form a "book mile."

Solution Notes

1. If necessary refer students to p. 97 to see the pattern again. (7 in 7th row, etc.)

2. Help students to see that the pattern is a 90 degree rotation, plus an added dot. Students may find it helpful to draw the figure and act this problem out.

3. Have students multiply 6 × 6 × 6 and 6 × 6 × 6 × 6 and continue until they see the pattern of 6 being in ones place in each product and therefore 6 ones are multiplied by 6 again and again.

4. Students should see that the pattern is each addend being 1 more than the previous one.

Higher-Order Thinking Skills

5. [Evaluation] A possible answer is to show the pattern:
+2 +3 +4 +5 +6 +7 +8 +9 +10
1, 3, 6, 10, 15, 21, 28, 36, 45, 55.
Another possible solution is to actually show the first 10 rows of Pascal's triangle and count.

6. [Synthesis] Multiply the number in any row by 11 to get the number in the next row.

7. [Analysis] It is the same in that the end numbers in each row are the same, the middle numbers are sums of the two numbers above, and each row has one more digit than the row above. It is different in that the digits in the rows are different.

8. [Analysis] Add the two preceding numbers to get the next number.

Calculator Use

pages 99-100

Objective

To use a calculator to balance a bank account

Materials

play or real money
sample checks
calculators

Mental Math

Have students round each to nearest ten and hundred:

1. 4,762 (4,760, 4,800)
2. 791 (790, 800)
3. 1,506 (1,510, 1,500)

Round to nearest hundred and thousand:

4. 19,876 (19,900, 20,000)
5. 51,604 (51,600, 52,000)
6. 77,499 (77,500, 77,000)

Skill Review

Write **$150.61** on the board. Tell students to use their calculators to add $64.65 to this amount. ($215.26) Have students subtract $83.70 from the $215.26 and tell the difference. ($131.56) Continue to have students add or subtract amounts of money to the sum or difference of the previous problem.

Calculators and Bank Accounts

Lorie is the student body treasurer. She writes **checks, deposits** money, and keeps the records for **expenses.** After a check is written, Lorie **balances the account.** How much does the student body have in its account?

We want to find the amount of money in the account. The amount after the last check was written

is _$85.35_ . The check is written for _$17.80_ .

To find the balance, we subtract _$17.80_ from

_$85.35_ . Enter this code into your calculator and write the answer on the screen.

85 $\cdot$ 35 $-$ 17 $\cdot$ 80 $=$ (67.55)

The student body has _$67.55_ in its account.

Some checkbooks have a **register** to keep records. How much is in the student body account after the $45.50 deposit? Remember to add deposits and subtract amounts of checks. Write the answer in the balance column.

CHECK NO.	DATE	CHECKS ISSUED TO OR DESCRIPTION OF DEPOSIT	AMOUNT OF CHECK		✔	AMOUNT OF DEPOSIT		BALANCE	
								85	35
102	NOV 1	Ace Ice Cream Store	17	80				67	55
	NOV 2	Deposit				45	50	113	05
103	NOV 7	Paul's Paper Company	13	68				99	37
	NOV 8	Deposit				23	58	122	95

Use the register to show the balance after each of the following:

1. Check 103 on November 7, to Paul's Paper Company, for $13.68
2. Deposit on November 8 of $23.58

99

Teaching the Lesson

Introducing the Problem Tell students that the picture shows a girl's record keeping for a checking account. Have a student read the problem aloud and tell what is to be found. (How much money is in the student body's account?) Tell students that when they open a checking account, money must be deposited in a bank and then they can write checks to use that money. Tell students that each check uses up some of the money deposited and therefore we subtract each check amount to see how much money is left in the account. Discuss why people have checking accounts. (convenience, safety, etc.) Have students complete the sentences and the calculator codes to solve the problem. Talk through each of the 2 problems at the bottom of the page, and have students complete the register.

Developing the Skill Have a student be a banker with play money. Have a second student go to the banker to deposit $10. Have the banker give the second student some slips of paper to resemble **checks.** Write **Deposit $10.00** on the board. Tell the second student to pay a third student $4.87 by completing a sample check made out to the third student. Write **−$4.87** on the board beside the $10.00. Tell students the $4.87 is an **expense.** Have students find the difference to see how much money the second student still has in his bank account. ($5.13) Act out more deposits and check payments as additions and subtractions are shown on the board.

The thought bubble check register at top right:

No. 102
19 __
To Ace Ice Cream Store

	DOLLARS	CENTS
Balance Forward	85	35
Deposits		
Total	85	35
This Check	17	80
Balance		
Deductions		
Balance Forward		

Practice

Complete the register using the information below.
Find each new balance.

Check Number	Date	Check or Deposit	Amount
201	Dec. 1	Bi/More Grocery Store	$36.18
202	Dec. 3	United Gas and Electrical	$45.73
203	Dec. 5	Sam's Shoes	$39.88
	Dec. 8	Deposit	$215.75
204	Dec. 8	Savings Account	$80.00
205	Dec. 9	Home Insurance Co.	$135.05
206	Dec. 16	Paul's Pizza Parlor	$16.50
207	Dec. 17	Dr. E. J. Goode	$49.78
	Dec. 22	Deposit	$205.75
208	Dec. 23	Alice Carson (rent)	$150.00

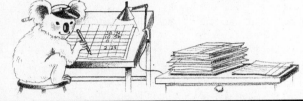

CHECK NO.	DATE	CHECKS ISSUED TO OR DESCRIPTION OF DEPOSIT	AMOUNT OF CHECK		✔	AMOUNT OF DEPOSIT		BALANCE	
								306	75
201	DEC 1	Bi/More Grocery Store	36	18				270	57
202	DEC 3	United Gas and Electrical	45	73				224	84
203	DEC 5	Sam's Shoes	39	88				184	96
	DEC 8	Deposit				215	75	400	71
204	DEC 8	Savings Account	80	00				320	71
205	DEC 9	Home Insurance Co.	135	05				185	66
206	DEC 16	Paul's Pizza Parlor	16	50				169	16
207	DEC 17	Dr. E. J. Goode	49	78				119	38
	DEC 22	Deposit				205	75	325	13
208	DEC 23	Alice Carson (rent)	150	00				175	13

100

Correcting Common Errors

Some students may have trouble distinguishing between checks and deposits when balancing a bank account. Have them imagine the account as a number line and think of a deposit as a move to the right and a check as a move to the left.

Enrichment

Tell students to work with a friend to make a check register with the following: a beginning balance of $26.23, 3 checks written before a deposit of $20 and a current balance of $26.46.

Practice

Help students record the first check amount in the register. Using their calculators have students find and record the balance independently. Continue to help students record the next 2 checks and the first deposit. Have students balance the register after each transaction. Remind students that check amounts are subtracted and deposit amounts are added. Have students complete the register independently.

Extra Credit *Statistics*

Have students record the highest daily temperature in their city for a week, using the Fahrenheit reading. Provide graph paper. Ask students to make a line graph, starting by listing the days of the week across the bottom, and the temperature degrees, in 5 degree increments, up the side. Have students record each day's temperature with a point. When students have marked each of the seven temperatures, have them connect the points to show any trends in the week's temperatures. Ask students if their graph would look different if they had recorded Celsius temperatures.

Chapter Test

page 101

Item	Objective
1-35	Recall multiplication facts through 9 (See pages 83-94)
36-41	Find value of expression with parentheses (See pages 89-90)
42-47	Find missing factors (See pages 95-96)

Multiply.

1. $\begin{array}{r} 3 \\ \times 2 \\ \hline 6 \end{array}$	**2.** $\begin{array}{r} 8 \\ \times 5 \\ \hline 40 \end{array}$	**3.** $\begin{array}{r} 7 \\ \times 9 \\ \hline 63 \end{array}$	**4.** $\begin{array}{r} 4 \\ \times 3 \\ \hline 12 \end{array}$	**5.** $\begin{array}{r} 2 \\ \times 8 \\ \hline 16 \end{array}$	**6.** $\begin{array}{r} 9 \\ \times 6 \\ \hline 54 \end{array}$	**7.** $\begin{array}{r} 5 \\ \times 8 \\ \hline 40 \end{array}$
8. $\begin{array}{r} 5 \\ \times 5 \\ \hline 25 \end{array}$	**9.** $\begin{array}{r} 6 \\ \times 7 \\ \hline 42 \end{array}$	**10.** $\begin{array}{r} 5 \\ \times 9 \\ \hline 45 \end{array}$	**11.** $\begin{array}{r} 8 \\ \times 7 \\ \hline 56 \end{array}$	**12.** $\begin{array}{r} 6 \\ \times 0 \\ \hline 0 \end{array}$	**13.** $\begin{array}{r} 6 \\ \times 5 \\ \hline 30 \end{array}$	**14.** $\begin{array}{r} 8 \\ \times 9 \\ \hline 72 \end{array}$
15. $\begin{array}{r} 7 \\ \times 8 \\ \hline 56 \end{array}$	**16.** $\begin{array}{r} 3 \\ \times 9 \\ \hline 27 \end{array}$	**17.** $\begin{array}{r} 9 \\ \times 7 \\ \hline 63 \end{array}$	**18.** $\begin{array}{r} 6 \\ \times 6 \\ \hline 36 \end{array}$	**19.** $\begin{array}{r} 8 \\ \times 6 \\ \hline 48 \end{array}$	**20.** $\begin{array}{r} 4 \\ \times 5 \\ \hline 20 \end{array}$	**21.** $\begin{array}{r} 6 \\ \times 9 \\ \hline 54 \end{array}$
22. $\begin{array}{r} 1 \\ \times 5 \\ \hline 5 \end{array}$	**23.** $\begin{array}{r} 7 \\ \times 7 \\ \hline 49 \end{array}$	**24.** $\begin{array}{r} 8 \\ \times 8 \\ \hline 64 \end{array}$	**25.** $\begin{array}{r} 5 \\ \times 7 \\ \hline 35 \end{array}$	**26.** $\begin{array}{r} 0 \\ \times 4 \\ \hline 0 \end{array}$	**27.** $\begin{array}{r} 9 \\ \times 9 \\ \hline 81 \end{array}$	**28.** $\begin{array}{r} 7 \\ \times 5 \\ \hline 35 \end{array}$
29. $\begin{array}{r} 9 \\ \times 8 \\ \hline 72 \end{array}$	**30.** $\begin{array}{r} 4 \\ \times 6 \\ \hline 24 \end{array}$	**31.** $\begin{array}{r} 7 \\ \times 6 \\ \hline 42 \end{array}$	**32.** $\begin{array}{r} 3 \\ \times 2 \\ \hline 6 \end{array}$	**33.** $\begin{array}{r} 9 \\ \times 5 \\ \hline 45 \end{array}$	**34.** $\begin{array}{r} 5 \\ \times 6 \\ \hline 30 \end{array}$	**35.** $\begin{array}{r} 6 \\ \times 8 \\ \hline 48 \end{array}$

Solve for n.

36. $(5 \times 3) + 6 = n$ $n = \underline{21}$

37. $(8 - 4) \times 5 = n$ $n = \underline{20}$

38. $9 \times (3 + 5) = n$ $n = \underline{72}$

39. $45 + (9 \times 5) = n$ $n = \underline{90}$

40. $36 - (9 \times 4) = n$ $n = \underline{0}$

41. $(8 \times 7) - 38 = n$ $n = \underline{18}$

42. $7 \times n = 49$ $n = \underline{7}$

43. $n \times 3 = 15$ $n = \underline{5}$

44. $n \times 8 = 64$ $n = \underline{8}$

45. $n \times 4 = 36$ $n = \underline{9}$

46. $6 \times n = 42$ $n = \underline{7}$

47. $6 \times n = 0$ $n = \underline{0}$

101

Circle the letter of the correct answer.

1 What is the value of the 9 in 9,058?
a tens
b hundreds
c thousands
d NG

2 3,651 ◯ 3,615
a >
b <

3 Round 850 to the nearest hundred.
a 800
b 900
c NG

4 Round 6,786 to the nearest thousand.
a 8,000
b 7,000
c NG

5 What is the value of the 3 in 632,461?
a tens
b hundreds
c thousands
d NG

6
  57
+ 86
a 133
b 143
c 1,313
d NG

7 359 + 283
a 532
b 542
c 552
d NG

8
  29,468
+ 36,875
a 66,343
b 66,433
c 67,343
d NG

9
  $16.48
+ 37.19
a $43.67
b $53.57
c $53.67
d NG

10 83 − 27
a 56
b 64
c 66
d NG

11 926 − 458
a 432
b 468
c 532
d NG

12
  40,276
− 29,867
a 10,409
b 20,409
c 29,611
d NG

13
  9
× 6
a 54
b 56
c 63
d NG

◻ score

102

Cumulative Review

page 102

Item	Objective
1	Identify place value of digit in a number less than 10,000 (See pages 27-28)
2	Compare, order numbers less than 10,000 (See pages 29-30)
3	Round numbers to nearest 100 (See pages 31-34)
4	Round numbers to nearest 1,000 (See pages 33-34)
5	Identify place value of digit in a number less than 1,000,000 (See pages 35-36)
6	Add two 2-digit numbers (See pages 43-44)
7	Add two 3-digit numbers (See pages 45-46)
8	Add two 4-digit numbers (See pages 47-48)
9	Add money amounts (See pages 53-54)
10	Subtract two 2-digit numbers (See pages 61-62)
11	Subtract two 3-digit numbers (See pages 63-64)
12	Subtract two 5-digit numbers (See pages 67-68)
13	Multiply using 9 as factor (See pages 93-94)

Alternate Cumulative Review

Circle the letter of the correct answer.

1 What is the value of the 6 in 4,692?
a tens
b hundreds
c thousands
d NG

2 4,321 ◯ 4,231
a >
b <
c =

3 Round 731 to the nearest hundred.
a 600
b 700
c 800
d NG

4 Round 8,463 to the nearest thousand.
a 8,000
b 8,400
c 9,000
d NG

5 What is the value of the 4 in 421,653?
a hundreds
b thousands
c ten thousands
d NG

6
  38
+ 65
a 913
b 93
c 103
d NG

7 647 + 274 =
a 821
b 921
c 811
d NG

8
  46,365
+ 13,947
a 32,418
b 59,202
c 60,202
d NG

9
  $25.67
+ 46.24
a $71.91
b $61.91
c $61.81
d NG

10 94 − 37 =
a 67
b 63
c 57
d NG

11 632 − 467 =
a 165
b 235
c 275
d NG

12
  50,385
− 18,568
a 42,827
b 41,827
c 48,223
d NG

13
  8
×7
a 48
b 54
c 56
d NG

10

Multiples

pages 103-104

Objective

To name multiples of whole numbers

Materials

*multiplication fact cards

Mental Math

Have students tell the number that comes after:

1. $47 - 19$ (29)
2. 1,719,000 (1,719,001)
3. $28 \times \frac{1}{2}$ (15)
4. $356 - 4$ tens (317)
5. 7×20 (141)
6. $26 \div 2$ (14)
7. $\frac{1}{4}$ of 200 (51)
8. 20 tens $+ 268$ (469)

Skill Review

Have students count by 2's and 5's to 100 and then write those sequences on the board.

Multiples

The 25 students in Miss Lane's class are counting off to form squares for square dancing. Every fifth person will stand in the center of a square. What numbers will the students in the centers have?

We want to know the numbers of the students who will stand in the centers of the squares.

We know there are ___25___ students in Miss Lane's class.

We know that every ___5th___ student will stand in the center of a square.

We can count from 1 to 25, marking off every fifth number.

1, 2, 3, 4,(5)   6, 7, 8, 9,(10)   11, 12, 13, 14,(15)

16, 17, 18, 19,(20)   21, 22, 23, 24,(25)

We say ___5___, ___10___, ___15___, ___20___ and ___25___ are multiples of 5. A **multiple** of a number is a product that has that number for at least one of its factors.

✔ The least multiple of any number is 0 because any number times 0 equals 0. Naming multiples of a number is called **skip-counting**. We skip-count by 5's by saying 0, 5, 10, 15, 20, 25, etc.

Students with the numbers ___5___, ___10___, ___15___, ___20___ and ___25___ will stand in the centers of the squares.

Getting Started

Write the first nine multiples of each of these numbers.

6   _0_, _6_, _12_, _18_, _24_, _30_, _36_, _42_, _48_

8   _0_, _8_, _16_, _24_, _32_, _40_, _48_, _56_, _64_

103

Teaching the Lesson

Introducing the Problem Have a student read the problem and tell about the picture. Tell students they need to first decide what information is known. (There are 25 students and every fifth person will stand in the middle of a square of 4.) Have students complete the sentences and skip counting to solve the problem. Students may want to check their solution by acting it out.

Developing the Skill Place the multiplication fact cards for 4×0 through 4×9 in order across the chalktray. Have students write the product of each fact above it. Tell students these products are called **multiples** of 4. Have students say the multiples of 4 in order. Have students write the numbers 1, 2, and 3 between the first 2 fact cards to show that these numbers were skipped. Continue to have students write the skipped numbers between the multiples. Tell students that **skip counting** is saying the multiples of a number in order. Place the fact cards for 6×0 through 6×9 across the chalktray and repeat the activity. Group students in pairs to continue the activity for more skip counting.

Practice

Write the first nine multiples of each of these numbers.

2 _0_ , _2_ , _4_ , _6_ , _8_ , _10_ , _12_ , _14_ , _16_

5 _0_ , _5_ , _10_ , _15_ , _20_ , _25_ , _30_ , _35_ , _40_

7 _0_ , _7_ , _14_ , _21_ , _28_ , _35_ , _42_ , _49_ , _56_

4 _0_ , _4_ , _8_ , _12_ , _16_ , _20_ , _24_ , _28_ , _32_

9 _0_ , _9_ , _18_ , _27_ , _36_ , _45_ , _54_ , _63_ , _72_

Skip-count by 7.

56, _63_ , _70_ , _77_ , _84_ , _91_ , _98_ , _105_ , 112

Skip-count by 4.

32, _36_ , _40_ , _44_ , _48_ , _52_ , _56_ , _60_ , 64

Skip-count by 9.

72, _81_ , _90_ , _99_ , _108_ , _117_ , _126_ , _135_ , 144

Skip-count by 2.

16, _18_ , _20_ , _22_ , _24_ , _26_ , _28_ , _30_ , 32

Skip-count by 6.

48, _54_ , _60_ , _66_ , _72_ , _78_ , _84_ , _90_ , 96

Skip-count by 5.

40, _45_ , _50_ , _55_ , _60_ , _65_ , _70_ , _75_ , 80

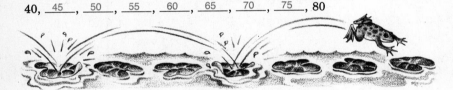

104

Practice

Remind students that the least multiple of any number is zero. Have students complete the page independently.

Correcting Common Errors

Some students may have difficulty with skip-counting. Draw a number line across the chalkboard and show the numbers from 0 through 81. Have the students work together in a group. Students may take turns being a rabbit who stands at the chalkboard and places a finger at 0. As the other students say ten multiples of a number in order, the rabbit hops with his or her finger to the multiples on the number line. Each time a new rabbit goes to the board, a new set of multiples should be used.

Enrichment

Tell students to skip count the books on several classroom or library shelves. Have them record their totals and compare with a friend's work.

Extra Credit *Logic*

Dictate this problem to students to build skills in multiplication, division and factoring. It can be solved by large or small groups.

Mrs. Carson, the sixth grade science teacher, kept a cage full of gerbils in her homeroom. One morning she and her students found that someone had left the cage door open and all the gerbils had escaped. The students volunteered to find them and asked how many were missing. Mrs. Carson didn't remember how many gerbils there had been. She did remember that when she counted them by two's, three's, or by four's, she always had one left over. When she counted them by five's, however, her count came out even. Her students quickly calculated how many gerbils were missing and returned them all safely to the cage. What is the smallest possible number of gerbils in the cage? (25)

Multiplication Properties

pages 105-106

Objective

To understand the grouping, multiplication-addition and multiplying-by-10 properties

Materials

small paper pieces

Mental Math

Have students tell the best buy:

1. 3/89¢ or 4/$1 (4/$1)
2. 6/$9 or 3/$3 (3/$3)
3. 2/$1 or 45¢ each (45¢)
4. 8/$2 or 10/$3 (8/$2)
5. 4/20¢ or 6/60¢ (4/20¢)
6. 3/99¢ or 8/$2 (8/$2)
7. 10¢ each or 12/$1 (12/$1)
8. 2/$9 or $5 each (2/$9)

Skill Review

Write the numbers 1 through 9 across the board. Have students skip count by 10's as you write the numbers 10 through 90 under 1 through 9. Have students write the facts for 10 on the board in a column. ($1 \times 10 = 10$ through $9 \times 10 = 90$) Leave this work on the board for later use.

Understanding Multiplication Properties

Besides the properties we have already studied, there are others that help us to find shortcuts in multiplication.

I can group any way I want in multiplication.

Multiplying by 10 is easy.

The Grouping Property
Factors can be grouped in anyway.

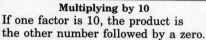
$(3 \times 2) \times 5 = 3 \times (2 \times 5)$

$\underline{6} \times 5 = 3 \times \underline{10}$

$\underline{30} = \underline{30}$

The Multiplication-Addition Property
Multiplication can be distributed over addition.

$5 \times (4 + 6) = (5 \times 4) + (5 \times 6)$

$5 \times \underline{10} = \underline{20} + \underline{30}$

$\underline{50} = \underline{50}$

Multiplying by 10
If one factor is 10, the product is the other number followed by a zero.

$2 \times 10 = \underline{20}$

$10 \times 7 = \underline{70}$

Getting Started

Solve for n. Use the properties to help you.

1. $4 \times (7 + 3) = n$
$n = \underline{40}$

2. $8 \times 10 = n$
$n = \underline{80}$

3. $6 \times (4 \times 10) = n$
$n = \underline{240}$

4. $9 \times 10 = n$
$n = \underline{90}$

5. $(2 \times 3) \times 10 = n$
$n = \underline{60}$

6. $(7 \times 2) + (7 \times 8) = n$
$n = \underline{70}$

105

Teaching the Lesson

Introducing the Problem Have a student read aloud the thoughts of the 2 students pictured. Have a student read the introductory paragraph. Guide them through each property definition as they fill in the examples.

Developing the Skill Refer students to the work on the board. Help students see that when a number is multiplied by 10, the product is that number followed by a zero. Write **(2 × 4) × 6** on the board and remind students that the operation in the parentheses is done first. Have a student write the product on the board. (48) Now write **2 × (4 × 6)** on the board and have a student write the product. (48) Ask if the product changed when the factors were grouped differently. (no) Have a student write the problem to show another way to regroup the factors. ((6 × 2) × 4) or (6 × 4) × 2) Tell students that they can use the multiplying-by-10 and the multiplication-addition properties to find a shortcut

in multiplication. Write on the board: **(9 × 2) + (9 × 8) = 9 × (2 + 8) = 9 × 10 = 90.** Have students read the problem with you. Write **(8 × 3) + (8 × 7)** on the board and have students follow the previous examples to work the problem on the board.

105

Practice

Solve for n. Use the properties to help you.

1. $3 \times (0 \times 5) = n$
$n = \underline{0}$

2. $5 \times (4 + 3) = n$
$n = \underline{35}$

3. $(6 \times 3) + (6 \times 7) = n$
$n = \underline{60}$

4. $7 \times 10 = n$
$n = \underline{70}$

5. $2 \times (5 \times 8) = n$
$n = \underline{80}$

6. $9 \times (0 \times 3) = n$
$n = \underline{0}$

7. $4 \times 10 = n$
$n = \underline{40}$

8. $8 \times (3 \times 1) = n$
$n = \underline{24}$

9. $(8 \times 5) \times 0 = n$
$n = \underline{0}$

10. $(1 \times 1) \times 1 = n$
$n = \underline{1}$

11. $10 \times 9 = n$
$n = \underline{90}$

12. $(3 \times 4) + (3 \times 6) = n$
$n = \underline{30}$

13. $7 \times (9 + 1) = n$
$n = \underline{70}$

14. $0 \times (8 \times 10) = n$
$n = \underline{0}$

15. $(4 \times 0) + (4 \times 0) = n$
$n = \underline{0}$

Add, subtract or multiply.

16. $6 + 9 = \underline{15}$
17. $8 \times 7 = \underline{56}$
18. $15 - 9 = \underline{6}$
19. $3 + 7 = \underline{10}$
20. $5 \times 8 = \underline{40}$
21. $6 - 6 = \underline{0}$

22. $4 \times 7 = \underline{28}$
23. $9 \times 6 = \underline{54}$
24. $17 - 8 = \underline{9}$
25. $7 + 6 = \underline{13}$
26. $0 \times 8 = \underline{0}$
27. $5 + 8 = \underline{13}$

28. $13 - 6 = \underline{7}$
29. $9 - 0 = \underline{9}$
30. $7 \times 1 = \underline{7}$
31. $6 \times 6 = \underline{36}$
32. $4 \times 3 = \underline{12}$
33. $8 + 1 = \underline{9}$

34. $5 + 5 = \underline{10}$
35. $5 \times 5 = \underline{25}$
36. $5 - 5 = \underline{0}$
37. $14 - 9 = \underline{5}$
38. $8 \times 6 = \underline{48}$
39. $8 - 3 = \underline{5}$

40. $8 \times 8 = \underline{64}$
41. $8 - 8 = \underline{0}$
42. $8 + 8 = \underline{16}$
43. $16 - 7 = \underline{9}$
44. $3 + 9 = \underline{12}$
45. $9 - 6 = \underline{3}$

106

Multiplying Tens and Ones

pages 107-108

Objective

To multiply 2-digit numbers by 1-digit numbers, no trading

Materials

page of 100 basic multiplication facts

Mental Math

Ask students if A = 1, and Z = 26, what is the number value of:

1. C × J (30)
2. Z ÷ B (13)
3. ½ of X (12)
4. A + Z + D (31)
5. 100 × (B + C) (500)
6. sum of letters in your first name
7. Y − (D × E) (5)

Skill Review

Give students a timed test of the 100 basic multiplication facts. It is a good idea to repeat these tests often throughout the year.

Multiplying Tens and Ones

River Road Foods processes and packs fresh fruits. How many cans of tomatoes will River Road pack in 4 minutes?

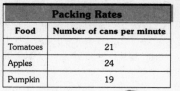

Packing Rates	
Food	Number of cans per minute
Tomatoes	21
Apples	24
Pumpkin	19

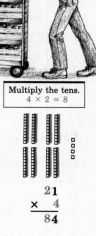

We want to know the number of cans of tomatoes packed in ___4___ minutes.

We know that ___21___ cans of tomatoes are packed in one minute. To find the number of cans packed in

4 minutes, we multiply ___21___ by ___4___.

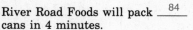

21 × 4 = ?

Multiply the ones.
4 × 1 = 4

Multiply the tens.
4 × 2 = 8

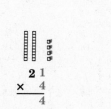

```
   2 1        2 1          2 1
 ×   4      ×   4        ×   4
             ———          ———
   4           4           84
```

River Road Foods will pack ___84___ cans in 4 minutes.

Getting Started

Multiply.

```
1.    2 4        2.    4 0        3.    3 1
    ×   2            ×   2            ×   3
    ———            ———            ———
      48              80              93
```

Copy and multiply.

```
4. 2 × 13        5. 4 × 11        6. 4 × 22
      26              44              88
```

107

Teaching the Lesson

Introducing the Problem Have a student read the problem aloud. Ask what they must find. (the number of cans of tomatoes packed in 4 minutes) Ask students if they will need all the information in the table. (no) Have students read and complete the information sentences. Guide them through the multiplication in the model, describing the trade to solve the problem. Tell students to round 21 cans to the nearest ten and estimate to check their solution.

Developing the Skill Write on the board:

$$42 \times 2 = (40 + 2) \times 2$$
$$= (40 \times 2) + (2 \times 2)$$
$$= 2 \text{ forties} + 2 \text{ twos}$$
$$= 2 \times (4 \text{ tens}) + 2 \times (2 \text{ ones})$$

Read through the problem with the students. Remind students that the multiplication-addition property helps us see that 42 × 2 is the same as 2 times 4 tens plus 2 times 2 ones. Write **42 × 2** vertically on the board. Tell students that we multiply the ones column first to find the product of 2 times 2 ones. Write **4** in the answer. Remind students that there are 4 tens in 40 and 2 times 4 tens equals 8 tens as you write **8** in the tens column. Repeat the procedure for 31 × 3 and 24 × 2.

Practice

Multiply.

1. $\;\;13$ $\underline{\times\;2}$ $\;\;26$	2. $\;\;30$ $\underline{\times\;2}$ $\;\;60$	3. $\;\;58$ $\underline{\times\;1}$ $\;\;58$	4. $\;\;40$ $\underline{\times\;2}$ $\;\;80$	5. $\;\;21$ $\underline{\times\;3}$ $\;\;63$
6. $\;\;42$ $\underline{\times\;2}$ $\;\;84$	7. $\;\;34$ $\underline{\times\;2}$ $\;\;68$	8. $\;\;11$ $\underline{\times\;7}$ $\;\;77$	9. $\;\;24$ $\underline{\times\;2}$ $\;\;48$	10. $\;\;33$ $\underline{\times\;3}$ $\;\;99$
11. $\;\;12$ $\underline{\times\;4}$ $\;\;48$	12. $\;\;11$ $\underline{\times\;5}$ $\;\;55$	13. $\;\;44$ $\underline{\times\;2}$ $\;\;88$	14. $\;\;31$ $\underline{\times\;3}$ $\;\;93$	15. $\;\;32$ $\underline{\times\;3}$ $\;\;96$

Copy and Do

16. 10×5
50
17. 36×1
36
18. 23×3
69
19. 4×12
48

20. 3×13
39
21. 43×2
86
22. 2×22
44
23. 9×11
99

24. 41×2
82
25. 3×12
36
26. 2×33
66
27. 2×31
62

Apply

Solve these problems.

28. Manuel bought 4 dozen eggs to put into the cakes he was making for the church bake sale. How many eggs did Manuel buy?
48 eggs

29. Beth bought 4 stamps. Each stamp cost 22¢. How much did Beth pay for the stamps?
88¢

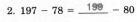
EXCURSION

Fill in the blanks so that both sides of the sentences are equal.

1. $145 - 12 = 143 - \underline{\;\;10\;\;}$
2. $197 - 78 = \underline{\;\;199\;\;} - 80$
3. $41 - 13 = 48 - \underline{\;\;20\;\;}$
4. $73 - 29 = \underline{\;\;74\;\;} - 30$
5. $187 - 65 = 182 - \underline{\;\;60\;\;}$
6. $395 - 264 = \underline{\;\;335\;\;} - 204$
7. $359 - 126 = 363 - \underline{\;\;130\;\;}$
8. $254 - 181 = \underline{\;\;274\;\;} - 201$

108

Practice

Remind students to multiply the ones first. Have students complete the problems independently.

Excursion

Have students first compare known minuends or subtrahends in the exercises and explain how they differ. Stress that changes made in the minuend must be repeated in the subtrahend.

Discuss how changing the minuend and subtrahend through addition or subtraction can help to simplify problems such as in $837 - 685 = (837 + 20) - (685 + 20) = 857 - 705$.

Extra Credit *Biography*

Put the following on the board or duplicate for students.
Directions: Fill in the blanks below. The first letters, reading down, spell the name of a famous person in mathematics, who is described here.

She was the first woman mathematician. She lived in the fifth century A.D. and studied medicine as well as mathematics. She was murdered by a Christian mob who thought she was a devil.

(H̲ A L F̲) 6 is what part of 12?
(Y̲ E S̲) Is is true that $42 \times 7 = 294$?
(P̲ R̲ O̲ D̲ U̲ C̲ T) What is the result of a multiplication called?
(A̲ D̲ D) How do you find the sum of 5 and 7?
(T̲ R̲ I̲ A̲ N̲ G̲ L̲ E̲) What is the name of this figure? △
(I̲) What was the Roman symbol for the number 1?
(A̲ N̲ G̲ L̲ E̲) What is this figure? ∠

(Hypatia. Ask a student to find additional information about Hypatia, and report to the class.)

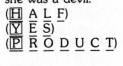

108

Multiplying, Trading Ones

pages 109-110

Objective

To multiply 2-digit numbers by 1-digit numbers with trading

Materials

place value materials

Mental Math

Dictate the following:

1. $2 \times 6 \times 3$ (36)
2. $10 \times (8 + 2)$ (100)
3. $(3 \times 6) + (4 \times 6)$ (42)
4. $(8 \times 6) - (3 \times 6)$ (30)
5. $(14 - 7) \div 7$ (1)
6. $20 \times 6 \times 1$ (120)
7. $8 \times 0 \times 10$ (0)
8. $(18 - 6) \div 3$ (4)

Skill Review

Have students work in pairs to write multiplication facts which have products of 10 or more, and then rename each product in tens and ones.

Multiplying, Trading Ones

The fourth grade science lab has 16 stations. Mr. Owens needs to make a battery and bulb hook-up for each station. How many batteries will Mr. Owens need?

We want to find the number of batteries Mr. Owens needs.

There are ___16___ stations in the science lab.

Each station requires ___2___ batteries.

To find the number of batteries needed,

we multiply ___16___ by ___2___.

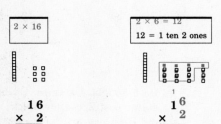

2×16	$2 \times 6 = 12$ $12 = 1$ ten 2 ones	$2 \times 1 = 2$ 2 tens $+ 1$ ten $= 3$ tens

$$\begin{array}{r} 1\,6 \\ \times\ \ 2 \\ \hline ? \end{array} \qquad \begin{array}{r} {}^{1}1\,6 \\ \times\ \ 2 \\ \hline 2 \end{array} \qquad \begin{array}{r} {}^{1}1\,6 \\ \times\ \ 2 \\ \hline 3\,2 \end{array}$$

Mr. Owens needs ___32___ batteries.

Getting Started

Multiply.

1. $\begin{array}{r} 27 \\ \times\ 3 \\ \hline 81 \end{array}$
2. $\begin{array}{r} 45 \\ \times\ 2 \\ \hline 90 \end{array}$
3. $\begin{array}{r} 34 \\ \times\ 2 \\ \hline 68 \end{array}$
4. $\begin{array}{r} 25 \\ \times\ 3 \\ \hline 75 \end{array}$

Copy and multiply.

5. 48×2 (96)   6. 5×19 (95)   7. 38×2 (76)

109

Teaching the Lesson

Introducing the Problem Have a student read the problem and tell what is to be found. (how many batteries Mr. Owens will need) Ask students what information is given in the problem. (There are 16 stations.) Ask what information is needed yet. (number of batteries per station) Have students complete the sentences. Talk through the model multiplication with them and have students complete the solution statement.

Developing the Skill Write on the board:

$39 \times 2 = (30 + 9) \times 2$

$\qquad = (30 \times 2) + (9 \times 2)$

$\qquad = 2$ thirties $+ 9$ twos

$\qquad = 2 \times (3$ tens$) + 9 \times (2$ ones$)$

$\qquad = 6$ tens $+ 18$ ones

$\qquad = 7$ tens $+ 8$ ones

$\qquad = 78$

Help students see that 10 ones are traded when adding 6 tens and 18 ones. Now write **39 × 2** vertically on the board and have students tell the product of 9×2. (18) Refer students to the 6 tens + 18 ones step to see that the 10 ones are traded. Ask students the product of 3 tens $\times$ 2. (6 tens) Tell them now to add the 1 ten as you refer to the 7 tens + 8 ones step above. Complete the problem and then repeat the procedure for 17×3 and then 43×2, where no trade is needed.

Practice

Multiply.

1. $\begin{array}{r} 25 \\ \times\ 3 \\ \hline 75 \end{array}$
2. $\begin{array}{r} 16 \\ \times\ 5 \\ \hline 80 \end{array}$
3. $\begin{array}{r} 34 \\ \times\ 2 \\ \hline 68 \end{array}$
4. $\begin{array}{r} 19 \\ \times\ 3 \\ \hline 57 \end{array}$
5. $\begin{array}{r} 27 \\ \times\ 3 \\ \hline 81 \end{array}$

6. $\begin{array}{r} 24 \\ \times\ 4 \\ \hline 96 \end{array}$
7. $\begin{array}{r} 17 \\ \times\ 5 \\ \hline 85 \end{array}$
8. $\begin{array}{r} 23 \\ \times\ 3 \\ \hline 69 \end{array}$
9. $\begin{array}{r} 23 \\ \times\ 4 \\ \hline 92 \end{array}$
10. $\begin{array}{r} 46 \\ \times\ 2 \\ \hline 92 \end{array}$

11. $\begin{array}{r} 12 \\ \times\ 4 \\ \hline 48 \end{array}$
12. $\begin{array}{r} 29 \\ \times\ 3 \\ \hline 87 \end{array}$
13. $\begin{array}{r} 17 \\ \times\ 4 \\ \hline 68 \end{array}$
14. $\begin{array}{r} 12 \\ \times\ 8 \\ \hline 96 \end{array}$
15. $\begin{array}{r} 13 \\ \times\ 7 \\ \hline 91 \end{array}$

Copy and Do

16. 2×36 — 72
17. 18×4 — 72
18. 26×3 — 78
19. 15×4 — 60
20. 28×2 — 56

21. 12×7 — 84
22. 2×43 — 86
23. 38×2 — 76
24. 24×3 — 72
25. 16×3 — 48

26. 2×19 — 38
27. 5×12 — 60
28. 18×3 — 54
29. 3×29 — 87
30. 4×19 — 76

Apply

Solve these problems.

31. A tablet of colored paper costs 29¢. How much will 3 tablets cost?
87¢

32. One pair of jeans costs $23.45. A shirt costs $16.79. The jeans cost how much more than a shirt?
$6.66

33. A tape costs $5. How much will 16 tapes cost?
$80

34. Mark has 47 cents. Jan has twice as much as Mark. How much does Jan have?
94¢

110

More Multiplying, Trading Ones

pages 111-112

Objective

To multiply 2-digit numbers by 1-digit numbers with trading

Materials

*place value materials

Mental Math

Ask students to name the number that tells:

1. days in 5 weeks. (35)
2. weeks in a decade. (520)
3. milliliters in 10 ½ liters. (10,500)
4. hours in 3 days. (72)
5. tenths in 1 whole. (10)
6. years in a millennium. (1,000)
7. months in 9 years. (108)
8. quarts in 50 gallons. (200)

Skill Review

Write on the board **16 tens + 28 ones** and have students tell how to trade as they find the total tens and ones and the sum. (188) Repeat more examples for sums less than 1,000.

Multiplying 2-Digit Numbers

Paul is stocking the produce bins of his grocery store. He wants to make 6 equal columns of tomatoes. How many tomatoes can Paul display?

We need to find how many tomatoes Paul will display.

There will be ___24___ tomatoes in each column.

Paul will pack ___6___ columns of tomatoes.

To find the number of tomatoes Paul can pack, we multiply ___24___ by ___6___.

Multiply the ones. Trade if needed.

$$\begin{array}{r} \overset{2}{2}4 \\ \times\ 6 \\ \hline 4 \end{array}$$

Multiply the tens. Add any extra tens.

$$\begin{array}{r} \overset{2}{2}4 \\ \times\ 6 \\ \hline 144 \end{array}$$

Paul can pack ___144___ tomatoes.

Getting Started

Multiply.

1. $\begin{array}{r} 25 \\ \times\ 7 \\ \hline 175 \end{array}$
2. $\begin{array}{r} 32 \\ \times\ 9 \\ \hline 288 \end{array}$
3. $\begin{array}{r} 50 \\ \times\ 6 \\ \hline 300 \end{array}$
4. $\begin{array}{r} 49 \\ \times\ 8 \\ \hline 392 \end{array}$

Copy and multiply.

5. 62×8
   496
6. 27×4
   108
7. 8×36
   288

111

Teaching the Lesson

Introducing the Problem Have students describe the picture. Have a student read the problem and tell what is to be found. (the total number of tomatoes to be displayed) Ask what information is given in the problem and the picture. (There are 24 in each column and there will be 6 columns.) Have students complete the sentences and guide them through the multiplication in the model. Tell students they may want to think of the total pennies in 6 quarters to estimate the product of 6 twenty-fours to check their solution. Have students complete the solution sentence.

Developing the Skill Write **20 × 6** vertically on the board and ask students to tell the product. (120) Write **120** and remind students that 12 tens is the same as 1 hundred 2 tens. Now write **28 × 6** vertically on the board and ask students the product of 6 times 8. (48) Remind students that the 4 tens in 48 are extra tens and need to be added to the total tens in 6 times 2 tens. Write **16** in the answer and remind students that 16 tens equal 1 hundred 6 tens. Repeat for 40 × 8 and 49 × 8.

Practice

Multiply.

1. 26 × 5 130	2. 64 × 6 384	3. 39 × 3 117	4. 29 × 2 58	5. 48 × 4 192
6. 53 × 8 424	7. 96 × 2 192	8. 88 × 3 264	9. 21 × 9 189	10. 32 × 7 224
11. 73 × 4 292	12. 80 × 9 720	13. 98 × 2 196	14. 81 × 8 648	15. 42 × 5 210

Copy and Do

16. 2 × 56
112
17. 3 × 97
291
18. 7 × 23
161
19. 4 × 59
236
20. 6 × 19
114

21. 5 × 73
365
22. 7 × 24
168
23. 6 × 28
168
24. 2 × 78
156
25. 9 × 19
171

26. 8 × 88
704
27. 5 × 68
340
28. 6 × 37
222
29. 3 × 46
138
30. 7 × 35
245

Apply

Solve these problems.

31. The Petersons drove 317 miles the first day of their vacation. They drove 287 miles the second day. How far did the Petersons drive in two days?
604 miles

32. A washing machine uses 18 gallons of water for each load. How many gallons of water are used for 6 loads of wash?
108 gallons

33. Each tablet contains 32 pieces of paper. How many tablets do you need to buy to get at least 100 pieces of paper?
4 tablets

34. Each glass holds 6 ounces of juice. How many ounces of juice is needed to fill 48 glasses?
288 ounces

112

Correcting Common Errors

Students may add the regrouped digit to the tens before they multiply.

INCORRECT   CORRECT

4 27 × 6 362	4 27 × 6 162

Have students use a place-value form like the following to chart the steps.

	100's	10's	1's
6 × 7 ones		4	2
6 × 2 tens	1	2	0
	1	6	2

Enrichment

Have students make a table showing the number of rows needed to seat 400 people if: each row had 8, or 10, or 25 or 50 seats.

Practice

Remind students they may not need to trade in every problem. Have students complete the page independently.

Extra Credit   *Statistics*

Have students keep a record of how many minutes they spend eating, sleeping, playing and watching television each day, for one week. At the end of the week have them convert the minutes to hours and make a bar graph showing the time spent at each of these activities. Display the bar graphs, and have students compare their daily routines. As an extension, have students find the average time they spent, per day, on each activity.

Multiplying Money

pages 113-114

Objective

To multiply money

Materials

*dollars, dimes, pennies

Mental Math

Dictate the following:
1. 68 × 2 (136)
2. 68 ÷ 2 (34)
3. (68 ÷ 2) × 2 (68)
4. (68 × 2) ÷ 2 (68)
5. (68 ÷ 2) ÷ 2 (17)
6. (68 × 2) × 2 (272)
7. 1/2 of 68 (34)
8. 2 × (1/2 of 68) (68)

Skill Review

Pair students at the board. Ask one student to write the number 236 in tens and ones and the second student to write the same number in dimes and pennies. Have students check their partner's work and then change roles for another problem. Continue until students can readily name 2- and 3-digit numbers in tens and ones and in dimes and pennies.

Multiplying Money

Eric is buying 7 folders at the school store. How much does this cost him?

We want to know the cost of Eric's purchases.

He bought __7__ folders.

Each folder cost __34¢__.

We find the total cost by multiplying

__34¢__ by __7__.

✔ Remember that $0.34 is another way of writing 34¢.

Multiply the pennies. Trade if needed.	Multiply the dimes. Add any extra dimes. Place the dollar sign and decimal point.

$$\begin{array}{r} {}^{2} \\ \$0.3\overset{4}{} \\ \times 7 \\ \hline 8 \end{array} \qquad \begin{array}{r} {}^{2} \\ \$0.\overset{3}{}4 \\ \times 7 \\ \hline \$2.38 \end{array}$$

Eric must pay __$2.38__ to the school store.

Getting Started

Multiply.

1. $0.42
    ×   6
    $2.52

2. $0.87
    ×   3
    $2.61

3. $0.38
    ×   9
    $3.42

4. $0.45
    ×   5
    $2.25

Solve these problems using the information above.

5. Robin buys 5 pencils and 3 erasers. How much does Robin spend?
85¢

6. How much more are 3 pennants than 4 tablets?
81¢

113

Teaching the Lesson

Introducing the Problem Have a student describe the picture. Have a student read the problem and tell what needs to be found solved. (the cost of 7 folders) Ask students if all the necessary information is given in the problem. (no) Ask students what information is needed from the picture. (A folder costs 34¢.) Have students complete the sentences and guide them through the multiplication in the model. Remind them to place the dollar sign and decimal point. Ask students how they might check their solution to see if it makes sense. (7 folders at 30¢ each would be $2.10, and 7 more 4¢ amounts would be 28¢.)

Developing the Skill Write **80 × 4** vertically on the board and have a student find the product. (320) Place dollar signs and decimal points in the problem to show money. Ask how many dollars, dimes and pennies in $3.20. (3, 2, 0) Now write **$0.79 × 6** vertically on the board and have a student talk through the problem, showing the trades, to find the product. ($4.74) Ask students to tell the number of dollars, dimes and pennies in the answer. (4, 7, 4) Repeat for $0.89 × 7.

Practice

Multiply.

1. $0.63
× 3
———
$1.89

2. $0.72
× 8
———
$5.76

3. $0.16
× 4
———
$0.64

4. $0.49
× 2
———
$0.98

5. $0.84
× 3
———
$2.52

6. $0.67
× 7
———
$4.69

7. $0.89
× 9
———
$8.01

8. $0.85
× 6
———
$5.10

9. $0.53
× 6
———
$3.18

10. $0.09
× 8
———
$0.72

11. $0.75
× 7
———
$5.25

12. $0.67
× 9
———
$6.03

Apply

Solve these problems.

13. Adam buys a folder and a pen. He gives the storekeeper a dollar bill. How much change should Adam receive?
7¢

14. Nancy has $4.25. She wants to buy 7 pennants and 3 tablets. How much more money does Nancy need?
$2.45

15. Todd buys 4 pencils and 5 folders. He gives the clerk a five dollar bill. How much change should Todd receive?
$2.98

16. Mrs. Lopez decides to use the rapid train to go to and from work for 4 days. How much will she save if she usually takes the bus?
$1.20

SCHOOL SUPPLIES

PENCILS 8¢ ERASERS 15¢

PENS 59¢ TABLETS 39¢

PENNANTS 79¢ FOLDERS 34¢

COMMUTER SPECIALS

CITY BUS 75¢ ONE WAY RAPID TRAIN 60¢ ONE WAY

114

Practice

Have students work the first 2 rows of problems independently. Help students develop a plan for each of the word problems and then have students complete the DO and CHECK stages independently.

Mixed Practice

1. 18,027 − 6,510 (11,517)
2. 6 × 43 (258)
3. (3 × 7) + (4 × 9) (57)
4. 32,158 + 19,765 (51,923)
5. $728.95 − 374.27 ($354.68)
6. 27 + 3 × 8 (51)
7. $.87 × 7 ($6.09)
8. 361 + 408 + 294 (1,063)
9. 13 × 8 (104)
10. 7,656 − 4,293 (3,363)

Extra Credit   *Measurement*

Conduct a Class Olympics. Have students create and write rules for several skill games that could be conducted inside the classroom, and whose results could be measured by length. Encourage the group to be creative in choosing appropriate events. Examples might be building a tower of milk cartons, an eraser toss, or a broom jump. Have students measure results in inches, feet, yards and meters and record them on a classroom chart. Continue the Olympics for several days, changing the activities daily and recording all data.

Problem Solving, Make a Table

pages 115-116

Objective

To make a table to solve problems

Materials

Mental Math

Tell students to multiply by 6, then divide by 3:

1. 10 (20)
2. 15 (30)
3. 200 (400)
4. 11 (22)
5. 70 (140)
6. 22 (44)
7. 13 (26)
8. 40 (80)

Make a Table

Our class timed some students walking down the hallway. They walked 6 meters in 9 seconds. At this rate how far would they walk in 81 seconds?

★ **SEE**

We want to find out how far the students will walk in 81 seconds.

They walk __6__ meters in __9__ seconds.

★ **PLAN**

We can make a table comparing the number of seconds to the distance the students walk.

★ **DO**

Seconds	9	18	27	36	45	54	63	72	81
Distance (in meters)	6	12	18	24	30	36	42	48	54

The students will walk __54__ meters in 81 seconds.

★ **CHECK**

We know the students walk 6 meters in 9 seconds.

We know that in 81 seconds there are 9 groups of 9 seconds each.

If we multiply 6 meters by the number of groups, we will get __54__ meters.

115

Teaching the Lesson

Remind students that a table is an organized way to see patterns in data. Tell students that rows or columns of a table must be labeled so that anyone else looking at it can clearly see the information.

Have a student read the problem and the SEE and PLAN stages. Tell students there are several ways to solve this problem, but a table will help them see the pattern as they compare the distance students walk to the amount of time they walk. Have students complete the table to solve the problem. Discuss why they needed to count by 9's in the Seconds column and by 6's in the Distance column. Help students see that a table can be extended to find other information. Have students find the distance walked in 135 seconds or 900 seconds. Have students find the time it would take to walk 90, 150 or 300 meters. Have students complete the check stage of the problem.

Apply

Make a table to help you solve these problems.

1. Laura swims 7 laps on the first day of swim practice. She increases her distance by 2 laps each day. On the last day of practice she swims 25 laps. How many days does practice last?
10 days

2. It costs $0.29 to mail a letter and $0.19 to mail a postcard. Betty wrote to 11 friends and spent $2.89 on postage. How many letters and postcards did she write?
8 letters, 3 postcards

3. While looking out the window, I saw some boys and some dogs walk by the house. I counted 22 heads and 68 legs. How many boys and how many dogs passed by?
12 dogs; 10 boys

4. Make a table of at least 4 other length and width measures for a 36-square unit rectangle.
See Solution Notes.

5. The fourth graders are collecting aluminum cans to be recycled. They receive $0.15 a pound for the cans. It takes 25 cans to make a pound. How many empty cans will they need to collect to earn $1.35?
225 cans

6. To produce a special shade of paint, we use 4 drops of dark green to 1 drop of white to 5 drops of light green. If we use 20 drops of light green, how many drops of dark green and how many drops of white must we use?
16 drops of dark green, 4 drops of white

7. Read Exercise 3 again. What if I saw 23 heads and 70 legs? Now how many boys and how many dogs passed by?
11 boys, 12 dogs

8. Peppy the puppy looked in the pond and saw 3 times as many gold fish as blue fish and 2 times as many silver fish as blue fish. Does Peppy see more gold fish or more silver fish? Explain how you know.
More gold fish

116

Solution Notes

1.

Day	1	2	3	. . .	10
Laps	7	9	11	. . .	25

2. The table will vary depending on whether students begin with 1 letter and 10 postcards or 1 postcard and 10 letters.

Cards	Cost	Letters	Cost	Total
10	$1.90	1	$ .29	$2.19
9	$1.71	2	$ .58	$2.29
.	.	.	.	.
.	.	.	.	.
3	$ .57	8	$2.32	$2.89
2	$ .38	9	$2.61	$2.99
1	$ .19	10	$2.90	$3.04

3.

Dogs	Boys	Heads	Legs
11	11	22	66
10	12	22	64
12	10	22	68

4. Have students think of all factors of 36. Discuss how 1×36 and 36×1 rectangles differ.

Length	4	1	36	18	2	3	12	6
Width	9	36	1	2	18	12	3	6
Area	36	36	36	36	36	36	36	36

5.

Cans	Pounds	Money Earned
25	1	$ .15
50	2	$ .30
75	3	$ .45
100	4	$ .60
200	8	$1.20
225	9	$1.35

6. Students will see how the numbers for 3 items change proportionally.

light green	5	10	15	20
green	4	8	12	16
white	1	2	3	4

Higher-Order Thinking Skills

7. [Analysis] There is no need to do the whole problem over again; 1 more head and 2 more legs means that 1 more boy passed by.

8. [Analysis] Regardless of how many blue fish Peppy sees, 3 times the number is always more than 2 times the number.

116

Calculator Multiplication

pages 117-118

Objective

To use a calculator to multiply

Materials

calculators

Mental Math

Have students work the following on their calculators:

1. $16 \times 2 \times 0$ (0)
2. $16 \times 2 \times 1$ (32)
3. $2,000 + 200 - 800$ (1,400)
4. 5×111 (555)
5. $486 - 6$ tens $+ 2$ hundreds (626)
6. $37,260 - 1,001$ (36,259)
7. $250 \times 10 - 0$ (2,500)
8. $9 \times 9 \times 2$ (162)

Skill Review

Have students find the total of $110 + 16 - 21$ on their calculators. (105) Give students more addition and subtraction problems to work on their calculators. Give some problems of money amounts and have students write the calculator codes for each. Be sure students remember the use of the decimal point.

Calculators, the Multiplication Key

Mary is making a table to show how many eggs there are in any number of cartons. How many eggs are there in 6 cartons?

DOZENS	1	2	3	4	5
EGGS	12	24	36	48	60

Mary found three ways to find the number of eggs in 6 cartons. Complete the codes to see what Mary found.

The addition code

12 [+] 12 [+] 12 [+] 12 [+] 12 [+] 12 [=] (72)

The equals code

12 [+] 12 [=] [=] [=] [=] [=] (72)

The multiplication code

12 [×] 6 [=] (72)

There are __72__ eggs in 6 cartons.

Complete these calculator codes.

45 [+] 45 [+] 45 [+] 45 [=] (180)

45 [+] 45 [=] [=] [=] (180)

45 [×] 4 [=] (180)

117

Teaching the Lesson

Introducing the Problem Have a student read the problem and tell what is to be solved. (number of eggs in 6 cartons) Ask students if all the necessary information is given in the problem. (no) Ask students what information is needed yet. (number of eggs in each carton) Tell students to look at the table in the picture, and ask them what key word tells the number of eggs in each carton. (dozens) Ask how many are in a dozen. (12) Ask students how they would find the number of eggs in 2 dozen. (multiply 12×2) Have students complete the three different codes with you to solve the problem. Have students check their work by thinking of $(6 \times 10) + (6 \times 2)$. Have students then complete the calculator codes at the bottom of the page.

Developing the Skill Have students clear their calculator screens. Tell students to enter 10 on their calculators and press the + key, enter 10 again and press + again. Have students add 2 more 10's and press the = key and tell what 4×10 equals. (40) Tell students this is called an **addition code** as you write $10 + 10 + 10 + 10 =$ on the board. Tell students another way to find what 4 times 10 equals is to press 10, the × key and 4 and then the = key. Have students clear their screens and work the multiplication code of 10×4. (40) Have students clear their screens again to work the same problem in yet another way. Write $10 + 10 =$ on the board. Tell students to enter the code and tell the answer. (40) Write 21×6 on the board. Have students write the 3 calculator codes to solve this on the board and then use their calculators to work the codes.

Practice

Enter these codes. Show the results on the screens.

1. 25 [+] 25 [+] 25 [=] (75)   2. 3 [×] 25 [=] (75)

3. 49 [+] 49 [=] [=] [=] (196)   4. 49 [×] 4 [=] (196)

5. 3 [×] 5 [×] 6 [=] (90)   6. 115 [−] 37 [×] 9 [=] (702)

Use your calculator to find each product.

7. 37 × 9	8. 28 × 7	9. 34 × 8	10. 79 × 6
333	196	272	474

11. 84 × 7	12. 52 × 4	13. $0.85 × 8	14. $0.96 × 5
588	208	$6.80	$4.80

EXCURSION

Complete these calculator codes.
What conclusion can you make from the answers?

1 [+] 3 [=] (4)   2 [×] 2 [=] (4)

1 [+] 3 [+] 5 [=] (9)   3 [×] 3 [=] (9)

1 [+] 3 [+] 5 [+] 7 [=] (16)   4 [×] 4 [=] (16)

1 [+] 3 [+] 5 [+] 7 [+] 9 [=] (25)   5 [×] 5 [=] (25)

1 [+] 3 [+] 5 [+] 7 [+] 9 [+] 11 [=] (36)   6 [×] 6 [=] (36)

1 [+] 3 [+] 5 [+] 7 [+] 9 [+] 11 [+] 13 [=] (49)

                                 7 [×] 7 [=] (49)

1 [+] 3 [+] 5 [+] 7 [+] 9 [+] 11 [+] 13 [+] 15 [=] (64)

                                 8 [×] 8 [=] (64)

Conclusion: Square numbers are the sum of consecutive odd numbers beginning

with one.

118

Correcting Common Errors

Students may have difficulty understanding how to multiply different ways. Have them work in groups of three to multiply 6 × 14, where one student uses the addition code, another the equals code, and the third the multiplication code. Then they should compare their answers to see that, regardless of the code used, 6 × 14 = 84. Have them do more problems so that students gain practice with all three codes.

Enrichment

Tell students that each class in their school has 25 students. Ask them to write and work on their calculators the addition, equals and multiplication codes to find an estimated total number of students in the whole school.

Practice

Ask students what key must be pressed before each new problem. (C) Remind students that in money problems a 7.2 answer on the calculator screen would be $7.20 since zeros to the far right of the decimal point are not entered nor shown in answers. Have students complete the problems independently.

Excursion

Tell students that when they think they have discovered the pattern (The sum of the addends is equal to the product of the number of addends times itself.), they should do the operations without using the calculator. Another way to test the pattern is to extend it beyond 9 × 9.

Extra Credit   *Logic*

Read or duplicate the following for the students to solve: If your doctor gives you three pills, and instructs you to take one every half hour, how long will your pills last? (One hour. You take one now, one a half hour from now and one a half hour later. Total time lapse is 1 hour)

Chapter Test

page 119

Item	Objective
1-3, 5	Recall basic multiplication facts (See page 104)
6-11	Solve for n while following rule of order (See pages 89-90)
12-15	Multiply 2-digit number by 1-digit number without trading (See pages 107-108)
4, 16-19	Multiply 2-digit number by 1-digit number with one regrouping (See pages 109-110)
20-23	Multiply 2-digit number by 1-digit number with two regroupings (See pages 111-112)

Write the first nine multiples of each of these numbers.

1. 3 $\underline{0}$, $\underline{3}$, $\underline{6}$, $\underline{9}$, $\underline{12}$, $\underline{15}$, $\underline{18}$, $\underline{21}$, $\underline{24}$

2. 7 $\underline{0}$, $\underline{7}$, $\underline{14}$, $\underline{21}$, $\underline{28}$, $\underline{35}$, $\underline{42}$, $\underline{49}$, $\underline{56}$

3. 9 $\underline{0}$, $\underline{9}$, $\underline{18}$, $\underline{27}$, $\underline{36}$, $\underline{45}$, $\underline{54}$, $\underline{63}$, $\underline{72}$

Skip-count by 8.

4. 64, $\underline{72}$, $\underline{80}$, $\underline{88}$, $\underline{96}$, $\underline{104}$, $\underline{112}$, $\underline{120}$, 128

Skip-count by 4.

5. 12, $\underline{16}$, $\underline{20}$, $\underline{24}$, $\underline{28}$, $\underline{32}$, $\underline{36}$, $\underline{40}$, 44

Solve for n.

6. $6 \times (0 \times 3) = n$ 7. $8 \times 10 = n$ 8. $(7 \times 2) + (7 \times 3) = n$

$n = \underline{0}$ $n = \underline{80}$ $n = \underline{35}$

9. $9 \times (4 \times 2) = n$ 10. $(5 \times 0) + (5 \times 0) = n$ 11. $10 \times 9 = n$

$n = \underline{72}$ $n = \underline{0}$ $n = \underline{90}$

Multiply.

12.
$$\begin{array}{r} 21 \\ \times\ 3 \\ \hline 63 \end{array}$$

13.
$$\begin{array}{r} 11 \\ \times\ 7 \\ \hline 77 \end{array}$$

14.
$$\begin{array}{r} 42 \\ \times\ 2 \\ \hline 84 \end{array}$$

15.
$$\begin{array}{r} 12 \\ \times\ 4 \\ \hline 48 \end{array}$$

16.
$$\begin{array}{r} 36 \\ \times\ 2 \\ \hline 72 \end{array}$$

17.
$$\begin{array}{r} 29 \\ \times\ 3 \\ \hline 87 \end{array}$$

18.
$$\begin{array}{r} 16 \\ \times\ 4 \\ \hline 64 \end{array}$$

19.
$$\begin{array}{r} 47 \\ \times\ 2 \\ \hline 94 \end{array}$$

20.
$$\begin{array}{r} \$0.37 \\ \times\ 6 \\ \hline \$2.22 \end{array}$$

21.
$$\begin{array}{r} \$0.49 \\ \times\ 4 \\ \hline \$1.96 \end{array}$$

22.
$$\begin{array}{r} \$0.73 \\ \times\ 9 \\ \hline \$6.57 \end{array}$$

23.
$$\begin{array}{r} \$0.87 \\ \times\ 8 \\ \hline \$6.96 \end{array}$$

119

Circle the letter of the correct answer.

1 What is the value of the 5 in 4,576?
- a thousands
- (b) hundreds
- c tens
- d NG

2 5,219 ◯ 5,129
- a <
- (b) >

3 Round 735 to the nearest hundred.
- (a) 700
- b 800
- c NG

4 Round 7,295 to the nearest thousand.
- a 8,000
- (b) 7,000
- c NG

5 What is the value of the 0 in 302,926.
- a tens
- b hundreds
- c thousands
- (d) NG

6 49 + 73
- a 23
- b 112
- c 1,112
- (d) NG

7 659
+ 268
- a 817
- b 917
- (c) 927
- d NG

8 3,629
+ 4,381
- a 7,000
- b 8,000
- (c) 8,010
- d NG

9 54,275
+ 27,196
- a 71,371
- b 81,461
- (c) 81,471
- d NG

10 51 − 26
- (a) 25
- b 36
- c 45
- d NG

11 817
− 298
- (a) 519
- b 619
- c 681
- d NG

12 61,083
− 28,596
- (a) 32,487
- b 32,587
- c 47,513
- d NG

13 24 × 4
- a 86
- (b) 96
- c 816
- d NG

◻ score

120

Cumulative Review

page 120

Item	Objective
1	Identify place value through thousands (See pages 27-28)
2	Compare, order numbers through thousands (See pages 29-30)
3	Round numbers to nearest 100 (See pages 31-34)
4	Round numbers to nearest 1,000 (See pages 33-34)
5	Identify place value through hundred thousands (See pages 37-38)
6	Add two 2-digit numbers (See pages 43-44)
7	Add two 3-digit numbers (See pages 45-46)
8	Add two 4-digit numbers (See pages 47-48)
9	Add two 5-digit numbers (See pages 53-54)
10	Subtract two 2-digit numbers (See pages 61-62)
11	Subtract two 3-digit numbers (See pages 63-64)
12	Subtract two 5-digit numbers with zero in minuend (See pages 67-68)
13	Multiply 2-digit number by 1-digit number with two regroupings (See pages 111-112)

Alternate Cumulative Review

Circle the letter of the correct answer.

1 What is the value of the 3 in 3,691?
- (a) thousands
- b hundreds
- c tens
- d NG

2 7,364 ◯ 7,435
- a >
- (b) <
- c =

3 Round 472 to the nearest hundred.
- a 400
- (b) 500
- c NG

4 Round 9,327 to the nearest thousand.
- a 8,000
- (b) 9,000
- c 10,000
- d NG

5 What is the value of the 6 in 625,430?
- a hundreds
- b thousands
- c ten thousands
- (d) NG

6 65 + 78 =
- a 1313
- (b) 143
- c 133
- d NG

7 572
+ 368
- a 830
- b 840
- c 930
- (d) NG

8 6,947
+ 2,164
- a 8,012
- (b) 9,111
- c 9,011
- d NG

9 37,621
+ 45,489
- a 72,111
- b 82,110
- (c) 83,110
- d NG

10 86 − 49 =
- (a) 37
- b 43
- c 47
- d NG

11 925
− 578
- a 453
- (b) 347
- c 456
- d NG

12 73,071
− 34,294
- a 41,223
- b 39,877
- c 38,877
- (d) NG

13 34
× 3
- a 92
- (b) 102
- c 67
- d NG

Dividing By 2 or 3

pages 121-122

Objective

To review division facts with 2 or 3 as the divisor

Materials

books and other objects

Mental Math

Have students complete each comparison: Quartet is to 4 as:
1. triplet is to ___. (3)
2. unison is to ___. (1)
3. hexagonal is to ___. (6)
4. bi-weekly is to ___. (2)
5. baker's dozen is to ___. (13)
6. couple is to ___. (2)
7. gross is to ___. (144)

Skill Review

Draw 9 groups of 2 X's on the board. Ask how many groups of 2. (9) Ask students how many X's in all. (18) Mark through 1 group of X's as you write one tally mark on the board and tell students you've subtracted 1 group of 2. Ask how many X's are left. (16) Continue to mark through one group at a time, record a tally mark and ask students how many X's are left. When all X's are subtracted, ask students to count the tally marks to tell how many groups of 2 X's were subtracted in all. (9) Repeat for 8 groups of 3 X's.

Dividing by 2 or 3

Lori is fixing lunches and is putting 2 cookies in each lunch box. How many lunch boxes can she supply with cookies?

We want to know how many lunch boxes Lori can supply with 2 cookies each.

We know she has __12__ cookies to share.

There are __2__ cookies for each lunch.
To find how many lunches can be supplied with cookies, we

divide __12__ by __2__.

In all	In each	Boxes
12	÷ 2	= 6

$$12 \div 2 = 6 \quad \text{or} \quad 2\overline{)12}$$

dividend   divisor   quotient   divisor

6 quotient
2)12 dividend

Lori can supply 2 cookies each to __6__ lunches.

Getting Started

Divide.

1. $6 \div 3 =$ __2__   2. $14 \div 2 =$ __7__   3. $8 \div 2 =$ __4__

4. $24 \div 3 =$ __8__   5. $15 \div 3 =$ __5__   6. $6 \div 2 =$ __3__

7. $2\overline{)16}$ __8__   8. $3\overline{)27}$ __9__   9. $3\overline{)18}$ __6__

121

Teaching the Lesson

Introducing the Problem Have a student read the problem and tell what is asked. (how many lunches can have 2 cookies) Ask students if all the information they will need is given in the problem. (no) Ask what information is needed. (number of cookies in all) Have students read and complete the information sentences. Point out the terms used in the model as students complete the solution sentence. Tell students to check their solution by asking what number times 2 equals 12.

Developing the Skill Tell students that division is related to multiplication as you write $3 \times 6 = 18$ on the board. Tell students that if the product 18 is **divided** by 6 the answer will always be 3 because $3 \times 6 = 18$. Repeat for 18 divided by 3. Draw 18 X's on the board and ask students to circle each group of 3. Draw 18 X's again and repeat for groups of 6. Write $18 \div 3 = 6$ and $18 \div 6 = 3$ on the board and tell students there is a name for each part of a division problem as you write the words **dividend, divisor** and **quotient** under the respective number in each problem. Now write $3\overline{)18}$ and $6\overline{)18}$ on the board as you tell students that this is another way to write a division problem. Tell students that to check a division problem using multiplication, the divisor times the quotient must equal the dividend. Provide additional division problems.

Practice

Divide.

1. $10 \div 2 = \underline{5}$ 2. $12 \div 3 = \underline{4}$ 3. $24 \div 3 = \underline{8}$ 4. $18 \div 2 = \underline{9}$

5. $21 \div 3 = \underline{7}$ 6. $18 \div 3 = \underline{6}$ 7. $27 \div 3 = \underline{9}$ 8. $15 \div 3 = \underline{5}$

9. $3\overline{)12}$ — 4   10. $3\overline{)6}$ — 2   11. $2\overline{)14}$ — 7   12. $3\overline{)27}$ — 9

13. $2\overline{)16}$ — 8   14. $2\overline{)10}$ — 5   15. $3\overline{)9}$ — 3   16. $2\overline{)6}$ — 3

Apply

Solve these problems.

17. Joan and Angie each have 12 sweaters to store in boxes. Joan will store 3 sweaters in each box. Angie can only put 2 of her sweaters in a box. How many boxes will Joan and Angie both need for all their sweaters?
10 boxes

18. Pat is putting marbles into bags. He has 24 marbles and wants to put 3 marbles into each bag. How many bags will Pat need?
8 bags

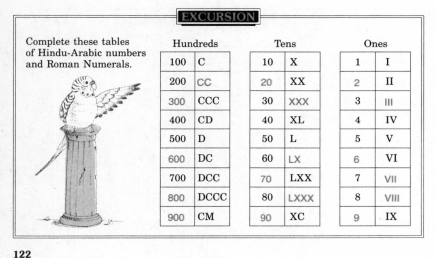
122

Correcting Common Errors

Some students may have difficulty finding the quotient when they are dividing by 2 or 3. For a division sentence such as $12 \div 3 = \square$, have students write a related multiplication sentence, $\square \times 3 = 12$, to help them see that 4 is the quotient. For more help, have students think $1 \times 3 = 3$, $2 \times 3 = 6$, $3 \times 3 = 9$, $4 \times 3 = 12$, until they find the correct factor.

Enrichment

Tell students to write a related division problem for each multiplication fact for the 2's and 3's.

Practice

Remind students to check each problem by multiplying. Tell students to make a plan to solve each of the word problems. Have students complete the page independently.

Excursion

Ask students to identify the patterns that exist between the three columns such as using three letters to represent the Roman numerals for 3, 30, and 300.
Have students write expanded notations such as 23 = 20 + 3, XXIII = XX + III, 719 = 700 + 10 + 9, and DCCXIX = DCC + X + IX.

Extra Credit *Applications*

Tell students they can find the day of the week for any given date, by using these codes:

January	1	July	0	Sunday	1
February	4	August	3	Monday	2
March	4	September	6	Tuesday	3
April	0	October	1	Wednesday	4
May	2	November	4	Thursday	5
June	5	December	6	Friday	6
				Saturday	0

Write these directions and a month, day and year on the board for students to use as an example.

1. Divide the last two digits of the year by 4, dropping remainders.
2. Add the code number of the month.
3. Add the day's date.
4. Divide by 7, keeping the remainder. Tell students this remainder is the code number for the day of the week, of the date they started with.

Dividing By 4 or 5

pages 123-124

Objective

To review division facts with 4 or 5 as the divisor

Materials

Mental Math

Have students name a fact to check:
1. 7×2 ($14 \div 2$, $14 \div 7$)
2. $24 \div 6$ (4×6)
3. $15 - 7$ ($7 + 8$)
4. $36 \div 4$ (9×4)
5. $17 - 9$ ($8 + 9$)
6. 1/2 of 16 (8×2, $16 \div 2$)
7. 1/5 fo 35 (7×5, $35 \div 5$)

Skill Review

Pair students up. Write **20** on the board and ask students to show as many different ways as they can to make 20 using addition or subtraction. ($18 + 2$, $25 - 5$, etc.) Give students 5 minutes and then have them list their solutions in columns on the board. Ask if there are other ways not yet given. (Yes, there are an infinite number of solutions since $2,010 - 1,980 = 20$, etc.)

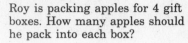

Dividing by 4 or 5

Roy is packing apples for 4 gift boxes. How many apples should he pack into each box?

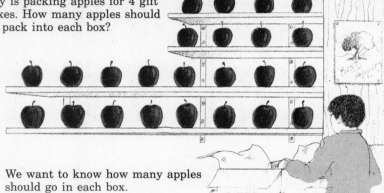

We want to know how many apples should go in each box.

There are ___24___ apples.

Roy is packing ___4___ boxes.

To find the number of apples for each box,

we divide ___24___ by ___4___.

In all		Boxes		In each

$$24 \div 4 = \underline{6} \quad \text{or} \quad 4\overline{)24} \begin{smallmatrix} 6 \text{ quotient} \\ \text{dividend} \end{smallmatrix}$$

dividend · divisor · quotient · · divisor

Roy should put ___6___ apples into each box.

Getting Started

Divide.

1. $36 \div 4 = \underline{9}$ 2. $45 \div 5 = \underline{9}$ 3. $16 \div 4 = \underline{4}$ 4. $20 \div 5 = \underline{4}$

5. $5\overline{)30}^{\,6}$ 6. $4\overline{)12}^{\,3}$ 7. $4\overline{)36}^{\,9}$ 8. $5\overline{)10}^{\,2}$

123

Teaching the Lesson

Introducing the Problem Have students read the problem and use the picture to tell what is to be solved. (number of apples in each box) Ask students what is known already. (24 apples are to go into 4 boxes.) Ask if there is any unnecessary information given. (no) Have students complete the sentences. Guide them through the division shown in the model, and have students complete the solution sentence. Ask students how they can use multiplication to check their solution. (The product of their solution and 4 should equal 24.)

Developing the Skill Have 16 students stand. Tell students we need to find a number which multiplied by 4 will equal 16. Ask students to think of a multiplication fact where 4 multiplied by another number equals 16. ($4 \times 4 = 16$) Have 4 students go to each of the 4 corners of the room and ask all students if $16 \div 4 = 4$. (yes) Now have the 16 students sit down. Tell students that another way to solve this problem is to place 1 student at a time in each of the 4 areas until all 16 students are moved. Have students act this out and check to see if 4 groups of 4 equal 16 in all. (yes) Repeat the activity for $45 \div 5$ and $28 \div 4$.

123

Practice

Divide.

1. $25 \div 5 = \underline{5}$ 2. $10 \div 5 = \underline{2}$ 3. $8 \div 4 = \underline{2}$ 4. $28 \div 4 = \underline{7}$

5. $24 \div 4 = \underline{6}$ 6. $18 \div 3 = \underline{6}$ 7. $20 \div 5 = \underline{4}$ 8. $15 \div 5 = \underline{3}$

9. $16 \div 4 = \underline{4}$ 10. $32 \div 4 = \underline{8}$ 11. $24 \div 3 = \underline{8}$ 12. $30 \div 5 = \underline{6}$

13. $35 \div 5 = \underline{7}$ 14. $12 \div 4 = \underline{3}$ 15. $20 \div 4 = \underline{5}$ 16. $36 \div 4 = \underline{9}$

17. $4\overline{)12}$ $\overset{3}{}$ 18. $5\overline{)25}$ $\overset{5}{}$ 19. $5\overline{)30}$ $\overset{6}{}$ 20. $2\overline{)18}$ $\overset{9}{}$

21. $4\overline{)20}$ $\overset{5}{}$ 22. $4\overline{)16}$ $\overset{4}{}$ 23. $3\overline{)27}$ $\overset{9}{}$ 24. $4\overline{)28}$ $\overset{7}{}$

25. $5\overline{)10}$ $\overset{2}{}$ 26. $5\overline{)20}$ $\overset{4}{}$ 27. $5\overline{)35}$ $\overset{7}{}$ 28. $4\overline{)8}$ $\overset{2}{}$

29. $2\overline{)12}$ $\overset{6}{}$ 30. $4\overline{)24}$ $\overset{6}{}$ 31. $5\overline{)45}$ $\overset{9}{}$ 32. $5\overline{)40}$ $\overset{8}{}$

Apply

Solve these problems.

33. There are 24 children in Miss Chen's class. The children sit at 4 tables. How many children are at each table?
6 children

34. There are 20 children playing soccer. There are 5 teams. How many children are on each team?
4 children

35. There are 28 children in Mr. Orr's class. 4 children were absent on Tuesday. How many children were present in Mr. Orr's class?
24 children

36. Kay is sewing 5 buttons on each blouse. On Friday, Kay sewed on 30 buttons. How many blouses did Kay put buttons on?
6 blouses

124

124

Dividing By 6 or 7

pages 125-126

Objective

To review division facts with 6 or 7 as the divisor

Materials

multiplication facts for 2's through 7's

Mental Math

Have students name what two different 1-digit numbers have a:

1. sum = to 15 + 2. (8 and 9)
2. quotient of 4. (8 and 2, 4 and 1)
3. product = to 60 + 3. (7 and 9)
4. sum = to 30 ÷ 2. (7 and 8)
5. difference = to 81 ÷ 9. (9 and 0)
6. product = to 15 × 2. (6 and 5)
7. quotient = to 15 ÷ 3. (5 and 1)
8. product > 70. (9 and 8)

Skill Review

Write **4 × 5 = 20** on the board and have students write a related division problem. (20 ÷ 4 = 5 or 20 ÷ 5 = 4) Continue for other multiplication facts of 2's, 3's, 4's or 5's.

Dividing by 6 or 7

Mr. Lopez spent $24 on concert tickets for his family. How many tickets did he buy?

We want to know the number of tickets Mr. Lopez bought.

He spent __$24__ on all the tickets.

Each ticket cost __$6__.
To find the number of tickets,

we divide __$24__ by __$6__.

| ___ × 6 = 24? | 6 × ___ = 24? |

$24 \div 6 =$ __4__ or $6\overline{)24}$ (quotient 4)

Mr. Lopez bought __4__ tickets.

Getting Started

Divide.

1. $54 \div 6 =$ __9__
2. $28 \div 7 =$ __4__
3. $42 \div 6 =$ __7__
4. $14 \div 7 =$ __2__
5. $56 \div 7 =$ __8__
6. $18 \div 6 =$ __3__
7. $30 \div 6 =$ __5__
8. $42 \div 7 =$ __6__

9. $6\overline{)12}$ (2)
10. $6\overline{)24}$ (4)
11. $7\overline{)21}$ (3)
12. $7\overline{)35}$ (5)

13. $7\overline{)49}$ (7)
14. $6\overline{)36}$ (6)
15. $7\overline{)63}$ (9)
16. $6\overline{)48}$ (8)

125

Teaching the Lesson

Introducing the Problem Have students describe the picture and read the problem. Ask students what is to be found. (number of tickets Mr. Lopez bought) Ask what information will be used to solve the problem. (He spent $24 and 1 ticket costs $6.) Have students read through the plan, complete the sentences and solve the problem. Have students check their solutions by multiplying.

Developing the Skill Write **6 × ___ = 54** on the board. Have a student complete the problem. (9) Tell students we can also write this as a division problem as you write **54 ÷ 6 =** on the board. Have a student complete the problem. (9) Have students think of a multiplication fact for the 6's or 7's and write the division problem for the fact on the board. Continue until all facts for 6's and 7's have been written.

Practice

Divide.

1. $24 \div 3 = \underline{8}$ 2. $28 \div 4 = \underline{7}$ 3. $35 \div 7 = \underline{5}$ 4. $30 \div 6 = \underline{5}$

5. $12 \div 6 = \underline{2}$ 6. $16 \div 2 = \underline{8}$ 7. $14 \div 7 = \underline{2}$ 8. $56 \div 7 = \underline{8}$

9. $54 \div 6 = \underline{9}$ 10. $63 \div 7 = \underline{9}$ 11. $24 \div 6 = \underline{4}$ 12. $28 \div 7 = \underline{4}$

13. $18 \div 6 = \underline{3}$ 14. $28 \div 4 = \underline{7}$ 15. $42 \div 7 = \underline{6}$ 16. $36 \div 6 = \underline{6}$

17. $6\overline{)30}$ 5   18. $6\overline{)36}$ 6   19. $7\overline{)21}$ 3   20. $7\overline{)49}$ 7

21. $6\overline{)24}$ 4   22. $7\overline{)42}$ 6   23. $6\overline{)18}$ 3   24. $7\overline{)28}$ 4

25. $7\overline{)63}$ 9   26. $5\overline{)40}$ 8   27. $6\overline{)54}$ 9   28. $6\overline{)48}$ 8

29. $7\overline{)35}$ 5   30. $4\overline{)32}$ 8   31. $6\overline{)42}$ 7   32. $7\overline{)56}$ 8

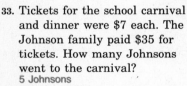

Apply

Solve these problems.

33. Tickets for the school carnival and dinner were $7 each. The Johnson family paid $35 for tickets. How many Johnsons went to the carnival?
5 Johnsons

34. Ruth bought 6 albums for $8 each. Randy bought 7 albums for $7 each. How much more did Randy spend for his albums?
$1

35. Danny bought 3 concert tickets for $5 each. Marta bought 2 tickets at $6 each. How much did they pay for all the tickets?
$27

36. All children's books were on sale for $6 each. Rene bought books worth $30. How many books did Rene buy?
5 books

126

126

Dividing By 8 or 9

pages 127-128

Objective

To review division facts with 8 or 9 as the divisor

Materials

counters and cups

Mental Math

Have students name the number referred to in the word:

1. decade. (10)
2. quadrilateral. (4)
3. triplets. (3)
4. octagonal. (8)
5. sextet. (6)
6. centennial. (100)
7. millionaire. (1,000,000)
8. couplet. (2)

Skill Review

Review the multiplication facts for 8's and 9's by having students make number lines on the board to show the products of 8 × 1, through 8 × 9, and similarly for the 9's. Have students then practice skip-counting by 8's and 9's.

Dividing by 8 or 9

Mrs. Ferris is buying one place setting of dinnerware at a time to complete her set. How much will she pay for one place setting?

We want to find the cost of 1 place setting.

The sale price on the place setting is $\underline{\$48}$

for $\underline{8}$ of them.

To find the cost of one place setting, we divide

$\underline{\$48}$ by $\underline{8}$.

$\boxed{\underline{\quad} \times 8 = \$48?}$ $\boxed{8 \times \underline{\quad} = \$48?}$

$\$48 \div 8 = \underline{6}$ or $8\overline{)\$48}$ with quotient 6

Mrs. Ferris will pay $\underline{\$6}$ for one place setting.

Getting Started

Divide.

1. $64 \div 8 = \underline{8}$ 2. $36 \div 9 = \underline{4}$ 3. $18 \div 9 = \underline{2}$ 4. $45 \div 9 = \underline{5}$

5. $56 \div 8 = \underline{7}$ 6. $72 \div 8 = \underline{9}$ 7. $48 \div 8 = \underline{6}$ 8. $27 \div 9 = \underline{3}$

9. $3\overline{)15}$ → 5 10. $8\overline{)32}$ → 4 11. $9\overline{)72}$ → 8 12. $9\overline{)54}$ → 6

13. $8\overline{)72}$ → 9 14. $9\overline{)63}$ → 7 15. $8\overline{)24}$ → 3 16. $8\overline{)40}$ → 5

127

Teaching the Lesson

Introducing the Problem Have students describe the picture. Have students discuss sales and what price might have originally been charged for the 8 place settings. Ask a student to read the problem and tell what is being asked. (cost of 1 place setting) Ask students where the necessary information can be found. (in the picture) Have students complete the sentences and the division in the model with you. Have students complete the solution sentence. Have students check by multiplying.

Developing the Skill Write **42 ÷ 7 =** and **7 × __=42** on the board. Have students complete the problems. (6,6) Remind students that the missing-factor multiplication problem helps us to solve the division problem. Write more division problems on the board and have students write missing-factor problems for each and then solve both.

127

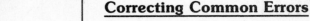

Practice

Divide.

1. $36 \div 9 = \underline{4}$ 2. $42 \div 7 = \underline{6}$ 3. $48 \div 8 = \underline{6}$ 4. $64 \div 8 = \underline{8}$

5. $27 \div 9 = \underline{3}$ 6. $81 \div 9 = \underline{9}$ 7. $56 \div 8 = \underline{7}$ 8. $16 \div 8 = \underline{2}$

9. $18 \div 9 = \underline{2}$ 10. $45 \div 9 = \underline{5}$ 11. $25 \div 5 = \underline{5}$ 12. $63 \div 9 = \underline{7}$

13. $48 \div 6 = \underline{8}$ 14. $24 \div 8 = \underline{3}$ 15. $72 \div 8 = \underline{9}$ 16. $32 \div 8 = \underline{4}$

17. $8\overline{)32}$ ^{4} 18. $8\overline{)16}$ ^{2} 19. $6\overline{)36}$ ^{6} 20. $9\overline{)45}$ ^{5}

21. $9\overline{)63}$ ^{7} 22. $9\overline{)27}$ ^{3} 23. $8\overline{)72}$ ^{9} 24. $8\overline{)24}$ ^{3}

25. $9\overline{)36}$ ^{4} 26. $8\overline{)48}$ ^{6} 27. $4\overline{)16}$ ^{4} 28. $8\overline{)56}$ ^{7}

29. $3\overline{)9}$ ^{3} 30. $9\overline{)54}$ ^{6} 31. $9\overline{)18}$ ^{2} 32. $9\overline{)81}$ ^{9}

Apply

Solve these problems.

33. Katy paid $24 for 8 scarves. How much did each scarf cost?
$3

34. Phil paid 54¢ for 9 nails. How much did each nail cost?
6¢

35. Leigh paid $40 for 8 baseballs. Later, she sold the baseballs for $7 each. How much profit did Leigh make on each baseball?
$2

36. There are 9 players on a softball team. All 45 people who came to practice were put on teams. Each team paid a fee of $8 to use the field. How much was collected from the teams?
$40

128

128

0 and 1 in Division

pages 129-130

Objective

To understand 0 and 1 in division

Materials

9 counters and 9 cups

Mental Math

Tell students to round each to nearest hundred and solve:

1. $2,615 - 1,398$ (1,200)
2. $294 + 606$ (900)
3. 719×2 (1,400)
4. $1,017 \div 2$ (500)
5. 85×3 (300)
6. $\frac{1}{3}$ of 876 (300)
7. $\frac{1}{10}$ of 979 (100)
8. $15,982 \div 2$ (8,000)

Skill Review

Have students tell the product of 1×0 through 1×9 to review that 1 times any number is that number itself. Ask students the product of zero times other numbers, to review that the product will be zero.

Understanding 1 and 0 in Division

One and zero are special numbers in division. Understanding these rules will help you be a better mathematician.

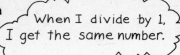

> When I divide by 1, I get the same number.

> When I divide a number by itself, I get 1.

If the divisor is 1, the quotient is equal to the dividend.

$$6 \div 1 = \underline{\quad 6 \quad} \qquad 1\overline{)6}^{\,6}$$

If the divisor and the dividend are the same number, the quotient is 1.

$$8 \div 8 = \underline{\quad 1 \quad} \qquad 8\overline{)8}^{\,1}$$

If the dividend is 0, the quotient is 0.

$$0 \div 4 = \underline{\quad 0 \quad} \qquad 4\overline{)0}^{\,0}$$

NEVER divide by zero. The divisor can never be zero.

Getting Started

Divide.

1. $9\overline{)9}^{\,1}$ 2. $1\overline{)7}^{\,7}$ 3. $5\overline{)0}^{\,0}$ 4. $6\overline{)0}^{\,0}$

5. $8\overline{)8}^{\,1}$ 6. $1\overline{)1}^{\,1}$ 7. $7\overline{)0}^{\,0}$ 8. $1\overline{)3}^{\,3}$

9. $4 \div 4 = \underline{\,1\,}$ 10. $8 \div 1 = \underline{\,8\,}$ 11. $0 \div 1 = \underline{\,0\,}$ 12. $2 \div 2 = \underline{\,1\,}$

129

Teaching the Lesson

Introducing the Problem   Have a student read the thought bubbles in the picture. Read the paragraph to the students. Have students fill in the examples as you read the rules. Emphasize these are basic rules of division.

Developing the Skill   Write **$7 \div 1 = 7$** on the board. Ask students to tell which number in this problem is the divisor. (1) Write the word under the 1. Repeat for the dividend and quotient. Now write $1\overline{)7}$ on the board and have students name each number as the quotient, divisor and dividend. Remind students that the quotient times the divisor must equal the dividend as you write **$7 \times 1 = 7$** on the board. Remind students that in reviewing the facts of 1 they found that **1 times any number is that number itself.** Write **$9 \div 1 =$** on the board and ask students what number times 1 equals 9. (9) Tell students that **when any dividend is divided by 1, the quotient is the same as the dividend.** Now write **$9 \div 9 =$** on the board and ask students

what number times 9 equals 9. (1) Continue for $8 \div 8$, $7 \div 7$, etc. Tell students **a number can never be divided by zero.** Tell students that $5 \div 0$ cannot be worked because its related multiplication fact, $\_\_ \times 0 = 5$, cannot be solved. Have students experiment with other problems where zero is the dividend.

129

Practice

Divide.

$\overset{?}{3\overline{)3}}$ $\overset{?}{1\overline{)3}}$

1. $2\overline{)2}^{\,1}$ 2. $5\overline{)10}^{\,2}$ 3. $4\overline{)4}^{\,1}$ 4. $7\overline{)14}^{\,2}$ 5. $1\overline{)1}^{\,1}$

6. $2\overline{)0}^{\,0}$ 7. $1\overline{)4}^{\,4}$ 8. $2\overline{)6}^{\,3}$ 9. $9\overline{)9}^{\,1}$ 10. $3\overline{)18}^{\,6}$

11. $1\overline{)7}^{\,7}$ 12. $1\overline{)8}^{\,8}$ 13. $2\overline{)12}^{\,6}$ 14. $4\overline{)12}^{\,3}$ 15. $2\overline{)4}^{\,2}$

16. $5\overline{)5}^{\,1}$ 17. $9\overline{)27}^{\,3}$ 18. $1\overline{)9}^{\,9}$ 19. $7\overline{)49}^{\,7}$ 20. $9\overline{)0}^{\,0}$

21. $7\overline{)35}^{\,5}$ 22. $3\overline{)27}^{\,9}$ 23. $8\overline{)8}^{\,1}$ 24. $7\overline{)63}^{\,9}$ 25. $3\overline{)0}^{\,0}$

26. $7\overline{)56}^{\,8}$ 27. $5\overline{)35}^{\,7}$ 28. $9\overline{)72}^{\,8}$ 29. $3\overline{)3}^{\,1}$ 30. $2\overline{)8}^{\,4}$

31. $1\overline{)2}^{\,2}$ 32. $8\overline{)40}^{\,5}$ 33. $1\overline{)0}^{\,0}$ 34. $6\overline{)6}^{\,1}$ 35. $4\overline{)0}^{\,0}$

36. $18 \div 9 = \underline{2}$ 37. $6 \div 1 = \underline{6}$ 38. $6 \div 3 = \underline{2}$ 39. $28 \div 7 = \underline{4}$

40. $3 \div 1 = \underline{3}$ 41. $18 \div 6 = \underline{3}$ 42. $5 \div 1 = \underline{5}$ 43. $7 \div 7 = \underline{1}$

44. $48 \div 6 = \underline{8}$ 45. $54 \div 6 = \underline{9}$ 46. $0 \div 9 = \underline{0}$ 47. $20 \div 5 = \underline{4}$

48. $40 \div 5 = \underline{8}$ 49. $1 \div 1 = \underline{1}$ 50. $0 \div 6 = \underline{0}$ 51. $45 \div 9 = \underline{5}$

130

Correcting Common Errors

Some students may think the answer to both of these problems is 1.

INCORRECT	CORRECT
$6 \div 1 = 1$	$6 \div 6 = 1$

Correct by having each student work with a partner, using counters to model the problem. Have them deal out 6 counters into 6 cups to see that when the counters have been dealt, there is 1 counter in each cup. So, $6 \div 6 = 1$. Then have them deal out 6 counters into 1 cup to see that $6 \div 1 = 6$.

Enrichment

Tell students to draw a picture showing how sharing and $3 \div 3 = 1$ are alike. Then have them draw a second picture which shows how selfishness and $4 \div 1 = 4$ are alike. Ask them to explain their drawings to the class.

Practice

Remind students to refer to the rules for 1 and 0 in division as they work all the problems on this page independently.

Mixed Practice

1. $438 + 15{,}193$ (15,631)
2. $520 - 393$ (127)
3. $7 \times \$.95$ ($6.65)
4. $652 + 729 + 803$ (2,184)
5. $63 \div 9$ (7)
6. 78×6 (468)
7. $2{,}473 - 658$ (1,815)
8. $8 \div 1$ (8)
9. $(2 \times 5) + 18$ (28)
10. $0 \div 5$ (0)

Extra Credit   *Geometry*

Have students research and report on geodesic domes. Ask what kinds of geometric shapes are used to construct such a structure? Have the students construct their own domes as an extended project, working independently or in groups. Possible materials to use would be straws (paper straws work best), toothpicks and paper cut into the geometric shapes and glued together.

Remainders

pages 131-132

Objective

To divide by a 1-digit number with remainders

Materials

10 pencils

Mental Math

Ask students how much money in:

1. (¹⁄₁₀ of $1) + (¼ of $1) (35¢)
2. 2 × ¼ of $1 (50¢)
3. ¹⁄₂₀ of $1 × 6 (30¢)
4. 25 dimes − 4 × 10¢ ($2.10)
5. ¼ of $10 ($2.50)
6. 7 quarters and 6 dimes ($2.35)
7. 8 dimes × 9 ($7.20)
8. (⅕ of $5) ÷ 5 (20¢)

Skill Review

Write **3, 6, 9, 12** on the board. Review skip counting by having a student continue to write the multiples of 3 through 27. Repeat for multiples of other numbers.

Working with Remainders

Keith wants to buy as many pretzels as he can for 43¢. How many can he buy? How much money will he have left?

We want to know the number of pretzels Keith will buy and the amount he will have left.

We know Keith has ___43¢___ to spend.

Each pretzel costs ___5¢___.

To find the number of pretzels that can be bought, and the left-over money, we divide ___43¢___ by ___5¢___.

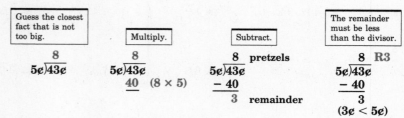

Guess the closest fact that is not too big.		Multiply.		Subtract.		The remainder must be less than the divisor.

```
      8            8              8   pretzels        8  R3
 5¢)43¢       5¢)43¢         5¢)43¢              5¢)43¢
               40  (8 × 5)     − 40                − 40
                                 3   remainder        3
                                                  (3¢ < 5¢)
```

Keith can buy ___8___ pretzels. He will have ___3¢___ left.

Getting Started

Divide. Show your work.

1. 6)13̄ → 2 R1
2. 4)23̄ → 5 R3
3. 2)19̄ → 9 R1
4. 7)32̄ → 4 R4
5. 8)23̄ → 2 R7

6. 9)30̄ → 3 R3
7. 5)37̄ → 7 R2
8. 3)25̄ → 8 R1
9. 8)58̄ → 7 R2
10. 6)39̄ → 6 R3

131

Teaching the Lesson

Introducing the Problem   Have a student describe the picture. Have another student read the problem and tell what two problems are to be solved. (number of pretzels Keith can buy and amount of money he will have left) Ask students what information is needed from the problem and the picture. (Keith has 43¢ and 1 pretzel costs 5¢.) Have students complete the sentences. Guide them through the division steps in the model, to solve the problem. Tell students they can check their answer by multiplying the quotient by the divisor and adding the remainder.

Developing the Skill   Have 3 students stand. Ask another student to distribute 10 pencils, one at a time, to the 3 students so that all students have the same number. When 3 pencils are given to each of the 3 students, ask if any pencils are left over. (yes) Tell students the leftover is called a **remainder** because it remains after dividing the pencils equally. Repeat the activity with 8 pencils and 3 students for a remainder of 2. Write **16 ÷ 3 =** on the board and tell students we need to think of a number that when multiplied by 3, will be almost 16. (5) Ask students if there will be a remainder. (yes) Have a student write 5R1 in the answer. Repeat for 19 ÷ 2, 21 ÷ 4 and 37 ÷ 6.

Practice

Divide. Show your work.

1. $\overset{5\ R3}{8)\overline{43}}$
2. $\overset{7\ R4}{5)\overline{39}}$
3. $\overset{8\ R2}{4)\overline{34}}$
4. $\overset{9\ R5}{9)\overline{86}}$
5. $\overset{6\ R3}{7)\overline{45}}$

6. $\overset{6\ R1}{2)\overline{13}}$
7. $\overset{4\ R3}{6)\overline{27}}$
8. $\overset{9\ R1}{3)\overline{28}}$
9. $\overset{5\ R4}{6)\overline{34}}$
10. $\overset{7\ R2}{4)\overline{30}}$

11. $\overset{7\ R4}{5)\overline{39}}$
12. $\overset{5\ R6}{9)\overline{51}}$
13. $\overset{3\ R7}{8)\overline{31}}$
14. $\overset{4\ R1}{7)\overline{29}}$
15. $\overset{5\ R1}{3)\overline{16}}$

EXCURSION

Complete the tables.

Multiply	Add	Find Missing Factor	Find Missing Addend
$1 \times 9 = \underline{9}$	$0 + 9 = \underline{9}$	$9 \times \underline{1} = 9$	$0 + \underline{9} = 9$
$2 \times 9 = \underline{18}$	$1 + 8 = \underline{9}$	$9 \times \underline{2} = 18$	$1 + \underline{8} = 9$
$3 \times 9 = \underline{27}$	$2 + 7 = \underline{9}$	$9 \times \underline{3} = 27$	$2 + \underline{7} = 9$
$4 \times 9 = \underline{36}$	$3 + 6 = \underline{9}$	$9 \times \underline{4} = 36$	$3 + \underline{6} = 9$
$5 \times 9 = \underline{45}$	$4 + 5 = \underline{9}$	$9 \times \underline{5} = 45$	$4 + \underline{5} = 9$
$6 \times 9 = \underline{54}$	$5 + 4 = \underline{9}$	$9 \times \underline{6} = 54$	$5 + \underline{4} = 9$
$7 \times 9 = \underline{63}$	$6 + 3 = \underline{9}$	$9 \times \underline{7} = 63$	$6 + \underline{3} = 9$
$8 \times 9 = \underline{72}$	$7 + 2 = \underline{9}$	$9 \times \underline{8} = 72$	$7 + \underline{2} = 9$
$9 \times 9 = \underline{81}$	$8 + 1 = \underline{9}$	$9 \times \underline{9} = 81$	$8 + \underline{1} = 9$

132

Correcting Common Errors

Some students will forget to write the remainder in a division problem. Have each student work with a partner, using counters and cups to model problems such as $6)\overline{25}$. Ask them to deal out 25 counters equally into 6 cups. They will see that each cup contains 4 counters and they have 1 counter left over.

$$\overset{4\ R1}{6)\overline{25}}$$

Enrichment

Tell students to draw a picture that shows a quotient of 9R7.

Practice

Remind students to use R and the number to write the remainder beside the quotient. Have students complete the page independently.

Excursion

Have students fill in the tables. When they have completed the excursion, ask if they can see any patterns in the answers. (In column one, the sum of the digits in each product is 9.) Help them to see that the products in column 1 can be compared to the addend pair in column 2. (18 compared to 1 + 8) In column three, help them see that the missing factor is one number greater than the first digit of the product.

Mixed Practice

1. $623 + 407 + 748$ (1,778)
2. $\$38.06 - 12.47$ ($25.59)
3. $38 \div 4$ (9 R2)
4. $13,275 - 9,027$ (4,248)
5. 4×73 (292)
6. $32,458 + 16,369$ (48,827)
7. $51 \div 8$ (6 R3)
8. $5 + 7 + 3 + 8 + 5$ (28)
9. $6 + 3 \times 9$ (33)
10. 17×4 (68)

Extra Credit *Numeration*

Provide each pair of students with 18 counters and 3 or more sorting containers, such as paper cups or egg cartons. Tell students they are to take turns sorting any or all of the counters into any number of equal piles to form multiplication sentences. One student does the sorting, the other records the number of piles, number in each pile and total counters used, on a chart. Have students write the correct multiplication sentence for each combination. Increase the number of counters, and repeat the activity.

2-Digit Quotient

pages 133-134

Objective

To divide by a 1-digit number for a 2-digit quotient with no remainders

Materials

*multiplication facts
ten-strips and singles

Mental Math

Tell student to start at 10 and:

1. add 2, divide by 6. (2)
2. multiply by 9, add 6. (96)
3. divide by 5, subtract 2. (0)
4. add ½ of 40. (30)
5. triple the number. (30)
6. multiply by 6, divide by 5. (12)
7. quadruple the number. (40)
8. add 4 × 9. (46)

Skill Review

Show multiplication facts through 9 and have students write and solve the related division problems on the board.

Dividing, 2-digit Quotients

The Girl Scouts have to carry 52 pieces of kindling to their campsite. How many pieces of wood will each scout have to carry?

We are looking for the number of pieces of wood each scout will carry. Altogether, there are

  52 pieces of kindling wood to be carried.

There are   4   scouts.

To find the number of pieces of wood each scout will carry, we divide   52   by   4 .

Divide the 5 tens by 4. Each scout will carry ten pieces with 1 ten left over.

$$\begin{array}{r} 1 \\ 4\overline{)52} \\ \underline{4} \\ 1 \end{array}$$

Divide. $4\overline{)5}$
Multiply. $4 \times 1 = 4$
Subtract. $5 - 4 = 1$
Compare. $1 < 4$

Trade the 1 ten for ten ones.

$$\begin{array}{r} 1 \\ 4\overline{)52} \\ \underline{4\downarrow} \\ 12 \end{array}$$

Bring down the ones digit. There are now 12 ones.

Divide the 12 ones by 4. Each scout will carry 3 more pieces for a total of 13 each.

$$\begin{array}{r} 13 \\ 4\overline{)52} \\ \underline{4} \\ 12 \\ \underline{12} \\ 0 \end{array}$$

Divide. $4\overline{)12}$
Multiply. $4 \times 3 = 12$
Subtract. $12 - 12 = 0$

Each scout must carry   13   pieces of wood.

Getting Started

Divide. Show your work.

1. $\begin{array}{r} 14 \\ 3\overline{)42} \end{array}$ 2. $\begin{array}{r} 23 \\ 4\overline{)92} \end{array}$

Copy and divide.

3. 50 ÷ 5
   10

4. 96 ÷ 8
   12

133

Teaching the Lesson

Introducing the Problem Have a student read the problem and tell what is to be found. (the number of pieces of wood each Girl Scout will carry) Ask students what information is needed. (number of pieces of wood in all and number of girls) Ask if all necessary information is given in the written problem. (no) Have students read and complete the sentences. Guide them through the division problem in the model, and have students complete the solution sentences.

Developing the Skill Write $2\overline{)56}$ on the board. Have students put out 5 ten-strips and 6 singles. Tell students that we first divide the 5 tens by 2 by asking what number times 2 is 5, or slightly less than 5. (2) Write **2** above the 5 and tell students we must then subtract 4 tens from the 5 to see how many tens are left. (1) Draw a line under the 4 and write **1** as you tell students that 1 ten is left. Tell students that the number of tens left must be less than the divisor. Tell students that to divide the ones, we trade the 1 ten for ones and add the 6 ones to make 16 ones in all. Record the 6 beside the 1 as students make the trade with their manipulatives. Ask students what number times 2 equals 16. (8) Record 8 in the quotient and tell students we now subtract the 16 ones from the 16 to get zero. Repeat the procedure for 90 ÷ 5 and 76 ÷ 4.

133

Practice

Divide. Show your work.

1. $\overset{15}{5\overline{)75}}$ 2. $\overset{21}{3\overline{)63}}$ 3. $\overset{20}{4\overline{)80}}$ 4. $\overset{24}{4\overline{)96}}$ 5. $\overset{28}{2\overline{)56}}$

6. $\overset{24}{3\overline{)72}}$ 7. $\overset{12}{7\overline{)84}}$ 8. $\overset{16}{5\overline{)80}}$ 9. $\overset{16}{6\overline{)96}}$ 10. $\overset{10}{9\overline{)90}}$

11. $\overset{11}{8\overline{)88}}$ 12. $\overset{27}{3\overline{)81}}$ 13. $\overset{14}{4\overline{)56}}$ 14. $\overset{13}{7\overline{)91}}$ 15. $\overset{14}{6\overline{)84}}$

16. $\overset{47}{2\overline{)94}}$ 17. $\overset{25}{3\overline{)75}}$ 18. $\overset{14}{7\overline{)98}}$ 19. $\overset{30}{3\overline{)90}}$ 20. $\overset{19}{4\overline{)76}}$

Copy and Do

21. $78 \div 6$
13
22. $90 \div 2$
45
23. $85 \div 5$
17
24. $72 \div 6$
12

25. $60 \div 5$
12
26. $60 \div 2$
30
27. $72 \div 4$
18
28. $88 \div 4$
22

29. $66 \div 3$
22
30. $76 \div 2$
38
31. $87 \div 3$
29
32. $96 \div 8$
12

Apply

Solve these problems.

33. Gene bought 3 thank you notes for 75¢. How much did each card cost?
25¢

34. Lemons cost 8¢ each. Sally has 96¢. How many lemons can Sally buy?
12 lemons

35. Rhoda fed her horse 3 apples a day for lunch. How many apples did the horse eat in 27 days?
81 apples

36. The tennis club used 45 tennis balls in a tournament. Tennis balls are sold 3 in a can. How many cans did the club use?
15 cans

134

Correcting Common Errors

Students may place the first digit of the quotient incorrectly when tens are involved, as in $3\overline{)48}$.

INCORRECT

$$\overset{1}{3\overline{)48}}$$
$$\underline{3}$$
$$5$$

Have students work in cooperative groups of 3 using 48 cards. Have them group the cards into 4 tens and 8 ones, putting rubber bands around each set of ten. Ask: How many tens can you give to each person in your group? (1) Have them write a 1 above the 4 tens in the dividend. Ask: What will you do with the 1 ten left over. (Make it 10 ones to put with the others to make 18 ones.) Ask: How many ones can you give to each person in your group? (6) Write a 6 above the ones in the dividend. Work other division problems in a similar way.

Enrichment

Tell students to find the three 1-digit numbers which can be divided into 99 evenly and show their work.

Practice

Work through the first few problems on the board to provide models for students to follow if needed as they complete this page. Have students complete the page independently.

Extra Credit *Logic*

Read or duplicate the following problem for the students to solve: A chemist rushed from his laboratory carrying a flask of steaming, purple liquid. "Look what I have invented," he said proudly to a fellow chemist. "The liquid in this bottle is such a powerful acid that it will dissolve anything it touches." "You're wrong," said the second chemist immediately. How did the second chemist know that the first was wrong? (If the liquid would dissolve anything it touched, it would have dissolved the flask in which it was being carried.)

2-Digit Quotient With Remainder

pages 135-136

Objective

To divide by a 1-digit number for a 2-digit quotient with remainders

Materials

*counters

Mental Math

Have students give a calculator code to find:

1. 7 + 2 + 10 (7 + 2 + 10 =)
2. $60 + 38.25 (60. + 38.25 =)
3. 7 eights (7 × 8 =)
4. 800 − 46 (800 − 46 =)
5. $7 + 16 − 80 (7. + .16 − .8 =)
6. $2.06 − .90 (2.06 − .9 =)
7. $4.00 × 3 (4. × 3 =)
8. 7 × 9 × 2¢ (7 × 9 × .02 =)

Skill Review

Write 3)16 on the board and have a student work the problem to show a quotient of 5R1. Provide more problems for 1-digit quotients with remainders.

Dividing, 2-digit Quotients with Remainders

Scott, Sam and Tom were quizzing each other with 52 multiplication fact cards. Scott dealt all the cards so that each boy received the same number. He put the extra cards in the center. How many cards will each boy receive? How many will be put in the center?

We want to know the number of cards dealt to each boy and the number left over for the center.

There are __52__ cards all together.

The cards are dealt out to __3__ boys.
To find the number of cards each boy will get and

the number in the center, we divide __52__ by __3__.

Divide the tens. Guess the closest fact that is not too big. Multiply. 3 × 1 = 3 Subtract. 5 − 3 = 2 Compare. 2 < 3	Bring down the ones. Divide the ones. Guess the closest fact that is not too big. Multiply. 3 × 7 = 21	Subtract. 22 − 21 = 1 Compare. 1 < 3 Write the remainder.

```
    1              17             17 R1
3)52           3)52           3)52
  3              3              3
  2             22             22
               21             21
                               1
```

Each boy will get __17__ cards. There will be __1__ card left over.

Getting Started

Divide. Show your work.

1. 3)19 → 6 R1

2. 6)65 → 10 R5

Copy and divide.

3. 76 ÷ 2
   38

4. 90 ÷ 4
   22 R2

135

Teaching the Lesson

Introducing the Problem Have a student read the problem and tell what they need to find. (how many cards each boy will receive and how many cards will be left) Ask students what necessary information they will use to solve the 2 problems. (There are 52 cards and 3 boys.) Have students complete the sentences. Guide them through the division problem in the model, explaining each step through the remainder. Have students compute the solution sentence. Students may want to act out the problem to check their solution.

Developing the Skill Write 4)94 on the board. Ask students how many of the 9 tens can be given to each of 4 people. (2) Record the 2 and ask students how many tens would be left as you subtract 8 from 9. (1) Remind students that the 1 < 4. Ask students how 1 ten can be divided among 4 people. (change to 10 ones) Bring down the 4 ones as you tell students that there are 14 ones to be di-

vided by 4. Ask students to think of a multiplication fact for 4 whose product is 14 or almost 14. (3 × 4) Record the 3 and ask students how many ones are left. (2) Tell students to compare the 2 and 4 to be sure the left-over ones are less than the divisor. Record the remainder. Write 5)87 on the board and have a student work through the problem, talking through each step. Now write 9)98 on the board to show students the placement of the zero in ones place in the quotient.

135

Practice

Divide. Show your work.

1. $\overset{9\ R1}{4\overline{)37}}$ 2. $\overset{12\ R1}{7\overline{)85}}$ 3. $\overset{28\ R1}{2\overline{)57}}$ 4. $\overset{15\ R1}{6\overline{)91}}$ 5. $\overset{14\ R2}{5\overline{)72}}$

6. $\overset{10\ R3}{9\overline{)93}}$ 7. $\overset{8\ R3}{8\overline{)67}}$ 8. $\overset{25\ R1}{3\overline{)76}}$ 9. $\overset{17\ R3}{4\overline{)71}}$ 10. $\overset{9\ R1}{2\overline{)19}}$

11. $\overset{32\ R1}{3\overline{)97}}$ 12. $\overset{10\ R8}{9\overline{)98}}$ 13. $\overset{11\ R1}{8\overline{)89}}$ 14. $\overset{14\ R1}{6\overline{)85}}$ 15. $\overset{13\ R5}{7\overline{)96}}$

Copy and Do

16. $47 \div 3$
15 R2

17. $54 \div 5$
10 R4

18. $86 \div 4$
21 R2

19. $62 \div 6$
10 R2

20. $92 \div 7$
13 R1

21. $91 \div 2$
45 R1

22. $98 \div 8$
12 R2

23. $59 \div 3$
19 R2

24. $71 \div 4$
17 R3

25. $87 \div 5$
17 R2

26. $76 \div 7$
10 R6

27. $80 \div 6$
13 R2

Apply

Solve these problems.

28. Margaret has 76¢ to buy stickers. Stickers cost 5¢ each. How many stickers can Margaret buy? How much will she have left over?
15 stickers   1¢ left over

29. Yoko has divided her 51 books equally on 4 shelves. How many books go on each shelf? How many are left over?
12 books   3 books left over

30. John has 65 flowers that he wants to plant in 6 rows. How many flowers will go into each row. How many will be left over?
10 flowers   5 flowers left over

31. Thomas is putting 5 flowers into each vase. He is filling 16 vases. How many flowers does Thomas need to put into the vases?
80 flowers

136

Mixed Practice

1. $28,127 + 34,392$ (62,519)
2. 57×9 (513)
3. $7 + 4 \times 8$ (39)
4. $51 \div 3$ (17)
5. $\$70.00 - 27.68$ ($42.32)
6. $5 \times \$.25$ ($1.25)
7. $396 + 275 + 427$ (1,098)
8. $95 \div 6$ (15 R5)
9. $64,326 - 19,610$ (44,716)
10. $84 \div 8$ (10 R4)

136

Problem Solving, Draw a Picture

pages 137-138

Objective

To draw a picture or diagram to solve a problem

Materials

Mental Math

Tell students to divide the number by 60:

1. 80 + 220 (5)
2. 100 + 1,700 (30)
3. 60 × 40 (40)
4. 24 × 200 (80)

and divide the number by 6:

5. 80 + 220 (50)
6. 100 + 1,700 (300)
7. 60 × 40 (400)
8. 24 × 200 (800)

Drawing a Picture or Diagram

Five girls are running a race. Elaine is 50 yards behind Diane. Cassie is 10 yards behind Elaine. Barb is 50 yards ahead of Cassie. Ann is 20 yards behind Barb. Ann is 30 yards behind Diane.
What is the order of the girls at this time?

★ SEE

We want to know the order of the five girls in the race.

We know Elaine is __50__ yards behind Diane.

__Cassie__ is 10 yards behind __Elaine__. Cassie

is __50__ yards behind Barb. Ann is __20__ yards

behind Barb. Ann is __30__ yards behind Diane.

★ PLAN

We draw a diagram of the race, showing positions 10 yards apart. Using the facts, we place the initial of each girl to show her position.

★ DO

```
•   •   •   •   •   •   •   •
    C   E       A       B   D
```

The order of the girls from first to last is

__Diane__, __Barb__, __Ann__, __Elaine__ and __Cassie__.

★ CHECK

We compare the results of the race with the facts in the original problem to see if each girl is in position.

137

Teaching the Lesson

Ask students if they have ever tried to describe something to someone and found the best way was to draw a picture or diagram to show what they meant. Discuss with students how we draw a map to tell someone how to get to a specific place. Tell students that we often draw a picture or diagram to solve such problems because it can make a problem easier for us to understand.

Have a student read the problem. Tell students that in order to make sense of all this information, we need to organize it by drawing the position of each runner. Have a student read the SEE step. Tell students that sometimes we need to reword the information in the SEE step to make it more clear. Have students note that some of the sentences have been reworded so that each tells who is behind whom. Have a student read the PLAN step. Tell students that although the problem does not ask the distance between each runner, it will help to draw the spaces marked off at 10-yard intervals. Have students locate each runner to complete the

DO step, and then see if each statement in the CHECK step can be answered yes. Ask students if any of the facts were unnecessary to solve the problem. (Yes, Ann is 30 yards behind Diane.) Have students now use their diagrams to find the distance between every 2 runners.

Apply

Draw a picture to help you solve each problem.
Remember to use the four-step plan.

1. Sal is waiting to buy tickets for a movie. There are 5 people waiting ahead of him and 7 people waiting behind him. How many people are in line for tickets?
13 people

2. If it takes 3 minutes to make a cut, how long will it take to cut a log that is 20 feet long into 10 equal lengths?
27 minutes

3. If you count the number of pickets and spaces along a fence, you will find that the number of spaces is always one less than the number of pickets. How many spaces will a bicycle wheel with 15 spokes contain?
15 spaces

4. Valleyview is south of Columbus but north of Lexington. Trenton is south of Valleyview but north of Lexington. Mintown is north of Valleyview but south of Columbus. Which town is the second most southerly town of the five?
Trenton

5. One dozen eggs and 2 half gallons of milk cost $3.35. Two dozen eggs and 2 half gallons of milk cost $4.30. How can you use subtraction to find which costs more, 1 dozen eggs or 1 half gallon of milk?
See Solution Notes.

6. Five dogs are chasing a cat. The dogs are 3 yards behind the cat and running hard. The cat is 12 feet from a tree. The cat and the dogs are running at the same rate. Will the cat reach the tree and safety before the dogs catch the cat?
Yes

7. A brown, a gray, and a white horse were in a race on a rainy day. Draw a picture to show how many different ways the horses can finish if the gray never wins on rainy days.
See Solution Notes.

138

Extra Credit   *Geometry*

Have students create tangram designs to share with others in the class. A tangram is a Chinese puzzle made by cutting a square of paper into five triangles, a square and a rhomboid, and arranging them into many different figures. Have them trace their design on paper and ask them to exchange with another student. Then have students calculate the perimeter of each of the geometric shapes in the tangram design.

Solution Notes

1. Remind students to draw a diagram.
2. Have students draw a board cut into 10 pieces to help them see they only need 9 cuts, which will take 27 minutes.
3. Have students draw the wheel with 15 spokes and count the 15 spaces. Help students see that in counting the spaces, they can avoid confusion if they mark the space where they begin. Have students draw a picket fence, count the spaces and then discuss the difference in the number of spaces on the fence and the wheel. You may want to have students extend this activity by finding other examples.
4. The complete order of the towns is: Columbus, Mintown, Valleyview, Trenton, Lexington.

Higher-Order Thinking Skills

5. [Analysis and Evaluation] Since the second price involves only 1 more dozen of eggs, a dozen costs the difference between $4.30 and $3.35, or $0.95. Two half-gallons of milk costs the difference between $3.35 and $0.95, or $2.40, which means that 1 half-gallon costs more than $0.95; so milk costs more.
6. [Synthesis] Students should deduce that if both the cat and the dogs are running at about the same rate, the cat will reach and run up the tree before the dogs reach the cat regardless of the distances.
7. [Analysis] Any picture will do that shows the following arrangements:
Brown, Gray, White
Brown, White, Gray
White, Gray, Brown
White, Brown, Gray.

Calculator Use, Division

pages 139-140

Objective

To use a calculator to divide

Materials

calculators
multiplication fact cards

Mental Math

Have students name the fact to check:

1. $16 \div 4$ (4×4)
2. $32 \div 8$ (8×4)
3. $12 \div 6$ (2×6)
4. $56 \div 7$ (8×7)
5. $81 \div 9$ (9×9)
6. $36 \div 6$ (6×6)
7. $54 \div 9$ (9×6)
8. $42 \div 7$ (6×7)

Skill Review

Have students find the product of 86 and 7 on their calculators. (602) Give other multiplication problems for students to practice multiplying a 2-digit by a 1-digit number.

Calculators, the Division Key

Natalie is packing lunches for a picnic. She needs to buy 5 apples. How much will Natalie pay for the 5 apples?

We want to know the price for 5 apples.

We know that ___3___ apples cost ___51¢___ .

To find the cost of 5 apples, we first find the cost of 1 by dividing ___$0.51___ by ___3___ . Then, we multiply the cost of 1 apple by ___5___ .

This can be done on the calculator in one code.

[·] 51 [÷] 3 [×] 5 [=] (0.85)

Natalie will pay ___$0.85___ for 5 apples.

Complete these calculator codes.

1. 42 [÷] 7 [=] (6)   2. 76 [÷] 2 [=] (38)
3. 96 [÷] 4 [=] (24)   4. 52 [÷] 4 [=] (13)
5. 36 [÷] 9 [×] 7 [=] (28)   6. 84 [÷] 4 [×] 3 [=] (63)
7. 72 [÷] 6 [×] 9 [=] (108) 8. 75 [÷] 5 [×] 9 [=] (135)

139

Teaching the Lesson

Introducing the Problem Ask students what is shown in the picture. (A girl is looking at apples in a store.) Have a student read the problem aloud and tell what they are to find. (the cost of 5 apples) Ask students what information is needed to solve this problem. (the cost of 1 apple) Ask if this information is given. (no) Ask what is given that may be helpful. (The cost of 3 apples is 51¢.) Have students complete the sentences and the calculator code with you to solve the problem. Have students decide if their solution makes sense.

Developing the Skill Write $6 \div 2 \times 3 =$ on the board. Have students work the problem on their calculators. Ask students why this calculator code is easier than $6 \div 2 =$ and $3 \times 3 =$. (There are fewer steps or fewer keys to punch.) Give more practice with $20 \div 4 \times 7 =$, $6 \times 5 \div 4 =$ and $72 \div 6 - 3 =$. Have students complete the codes at the bottom of the page.

Practice

Complete these calculator codes.

1. 85 ÷ 5 = (17)
2. 57 ÷ 3 = (19)
3. 91 ÷ 7 = (13)
4. 96 ÷ 6 = (16)
5. 88 ÷ 2 = (44)
6. 90 ÷ 9 = (10)
7. 48 ÷ 4 × 9 = (108)
8. 63 ÷ 7 × 8 = (72)
9. 63 ÷ 9 × 7 = (49)
10. 75 ÷ 5 × 6 = (90)
11. 62 + 26 ÷ 4 = (22)
12. 216 − 158 ÷ 2 = (29)

Apply

Solve these problems.

13. Jeff can clean 4 rugs every 76 minutes. How long will it take Jeff to clean 9 rugs?
171 minutes

14. Nathan can jog 5 miles in 65 minutes. How long will it take Nathan to jog 8 miles?
104 minutes

15. Sandi earns $32 every 4 days selling flowers. How much will Sandi earn in 7 days?
$56

16. Bananas are on sale at 6 for 96¢. How much do 8 bananas cost?
$1.28

EXCURSION

Find the missing numbers.

1. The product of 2 numbers is 45. Their sum is 14. What are the numbers?
5, 9

2. The sum of 2 numbers is 60. Their difference is 12. What are the numbers?
24, 36

3. The quotient of two numbers is 15. Their sum is 144. What are the numbers?
135, 9

4. Five times one number is three more than six times another number. The difference between the numbers is 1. What are the numbers?
3, 2

140

Correcting Common Errors

Some students may not understand the calculator code for finding the cost of a number of objects. Write this calculator code on the chalkboard. 84 ÷ 3 × 5 = ( ). Ask: If 3 whistles cost 84¢, how do you find the cost of 1 whistle? (Divide by 3.) Point to the calculator code for this operation. Have students find the cost of 1 whistle. Ask: How can you find the cost of 5 whistles? (Multiply 28¢ by 5.) Point to the calculator code for this operation. Then have them use their calculators to find the cost. ($1.40)

Enrichment

Tell students to use their calculators to find 9 times the sum of the digits 0 through 9.

Practice

Remind students to clear their calculators before each new problem. Tell students to make a plan of one calculator code to work each story problem. Have students complete the page independently.

Excursion

Tell students it will be helpful for them to set up each problem using a solve for n format. Have them use their calculators and remind students to read the operation words carefully.

Extra Credit *Applications*

Incorporate the students' interest in sports to give them practice in division. Have students list their favorite team sports. Next to each sport, have them list the number of players on a team. Now have students create mathematical problems, involving any combination of operations, using this list. Have them solve these examples to get them started:

1. How many soccer teams can be formed from 36 players? (36 players ÷ 6 per team = 6 teams)
2. How many players would you need for 4 baseball teams? (4 teams × 9 players = 36 players)
3. How many players are on a team if you have 45 players and 9 teams? (45 players ÷ 9 teams = 5 players per team.)

Chapter Test

page 141

Item	Objective
1-18	Master division facts using 2-9 as divisor (See pages 121-128)
3, 8	Master special division properties of 1 and 0 (See pages 129-130)
19-22	Solve basic division problems with remainders (See pages 131-132)
23-26	Divide 2-digit number by 1-digit quotient without remainder (See pages 133-134)
27-34	Divide 2-digit number by 1-digit number to get 2-digit quotient with remainder (See pages 135-136)

Divide.

1. $5\overline{)30}$ 6

2. $7\overline{)49}$ 7

3. $1\overline{)6}$ 6

4. $8\overline{)56}$ 7

5. $4\overline{)4}$ 1

6. $3\overline{)9}$ 3

7. $6\overline{)42}$ 7

8. $9\overline{)0}$ 0

9. $2\overline{)18}$ 9

10. $6\overline{)36}$ 6

11. $45 \div 5 = \underline{9}$

12. $21 \div 7 = \underline{3}$

13. $36 \div 4 = \underline{9}$

14. $18 \div 9 = \underline{2}$

15. $24 \div 8 = \underline{3}$

16. $24 \div 6 = \underline{4}$

17. $49 \div 7 = \underline{7}$

18. $25 \div 5 = \underline{5}$

Divide. Show your work.

19. $4\overline{)17}$ 4 R1

20. $7\overline{)50}$ 7 R1

21. $3\overline{)26}$ 8 R2

22. $9\overline{)46}$ 5 R1

23. $3\overline{)36}$ 12

24. $5\overline{)50}$ 10

25. $2\overline{)84}$ 42

26. $4\overline{)76}$ 19

27. $6\overline{)73}$ 12 R1

28. $8\overline{)94}$ 11 R6

29. $2\overline{)93}$ 46 R1

30. $7\overline{)70}$ 10

31. $4\overline{)82}$ 20 R2

32. $5\overline{)79}$ 15 R4

33. $9\overline{)97}$ 10 R7

34. $3\overline{)88}$ 29 R1

Circle the letter of the correct answer.

1 Round 809 to the nearest hundred.
- a 700
- (b) 800
- c NG

2 Round 8,575 to the nearest thousand.
- a 7,000
- b 8,000
- (c) NG

3 What is the value of the 3 in 423,916?
- a tens
- b hundreds
- (c) thousands
- d NG

4 587
+ 274
- a 751
- b 761
- (c) 861
- d NG

5 4,297
+ 3,486
- a 7,683
- (b) 7,783
- c 8,783
- d NG

6 57,198
+ 23,488
- a 70,686
- (b) 80,686
- c 81,686
- d NG

7 73
− 29
- a 52
- b 56
- c 102
- (d) NG

8 603
− 256
- (a) 347
- b 357
- c 453
- d NG

9 42,823
− 29,647
- (a) 13,176
- b 13,276
- c 27,224
- d NG

10 8 × 3
- (a) 24
- b 25
- c 32
- d NG

11 37
× 2
- a 64
- (b) 74
- c 614
- d NG

12 59
× 6
- a 304
- (b) 354
- c 356
- d NG

13 4)93
- a 2 R1
- b 23
- (c) 23 R1
- d NG

☐ score

142

Cumulative Review

Item	Objective
1	Round numbers to nearest 100 (See pages 31-34)
2	Round numbers to nearest 1,000 (See pages 33-34)
3	Identify place value through hundred thousands (See pages 27-28)
4	Add two 3-digit numbers (See pages 45-46)
5	Add two 4-digit numbers (See pages 47-48)
6	Add two 5-digit numbers (See pages 53-54)
7	Subtract two 2-digit numbers (See pages 61-62)
8	Subtract two 3-digit numbers with zero in minuend (See pages 65-66)
9	Subtract two 5-digit numbers (See pages 67-68)
10	Multiply using factor 8 (See pages 93-94)
11-12	Multiply 2-digit number by 1-digit number with two regroupings (See pages 111-112)
13	Divide using 1-digit divisor (See pages 123-124

Alternate Cumulative Review

Circle the letter of the correct answer.

1 Round 659 to the nearest hundred.
- a 500
- b 600
- (c) 700
- d NG

2 Round 4,306 to the nearest thousand.
- a 3,000
- (b) 4,000
- c 5,000
- d NG

3 What is the value of the 9 in 659,243?
- a hundreds
- (b) thousands
- c ten thousands
- d NG

4 376
+ 547
- a 813
- b 814
- c 913
- (d) NG

5 3,163
+ 2,498
- a 5,552
- b 5,551
- (c) 5,661
- d NG

6 65,376
+ 16,596
- a 71,973
- b 71,862
- (c) 81,972
- d NG

7 85
− 48
- a 43
- b 47
- c 33
- (d) NG

8 808
− 249
- (a) 559
- b 569
- c 641
- d NG

9 36,715
− 17,426
- (a) 19,289
- b 19,299
- c 21,311
- d NG

10 7 × 6 =
- a 35
- b 36
- (c) 42
- d NG

11 26
× 4
- (a) 104
- b 824
- c 84
- d NG

12 75
× 5
- a 3,525
- (b) 375
- c 130
- d NG

13 6)85
- a 14
- (b) 14 R1
- c 10
- d NG

142

Measuring Time

Objective

To tell time on a standard or digital clock

To tell time when hours and minutes are added or subtracted

Materials

*demonstration clock

Mental Math

Ask students what operation(s) can be used to get from:

1. 16 to 10 (−)
2. 200 to 4 (−, ÷)
3. ⅕ to ⅘ (+, ×)
4. 20 to 5 (−, ÷)
5. 5 to 20 (+, ×)
6. 1,000 to 10,000 (×, +)
7. 12 to 2 and then to 3 (− or ÷ and then +)
8. 45 to 9 and then to 27 (− or ÷ and then × or +)

Skill Review

Have students arrange the hands on the clock to show the times they rise and go to bed on school and weekend days, lunchtimes, etc. Have other students read the times shown. Include some times on the quarter-hour.

Telling Time

Mr. Jameson can drive home from work in 25 minutes. Draw both a standard clock and a digital clock to show the time when Mr. Jameson will arrive.

We want to know what time Mr. Jameson will get home.

The time now is <u>4:28</u> .

His trip home will take <u>25</u> minutes.

Arrival Time

Arrival Time

Mr. Jameson will get home at <u>4:53</u> or <u>7</u>

minutes to <u>5</u> .

Getting Started

Show the times.

1. five after twelve
   `12:05`

2. quarter to four
   `3:45`

3. 7:35

Write the times as you would say them.

4.
   three thirty

5.
   seven forty-five
   or
   quarter to eight

6.
   ten forty
   or
   twenty to eleven **143**

Teaching the Lesson

Introducing the Problem   Have a student read the problem aloud and tell what is to be solved. (the time when Mr. Jameson will arrive home from work) Ask students what information is needed to solve this problem. (time he leaves work and amount of time it takes him to get home) Tell students to write this information in the sentences. On a clock show students how they can determine the arrival time. Then have them supply the answers in the solution sentence. Tell students they can check their solution by adding or by counting on 25 minutes.

Developing the Skill   Show 3:15 on the clock. Ask how many minutes after 3:00. (15) Ask a student to move the minute hand to show 5 minutes later than 3:15. Ask students to tell and write ways to say this time. (3:20, twenty minutes after 3) Show 2:25 on the clock and slowly move the minute hand 40 minutes forward as you tell students that 40 minutes later than 2:25 is 3:05 or 5 minutes after 3. Write **12:55** on the board and tell students we can find the time that is 42 minutes later than 12:55 by adding 55 and 42. (97 minutes) Remind students that 60 of the 97 minutes equals 1 hour so the time would be 1 hour and 37 minutes later than 12:00 or 1:37. Repeat for subtracting minutes from a specific time to show the trade of 1 hour for 60 minutes. Give more examples of both. Do not use times that require use of AM and PM notations at this time.

Practice

Show the times.

1. 8:20

2. 3:50

3. noon

4. quarter to eleven
10:45

5. ten minutes after six
6:10

6. three twenty-five

Write the times as you would say them.

7.

_____
two fifteen
or
quarter after two

8.

_____
twenty to ten
or nine forty

9.

_____
two minutes after four

10.
12:30

_____
twelve thirty

11.

_____
one twenty-two

12.

_____
six forty-five
or quarter to seven

Apply

Solve these problems.

13. The donut shop opens at 8:30 AM. The baker needs 2 hours and 25 minutes to prepare bakery for his first customers. At what time should he start baking?
6:05 AM

14. At 11:50 AM, Mrs. Miller dismisses her morning kindergarten class. She has 1 hour and 50 minutes until the afternoon class arrives. What time will she greet her afternoon students?
1:40 AM

144

Practice

Have students complete the page independently.

Mixed Practice

1. 8,020 − 5,372 (2,648)
2. 73 × 6 (438)
3. $253.49 + 128.92 ($382.41)
4. 28 ÷ 8 (3 R4)
5. $100.00 − 64.38 ($35.62)
6. 75 ÷ 4 (18 R3)
7. 4 × $.68 ($2.72)
8. 363 + 257 + 1,806 (2,426)
9. 45 ÷ 6 (7 R3)
10. (6 × 8) + (7 × 7) (97)

Extra Credit *Numeration*

Tell students they will be doing an experiment to find out if it is easier to count objects in a row, or objects grouped in a pattern. Have each student draw several arrays of countable marks or letters on a paper. Tell them to put some in a straight row, and others forming patterns of some kind. Then, have them work in pairs. One student covers the worksheet with paper, revealing only one set of objects at a time for a brief moment. The second student writes an estimate of the number of objects he saw. Have them reverse roles. At the end of the activity, have them compare their estimates with the actual numbers. Ask students to conclude whether it is easier to count objects in a row or grouped in a pattern.

Telling Time
AM and PM

pages 145-146

Objective
To use AM and PM notations

Materials
*demonstration clock

Mental Math
Tell students to start at 400 and:

1. subtract their age
2. multiply by their age
3. divide by 4 (100)
4. find ½ that number (200)
5. multiply by 5 (2,000)
6. subtract 3.9 (396.1)
7. add 2 times the product of 10 × 6 (520)

Skill Review
Have students use the demonstration clock to show and compare their rising times, bedtimes, supper times, etc. Have students figure the amount of time between breakfast and lunch, lunch and supper, rising time and breakfast, etc. Discuss how different homes have different time schedules.

Telling Time, AM and PM

It takes 20 minutes to load the buses after school. What time did school end?

We want to find the time that school ends.

The buses leave the school at ___12:10 PM___.

It takes __20__ minutes to load the bus.
To find the time for school closing, we can

move the minute hand back __20__ minutes.

AM is from midnight to noon.

12	1	2	3	4	5	6	7	8	9	10	11
midnight						AM					

PM is from noon to midnight.

12	1	2	3	4	5	6	7	8	9	10	11
noon						PM					

It is 12:10 PM

Move the minute hand back 20 minutes.

School Closing Time

School ends at ___11:50 AM___.

Getting Started

Write the times.

1. 45 minutes later than 9:15 AM   ___10:00 AM___
2. 15 minutes earlier than 3:04 PM   ___2:49 PM___
3. 3 hours after 10:00 AM   ___1:00 PM___
4. 4 hours before 12:00 midnight   ___8:00 PM___
5. 1 hour and 20 minutes before 6:45 AM   ___5:25 AM___
6. 25 minutes after 12:15 AM   ___12:40 AM___

145

Teaching the Lesson

Introducing the Problem   Have a student read the problem aloud and tell what is to be solved. (the time school ends) Ask students what information is given. (Loading the buses takes 20 minutes. Buses leave at 12:10 PM.) Discuss AM and PM and work through the model of moving the clock back 20 minutes from PM to AM. Then have students complete the solution sentence.

Developing the Skill   Write **7:00** on the board and ask students what meal they might eat at this time. (breakfast or supper) Tell students there are two 7:00 times in each day since there are 24 hours in a day and a clock can only show 12 hours. Tell students that 12:00 midnight to 12:00 noon is noted by the use of AM. Write **6:15 AM** on the board and tell students this time occurs between midnight and noon. Tell students that the time from 12:00 noon to 12:00 midnight is noted by the use of PM. Write **10:00** on the board. Ask students to write AM or PM for class time (AM) or sleeping time. (PM) Now write **7:45 AM** and **2:15 PM** on the board and tell students we find the time between these times by counting to noon and beyond noon. (6 hours 30 minutes) Repeat for number of hours and minutes between breakfast at 6:30 and lunch at 12:09. (5 hours 39 minutes) Give other examples for students to practice.

Practice

Write the times.

1. 30 minutes later than 4:15 AM   <u> 4:45 AM </u>

2. 20 minutes earlier than 12:45 PM   <u> 12:25 PM </u>

3. 25 minutes before 2:15 PM   <u> 1:50 PM </u>

4. 40 minutes after 6:22 AM   <u> 7:02 AM </u>

5. 2 hours after noon   <u> 2:00 PM </u>

6. 1 hour and 15 minutes before 5:00 PM   <u> 3:45 PM </u>

Apply

Solve these problems.

7. What time is 2 hours and 25 minutes before 8:30 AM?
6:05 AM

8. What time is 1 hour and 52 minutes after 11:50 AM?
1:42 PM

9. School begins at 8:30 AM. Gina wakes up at 7:15 AM. How much time does Gina have before school?
1 hour and fifteen minutes

10. Ralph ran a marathon in 3 hours and 56 minutes. He began the race at 9:25 AM. When did Ralph finish the race?
1:21 PM

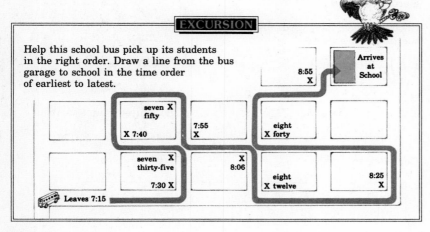

EXCURSION

Help this school bus pick up its students in the right order. Draw a line from the bus garage to school in the time order of earliest to latest.

Leaves 7:15 · seven thirty-five · 7:30 X · seven fifty · X 7:40 · 7:55 X · X 8:06 · eight X forty · eight X twelve · 8:25 X · 8:55 X · Arrives at School

146

Correcting Common Errors

Many students write the time before or after a given time incorrectly when first learning how to tell time. Have each student work with a partner, making a list of times they get up, go to school, go to bed, and so forth. Have them use a model of a clock to set these times, and then move the hands to show 30 minutes before and 45 minutes after the time. If they need more practice, have them repeat the activity with other lists of times.

Enrichment

Have students make a time line to include a minimum of 6 activities during a normal day in their lives. Then have them illustrate an activity under each time and find the number of hours and minutes between each 2 consecutive activities.

Practice

Remind students that time is divided into 2 twelve-hour groups, with AM meaning morning and PM meaning afternoon and evening. Have students complete the page independently.

Excursion

Tell students to first read all the times on the map. Make sure they see that the times are written in three different ways. It may help students to first number the times in order from earliest to latest. They can then easily draw the bus route.

Extra Credit   *Measurement*

Have pairs of students make cards to play measurement Concentration. Each card should have a measurement on it that will pair up with an equivalent measure on another card. Tell them to use pairs, such as 8 ounces and 1 cup, 12 inches and 1 foot, 2 pints and 1 quart, 365 days and 1 year, etc. Tell students to shuffle cards and arrange them face down. Have them take turns turning the cards over to find matching pairs, keeping the cards that match, and replacing cards if they don't match. The player with the most cards after all are matched wins.

Customary Length

Objective

To determine and use the appropriate unit of measure: inch, foot, yard and mile

Materials

*pictures of newborn baby, sitting or crawling baby and teenager

Mental Math

Tell students a gross is 12 dozen. Ask how much in:

1. 2 gross? (24 dozen)
2. ½ gross? (6 dozen)
3. 6 gross? (72 dozen)
4. ⅓ gross? (4 dozen)
5. ⅙ gross? (2 dozen)
6. 5 gross? (60 dozen)
7. 10 gross? (120 dozen)
8. 1½ gross? (18 dozen)

Skill Review

Review skip counting by 3's, 5's and 10's. Have students use addition to skip count by 12's and then use multiplication to check their answers. Repeat for skip counting by 24's and 36's.

Using Customary Units of Length

Some customary units of length are **inch, foot, yard** and **mile.** Is the height of the Chrysler Building in New York City about 1,000 inches or 1,000 feet?

We know:
This is an inch unit.

├─────────────┤

1 foot = 12 inches   1 yard = 3 feet   1 mile = 5,280 feet

__1__ ft = __12__ in.   __1__ yd = __3__ ft   __1__ mi = __5,280__ ft

                            __1__ yd = __36__ in.   __1__ mi = __1,760__ yd

The Chrysler Building is about 1,000 __feet__ tall.

Getting Started

Choose inches, feet, yards or miles to complete these sentences.

1. A jet flies about 6 __miles__ high.

2. A soccer field is 110 __yards__ long.

3. A hand is about 5 __inches__ wide.

4. The United States is about 3,000 __miles__ wide.

Complete the table.

5.

yards	1	2	3	4	5	6	7	8	9
inches	36	72	108	144	180	216	252	288	324

147

Teaching the Lesson

Introducing the Problem   Have a student read the problem and tell what is to be found. (if the Chrysler Building is about 1,000 inches or 1,000 feet tall) Work through the model with students as they fill in the equivalents. Have students tell the abbreviation for each of the 4 units of measure. (in., ft, yd, mi) Then have them complete the solution sentence.

Developing the Skill   Show the pictures of the children. Ask students which person might be 6 or 9 months old, 15 years old and 6 or 7 days old. Discuss with students how a baby's age is generally given in days until the baby is a week or two and then in weeks until the baby is a month and then in months. Discuss how ages are given in months until the age of 2 or so and then in years since days, weeks and months are cumbersome once years can be used. Now have a student draw on the board and label a line that is about 1 **inch** long. Have other students draw and label lines that are about 1 **foot** long and 1 **yard**. Ask students if a 1 **mile** long line can be drawn on the board. (no) Tell students to imagine 1,760 lines the length of the 1 yard line and that would be 1 mile. Hold up a book and tell students the book is about 9 units of measure in length. Ask if it would be 9 inches, 9 feet, 9 yards or 9 miles. (9 inches) Name some objects in and outside of the room and have students choose the best unit of measure to describe each.

147

Practice

Choose inches, feet, yards or miles to complete these sentences.

1. The front door is 3 __feet__ wide.

2. A tree is 10 __feet or yards__ tall.

3. Your nose is about 3 __inches__ long.

4. Your classroom is about 10 __yards__ wide.

5. A homerun is about 300 __feet__ long.

6. A pencil is about 7 __inches__ long.

7. A football field is 100 __yards__ long.

8. Your waist is about 25 __inches__ around.

9. The Mississippi River is about 2,000 __miles__ long.

10. The Empire State Building is about 1,000 __feet__ high.

Complete the table.

11.

yards	1	2	3	4	5	6	7	8	9	10
feet	3	6	9	12	15	18	21	24	27	30

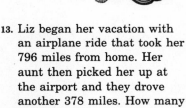

Apply

Solve these problems.

12. It is 27 miles from Ken's house to his grandparents' house and back. Last summer, Ken made 6 trips to see his grandparents. How many miles did Ken travel on all these trips?
162 miles

13. Liz began her vacation with an airplane ride that took her 796 miles from home. Her aunt then picked her up at the airport and they drove another 378 miles. How many miles did Liz travel to get to her vacation spot?
1,174 miles

148

Customary Capacity

Objective

To determine and use the appropriate unit of measure: cup, pint, quart and gallon

Materials

*measuring cup
*pint, quart, ½ gallon and gallon containers

Mental Math

Ask students to explain these figurative phrases:

1. to sum it up (a summary)
2. six of one and a half dozen of the other (all the same)
3. two steps forward and one step back (slow progress)
4. just a second (tiny bit of time)
5. second Tuesday of next week (never)
6. ounce of prevention is worth a pound of cure (better and easier to prevent)

Skill Review

Have students skip count by 2's, 4's and 6's and use multiplication or addition to check their answers.

Measuring Capacity, Customary Units

Some customary units of capacity are **cup, pint, quart** and **gallon**. Does the tub hold about 12 quarts or 12 gallons of water?

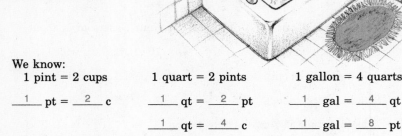

We know:

1 pint = 2 cups	1 quart = 2 pints	1 gallon = 4 quarts
__1__ pt = __2__ c	__1__ qt = __2__ pt	__1__ gal = __4__ qt
	__1__ qt = __4__ c	__1__ gal = __8__ pt

The bathtub probably holds about   __1__ gal = __16__ c

12 __gallons__ of water.

Getting Started

Circle the better estimate of capacity.

1. (2 cups)   2 quarts

2. (6 pints)   6 gallons

3. (2 quarts)   2 gallons

4. (1 pint)   1 quart

Complete the table.

5.

gallons	1	2	3	4	5	6	7	8	9	10
quarts	4	8	12	16	20	24	28	32	36	40

149

Teaching the Lesson

Introducing the Problem   Have students describe the picture. Have a student read the problem aloud and tell what is to be solved. (if the tub holds about 12 quarts or 12 gallons) Work through the model with students as they fill in the equivalents. Have students tell the abbreviation for each of the 4 units of measure. (c, pt, qt, gal) Then have them complete the solution sentence.

Developing the Skill   Tell students to imagine being sent to the store to buy milk. Ask students what they would need to know. (size of container wanted) Students may also question the kind of milk wanted, for example, 1%, 2%, whole, skim, etc. Tell students that milk is packaged in half-pint, pint, quart, half-gallon and gallon containers. Tell students these are measures of capacity and that a cup is another measure of capacity. Fill a gallon container with water and have students use the water to fill quart containers. (4) Have a student record the finding on the board. (4 quarts = 1 gallon) Continue to find the number of pints in a gallon, cups in a pint, pints in a quart, cups in a gallon, etc. Have students record all findings. Ask students to name items and products that can be purchased in these various units.

Practice

Circle the better estimate of capacity.

1.

   1 pint   (1 gallon)

2.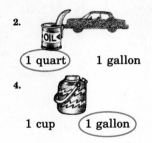

   (1 quart)   1 gallon

3.

   (8 cups)   8 quarts

4.

   1 cup   (1 gallon)

Complete each table.

5.
pints	1	2	3	4	5	6	7	8	9
cups	2	4	6	8	10	12	14	16	18

6.
quarts	1	2	3	4	5	6	7	8	9
pints	2	4	6	8	10	12	14	16	18

Apply

Solve these problems.

7. Ronnie bought 9 quarts of milk. Each quart cost 67¢. How much did Ronnie pay for milk?
$6.03

8. The dairy company produced 3,249 gallons of milk on Monday and 2,976 gallons on Tuesday. How many gallons of milk did the dairy produce in two days?
6,225 gallons

9. On Monday, a gas station had 16,075 gallons of gas delivered. By Wednesday 11,887 gallons were sold. How many gallons of gas were left?
4,188 gallons

10. Takashi drives a milk collection truck. After three stops, he has picked up 3,246 gallons. How many more gallons are needed to fill Takashi's milk truck if it's capacity is 6,000 gallons?
2,754 gallons

150

Customary Weight

Objective

To determine the appropriate unit of measure: ounce, pound and ton

Materials

*product containers having weight information on labels

Mental Math

Dictate the following problems:

1. $20 \times \frac{1}{2}$ of 8 (80)
2. $16 \div 2 \times 0$ (0)
3. $\$1.10 \times 4$ ($4.40)
4. $6\frac{1}{2} - 4\frac{1}{2}$ (2)
5. $56 \div (2 + 5)$ (8)
6. $60 \times (36 \div 6)$ (360)
7. $50\cancel{c} \times 10$ ($5)
8. $\$1 \times \frac{1}{2}$ of 6 ($3)

Skill Review

Have students work the following problems on the board:

16×4	$4\overline{)64}$	$4\overline{)67}$	$4\overline{)65}$
16×6	$6\overline{)96}$	$6\overline{)98}$	$6\overline{)99}$
16×5	$5\overline{)80}$	$5\overline{)84}$	$5\overline{)82}$

Have students compare the 3 division problems to each multiplication problem to see that the number of sixteens remains the same in each problem but there are different remainders.

Using Customary Units of Weight

Some customary units of weight are **ounces, pounds** and **tons.** Would the weight of a stick of butter more likely be 4 ounces or 4 pounds?

We know:

1 pound = 16 ounces	1 ton = 2,000 pounds
$\underline{1}$ lb = $\underline{16}$ oz	$\underline{1}$ T = $\underline{2,000}$ lb

Study these comparisons.

About 1 ounce   About 1 pound   About 1 ton

The stick of butter probably weighs about 4 $\underline{ounces}$.

Getting Started

Would these objects be weighed in ounces, pounds or tons?

1.   2.   3.

   ounces   tons   pounds

Use the table to help complete these measurements.

Pounds	1	2	3	4	5	6
Ounces	16	32	48	64	80	96

4. 50 oz = 3 lb $\underline{2}$ oz

5. 2 lb 5 oz = $\underline{37}$ oz

6. 4 lb 7 oz = $\underline{71}$ oz

7. 72 oz = $\underline{4}$ lb $\underline{8}$ oz

151

Teaching the Lesson

Introducing the Problem Have a student describe the picture. Have a student read the problem aloud and tell what is to be solved. (if a stick of butter would weigh 4 ounces or 4 pounds) Work through the model with students discussing the equivalents and comparisons. Have students tell the abbreviations for each of the 3 units of measure. (oz, lb, T) Then have them complete the solution sentence.

Developing the Skill Write on the board: **16 oz = 1 lb; 2,000 lb = 1T.** Ask students how many **tons** would equal 4,000 pounds. (2) Repeat for other multiples of 2,000. Ask students how many pounds would equal 32 ounces. (2) Repeat for other multiples of 16 through 96. Tell students that 2,100 pounds would equal 1 ton with 100 pounds left over. Ask students how many tons and left over pounds in 2,300 pounds. (1,300) Provide more examples for students to figure the tons and pounds left over. Repeat for pounds and ounces left over for 18 ounces, 37 ounces and 39 ounces. Have students list on the board the words ton, ounce and pound in order from least to greatest weight, and then from greatest to least weight. (ounce, pound, ton, and ton, pound, ounce)

Practice

Would these objects be weighed in ounces, pounds or tons?

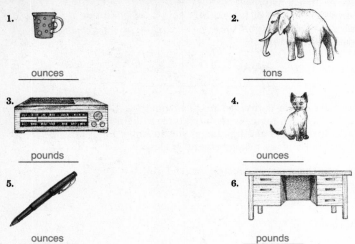

1. _____ ounces

2. _____ tons

3. _____ pounds

4. _____ ounces

5. _____ ounces

6. _____ pounds

Use the table on page 151 to help complete these measurements.

7. 53 oz = 3 lb ___5___ oz

8. 18 oz = ___1___ lb ___2___ oz

9. 4 lb 6 oz = ___70___ oz

10. 3 lb 4 oz = ___52___ oz

11. 85 oz = ___5___ lb ___5___ oz

12. 100 oz = ___6___ lb ___4___ oz

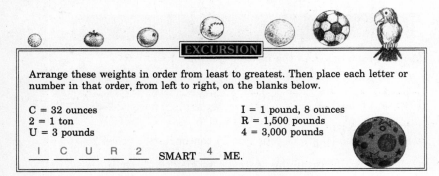

EXCURSION

Arrange these weights in order from least to greatest. Then place each letter or number in that order, from left to right, on the blanks below.

C = 32 ounces
2 = 1 ton
U = 3 pounds

I = 1 pound, 8 ounces
R = 1,500 pounds
4 = 3,000 pounds

__I__ __C__ __U__ __R__ __2__ SMART __4__ ME.

152

Practice

Tell students they are to write the most appropriate unit of measure for each object at the top of the page. Remind students that the table on page 151 will help them with the second exercise. Have them complete the page independently.

Excursion

Each letter in the exercise represents a word. If ordered correctly, the message will say,"I see you are too smart for me!"

Correcting Common Errors

Some students may have difficulty converting from ounces to pounds and vice versa. To give them more practice, have them work in groups with products from the grocery store. Have them make a list of all the containers with weights in pounds and use the table on page 151 to help convert them to ounces. Then have them convert the weights of containers expressed in ounces as pounds and ounces.

Enrichment

Have students write their birth weights and present weights in ounces. Then tell them to find their weight growth since birth in pounds and ounces.

Extra Credit *Statistics*

Extend the weather graphing activity introduced in the preceding extra credit. Have students create their own classroom weather station, equipping it with a thermometer, barometer and student-made rain gauge. Have them record weather statistics for a month, using their station equipment. Have students keep records by making graphs of the high and low temperatures for a week, make a similar graph for the temperatures for the same week a year ago and compare. Also have them research the record Hi and Low for each day's date, etc. See if students can think of other graphing activities.

Metric Length

Objective

To measure length in centimeters

Materials

rulers marked in centimeters
cards with lengths and widths
   varying in tenths of centimeters
   through 15.0 centimeters

Mental Math

Have students solve for n:

1. $48 - n = 32$ (16)
2. $5 \div n = 1$ (5)
3. $(n \times 7) + 16 = 65$ (7)
4. n is 1 more than $32 \div 2$ (17)
5. $n \times n = 60 + 4$ (8)
6. $\frac{1}{n}$ of $100 = 10$ (10)
7. $n \div 3 = 63 \div 7$ (27)
8. $n + n + n = 18$ (6)

Skill Review

Review rounding by drawing a number line from 0 through 100 with multiples of 10 noted. Ask students to name the multiple of 10 that is closest to 87. (90) Have students tell the multiple of 10 which is closest to other numbers you name.

Measuring Length in Centimeters

The **metric system** is another measurement system that we frequently use.
The **centimeter** is a unit of length in this system. Measure this golf tee to the nearest centimeter.

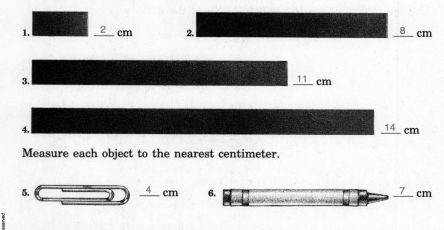

✔ This metric rule is divided into centimeters. A **cm** is the symbol for centimeter. The small line between centimeter markings helps you to know to which centimeter your measurement is closer.

Measured to the nearest centimeter, the golf tee

is __5__ centimeters long.

Getting Started

Measure each bar to the nearest centimeter.

1. __2__ cm   2. __8__ cm

3. __11__ cm

4. __14__ cm

Measure each object to the nearest centimeter.

5. __4__ cm   6. __7__ cm

7. Estimate the length of your arm in centimeters. ____ cm Answers will vary.

8. Measure your arm to the nearest centimeter. ____ cm

153

Teaching the Lesson

Introducing the Problem Have a student describe the picture and read the thought bubble. Read the problem to the students and ask what is to be found. (length of the golf tee to the nearest centimeter) Read and discuss the paragraph about centimeters with the students. Then have students complete the solution sentence.

Developing the Skill Remind students that there are 2 systems of measure used in the United States. Tell students they are going to work with the **metric** system now. Have students lay their **centimeter** rulers along the width of a card. Remind students to place the left edge of the ruler on the exact edge of the card. Suppose the card is 3.7 centimeters by 8.2 centimeters. Ask students if the card is exactly 3 centimeters in width. (no) Ask if it is closer to the 3 or the 4 centimeter mark. (4 centimeters) Tell students we would say the card's width is about 4 centimeters. Repeat the procedure for the length of the card to find it is about 8 centimeters in length. Have students check the measurements of other cards.

Practice

Measure each object to the nearest centimeter.

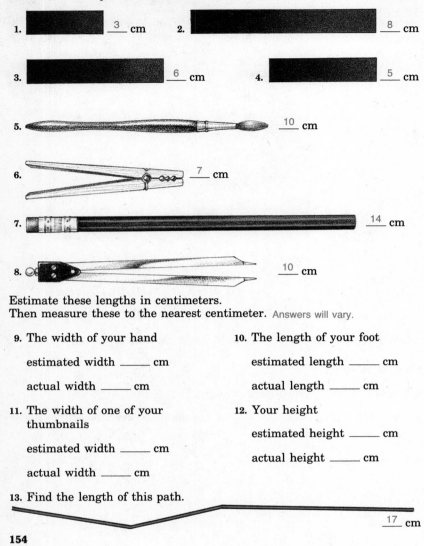

1. _3_ cm

2. _8_ cm

3. _6_ cm

4. _5_ cm

5. _10_ cm

6. _7_ cm

7. _14_ cm

8. _10_ cm

Estimate these lengths in centimeters.
Then measure these to the nearest centimeter. Answers will vary.

9. The width of your hand

   estimated width _____ cm

   actual width _____ cm

10. The length of your foot

    estimated length _____ cm

    actual length _____ cm

11. The width of one of your thumbnails

    estimated width _____ cm

    actual width _____ cm

12. Your height

    estimated height _____ cm

    actual height _____ cm

13. Find the length of this path.

    _17_ cm

154

154

More Metric Length

Objective

To determine and use the appropriate measure of length: centimeter, decimeter, meter and kilometer

Materials

meter sticks marked in centimeters and decimeters

Mental Math

Ask students which customary unit(s) of measure would they use for:

1. their height (foot, inch)
2. oil for car motor (quart)
3. newborn baby's length (inch)
4. buying fabric (yard)
5. speed limit (mile)
6. small amount of candy (ounce)
7. weight of tractor (ton)

Skill Review

Have students count by 10's through 2,000 in a relay, with each student counting the next 10 numbers from where the previous student left off. Remind students that when multiplying by 10, the product is that number with a zero to the right of it. Have students write the products of various 2-digit numbers multiplied by 10.

Understanding Decimeters, Meters and Kilometers

Some other units in length in the metric system are **decimeters**, **meters** and **kilometers**. How do these compare with centimeters?

 A crayon is about a decimeter long.

 A first-grader is about a meter tall.

We know:

$$1 \text{ decimeter} = 10 \text{ centimeters}$$

$$\underline{\quad 1 \quad} \text{ dm} = \underline{\quad 10 \quad} \text{ cm}$$

$$1 \text{ meter} = 10 \text{ decimeters} \qquad 1 \text{ kilometer} = 1{,}000 \text{ meters}$$

$$\underline{\quad 1 \quad} \text{ m} = \underline{\quad 10 \quad} \text{ dm} \qquad \underline{\quad 1 \quad} \text{ km} = \underline{\quad 1{,}000 \quad} \text{ m}$$

$$\underline{\quad 1 \quad} \text{ m} = \underline{\quad 100 \quad} \text{ cm}$$

Study these comparisons.

 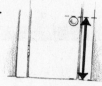 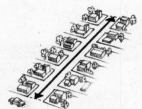

About 1 decimeter   About 1 meter   About 1 kilometer

It takes __10__ centimeters to make a decimeter, __100__ centimeters to make a meter and __1,000__ meters to make a kilometer.

Getting Started

Circle the best estimate of measurement.

1. the width of this page
   (21 cm)   21 m   21 km

2. the distance from Chicago to Miami
   2,000 dm   2,000 m   **(2,000 km)**

3. Complete this table.

meters	1	2	3	4	5	6	7	8
decimeters	10	20	30	40	50	60	70	80

Teaching the Lesson

Introducing the Problem Have a student describe the picture and read the thought bubbles. Read the problem to the students and ask what question is to be answered. (how decimeters, meters and kilometers compare with centimeters) Work through the equivalents with students and discuss the comparisons. Then have them complete the solution sentence.

Developing the Skill Ask students why we would not use centimeters to measure a street's length. (The numbers would be large and awkward to use.) Tell students that the metric system's units to measure longer lengths are the **decimeter, meter** and **kilometer**. Discuss the abbreviations for each. Tell students the metric system is based on units of 10. Have a student draw a 1 cm line on the board. Tell students that 10 cm is equal to 1 dm. Have a student draw a 10 cm line under the 1 cm line. Tell students that 10 dm is equal to 1 meter. Have a student draw a 1 m line under the

other 2 lines. Show a meter stick marked in cm and ask how many cm are in 1 m. (100) Have a student draw a 100 cm line. Ask students if 100 cm is equal to 10 dm. (yes) Tell students that a much longer line of 1,000 m is equal to 1 km. Tell students that the width of their fingernail gives an idea of 1 cm and a giant step gives an idea of the length of 1 m.

Practice

Circle the best estimate of measurement.

1. the length of 6 city blocks

   1 dm   (1 km)   1 m

2. the height of your classroom

   4 cm   (4 m)   4 km

3. the length of a paper clip

   (3 cm)   3 dm   3 m

4. the length of a school playground

   (75 m)   75 dm   75 km

5. the distance a car travels in 1 hour

   80 cm   80 m   (80 km)

6. the height of a box of cereal

   3 cm   (3 dm)   3 m

7. Complete this table.

meters	1	2	3	4	5	6	7	8
centimeters	100	200	300	400	500	600	700	800

8. Complete the table and use it for problems 9 and 10.

   ### Average Athlete's Pace

minutes	10	20	30	40	50	60
kilometers	3	6	9	12	15	18

9. Approximately how far can an athlete run in one hour?

   18 kilometers

10. What would be the approximate time for a 15-kilometer run?

    50 minutes

Apply

Solve these problems.

11. A fence is constructed of 25 sections. Each section is 6 meters long. How long is the fence?

    150 meters

12. The puppy is 38 centimeters long. The kitten is 3 decimeters long. Which pet is longer? How much longer is it?

    The puppy, 8 centimeters

156

Metric Capacity

pages 157-158

Objective

To determine and use the appropriate measure of capacity: liter and milliliter

Materials

*various small containers
*transparent liter containers
*4 drinking glasses
*eyedroppers with 1 milliliter marked
*teaspoons

Mental Math

Ask students how many hours and minutes from:

1. noon to 1:30 AM (13, 30)
2. 2:16 AM to 4:26 AM (2, 10)
3. 7:00 PM to 3:30 AM (8, 30)
4. 4:28 AM to 7:30 AM (3, 2)
5. midnight to 1:46 PM (13, 46)
6. 8:00 AM to 4:45 PM (8, 45)
7. 9:15 PM to 6:30 AM (9, 15)
8. 11:49 AM to 3:30 PM (3, 41)

Skill Review

Write on the board: **1/2 of 10 = .**
Have a student complete the problem.
(5) Repeat for 1/5 of 10, 1/2 of 100,
1/4 of 100, 1/5 of 100, 1/2 of 1,000,
1/4 of 1,000 and 1/5 of 1,000.

Understanding Metric Units of Capacity

Two of the most often used metric units of capacity are the **liter** and the **milliliter.** Would you probably drink 250 liters or 250 milliliters of milk at one meal?

We know:
1 liter = 1,000 milliliters

$\underline{\quad 1 \quad}$ L = $\underline{\quad 1,000 \quad}$ mL

Study these comparisons.

About 1 liter   About 250 milliliters   About 10 milliliters

A glass might hold about 250 ____milliliters____ of milk.

Getting Started

Circle the better estimate of capacity.

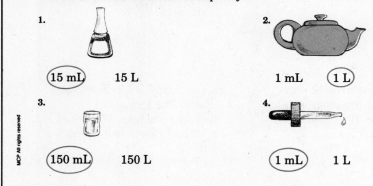

1.   (15 mL)   15 L

2.   1 mL   (1 L)

3.   (150 mL)   150 L

4.   (1 mL)   1 L

157

Teaching the Lesson

Introducing the Problem Have a student describe the picture. Have another student read the problem and tell what question is to be answered. (if a glass of milk contains 250 liters or 250 milliliters) Work through the equivalents with students and discuss the comparisons. Have students tell the abbreviations for liter and milliliter. (L, mL) Then have students complete the solution sentence.

Developing the Skill Fill a liter container with water. Tell students that the **liter** is a metric unit of capacity. Have a student pour the liter of water into the 4 glasses to show that a liter is about 4 glassfuls. Now fill the teaspoon with water and tell students the spoon holds about 4 milliliters of water. Tell students that one **milliliter** is a very small amount and an eye dropper works well to measure that quantity. Have students work with the eyedroppers and teaspoons to realize the comparable capacities of 1 milliliter and a teaspoon.

Practice

Circle the better estimate of capacity.

1.

25 mL   (25 L)

2.

1 mL   (1 L)

3.

(5 mL)   5 L

4.

(200 mL)   200 L

Apply

Solve these problems.

5. A pitcher holds 3 liters of juice. The fourth grade fills the pitcher 16 times at the class picnic. How many liters of juice does the fourth grade use?

48 liters

6. Marcie takes 8 milliliters of vitamin C, 6 milliliters of vitamin B and 9 milliliters of vitamin A everyday. How many milliliters of vitamins does Marcie take daily?

23 milliliters

Circle the smaller containers that can be filled by the larger container in each exercise.

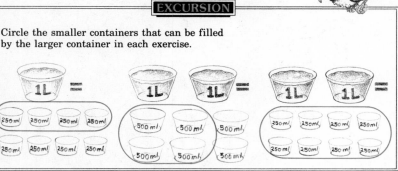

158

158

Metric Weight

pages 159-160

Objective

To determine and use the appropriate measure of weight: gram and kilogram

Materials

*items weighing about 1 gram
*items weighing about 1 kilogram
*scale that weighs in kilograms
*eyedroppers

Mental Math

Tell students if n = 4, and p = 6, find:

1. $2 \times n + 2 \times p$ (20)
2. $1/n$ of 40 (10)
3. $(2 \times p) - n$ (8)
4. $6 (n + p)$ (60)
5. $(2 \times p) \div n$ (3)
6. $5 \times p - 2 \times n$ (22)

Skill Review

Write on the board:

Capacity

Customary	Metric
cup (c)	**milliliter** (mL)
pint (pt)	**liter** (L)
quart (qt)	
gallon (gal)	

Have students write the abbreviations for each. Repeat for both customary and metric units of length.

Understanding Metric Units of Weight

Two of the most commonly used metric units of weight are **grams** and **kilograms**. Would a penny weigh about 3 grams or 3 kilograms?

We know:
1 kilogram = 1,000 grams

$\underline{1}$ kg = $\underline{1,000}$ g

Study these comparisons.

About 1,000 kg   About 1 kg   About 1 g

A penny weighs about 3 <u>grams</u>.

Getting Started

Circle the better estimate of weight.

1.

1,000 g   (1,000 kg)

2.

(5 g)   5 kg

3.

(30 g)   30 kg

4.

(350 g)   350 kg

159

Teaching the Lesson

Introducing the Problem Have a student describe the picture and read the thought bubbles. Read the problem to the students and ask what question is to be answered. (if a penny weighs about 3 grams or 3 kilograms) Work through the equivalents with students and discuss the comparisons. Have students tell the abbreviations for gram and kilogram. (g, kg) Then have students complete the solution sentence.

Developing the Skill Show students an array of items weighing about 1 gram or 1 kilogram each. Tell students that 1 **gram** is a very small amount of weight and that 1,000 grams equals 1 **kilogram.** Tell students they are to use this information to make an educated guess of the weight of each item and sort them into groups weighing about 1 gram or 1 kilogram. Have students use the scale to find the actual weight of each item to verify their guesses.

159

Practice

Circle the better estimate of weight.

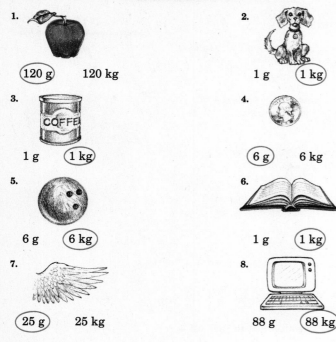

1. 120 g   120 kg

2. 1 g   1 kg

3. 1 g   1 kg

4. 6 g   6 kg

5. 6 g   6 kg

6. 1 g   1 kg

7. 25 g   25 kg

8. 88 g   88 kg

Apply

Solve these problems.

9. A large paper clip weighs 3 grams. Sean has a box of paper clips that has a net weight of 87 grams. How many paper clips are in the box?
29 paper clips

10. Oranges cost 68¢ a kilogram. Mary bought 6 kilograms of oranges. How much did Mary pay for the oranges?
$4.08

160

Problem Solving
Make a Model

pages 161-162

Objective

To make a model to solve problems

Materials

items such as pennies or similar circles,
  blocks, straws or toothpicks, etc.
graph paper

Mental Math

Have students tell the cost of:

1. 20 at 14¢ ($2.80)
2. 1 if 60/$1.80 (3¢)
3. 40 if 2/$1 ($20)
4. 9 if 5/$400 ($720)
5. 60 if 5/$1 ($12)
6. 800 at 10¢ ($80)
7. 4 if 8/$10 ($5)
8. 15 if 5/75¢ ($2.25)

Making a Model

How many ways can you tear off 4 attached
stamps from a sheet of postage stamps?

★ SEE
We want to know how many patterns of
4 attached stamps we can make.

★ PLAN
We can use a sheet of graph paper to
represent the stamps and cut out
different patterns of four squares.

★ DO

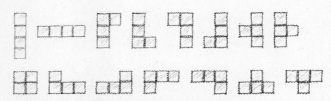

There are __15__ different ways to tear off 4
attached stamps.

★ CHECK

4 + 0 = 4   There are __2__ ways we can tear off
stamps with 4 in a row or column.

3 + 1 = 4   There are __12__ ways we can tear off
stamps with 3 in one row or column
and one in another row or column.

2 + 2 = 4   There is __1__ way we can tear off
stamps with 2 stamps in one row or
column and 2 stamps in another row
or column.

161

Teaching the Lesson

Remind students that they have learned to solve problems
by making drawings. Tell students that some problems are
easier for us to understand if we use actual items that can
be moved around and looked at from different sides. Tell
students we call this a **model** and it represents data in a
problem like a model airplane or car represents the real
vehicle. Tell students it is important when making a model
to choose objects which accurately represent the real thing.
Have a student read the problem and its SEE and PLAN
stages. Have students use graph paper to make a model if
stamps are not available. Tell them they can make the graph
paper more like stamps by drawing pictures on each square.
Have students experiment to draw each 4-stamp pattern.
Help students develop strategies to be sure they have included all possibilities and then fill in the solution sentences.
Have students complete the CHECK stage.

Apply

Make a model to help you solve each problem.

1. Move one block to another stack so the sums of all stacks are equal.

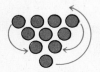

Move the 9-block to the column on the left.

2. It takes 27 sugar cubes to fill a cubical box. How many of the cubes will be touching the sides, top and bottom of the box?

26 cubes

3. Change the arrangement of the discs on the left to the arrangement on the right by moving only 3 of the discs.

4. Form 5 equilateral triangles using 9 sticks of equal length.

See Solution Notes.

5. Kevin makes a triangle with cubic blocks. He puts 1 block in the first row, 2 blocks in the second row, 3 blocks in the third row, and so on. Predict whether he will need more or fewer than 100 blocks to make 10 rows. Then prove that your prediction was correct.

See Solution Notes.

6. A lady bug walks around the edges of the top of a box of crackers. The top is shaped like a rectangle. When the bug gets back to where she started, she had walked 48 cm. If the top of the box is 6 cm wide, explain how you could find the length of the top of the box.

Answers will vary.

7. Read Exercise 2 again. What if it took 64 cubes to fill the box? Now how many of the cubes would be touching the sides, top, and bottom of the box?

56 cubes

8. There are 12 cubic blocks on the bottom layer of a box. The box is completely filled with blocks. Why could there not be 10 layers if there are fewer than 100 blocks in the box?

Answers will vary.

162

Solution Notes

1. Help students see to move one block at a time until they find that the 9 block needs to be moved to the far left column.

2. Remind students that a cube has equal edges. Help students search for a number that when multiplied by itself and again by itself, or cubed, is equal to 27. (3) The solution of 26 cubes is then found by thinking of a cube with 3 across, 3 up and 3 down so that only the 1 cube in the middle does not touch the sides, top or bottom.

4. Remind students that an equilateral triangle has 3 equal sides. Ask if all the triangles must be the same size. (no) Remind students to keep this in mind as they form their model, since the 5th triangle is the outline formed by the 4 smaller triangles.

Higher-Order Thinking Skills

5. [Analysis] Each row has the same number of blocks as the number of the row; the 10th row has 10 blocks and the greatest number of blocks in any of the 10 rows. So, there are fewer than ten 10-blocks in the 10 rows. Students can draw a picture or use blocks to prove their answer.

6. [Synthesis] Answers will vary but should include that there are 2 edges of 6 cm; so, the other 2 edges must have a total length of 48 cm − 12 cm, or 36 cm. Being equal in length, each edge is therefore 18 cm.

8. [Synthesis] A possible answer is that since there are 12 blocks in each layer, $10 \times 12 = 120$, which is more than 100.

Extra Credit   *Biography*

Mary Somerville was born in Scotland in 1780. Like other women mathematicians, Mary had to struggle against her parents' opposition to her studies. Intelligent women of the time were urged to hide their learning to avoid drawing envy and losing friends. Mary was sent to boarding school at age nine to learn to read and write. As a teenager, Mary discovered algebra and began to study it intensely. She read far into the night, using up the family's supply of candles. Her parents hid the candles from her, but she persisted, rising at dawn to study. Mary became a science writer, a person who could make complex, technical subjects understandable. She surpassed her male colleagues by the success of her work. Mary was one of the first women to become a member of the Royal Astronomical Society. Mary continued her studies and writing until she died at age 92.

Chapter Test

page 163

Item	Objective
1	Use AM and PM notation; find time by moving minute hand backward (See pages 145-146)
2	Use AM and PM notation; find time by moving hour and minute hands forward (See pages 145-146)
3-4	Determine appropriate customary unit of length (See pages 147-148)
5-6	Determine appropriate customary unit of capacity (See pages 149-150)
7-8	Determine appropriate customary unit of weight (See pages 151-152)
9-10	Determine appropriate metric unit of length (See pages 153-156)
11	Determine appropriate metric unit of capacity (See pages 157-158)
12	Determine appropriate metric unit of weight (See pages 159-160)

163

Write the times. Include AM or PM.

1. 45 minutes before midnight

11:15 PM

2. 3 hours and 15 minutes after 10:30 AM

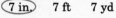

1:45 PM

Circle the best estimate for each measurement.

3.

6 in.   (6 ft)   6 mi

4.

(7 in.)   7 ft   7 yd

5.

1 pt   (1 qt)   1 gal

6.

(2 c)   2 pt   2 qt

7.

1 oz   (1 lb)

8.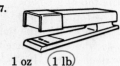

(15 oz)   15 lb

9.

(3 cm)   3 dm   3 m

10.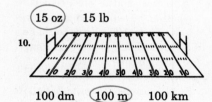

100 dm   (100 m)   100 km

11.

2 mL   (2 L)

12.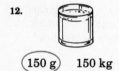

(150 g)   150 kg

163

Circle the letter of the correct answer.

1 Round 650 to the nearest hundred.
a 60
b 700
c NG

2 Round 4,296 to the nearest thousand.
a 4,000
b 5,000
c NG

3 What is the value of the 0 in 625,309?
a ten thousands
b thousands
c hundreds
d NG

4 3,475
   + 2,686
a 5,061
b 5,161
c 6,161
d NG

5 36,439
   + 17,806
a 43,235
b 54,215
c 54,245
d NG

6 822
   − 387
a 435
b 535
c 662
d NG

7 86,042
   − 41,385
a 44,357
b 45,343
c 54,357
d NG

8 6 × 8
a 48
b 54
c 63
d NG

9 26
  × 3
a 68
b 78
c 618
d NG

10 65 × 7
a 425
b 435
c 635
d NG

11 42 ÷ 7
a 5
b 7
c 9
d NG

12 4)48
a 1
b 2
c 12
d NG

13 86 ÷ 6
a 14 R2
b 14 R24
c 14
d NG

[ ] score

164

Cumulative Review

page 164

Item	Objective
1	Round numbers to nearest 100 (See pages 31-34)
2	Round numbers to nearest 1,000 (See pages 33-34)
3	Identify place value through hundred thousands (See pages 27-28)
4	Add two 4-digit numbers (See pages 47-48)
5	Add two 5-digit numbers (See pages 53-54)
6	Subtract two 3-digit numbers (See pages 63-64)
7	Subtract two 5-digit numbers with zero in minuend (See pages 67-68)
8	Multiply using factor 8 (See pages 93-94)
9-10	Multiply 2-digit number by 1-digit number with two regroupings (See pages 111-112)
11	Divide using 7 as divisor (See pages 125-126)
12	Divide 2-digit number by 1-digit number to get 2-digit quotient without remainder (See pages 133-134)
13	Divide 2-digit number by 1-digit number to get 2-digit quotient with remainder (See pages 135-136)

Alternate Cumulative Review

Circle the letter of the correct answer.

1 Round 836 to the nearest hundred
a 700
b 800
c 900
d NG

2 Round 5,687 to the nearest thousand.
a 5,000
b 6,000
c 7,000
d NG

3 What is the value of the 5 in 723,586?
a ten thousands
b thousands
c hundred thousands
d NG

4 4,953
   + 2,268
a 6,111
b 6,122
c 7,111
d NG

5 27,266
   + 24,817
a 41,073
b 52,083
c 52,183
d NG

6 718
   − 469
a 249
b 351
c 349
d NG

7 92,031
   − 51,654
a 40,387
b 41,487
c 41,623
d NG

8 7 × 9 =
a 49
b 56
c 63
d NG

9 19
  ×6
a 6543
b 114
c 75
d NG

10 37
  ×6
a 222
b 272
c 1,842
d NG

11 81 ÷ 9 =
a 8
b 9
c 10
d NG

12 2)46
a 2
b 20 R6
c 23
d NG

13 8)96
a 1
b 12
c 11 R8
d NG

Multiplying by Powers of 10

pages 165-166

Objective

To multiply l-digit numbers by a power of 10

Materials

Mental Math

Have students name the unit of measure:

1. L (liter)
2. ft (foot)
3. lb (pound)
4. kg (kilogram)
5. mL (milliliter)
6. c (cup)
7. dm (decimeter)
8. qt (quart)
9. oz (ounce)

Skill Review

Write on the board: **6 × 10** and **9 × 10.** Have students tell the products. (60, 90) Ask students to recall the rule for multiplying a number by 10. (The product is that number followed by a zero.) Have students find the products of more 1-digit numbers and 10.

MULTIPLYING WHOLE NUMBERS

Multiplying by Powers of 10

Ricky's job is shoveling snow from his front walk. The walk is 9 meters long. Find the length of Ricky's shoveling job in centimeters.

We want to know how many meters of walk Ricky has to shovel.

The walk is ___9___ meters long.

Each meter contains ___100___ centimeters. To find the length of the walk in centimeters,

we multiply ___9___ by ___100___.

Study these multiplications.

$3 \times 1 = 3$	$4 \times 1 = 4$	$6 \times 2 = 12$	$9 \times 6 = 54$
$3 \times 10 = 30$	$4 \times 10 = 40$	$6 \times 20 = 120$	$9 \times 60 = 540$
$3 \times 100 = 300$	$4 \times 100 = 400$	$6 \times 200 = 1,200$	$9 \times 600 = 5,400$
$3 \times 1,000 = 3,000$	$4 \times 1,000 = 4,000$	$6 \times 2,000 = 12,000$	$9 \times 6,000 = 54,000$

> Multiply the digits that are not zeros. The product has the same number of zeros as there are zeros in the factors.

$$9 \times 100 = \underline{900}$$

Ricky's front walk is ___900___ centimeters long.

Getting Started

Multiply.

1. $6 \times 100 =$ ___600___   2. $100 \times 5 =$ ___500___   3. $7 \times 1,000 =$ ___7,000___

4. $10 \times 9 =$ ___90___   5. $6,000 \times 7 =$ ___42,000___   6. $6 \times 9,000 =$ ___54,000___

165

10 cm = 1 dm
100 cm = 1 m

Teaching the Lesson

Introducing the Problem Have a student describe the picture. Ask students to tell the units of measure which are abbreviated on the sign. (centimeter, decimeter, meter) Ask a student to read the problem and tell what is to be found. (the length of the walk in centimeters) Have students complete the sentences to solve as they read with you. Point out the multiplication progression shown in the model. Have students complete the solution sentence.

Developing the Skill Write **100 × 6 = 600** on the board and tell students that a number times 100 is that number followed by 2 zeros. Write **2 × 200** on the board and ask students to apply the rule for multiplying by 100 to find the product. (2 × 1 = 2 and 2 followed by 2 zeros is

200) Have students work more problems of 1-digit numbers times 100. Write **1,000 × 6** on the board and help students develop a rule for multiplying by 1,000. (The product is that number followed by 3 zeros.) Tell students that **when multiplying by multiples of ten, the product is the product of the 2 numbers that are not zero, followed by the number of zeros in the factors.** Now write **200 × 9** on the board and tell students to multiply the 2 by the 9 for a product of 18. Ask students how many zeros are in the factors. (2) Write **00** after the 18 and have students read the number. (1,800) Repeat for more numbers.

Practice

Multiply.

1. 5 × 100 = _500_
2. 6 × 3,000 = _18,000_
3. 20 × 9 = _180_
4. 10 × 7 = _70_
5. 1,000 × 7 = _7,000_
6. 80 × 3 = _240_
7. 400 × 5 = _2,000_
8. 2 × 9,000 = _18,000_
9. 9 × 700 = _6,300_
10. 800 × 8 = _6,400_
11. 3 × 5,000 = _15,000_
12. 7,000 × 8 = _56,000_
13. 4 × 40 = _160_
14. 300 × 7 = _2,100_
15. 6,000 × 2 = _12,000_
16. 6 × 70 = _420_
17. 5 × 2,000 = _10,000_
18. 80 × 9 = _720_
19. 600 × 5 = _3,000_
20. 4,000 × 8 = _32,000_
21. 30 × 9 = _270_
22. 7 × 700 = _4,900_
23. 5,000 × 6 = _30,000_
24. 300 × 4 = _1,200_
25. 80 × 7 = _560_
26. 6,000 × 7 = _42,000_
27. 4 × 9,000 = _36,000_
28. 5 × 800 = _4,000_
29. 3 × 700 = _2,100_
30. 8 × 5,000 = _40,000_
31. 2 × 8,000 = _16,000_
32. 9 × 9,000 = _81,000_
33. 6 × 40 = _240_
34. 7,000 × 2 = _14,000_
35. 500 × 4 = _2,000_
36. 9 × 60 = _540_
37. 6,000 × 6 = _36,000_
38. 4 × 8,000 = _32,000_
39. 400 × 9 = _3,600_

Apply

Solve these problems.

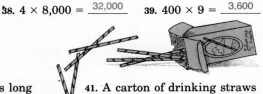

40. How many centimeters long is a table that is 9 decimeters in length?
90 centimeters

41. A carton of drinking straws contains 800 straws. How many straws are in 7 cartons?
5,600 straws

42. Computers cost $2,000 each. How much will the school pay for 7 computers?
$14,000

43. A small car weighs 3,000 pounds. How much do 8 small cars weigh?
24,000 pounds

166

Correcting Common Errors

Some students will write an incorrect number of zeros when multiplying. Have them complete the table shown below. For each answer, ask them to compare the number of zeros in the multiples of ten, hundred, and thousand with the number in the product.

	6	3	8	7
10	60	30	80	70
100	(600)	(300)	(800)	(700)
500	(3,000)	(1,500)	(4,000)	(3,500)
1,000	(6,000)	(3,000)	(8,000)	(7,000)
6,000	(36,000)	(18,000)	(48,000)	(42,000)
9,000	(54,000)	(27,000)	(72,000)	(63,000)

Enrichment

Tell students to make a 5-year table to show the total mileage on a car at the end of each year, if it is driven an average of 6,000 miles each year. Then have them show the mileage if the car is driven 9,000 or 7,000 miles each year.

Practice

Remind students to refer to the model on page 165 if necessary as they work to find the products in the 3 columns. Have students complete the page independently.

Mixed Practice

1. 38 × 5 (190)
2. 350 − 209 (141)
3. 17 ÷ 8 (2 R1)
4. (9 ÷ 9) × 9 (9)
5. 72,156 + 27,408 (99,564)
6. 6 × $.73 ($4.38)
7. $743.20 − 270.94 ($472.26)
8. 64 ÷ 5 (12 R4)
9. (8 × 7) + 4 (60)
10. 462 + 703 + 297 (1,462)

Extra Credit   *Measurement*

Have students estimate the weight, in grams, of various familiar objects. First give examples of items that weigh about a gram, such as a safety pin or paper clips. Next, name items that weigh about 1 kilogram, for example a baseball bat or 2 footballs. Have students gather 6–8 objects such as: a dime, comb, classroom globe, pair of tennis shoes or a math book. Have students estimate the weight of each item and gather the items into groups according to the following categories: under 10 grams, 11–100 grams, 101–500 grams, 501–999 grams, 1–5 kilograms, 5–10 kilograms and over 10 kilograms. Tell students to check their estimations for accuracy by weighing the lighter objects on a balance scale, using a set of metric weights, or heavier items on a metric bathroom scale.

3-digit Multiplicands

pages 167-168

Objective

To multiply 3-digit numbers by 1-digit numbers with trading

Materials

place value materials

Mental Math

Have students complete each comparison: 6 is to 36 as 3 is to 18 as:

1. 4 is to ___ (24)
2. 8 is to ___ (48)
3. 5 is to ___ (30)
4. 9 is to ___ (54)
5. 7 is to ___ (42)
6. 2 is to ___ (12)
7. 1 is to ___ (6)
8. 10 is to ___ (60)

Skill Review

Write problems such as **36 × 8** and **43 × 3** on the board for review of multiplying 2-digit numbers by 1-digit numbers, with or without trading. Have students talk through each problem as it is worked.

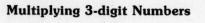

Multiplying 3-digit Numbers

Mach 1 is the unit used to measure the speed of sound. Major Joseph Rogers was the first person to fly a plane faster than Mach 2. How fast was Major Rogers flying when he reached Mach 2?

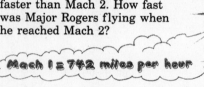

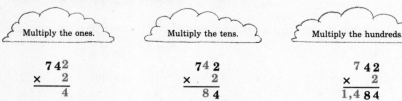

Mach 1 = 742 miles per hour

We want to know the miles per hour speed for Mach 2.

We know Mach 1 is ___742___ miles per hour.

Mach 2 is ___2___ times as fast.

To find Mach 2, we multiply ___2___ by ___742___.

Multiply the ones.	Multiply the tens.	Multiply the hundreds.
74**2** × __2__ 4	7**4**2 × __2__ 8 4	742 × __2__ 1,4 8 4

Major Rogers was flying ___1,484___ miles per hour when he reached Mach 2.

Getting Started

Multiply.

1. 223
    × __4__
    892

2. 112
    × __5__
    560

3. 730
    × __3__
    2,190

4. 411
    × __5__
    2,055

Copy and multiply.

5. 304 × 2
   608

6. 131 × 5
   655

7. 612 × 4
   2,448

8. 243 × 3
   729

167

Teaching the Lesson

Introducing the Problem Have a student tell what they see in the picture. Tell students that a Mach is a unit of measure to indicate the speed of sound and was named for the physicist who studied sound. Discuss how sound is produced as air vibrates. Have a student read the problem and tell what they are to find. (the speed of Mach 2) Ask what information is needed to solve the problem. (the speed of Mach 1, 742 miles per hour) Have students complete the sentences and guide them through the multiplication shown in the model. Have students round the speed of Mach 1 to the nearest hundred and multiply by 2 to check their solution.

Developing the Skill Draw the following place value grid on the board:

hundreds	tens	ones
1	3	2
x		3

Cover the 1 in hundreds place and remind students they already know how to multiply 32 × 3. Have a student multiply the tens and ones columns. Uncover the 1 and tell students they now need to multiply the hundreds by 3. Have a student tell what 3 × 100 equals. (300) Show students that the number of hundreds goes in hundreds place. Repeat for **423 × 4** and **642 × 4**.

167

Practice

Multiply.

1. 823
    × 3
    ———
    2,469

2. 510
    × 5
    ———
    2,550

3. 132
    × 4
    ———
    528

4. 721
    × 4
    ———
    2,884

5. 634
    × 2
    ———
    1,268

6. 124
    × 3
    ———
    372

7. 704
    × 2
    ———
    1,408

8. 911
    × 5
    ———
    4,555

Copy and Do

9. 2 × 734
   1,468

10. 3 × 204
    612

11. 620 × 4
    2,480

12. 814 × 2
    1,628

13. 433 × 3
    1,299

14. 311 × 5
    1,555

15. 4 × 323
    1,292

16. 5 × 921
    4,605

Apply

Solve these problems.

17. There are 212 rubber bands in each box. How many rubber bands are in 4 boxes?
    848 rubber bands

18. Marty ran 425 meters in 2 minutes. Todd ran 512 meters in the same time. How much farther did Todd run?
    87 meters

EXCURSION

What is a Roman emperor's favorite food? Find the correct letters by matching the numerals under each blank with the Roman Numeral in the chart.

S XXII	L XXXVII	A XI
E IV	C I	S LXVII
D LVI	A XLVI	A XXIX
A II	R XVI	S VII

C   A   E   S   A   R
1   2   4   7   11 16

S   A   L   A   D   S
22 29 37 46 56 67

168

Multiplying Money

pages 169-170

Objective

To multiply money through $9.99 by 1-digit numbers with trading

Materials

Mental Math

Have students name these equivalents:

1. 32 oz = (2) lb
2. 8,000 lb = (4) tons
3. 2,000 m = (2) km
4. 88 qt = (22) gal
5. 4 dm = (40) cm
6. 6 L = (6,000) mL
7. 300 cm = (3) m
8. 3 pt = (6) c

Skill Review

Write the following problems vertically on the board to have students review multiplication of money: **$0.76 × 3, $0.19 × 6, $0.47 × 6, $0.59 × 2.** Have students work the problems and then check by estimating the amounts to the nearest dime.

Multiplying Money

Mrs. Juarez is planning to cook dinner on the grill. She needs 4 pounds of steak. How much will it cost if she buys porterhouse steaks?

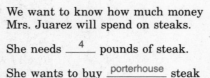

We want to know how much money Mrs. Juarez will spend on steaks.

She needs ___4___ pounds of steak.

She wants to buy __porterhouse__ steak

which sells for __$3.19__ a pound.

To find the total cost of the steak,

we multiply __$3.19__ by ___4___.

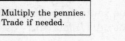

Multiply the pennies. Trade if needed.	Multiply the dimes. Add the extra dimes. Place the decimal point.	Multiply the dollars and add the dollar sign.
$$\begin{array}{r} \overset{3}{\$3.19} \\ \times \quad 4 \\ \hline 6 \end{array}$$	$$\begin{array}{r} \overset{3}{\$3.19} \\ \times \quad 4 \\ \hline 76 \end{array}$$	$$\begin{array}{r} \$3.19 \\ \times \quad 4 \\ \hline \$12.76 \end{array}$$

Mrs. Juarez will spend __$12.76__ on steak.

Getting Started

Multiply.

1. $$\begin{array}{r} \$6.13 \\ \times \quad 5 \\ \hline \$30.65 \end{array}$$
2. $$\begin{array}{r} \$8.26 \\ \times \quad 3 \\ \hline \$24.78 \end{array}$$
3. $$\begin{array}{r} \$7.18 \\ \times \quad 5 \\ \hline \$35.90 \end{array}$$
4. $$\begin{array}{r} \$4.37 \\ \times \quad 2 \\ \hline \$8.74 \end{array}$$

Copy and multiply.

5. 6 × $4.09
   $24.54
6. $3.29 × 2
   $6.58
7. 5 × $3.19
   $15.95
8. 2 × $6.01
   $12.02

169

Teaching the Lesson

Introducing the Problem Have a student describe the picture. Ask students how much each kind of steak costs per pound. (Sirloin is $3.29, T-Bone is $4.09 and Porterhouse is $3.19 per pound.) Have a student read the problem and tell what is to be solved. (the cost of 4 lbs of Porterhouse) Ask students if there is unnecessary information given. (yes) Have students read and complete the sentences to solve the problem. Guide students through the multiplication in the model, and have them complete the solution sentence. Have students estimate the cost of 1 steak to the nearest dime to check their solution.

Developing the Skill Write **326 × 4** and **$3.26 × 4** vertically on the board. Ask students how the 2 problems are similar. (Both have the same numbers.) Ask students how these problems differ. (One has a dollar sign and decimal point.) Have students talk through the problems as they work. Remind students that the decimal point goes before the dimes digit. Have a student place the dollar sign and decimal point. Repeat for **$7.86 × 2** and **$3.99 × 4.** Have students read the products to practice reading money amounts. Remind students that they say the word, **and** for the decimal in dollar amounts.

169

Practice

Multiply.

1. $6.23
   × 4
   ___
   $24.92

2. $7.17
   × 5
   ___
   $35.85

3. $5.38
   × 2
   ___
   $10.76

4. $9.15
   × 6
   ___
   $54.90

5. $4.08
   × 7
   ___
   $28.56

6. $8.14
   × 6
   ___
   $48.84

7. $3.49
   × 2
   ___
   $6.98

8. $6.26
   × 3
   ___
   $18.78

9. $2.25
   × 3
   ___
   $6.75

10. $9.09
    × 8
    ___
    $72.72

11. $7.29
    × 3
    ___
    $21.87

12. $3.17
    × 4
    ___
    $12.68

Copy and Do

13. 3 × $7.26
    $21.78

14. $4.12 × 8
    $32.96

15. $7.19 × 5
    $35.95

16. 2 × $8.37
    $16.74

17. $6.09 × 7
    $42.63

18. $8.13 × 6
    $48.78

19. 4 × $9.23
    $36.92

20. 3 × $4.29
    $12.87

Apply

Solve these problems.

21. Hal bought 5 gallons of paint. Each gallon costs $7.15. How much did Hal pay for the paint?
    $35.75

22. A fishing rod costs $8.97. A spinning reel costs $7.97. What is the total cost of the rod and reel?
    $16.94

23. Radial tires are $49.35 each. Regular tires are $34.89 each. How much more are the radial tires?
    $14.46

24. Oil filters are packed 6 to a box. One oil filter costs $1.09. How much does one box cost?
    $6.54

Use the data on page 169 to solve these problems.

25. How much more will 3 pounds of T-bone steaks cost than 3 pounds of sirloin steaks?
    $2.40

26. Mr. Kelly bought 2 pounds of T-bone and 2 pounds of porterhouse steaks. How much change did he receive from a $20 bill?
    $5.44

170

Correcting Common Errors

Some students may add the renamed number before they multiply.

INCORRECT   CORRECT
   1   1
$6.24 $6.24
×   3   ×   3
_____   _____
$18.92 $18.72

Have students work with partners to use the following form to compute.

$3.19
×   4
_____
.36 ← 4 × 9 pennies
.40 ← 4 × 1 dime
12.00 ← 4 × 3 dollars
_____
$12.76

Enrichment

Tell students to use newspapers or catalogs and cut two priced items they might buy in quantities of 2 or more. Then have them find the total cost if they bought 3 of one item and 4 of the other.

Practice

Remind students to place the dollar sign and decimal in each answer. Tell students to use the butcher shop data to solve the last 2 word problems. Have students complete the page independently.

Mixed Practice

1. 86 ÷ 2 (43)
2. 28,155 + 36,345 (64,500)
3. 500 × 3 (1,500)
4. 62,176 − 24,259 (37,917)
5. 72 − 8 × 8 (8)
6. 8 × 6,000 (48,000)
7. 376 + 1,204 + 314 (1,894)
8. $.85 × 4 ($3.40)
9. $50.70 − 38.29 ($12.41)
10. 72 ÷ 5 (14 R2)

Extra Credit   *Logic*

Tell students that you are going to write the numerals 0 through 9 on the board in a definite order. Ask them if they can figure out the logic represented by this order of numbers. Then write the following number sequence on the board for students to identify the pattern:

8, 5, 4, 9, 1, 7, 6, 3, 2, 0

(They are in alphabetical order according to the number word for each numeral.)

More 3-digit Multiplicands

pages 171-172

Objective

To multiply 3-digit numbers by 1-digit numbers with trading

Materials

Mental Math

Ask students which is greater:

1. 8 × 9 or ½ of 140. (8 × 9)
2. 42 ÷ 6 or 42 ÷ 7. (42 ÷ 6)
3. 35 ÷ 7 or 54 ÷ 9. (54 ÷ 9)
4. ¼ of 40 or ¼ of 36. (¼ of 40)
5. 26 − 18 or 18 ÷ 3. (26 − 18)
6. 64 × 1 or 20 × 4. (20 × 4)
7. 80 ÷ 8 or 99 ÷ 9. (99 ÷ 9)
8. ¹⁄₁₀ or ⅑. (⅑)

Skill Review

Write **7 × 2 tens** on the board. Have students tell how many tens in all (14 tens), and how many hundreds, tens and ones in the product. (1, 4, 0) Repeat for **7 ones × 9.** (6, 3) Present more problems for students to review the trades.

More Multiplying 3-digit Numbers

The distance from Denver to Seattle is approximately 3 times farther than the distance from Chicago to Pittsburgh. About how far is it from Seattle to Denver?

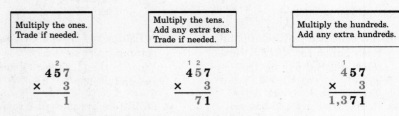

We want to know the approximate distance between Seattle and Denver.

The distance from Chicago to Pittsburgh

is __457__ miles.

The distance from Seattle to Denver

is approximately __3__ times that number.

To find the distance between Seattle and Denver,

we multiply __457__ by __3__.

Multiply the ones. Trade if needed.	Multiply the tens. Add any extra tens. Trade if needed.	Multiply the hundreds. Add any extra hundreds.

$$\begin{array}{r} \overset{2}{4}57 \\ \times\ \ 3 \\ \hline 1 \end{array} \qquad \begin{array}{r} \overset{1}{4}\overset{2}{5}7 \\ \times\ \ 3 \\ \hline 71 \end{array} \qquad \begin{array}{r} \overset{1}{4}57 \\ \times\ \ 3 \\ \hline 1{,}371 \end{array}$$

It is about __1,371__ miles from Seattle to Denver.

Getting Started

Multiply.

1. 525
    × 7
    ――――
    3,675

2. 289
    × 4
    ――――
    1,156

3. 328
    × 8
    ――――
    2,624

4. $2.48
    × 6
    ――――
    $14.88

Copy and multiply.

5. $5.96 × 9
   $53.64

6. 709 × 5
   3,545

7. 492 × 3
   1,476

8. 587 × 2
   1,174

171

Teaching the Lesson

Introducing the Problem Ask students what the map shows. (map with 4 cities and the approximate mileage from Chicago to Pittsburg) Have a student read the problem and tell what they are to find. (distance from Seattle to Denver) Ask students what information is given to help solve this problem. (It is about 457 miles from Pittsburg to Chicago and it is 3 times that distance from Seattle to Denver.) Have students read and complete the sentences. Guide them through the multiplication in the model. Have students complete the solution sentences.

Developing the Skill Remind students that they already know how to work this problem as you write **78 × 4** vertically on the board. Have students talk through the work, as you record their responses, to find the product. (312) Ask students why there are 31 tens instead of 28. (add the 3 tens traded from the ones) Now write **278 × 4** on the board and have students talk through the work to find the ones, then the tens. Ask students if some of the 31 tens can be traded for hundreds. (yes) Record the trade and ask students to find the total number of hundreds. (11 hundreds) Ask students if some hundreds can be traded for thousands. (yes) Record the trade and ask students to read the product. (1,112) Repeat for **69 × 3** and **469 × 3.**

Practice

Multiply.

1.   357
  ×   4
 1,428
	2. 292
	  × 8
	 2,336
		3. 537
		  × 5
		 2,685
			4. 673
			  × 2
			 1,346

5.   896
  ×   6
 5,376
	6. 383
	  × 7
	 2,681
		7. $4.77
		  × 3
		 $14.31
			8. 709
			  × 9
			 6,381

9.   372
  ×   8
 2,976
	10. 628
	  × 3
	 1,884
		11. 548
		  × 6
		 3,288
			12. 419
			  × 5
			 2,095

Copy and Do

13. 5 × 386
   1,930
14. 457 × 9
   4,113
15. $6.75 × 2
   $13.50
16. 3 × 628
   1,884

17. 727 × 8
   5,816
18. $2.94 × 5
   $14.70
19. 7 × 929
   6,503
20. 6 × 848
   5,088

Apply

Solve these problems.

21. The Cammero family drove 346 miles on their vacation. The Johnsons drove 4 times as far as the Cammeros. How far did the Johnsons drive?
1,384 miles

22. The distance from San Francisco to Los Angeles is 403 miles. Mr. Harris left San Francisco at 10:00 AM and drove 115 miles. How far does Mr. Harris still have to drive?
288 miles

23. A jet airliner can hold 186 people. How many people can 7 jet airliners carry?
1,302 people

24. A china platter costs $6.75 and cereal bowls cost $3.95 each. How much would Sharon pay for a platter and set of 4 bowls?
$22.55

172

4-digit Multiplicands

pages 173-174

Objective

To multiply 4-digit numbers by 1-digit numbers with trading

Materials

Mental Math

Have students name the century for the year:

1. 1977 (20th)
2. 2001 (21st)
3. 1776 (18th)
4. 1492 (15th)
5. 961 (10th)
6. 1382 (14th)
7. 2164 (22nd)

Skill Review

Ask students the total hundreds in $(9 \times 700) + (4 \times 100)$. (67) Have students tell the number of thousands in 1,300. (1) Repeat for $(200 \times 6) + (9 \times 100)$, $(900 \times 9) + (1 \times 100)$ and other similar problems.

Multiplying 4-digit Numbers

The course of the North Coast Cross Country Ski Race is 8 miles long. How many yards long is the race course?

We want to know how many yards long the ski race is.

The race is __8__ miles long.

There are __1,760__ yards in 1 mile.

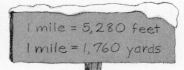

1 mile = 5,280 feet
1 mile = 1,760 yards

To find the total number of yards in the race, we multiply __1,760__ by __8__.

Multiply the ones. Trade if needed.	Multiply the tens. Add any extra tens. Trade if needed.	Multiply the hundreds. Add any extra hundreds. Trade if needed.	Multiply the thousands. Add any extra thousands.
$\begin{array}{r} 1,76\overset{4}{0} \\ \times\ \ \ \ 8 \\ \hline 0 \end{array}$	$\begin{array}{r} 1,7\overset{6}{6}\overset{4}{0} \\ \times\ \ \ \ 8 \\ \hline 80 \end{array}$	$\begin{array}{r} 1,\overset{6}{7}\overset{4}{6}0 \\ \times\ \ \ \ 8 \\ \hline 080 \end{array}$	$\begin{array}{r} \overset{6}{1},760 \\ \times\ \ \ \ 8 \\ \hline 14,080 \end{array}$

There are __14,080__ yards in the North Coast Ski Race.

Getting Started

Multiply.

1. $\begin{array}{r} 6,243 \\ \times\ \ \ \ 3 \\ \hline 18,729 \end{array}$
2. $\begin{array}{r} 4,086 \\ \times\ \ \ \ 3 \\ \hline 12,258 \end{array}$
3. $\begin{array}{r} 5,248 \\ \times\ \ \ \ 7 \\ \hline 36,736 \end{array}$
4. $\begin{array}{r} \$75.76 \\ \times\ \ \ \ 5 \\ \hline \$378.80 \end{array}$

Copy and multiply.

5. $4,273 \times 2$
8,546
6. $\$16.59 \times 9$
$149.31
7. $\$90.06 \times 8$
$720.48
8. $3,876 \times 6$
23,256

173

Teaching the Lesson

Introducing the Problem   Have a student read the problem and tell what they are to find. (the race course length in yards) Ask students what information is given in the problem. (The race course is 8 miles long.) Ask students what helpful information is given in the picture. (1 mile = 1,760 yards) Ask students why the number of feet in 1 mile is not useful. (do not need to answer the question in feet) Have students complete the sentences. Guide students through the multiplication in the model, and have them complete the solution. Have students round the number of yards in 1 mile to the nearest hundred and estimate to check their answer.

Developing the Skill   Write **789 × 4** on the board and have a student work the problem to review multiplying a 3-digit number. (3,156) Now write **4,789 × 4** on the board and tell students they will multiply the ones, tens and hundreds as in the first problem, but they will trade the hundreds for thousands and then multiply the thousands and add the traded thousands. Have a student work the problem. (19,156) Repeat for 816 × 9 and 3,816 × 9.

173

Practice

Multiply.

1. 3,216
   × 5
   —————
   16,080

2. 7,926
   × 8
   —————
   63,408

3. 2,079
   × 6
   —————
   12,474

4. 8,273
   × 2
   —————
   16,546

5. $37.86
   × 7
   —————
   $265.02

6. 9,376
   × 3
   —————
   28,128

7. 5,163
   × 4
   —————
   20,652

8. $48.06
   × 9
   —————
   $432.54

9. 1,673
   × 8
   —————
   13,384

10. $32.85
    × 5
    —————
    $164.25

11. 8,269
    × 2
    —————
    16,538

12. 4,675
    × 4
    —————
    18,700

Copy and Do

13. 8 × 4,271
    34,168

14. 3 × 6,129
    18,387

15. $63.38 × 9
    $570.42

16. 4,350 × 2
    8,700

17. 4,925 × 9
    44,325

18. 4 × $57.83
    $231.32

19. 5 × 6,256
    31,280

20. 9,816 × 6
    58,896

Apply

Solve these problems.

21. The highest mountain in Russia is 18,510 feet tall. The highest mountain in the United States is 1,810 feet higher than that. How tall is the highest mountain in the U.S.?
    20,320 feet

22. A computer with 1K memory can store 1,024 bits of information. How many bits of information can an 8K computer store?
    8,192 bits

23. The race cars drove a total of 4 miles in 9 laps. How many feet did they travel?
    21,120 feet

24. A dump truck can hold 1,426 pounds of dirt. How many pounds can the dump truck haul in 7 trips?
    9,982 pounds

25. Rebecca ran 450 yards less than 6 miles. How many yards did Rebecca run?
    10,110 yards

26. It costs $23.43 for a pair of jeans and $14.26 for a shirt. Pablo bought 2 pairs of jeans and 3 shirts. How much did Pablo spend?
    $89.64

174

Correcting Common Errors

Some students may make errors because they have difficulty writing numbers for the partial products in the correct place of the product. Have students work with partners and use the following form to compute.

```
    4,329
  ×     5
  ——————————
       45  ← 5 × 9
      100  ← 5 × 20
    1 500  ← 5 × 300
   20 000  ← 5 × 4,000
  ——————————
   21,645
```

Enrichment

Tell students to use a map or the World Almanac to find the distance they would travel if they drove 2 round trips from New York City to San Francisco.

Practice

Remind students to place a dollar sign and decimal point in the product for money problems. Tell students they will need to use the information on page 173 when solving one of the word problems. Remind students to watch for word problems that require 2 operations. Have students complete the page independently.

Extra Credit   *Statistics*

Using the newspaper weather page, have students make a bar graph showing the daily high temperatures of various cities around the world. Have them check a world map to make sure they have included cities in both hemispheres and temperate zones, tropic zones and polar zones. After two weeks of graphing, tell students to write 5 questions based on information shown on their graphs, for example: Which city has the lowest temperature? Which city has the most consistent temperature? etc. Have them exchange graphs with a partner who will answer the questions.

Multiplying By Multiples of 10

pages 175-176

Objective

To multiply 2- or 3-digit numbers by a 2-digit multiple of 10

Materials

Mental Math

Have students complete these equivalents.

1. (60) min = 1 hr
2. 1 gal = (8) pt
3. 3 hr = (180) min
4. (8) oz = ½ lb
5. 49 days = (7) weeks
6. 40 cm = (4) dec
7. (60) sec = 1 min
8. 2L = (2,000) mL

Skill Review

Ask students what 8 × 0 equals. (0) Repeat for 0 × 2 and 4 × 0. Write **24 × 0** vertically on the board. Ask students the product. (0) Ask students what any number times zero equals. (zero) Write **464 × 0** on the board and have a student write the product. (0) Repeat for more problems up to 4-digit numbers times zero.

Multiplying by Multiples of 10

Ronald is learning health skills in his CPR class. He is taking his pulse to find his heart rate for one minute. How many times will Ronald's heart beat in one hour?

...70, 71, 72

We want to know how often Ronald's heart beats hourly.

His heart beats ___72___ times in one minute,

and there are ___60___ minutes in one hour.

To find Ronald's hourly heart rate,

we multiply ___60___ by ___72___.

Multiply by the digit in the ones place.	Multiply by the digit in the tens place.
$\begin{array}{r} 72 \\ \times 60 \\ \hline 0 \end{array}$	$\begin{array}{r} {\scriptstyle 1} \\ 72 \\ \times 60 \\ \hline 4,320 \end{array}$

Ronald's heart beats ___4,320___ times in one hour.

Getting Started

Multiply.

1. $\begin{array}{r} 36 \\ \times 20 \\ \hline 720 \end{array}$ 2. $\begin{array}{r} 35 \\ \times 40 \\ \hline 1,400 \end{array}$ 3. $\begin{array}{r} 50 \\ \times 60 \\ \hline 3,000 \end{array}$ 4. $\begin{array}{r} 125 \\ \times 30 \\ \hline 3,750 \end{array}$

Copy and multiply.

5. 625 × 70
43,750

6. 820 × 80
65,600

7. 635 × 50
31,750

8. 708 × 90
63,720

175

Teaching the Lesson

Introducing the Problem   Have a student describe the picture and read the thought bubble. Ask a student to read the problem aloud and tell what is to be solved. (number of times Ronald's heart will beat in 1 hour) Ask students what information is given. (Ronald's heart beats 72 times in 1 minute.) Ask students what additional information is needed. (number of minutes in 1 hour) Have students read and complete the sentences and work through the model with them to solve the problem. Tell students to round 72 to the closest ten to check their answer.

Developing the Skill   Remind students that the rule for multiplying by a multiple of 10 is to multiply the numbers that are not zero, and write that product followed by the number of zeros in the factors. Write **84 × 90** on the board and have a student multiply 84 times 9. (756) Ask how many zeros are in the 2 factors. (1) Have the student write 1 zero after 756 and read the number. (7,560) Now write **84 × 90** vertically on the board and ask students what zero times 84 equals. (0) Write a zero in ones place. Tell students that we now multiply the numbers that are not zero. Have a student multiply 84 × 9. (756) Write **756** in front of the zero. Repeat for **96 × 40** and **38 × 70**.

175

Practice

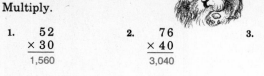

Multiply.

1. 52
   × 30
   1,560

2. 76
   × 40
   3,040

3. 27
   × 70
   1,890

4. 80
   × 90
   7,200

5. 48
   × 20
   960

6. 63
   × 60
   3,780

7. 400
   × 50
   20,000

8. 88
   × 80
   7,040

9. 153
   × 30
   4,590

10. 400
    × 70
    28,000

11. 94
    × 50
    4,700

12. 785
    × 20
    15,700

Copy and Do

13. 40 × 253
    10,120

14. 70 × 36
    2,520

15. 90 × 573
    51,570

16. 426 × 30
    12,780

17. 651 × 60
    39,060

18. 20 × 879
    17,580

19. 89 × 50
    4,450

20. 80 × 600
    48,000

Apply

Solve these problems.

21. How many minutes are there in 36 hours?
    2,160 minutes

22. How many seconds are there in 15 minutes?
    900 seconds

23. Charlotte can walk 1 kilometer in 9 minutes. One day, Charlotte walked for 90 minutes. How many kilometers did she walk?
    10 kilometers

24. A container holds 245 milliliters of juice. The cafeteria used 80 containers. How many milliliters of juice did the cafeteria use?
    19,600 milliliters

25. The school photographer took 875 pictures. Each picture takes 40 seconds to develop. How many seconds will it take to develop the pictures?
    35,000 seconds

26. Bobby read 16 chapters of a book. Each chapter was 30 pages long. Bobby took 50 minutes to read each chapter. How long did it take him to read the book?
    800 minutes

176

Correcting Common Errors

When multiplying by a multiple of 10, students may add the renamed number before they multiply.

INCORRECT	CORRECT
1	1
72	72
× 60	× 60
4,820	4,320

Correct by having students review renaming when multiplying 6 × 72 and apply it to multiplying by 60.

Enrichment

Tell students to make a table to show the total number of eggs if they had 10 dozen, 20 dozen, 40 dozen and 70 dozen.

Practice

Tell students there is unnecessary information in 1 word problem and another problem requires them to supply some of the information. Have students complete the page.

Mixed Practice

1. 9 + 5 + 6 + 3 + 1 (24)
2. 4 × 309 (1,236)
3. $8.27 × 2 ($16.54)
4. 6,210 − 3,008 (3,202)
5. 67 ÷ 4 (16 R3)
6. $563.17 + 327.93 ($891.10)
7. 47 × 6 (282)
8. 75 + 1,027 + 472 (1,574)
9. 39 ÷ 8 (4 R7)
10. 376 − 256 (120)

Extra Credit   *Statistics*

Give each pair of students a bag containing 25 assorted colored cubes. Ask them to pick out five of the cubes. Explain that there are a total of 25 cubes inside and ask them to guess at the contents based on their sample of 5. Tell students to put the cubes back in the bag and draw another 5. Ask them again to estimate the contents after each sample of 5. See if any group is able to guess the total contents after they have seen 20 of the 25 cubes.

2-digit Multipliers

pages 177-178

Objective

To multiply 2-digit numbers by 2-digit numbers, no trading

Materials

Mental Math

Dictate the following:

1. 5×50 (250)
2. $84 \div 2$ (42)
3. $6 \times 10 \times 2$ (120)
4. $(7 \times 5) \div 5$ (7)
5. $238 - 3$ tens (208)
6. 4 hundreds plus 28 ones (428)
7. $6,000 \times 9$ (54,000)
8. $43 \times 0 \times 8$ (0)

Skill Review

Write $(30 \times 84) + (4 \times 84)$ on the board and have students find the sum of the 2 multiplications. $(2,520 + 336 = 2,856)$ Repeat for $(26 \times 70) + (26 \times 9)$ and $(40 \times 18) + (18 \times 7)$.

Multiplying by 2-digit Numbers

January, usually the coldest month of the year, is named for the Roman god, Janus. How many hours are there in the month of January?

We want to know the total number of hours in January.

January has __31__ days.

There are __24__ hours in one day.

To find the total number of hours, we multiply __31__ by __24__.

Multiply by the ones.	Multiply by the tens.	Add the products.
$\begin{array}{r} 31 \\ \times\,24 \\ \hline 124 \end{array}$ ← 4 × 31	$\begin{array}{r} 31 \\ \times\,24 \\ \hline 124 \\ 620 \end{array}$ ← 20 × 31	$\begin{array}{r} 31 \\ \times\,24 \\ \hline 124 \\ 620 \\ \hline 744 \end{array}$ ← 24 × 31

There are __744__ hours in January.

Getting Started

Multiply.

1. $\begin{array}{r} 23 \\ \times\,32 \\ \hline 736 \end{array}$
2. $\begin{array}{r} 42 \\ \times\,24 \\ \hline 1,008 \end{array}$
3. $\begin{array}{r} 50 \\ \times\,35 \\ \hline 1,750 \end{array}$
4. $\begin{array}{r} 64 \\ \times\,22 \\ \hline 1,408 \end{array}$

Copy and multiply.

5. 12×14
168
6. 21×28
588
7. 53×23
1,219
8. 96×11
1,056

177

Teaching the Lesson

Introducing the Problem   Ask students what tool of measurement is shown in the picture. (calendar) Ask what a calendar measures. (time in days, weeks and months) Have a student read the problem and tell what they are to find. (the number of hours in the month of January) Ask students what information will be needed. (number of days in January and number of hours in a day) Ask which of that information is not given in the problem or picture. (number of hours in a day) Have students complete the sentences as you read with them. Guide students through the multiplication in the model, to solve the problem. Tell students to check their solution by rounding 31 to 30 and multiplying.

Developing the Skill   Write $61 \times 46 = (61 \times 40) + (61 \times 6)$ on the board and tell students the problem of 61×46 can be rewritten as 40 sixty-ones plus 6 sixty-ones. Now write on the board:

$$\begin{array}{r} 61 \\ \times\,46 \\ \hline 366 \\ 2,440 \\ \hline 2,806 \end{array}$$

(6 × 61)
(40 × 61)
(46 × 61)

Talk through multiplying of the ones, then the tens and adding the products. Repeat the exercise for 23×71 and 32×24.

Practice

Multiply.

1. $\begin{array}{r} 32 \\ \times 43 \\ \hline 1{,}376 \end{array}$	2. $\begin{array}{r} 71 \\ \times 56 \\ \hline 3{,}976 \end{array}$	3. $\begin{array}{r} 23 \\ \times 23 \\ \hline 529 \end{array}$	4. $\begin{array}{r} 75 \\ \times 11 \\ \hline 825 \end{array}$
5. $\begin{array}{r} 32 \\ \times 32 \\ \hline 1{,}024 \end{array}$	6. $\begin{array}{r} 22 \\ \times 43 \\ \hline 946 \end{array}$	7. $\begin{array}{r} 61 \\ \times 38 \\ \hline 2{,}318 \end{array}$	8. $\begin{array}{r} 80 \\ \times 49 \\ \hline 3{,}920 \end{array}$
9. $\begin{array}{r} 54 \\ \times 12 \\ \hline 648 \end{array}$	10. $\begin{array}{r} 62 \\ \times 33 \\ \hline 2{,}046 \end{array}$	11. $\begin{array}{r} 84 \\ \times 21 \\ \hline 1{,}764 \end{array}$	12. $\begin{array}{r} 32 \\ \times 42 \\ \hline 1{,}344 \end{array}$

Copy and Do

13. 41×38
1,558

14. 24×62
1,488

15. 63×33
2,079

16. 60×57
3,420

17. 76×81
6,156

18. 79×51
4,029

19. 96×11
1,056

20. 23×52
1,196

Apply

Solve these problems.

21. How many hours are there in 2 weeks and 5 days?
456 hours

22. How many inches are there in 25 feet and 11 inches?
311 inches

23. How many ounces are there in 3 pounds 9 ounces?
57 ounces

24. How many feet are there in 15 yards 2 feet?
47 feet

178

Correcting Common Errors

Students may forget to write a zero in the ones place of the second partial product when multiplying by the tens. Correct by having them rewrite the problems as shown below.
$21 \times 43 = (20 \times 43) + (1 \times 43)$
$860 + 43 = 903$

Enrichment

Have students find the combined weight, in ounces, of 2 babies if one weighs 11 pounds 6 ounces and the other weighs 13 pounds 10 ounces.

Practice

Remind students to think of the multiplier or bottom factor as a multiple of 10 plus some ones. Tell students they must supply some missing information to work each word problem. Have students complete the page.

Extra Credit   *Numeration*

Have students look up Egyptian hieroglyphics in an encyclopedia. Ask them to list the symbols for our numbers 1 through 10, for 20, 30, 40, 100, 1,000, 10,000, 100,000 and 1,000,000.

178

More 2-digit Multipliers

pages 179-180

Objective

To multiply 2-digit numbers by 2-digit numbers with trading

Materials

Mental Math

Have students tell the profit on each if they buy at:

1. 5/$1 and sell at 25¢ each. (5¢)
2. 40¢ each and sell at 6/$3. (10¢)
3. $1 each and sell at 2/$3. (50¢)
4. $35 each and sell at 2/$90. ($10)
5. 8/$20 and sell at $3.50 each. ($1)
6. $3.19 and sell at $4.50. ($1.31)
7. 4/88¢ and sell at 3/99¢. (11¢)
8. $4.80 for 2 and sell at $5.00 each. ($2.60)

Skill Review

Write **69 × 3** on the board and have a student talk through the multiplication and the trade. Repeat for more problems of 2-digit numbers by 1-digit numbers with trading.

More 2-digit Multiplying

Elmer's boss told him to pack as many bags of sugar as he could fit on the shelf. How many pounds of sugar does the store have on display?

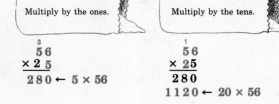

Capacity: (56) 25-pound bags

We want to find how much sugar is in the store's inventory.

Each sugar bag weighs ___25___ pounds.

There are ___56___ bags of sugar in stock.

To find the total number of pounds of sugar,

we multiply ___56___ by ___25___.

Multiply by the ones.	Multiply by the tens.	Add the products.

$$
\begin{array}{r}
\overset{3}{5}6 \\
\times\,2\,5 \\
\hline
280
\end{array}
\leftarrow 5 \times 56
$$

$$
\begin{array}{r}
\overset{1}{5}6 \\
\times\,2\,5 \\
\hline
280 \\
1120
\end{array}
\leftarrow 20 \times 56
$$

$$
\begin{array}{r}
56 \\
\times\,2\,5 \\
\hline
280 \\
1120 \\
\hline
1,400
\end{array}
\leftarrow 25 \times 56
$$

There are ___1,400___ pounds of sugar in the store's inventory.

Getting Started

Multiply.

1.
$$
\begin{array}{r} 36 \\ \times\,24 \\ \hline 864 \end{array}
$$

2.
$$
\begin{array}{r} 23 \\ \times\,42 \\ \hline 966 \end{array}
$$

3.
$$
\begin{array}{r} 39 \\ \times\,56 \\ \hline 2,184 \end{array}
$$

4.
$$
\begin{array}{r} 74 \\ \times\,38 \\ \hline 2,812 \end{array}
$$

Copy and multiply.

5. 86 × 58
4,988

6. 73 × 66
4,818

7. 48 × 93
4,464

8. 47 × 47
2,209

179

Teaching the Lesson

Introducing the Problem Have a student read the problem aloud and tell what is being asked. (total number of pounds of sugar) Discuss the concept of an inventory, using the number of books in the classroom, number of pencils each student has, etc. Ask students what information is needed to answer the problem. (number of bags and number of pounds per bag) Ask students where this information is given. (The sign says 56 bags can be stored and the problem tells that each bag holds 25 pounds.) Have students read with you as they complete the sentences. Guide students through the multiplication in the model, to solve the problem. Tell students they can estimate the correctness of their answer by finding the total pounds in 60 bags of sugar.

Developing the Skill Write **43 × 6** on the board and have a student talk through the trade and find the product. (258) Now write **43 × 56** on the board and have the same student multiply the 6 forty-three's, make the trade and write the product. (258) Remind students that the 5 in tens place means 5 tens or 50, and 50 forty-threes equals 2,150, as you write 2,150 under 258. Tell students we add the product of 6 forty-threes and the product of 50 forty-threes to find that 56 forty-threes equals 2,408 as you write the answer. Repeat for **32 × 9** and **32 × 69**.

179

Practice

Multiply.

1.. 26 × 34 884	2. 42 × 29 1,218	3. 85 × 37 3,145	4. 67 × 26 1,742
5. 73 × 48 3,504	6. 96 × 53 5,088	7. 59 × 74 4,366	8. 77 × 23 1,771
9. 89 × 76 6,764	10. 45 × 67 3,015	11. 28 × 98 2,744	12. 56 × 39 2,184
13. 35 × 35 1,225	14. 82 × 64 5,248	15. 19 × 56 1,064	16. 97 × 48 4,656

Copy and Do

17. 29 × 46
1,334
18. 85 × 37
3,145
19. 52 × 74
3,848
20. 65 × 98
6,370
21. 57 × 58
3,306
22. 26 × 76
1,976
23. 87 × 38
3,306
24. 79 × 89
7,031

Apply

Solve these problems.

25. Each package of trivia cards contains 75 questions. There are 25 packages in the game. How many trivia questions are there in the game?
1,875 questions

26. There are 28 chairs in each classroom in Ryan School. The school has 15 classrooms. How many classroom chairs are in the school?
420 chairs

27. Eric rode his bicycle 26 miles, and still has 17 miles to go. How far will Eric ride his bike?
43 miles

28. There are 96 oranges in each crate. Casey put 38 crates in inventory. How many oranges are in inventory?
3,648 oranges

180

Correcting Common Errors

Some students may confuse the regrouped digits from finding the first partial product with the regrouped digits from finding the second partial product. To avoid such errors, have them cross out the first set of regrouped digits when they have finished multiplying by the ones digit.

Enrichment

Tell students to look up the world's long-jump record, round the number to feet, and convert that measurement to inches.

Practice

Remind students to think of the bottom factor as a multiple of 10 plus ones. Have students complete the page independently.

Mixed Practice

1. 39 ÷ 7 (5 R4)
2. $468.27 + 693.84 ($1,162.11)
3. 362 + 507 + 221 (1,090)
4. 576 × 8 (4,608)
5. 1,875 − 658 (1,217)
6. 76 ÷ 6 (12 R4)
7. $19.03 − 7.98 ($11.05)
8. 8.42 × 7 ($58.94)
9. 46 − 3 × 9 (19)
10. 3 × 958 (2,874)

Extra Credit   *Numeration*

Have students use their table of Egyptian hieroglyphics or put a chart of the numbers 1 through 10, 100, 1,000, and 10,000 on the board. Explain that the Egyptians used hieroglyphics as a tallying system.

Have students write out the hieroglyphics for 59, 882, 4,205, and 552.

180

Multiplying Hundreds by Tens

pages 181-182

Objective

To multiply 3-digit numbers by 2-digit numbers with trading

Materials

Mental Math

Have students complete the comparison: 49 is to 7 as 14 is to 2 as:

1. 140 is to (20).
2. 7 is to (1).
3. $3.50 is to ($0.50).
4. (42) is to 6.
5. 210 is to (30).
6. (84) is to 12.
7. $1.75 is to ($0.25).
8. 56 is to (8).

Skill Review

Write several 2-digit by 2-digit multiplication problems on the board. Have students work in pairs at the board to solve the problems and check their solutions.

Multiplying Larger Numbers

Marcia is the summer hostess at the Village Lunch Shop. Her regular work week is 35 hours long. How much does Marcia earn each week?

Village Lunch Shop Wages	
Host/Hostess	$4.85
Cook	$4.16
Busperson	$3.78
Dishwasher	$3.52

We want to know Marcia's weekly wage.

We know Marcia works __35__ hours each week,

and earns __$4.85__ each hour.

To find her total weekly wages,

we multiply __$4.85__ by __35__.

Please wait for Hostess

Add the products. Place the dollar sign and decimal point.

Multiply by the ones.	Multiply by the tens.	

```
    4 2
  $4.85              $4.85                    $4.85
×    35            ×    35                  ×    35
  2425 ← 5 × 485     2425                     2425
                   14550 ← 30 × 485          14550
                                           $169.75
```

Marcia earns __$169.75__ each week.

Getting Started

Multiply.

1. $3.65
    ×   28
    $102.20

2.   575
    × 47
    27,025

3. $8.09
    ×   53
    $428.77

4.   639
    × 82
    52,398

Copy and multiply.

5. $0.85 × 39
   $33.15

6. $7.00 × 56
   $392.00

7. 825 × 78
   64,350

8. 960 × 62
   59,520

181

Teaching the Lesson

Introducing the Problem Have a student describe the picture and read the sign. Ask students what time period the wages would cover. (1 hour) Have a student read the problem and tell what is being asked. (how much Marcia earns per week) Ask students what information is needed to solve the problem. (number of hours she works per week and pay per hour) Ask what information will be used from the problem and picture. (35 hours per week at $4.85 per hour) Have students read and complete the sentences. Guide students through the multiplication steps in the model, to solve the problem. Ask students how they might check the solution. (round the pay to $5.00 or the hours to 40)

Developing the Skill Write **286 × 2, 286 × 23** and **$2.86 × 23** in vertical form across the board. Have a student work the first problem, talking through each trade. Review with students how the two 200's are multiplied. Have a student work the second problem and talk through each step as 3 times 286 is found and 20 times 286 is found. Ask students how the third problem differs from the second. (Dollar sign and decimal point are added.) Have a student talk through each step, work the problem and place the dollar sign and decimal point in the product. Repeat for 168 × 7, 168 × 97 and $1.68 × 97.

181

Practice

Multiply.

1.	216 × 27 = 5,832	**2.**	$9.45 × 46 = $434.70	**3.**	$8.73 × 39 = $340.47	**4.**	$0.46 × 78 = $35.88
5.	915 × 67 = 61,305	**6.**	707 × 54 = 38,178	**7.**	$6.38 × 85 = $542.30	**8.**	784 × 96 = 75,264
9.	$3.90 × 59 = $230.10	**10.**	458 × 66 = 30,228	**11.**	823 × 92 = 75,716	**12.**	$5.85 × 78 = $456.30
13.	867 × 84 = 72,828	**14.**	$6.81 × 49 = $333.69	**15.**	$9.06 × 37 = $335.22	**16.**	738 × 29 = 21,402

Copy and Do

17. 47 × 368
    17,296

18. $5.47 × 34
    $185.98

19. 296 × 88
    26,048

20. 56 × $8.28
    $463.68

21. $9.25 × 67
    $619.75

22. 96 × 428
    41,088

23. 72 × $7.93
    $570.96

24. 408 × 27
    11,016

Apply

Solve these problems. Use the chart on page 181.

25. Winston cooked for the Village Lunch Shop for 4 weeks. He worked 8 hours a day for 5 days each week. What did he earn during that time?
$665.60

26. Richard worked as a busperson for 40 hours. How much more did he earn than the dishwasher who worked 40 hours?
$10.40

27. Penny washed dishes at the Village Lunch Shop for 35 hours. How much did she earn?
$123.20

28. Lisa works 32 hours each week as a cook and 10 hours as a hostess. How much will Lisa earn in 3 weeks?
$544.86

182

Practice

Mixed Practice

1. 72,176 + 38,934 (111,110)
2. $781.27 − 350.89 ($430.38)
3. 473 × 40 (18,920)
4. ___ × 5 = 45 (9)
5. 795 − 432 (363)
6. (8 × 9) + (5 × 7) (107)
7. 2,567 × 7 (17,969)
8. 342 + 178 + 456 (976)
9. 91 ÷ 7 (13)
10. 80 × 500 (40,000)

Extra Credit   *Probability*

Explain that probabilities are frequently expressed as fractions. If there is one chance in two that something will happen, the probability is 1/2. Point out that this fraction is the number of specific ways the event can happen, over all possible ways. Show students a coin. Explain that there is one way of getting a specific result, for example, flipping a head, and there are two possible ways the flip can turn out. (heads or tails) The probability of flipping a head is 1/2. Have students express these probabilities as fractions:

1. of drawing an ace from a deck of 52 cards (4/52 or 1/13)
2. of getting a six on one roll of the dice (1/6)
3. of drawing a red cube out of a bag containing one red, one blue, and one yellow (1/3)

Estimation

Objective

To use estimation in multiplication

Materials

Mental Math

Have students tell their cumulative bank balance if they first:

1. deposit $26. ($26)
2. deposit $24.50. ($50.50)
3. write a check for $12.50. ($38)
4. write a check for $14.50. ($23.50)
5. earn interest of 7¢. ($23.57)
6. deposit $3.43. ($27)
7. write a check for $26.93. (7¢)
8. deposit $15.09. ($15.16)

Skill Review

Write on the board random numbers up to 100,000. Have students round each number to the nearest ten, hundred, thousand and ten-thousand.

Using Estimation

Estimation is used to see if an answer seems reasonable. How many days are in the 9 years shown here? The years 1984 and 1988 each have an extra day because they are leap years.

We want to find the number of days in the years 1981 through 1989.

There are __9__ years, and each year

has __365__ days.

1988 and 1984 have __2__ extra days in them

because they are __leap years__.

To find the total number of days in these years,

we multiply __365__ by __9__ and add __2__.

$$\begin{array}{r} {}^{5\ 4}\ \\ 3\,6\,5 \\ \times\quad 9 \\ \hline 3,285 \end{array}$$ __3,285__ + __2__ = __3,287__

There are __3,287__ days in 1981 through 1989.

Use estimation to check if your answer seems reasonable.

365 is about 370.

$$\begin{array}{r} 3\,7\,0 \\ \times\quad 9 \\ \hline 3,330 \end{array}$$

The answer seems reasonable.

Getting Started

Multiply. Use estimation to check your answers.

1.	2.	3.
68	94	27
× 39	× 8	× 58
2,652	752	1,566

Copy and multiply. Use estimation to check your answers.

4. 186 × 7   5. $4.37 × 55   6. 749 × 89
   1,302 $240.35   66,661

183

Teaching the Lesson

Introducing the Problem   Have students read the problem with you and tell what is to be answered. (the total days in the years 1981 through 1989) Tell students that a year is the time it takes the earth to revolve around the sun and is slightly over 365 days. Tell students that every 4 years, the extra bit of time over 365 days is made up by adding a day called a **leap day.** Thus the year is called a **leap year.** Ask students what information is given. (There are 9 years all together and 2 of the years have 1 extra day each.) Remind students that when they round a number, they are estimating. Have students complete the sentences to solve the problem with you.

Developing the Skill   Write **29 × 72** on the board. Have students find the product. (2,059) Tell students we can check this answer by rounding one of the numbers to the nearest ten before multiplying. Have students tell the nearest multiple of 10 for 29. (30) Have a student write the problem on the board and find its product. (2,100) Ask students if the product of 2,059 is reasonable. (yes) Repeat for 182 × 64 and $1.49 × 18.

Practice

Multiply. Use estimation to check your answers.

1. 77
    × 32
    2,464

2. 57
    × 89
    5,073

3. 92
    × 8
    736

4. 476
    × 67
    31,892

5. $5.09
    × 62
    $315.58

6. 895
    × 82
    73,390

Copy and Do

7. 83 × 7
   581

8. 6 × 247
   1,482

9. 39 × 81
   3,159

10. 73 × 24
    1,752

11. 75 × 87
    6,525

12. 13 × 98
    1,274

13. 9 × 576
    5,184

14. 37 × $6.58
    $243.46

Apply

Solve these problems. Use estimation to check your answers.

15. Brenda runs up the stadium steps each day. There are 89 steps in the stadium. How many steps does Brenda run in 38 days?
    3,382 steps

16. It takes 4 cups of whole wheat flour to make 1 loaf of bread. The City Bakery bakes 673 loaves of bread each day. How many cups of flour are used each day?
    2,692 cups

17. Juan has saved $136.46 to buy a color T.V. set. The T.V. set sells for $249.19. How much more does Juan need to save?
    $112.73

18. Paint sells for $7.79 a gallon. Mr. Jameson used 18 gallons of paint to paint his farm buildings. How much did the paint cost Mr. Jameson?
    $140.22

19. Annette gets 3¢ for each can she saves. One week Annette received 87¢. How many cans did she save?
    29 cans

20. Pete and his sister have the same birth date. When Pete is 12, his sister is 23. How many days older is Pete's sister?
    4,015 days

184

Problem Solving
Make a Tally

pages 185-186

Objective

To make a tally to solve problems

Materials

yellow pages of phone book
white pages of phone book
coins
tape measure marked in inches

Mental Math

Have students name all multiplication problems of whole numbers that have a product of:

1. 10 (10 × 1, 1 × 10, 2 × 5, 5 × 2)
2. 15 (15 × 1, 1 × 15, 3 × 5, 5 × 3)
3. 8 (8 × 1, 1 × 8, 2 × 4, 4 × 2)
4. 12 (12 × 1, 1 × 12, 2 × 6, 6 × 2, 3 × 4, 4 × 3)
5. 17 (1 × 17, 17 × 1)
6. 16 (16 × 1, 1 × 16, 2 × 8, 8 × 2, 4 × 4)
7. 20 (20 × 1, 1 × 20, 2 × 10, 10 × 2, 4 × 5, 5 × 4)
8. 35 (35 × 1, 1 × 35, 5 × 7, 7 × 5)

Making a Tally

How many times does each digit, 0 through 9, appear in the counting numbers 1 through 50?

★ SEE

We want to know how many times each digit appears in the counting numbers 1 through 50.

We know the digits are 0, 1, _2_, _3_, _4_, _5_, _6_, _7_, _8_ and _9_.

★ PLAN

We will make a list of the digits from 0 through 9. Then we will count from 1 through 50 and make a tally mark next to each of the digits in the numbers.

✔ Don't forget to cross the fifth tally mark, to make them easier to count.

★ DO

```
0  卌              5  卌 |
1  ||| |||| 卌     6  卌 |
2  卌 卌 卌         7  卌 |
3  卌 卌 卌         8  卌 |
4  卌 卌 卌         9  卌 |
```

(The numbers 1 through 20 have been tallied. Complete the tally for the numbers from 21 through 50.)

Digits	0	1	2	3	4	5	6	7	8	9
Times Used	5	15	15	15	15	6	5	5	5	5

★ CHECK

There are _9_ one-digit numbers and _41_ two-digit numbers from 1 through 50. 9 + (41 × 2) = 91 There are _91_ tallies.

185

Teaching the Lesson

Tell students that a tally is an organized way of counting things. Draw 3 columns on the board titled: brown, blue and green. Tell students they will make a tally of how many students have each color eyes. Have each student tell their eye color as you make a mark under that color word on the board. Tell students you are crossing every fifth tally mark, since you can then count by 5's to total each column. Have students total the columns and make a solution statement to report the tally. (There are __ students in our classroom with brown eyes, __ with blue eyes and __ with green eyes.) Remind students it is important to have a tally sheet with headings for the things being counted, so that anyone looking at it will clearly understand the method being used. Have a student read the thought bubble and the problem. Read through the SEE and PLAN stages with students and have them complete the tallies and the chart. Tell students to complete the sentences to check their solution.

185

Apply

Tally the data found in each situation.

1. Look in the White Pages of a phone book and pick ten random phone numbers. Tally the digits used in each phone number. How does this set of tallies compare to the one gathered in problem 3?
   Answers will vary.

2. Consider the vowels a, e, i, o and u. Count the number of times each vowel is used in the sentence:
   Now is the time for all good men to come to the aid of their country.
   See Solution Notes.

3. Look in the Yellow Pages of a phone book and pick ten random phone numbers. Tally the digits used in these numbers.
   Answers will vary.

4. Survey the students in your room and tally the number of pennies, nickels, dimes and quarters they have.
   Answers will vary.

5. Conduct a class survey and determine the most frequent date on all the coins in the room.
   Answers will vary.

6. Measure the distance around one wrist of each person in your classroom, to the nearest inch. Tally the findings.
   Answers will vary.

7. Suppose in Exercise 3 you had chosen 20 phone numbers at random from the same page. How do you think this would change your tally?
   Answers will vary.

8. Suppose you toss a penny 50 times. Predict the number of times it would land heads. Then toss a penny to check your prediction.
   See Solution Notes.

9. Douglas Dog ordered some cartons of canned dog food. The number of cartons he ordered is a two-digit number. The number of cans in each carton is a two-digit number. If you round to estimate the greatest number of cans that he could have ordered, what would be your estimate?
   See Solution Notes.

186

Solution Notes

Remind students to determine the headings of what will be counted before beginning to record tally marks. Help students see that some headings may have no tally marks and may be totaled as zero or be deleted when the counting is completed. Remind students to check to see that the sum of their tally totals equals the number of things in all, to be sure all information is included.

1. Answers will vary but students may find an abundance of zeros.
2.
Vowels	Counts	
a	//	= 2
e	⨋ /	= 6
i	////	= 4
o	⨋ ////	= 9
u	/	= 1
3. Answers will vary. Have students compare their results with answer to #1.
4. Answers will vary. Provide additional coins if students do not have a sufficient number to tally.
5. Provide coins if students do not have a sufficient number. Answers will vary.
6. Answers will vary. Be sensitive to any student measurement outside the class norm.

Higher-Order Thinking Skills

7. [Synthesis] Help students recognize that it is likely that each digit would receive twice as many tallies.
8. [Analysis] It is likely to land heads 25 times although the actual check might differ by 5 or more times either way.
9. [Synthesis] The greatest two-digit number is 99; so, the estimate would be 100×100, or 10,000.

Extra Credit *Logic*

Tell students to create the following figure divided into 5 rectangles, using 15 toothpicks.

Tell them to remove, not move, just 3 of the toothpicks to create a figure made up of 3 rectangles.

Calculator Work With Money

pages 187-188

Objective

To use a calculator to solve money problems

Materials

calculators

Mental Math

Ask students how old these people will be in 1992?

1. Adam is 7 now
2. Lil is 22 now
3. Rudy is 12 now
4. them
5. their dad
6. their mom
7. if born in 1944 (48)
8. if born in 1975 (17)

Skill Review

Have students work the following problems on the calculator: 26×8, $96¢ \times 4$, 12×6, $\$.87 \times 9$. Have students work in pairs to give each other review multiplication problems of 2-digit numbers times 1-digit numbers to solve.

Calculators, Working with Money

Mrs. Wallace bought 2 blouses and 4 ties at the Daisy Sale. She used her calculator to find the total cost of her purchases. Then she used estimation to see if that total made sense. How much did Mrs. Wallace pay for her purchases?

Daisy Sale
Belts ~ $8.25
Ties ~ $9.50
Blouses ~ $21.75

We want to know how much money Mrs. Wallace spent at the Daisy Sale.

She bought __2__ blouses for __$21.75__ each.

She also bought __4__ ties for __$9.50__ each. To find the total cost of each item, we multiply the cost of one by the number of items purchased. We use the $\boxed{\cdot}$ key to enter money amounts like $21.75. Complete these codes.

Blouses 21 $\boxed{\cdot}$ 75 $\boxed{\times}$ __2__ $\boxed{=}$ $\boxed{43.5}$

Ties 9 $\boxed{\cdot}$ 5 $\boxed{\times}$ __4__ $\boxed{=}$ $\boxed{38}$

✔ The cost of the blouses appears as 43.5. We have to write the dollar sign and zero for the answer to read $43.50. The cost of the ties is $38.00. When we enter the numbers on the calculator, we also can omit the zeros to the far right of the decimal point.
To find the total cost, complete this code.

43 $\boxed{\cdot}$ 5 $\boxed{+}$ 38 $\boxed{=}$ $\boxed{81.5}$

The total cost of the purchases is __$81.50__.

Estimate to see if the answer seems reasonable.

Blouses $21.75 → $22 × 2 = $44
Ties $ 9.50 → $10 × 4 = + 40
Total Cost $84
Estimate

187

Teaching the Lesson

Introducing the Problem Have a student read the problem aloud and tell what is to be solved. (the total cost of 2 blouses and 2 ties) Ask students what information is given. (Mrs. Wallace bought 2 blouses and 4 ties. Blouses cost $21.75 each while ties cost $9.50 each.) Have students complete the sentences as they read with you. Guide students through the calculator codes in the model to solve the problem. Remind students to clear their screens before each new problem. Have students estimate the cost to check their solution.

Developing the Skill Write **$12.98 × 6 =** on the board. Remind students there is no dollar sign on a calculator but they must enter the decimal point. Tell students we want to find the cost of 6 items at $12.98 each, so we enter 12.98, press the × key and 6 and the = key. Have a student write the product on the board. (77.88) Now write **$24.50 × 3 =** on the board and tell students to clear their calculator screens and enter this problem. Ask a student to write the product as it is shown on the screen. (73.5) Ask students what this answer means. ($73.50) Tell students we can find the total cost of the 6 items and 3 items by adding the 2 costs together. Have students help write the calculator codes for the 2 costs and enter the codes to find the total. (77.88 + 73.5 = 155.38) Now have students enter the code 13. × 6 = to find an estimate, clear their screens and enter 25. × 3 = . Tell students to clear their screens and enter 78. + 75. to find the estimated total. ($153) Repeat for the total cost of 5 items at $27.13 and 12 items at $19.85.

187

Practice

Use your calculator to find each answer.

1. $26.75
   × 3
   ——————
   $80.25

2. $18.79
   × 9
   ——————
   $169.11

3. $37.35
   × 12
   ——————
   $448.20

4. $57.38
   × 25
   ——————
   $1,434.50

5. $9.75
   × 36
   ——————
   $351.00

6. $11.75
   × 45
   ——————
   $528.75

7. $65.48
   × 82
   ——————
   $5,369.36

8. $49.75
   × 26
   ——————
   $1,293.50

Apply

Use the ad to help solve these problems. Use estimation to check your answers.

☆ Cassettes...$8.96
☆ Records...$3.49
☆ Albums...$12.69
★ SALE!★

9. What is the cost of 2 cassettes and 3 records?
$28.39

10. How much are 2 albums and 3 cassettes?
$52.26

11. How much more are 2 cassettes than 5 records?
$0.47

12. In another store, cassettes are on sale at 2 for $15. How much will you save if you buy 6 cassettes on sale?
$8.76

EXCURSION

Play this game with a friend. Pick 2 numbers from the apple and multiply them. If your product is a number on the board, put your initial on it. The winner is the first to have 3 numbers in a row.

9 18
16 8
15 4 12
19

228	224	216
270	144	240
171	120	108

188

Practice

Tell students to work the first group of problems on their calculators. Remind them to enter the decimal point in a money amount and it must have 2 digits to the right of the decimal point. Tell students they are to use the information in the ad to solve the word problems. Remind students to estimate to check their answers. Have students complete the page.

Excursion

Estimating products and predicting the units digit of the products, are two strategies which will help a student to be successful in this activity.

Extra Credit *Logic*

Read the following to students, or duplicate for them to solve: On her way to school, Heather counted 47 trees along the right side of the street. On the way home, she counted 47 trees on the left side of the street. How many trees did Heather count in all? (47 trees; The trees on her right, going to school are on her left when she comes home.)

188

Chapter Test

page 189

Item	Objective
1, 3, 4	Multiply 3-digit number by 1-digit number with trading (See pages 167-168)
2	Multiply 3-digit amount of money by 1-digit number (See pages 169-170)
5-8 22-23	Multiply 4-digit number or money by 1-digit number with or without trading ones, tens, hundreds (See pages 173-174)
9-12	Multiply two 2-digit numbers without trading (See pages 177-178)
13-16	Multiply two 2-digit numbers trading ones (See pages 179-180)
17-20 25, 26, 28	Multiply 3-digit number or money by 2-digit number (See pages 181-182)
21, 24, 27	Multiply money by 2-digit numbers (See pages 181-182)

Multiply. Use estimation to check your answers.

1.
$$\begin{array}{r} 224 \\ \times\ \ \ 3 \\ \hline 672 \end{array}$$

2.
$$\begin{array}{r} \$3.43 \\ \times\ \ \ \ 4 \\ \hline \$13.72 \end{array}$$

3.
$$\begin{array}{r} 679 \\ \times\ \ \ 7 \\ \hline 4,753 \end{array}$$

4.
$$\begin{array}{r} 425 \\ \times\ \ \ 9 \\ \hline 3,825 \end{array}$$

5.
$$\begin{array}{r} 3,208 \\ \times\ \ \ \ \ 5 \\ \hline 16,040 \end{array}$$

6.
$$\begin{array}{r} 5,728 \\ \times\ \ \ \ \ 2 \\ \hline 11,456 \end{array}$$

7.
$$\begin{array}{r} \$38.51 \\ \times\ \ \ \ \ \ 6 \\ \hline \$231.06 \end{array}$$

8.
$$\begin{array}{r} 7,286 \\ \times\ \ \ \ \ 8 \\ \hline 58,288 \end{array}$$

9.
$$\begin{array}{r} 21 \\ \times\ 32 \\ \hline 672 \end{array}$$

10.
$$\begin{array}{r} 43 \\ \times\ 23 \\ \hline 989 \end{array}$$

11.
$$\begin{array}{r} 57 \\ \times\ 11 \\ \hline 627 \end{array}$$

12.
$$\begin{array}{r} 72 \\ \times\ 43 \\ \hline 3,096 \end{array}$$

13.
$$\begin{array}{r} 76 \\ \times\ 29 \\ \hline 2,204 \end{array}$$

14.
$$\begin{array}{r} 59 \\ \times\ 75 \\ \hline 4,425 \end{array}$$

15.
$$\begin{array}{r} 83 \\ \times\ 46 \\ \hline 3,818 \end{array}$$

16.
$$\begin{array}{r} 67 \\ \times\ 96 \\ \hline 6,432 \end{array}$$

17.
$$\begin{array}{r} 245 \\ \times\ \ 36 \\ \hline 8,820 \end{array}$$

18.
$$\begin{array}{r} \$7.93 \\ \times\ \ \ \ 18 \\ \hline \$142.74 \end{array}$$

19.
$$\begin{array}{r} 425 \\ \times\ \ 86 \\ \hline 36,550 \end{array}$$

20.
$$\begin{array}{r} 729 \\ \times\ \ 79 \\ \hline 57,591 \end{array}$$

21.
$$\begin{array}{r} \$6.37 \\ \times\ \ \ \ 45 \\ \hline \$286.65 \end{array}$$

22.
$$\begin{array}{r} 4,759 \\ \times\ \ \ \ \ 8 \\ \hline 38,072 \end{array}$$

23.
$$\begin{array}{r} 6,904 \\ \times\ \ \ \ \ 9 \\ \hline 62,136 \end{array}$$

24.
$$\begin{array}{r} \$5.94 \\ \times\ \ \ \ 63 \\ \hline \$374.22 \end{array}$$

25.
$$\begin{array}{r} 748 \\ \times\ \ 59 \\ \hline 44,132 \end{array}$$

26.
$$\begin{array}{r} 286 \\ \times\ \ 43 \\ \hline 12,298 \end{array}$$

27.
$$\begin{array}{r} \$8.99 \\ \times\ \ \ \ 74 \\ \hline \$665.26 \end{array}$$

28.
$$\begin{array}{r} 987 \\ \times\ \ 39 \\ \hline 38,493 \end{array}$$

Circle the letter of the correct answer.

1 Round 723 to the nearest hundred.
(a) 700
b 800
c NG

2 Round 5,329 to the nearest thousand.
(a) 5,000
b 6,000
c NG

3 What is the value of the 5 in 315,603?
a tens
b hundreds
(c) thousands
d NG

4
  4,279
+ 3,651
a 7,820
(b) 7,930
c 8,930
d NG

5
  20,926
+ 50,285
a 70,211
b 71,101
(c) 71,211
d NG

6
  703
− 246
a 447
b 543
c 557
(d) NG

7
  52,186
− 19,429
a 32,657
(b) 32,757
c 47,363
d NG

8 63 ÷ 9
(a) 7
b 8
c 9
d NG

9 72 ÷ 3
a 20
b 22
(c) 24
d NG

10 4)93
a 2 R1
b 23
(c) 23 R1
d NG

11 Choose the best estimate of length.
((·chalk·))
(a) 7 cm
b 7 dm
c 7 m

12 Choose the best estimate of weight.
a 2 oz
(b) 2 lb

13
  4,032
×   3
a 1,296
(b) 12,096
c 12,396
d NG

14 27 × 46
a 1,222
b 1,322
c 1,442
(d) NG

☐ score

190

Cumulative Review

page 190

Item	Objective
1	Round numbers to nearest 100 (See pages 31-34)
2	Round numbers to nearest 1,000 (See pages 33-34)
3	Identify place value through hundred thousands (See pages 27-28)
4	Add two 4-digit numbers (See pages 47-48)
5	Add two 5-digit numbers (See pages 53-54)
6	Subtract two 3-digit numbers with zero in minuend (See pages 65-66)
7	Subtract two 5-digit numbers (See pages 67-68)
8	Divide using 9 as divisor (See pages 127-128)
9	Divide 2-digit number by 1-digit number to get 2-digit quotient without remainder (See pages 133-134)
10	Divide 2-digit number by 1-digit number to get 2-digit quotient with remainder (See pages 135-136)
11	Determine appropriate metric unit of length (See pages 153-154)
12	Determine appropriate customary unit of weight (See pages 149-150)
13	Multiply 4-digit number by 1-digit number (See pages 173-174)
14	Multiply two 2-digit numbers trading ones (See pages 179-180)

Alternate Cumulative Review

Circle the letter of the correct answer.

1 Round 621 to the nearest hundred.
a 500
(b) 600
c 700
d NG

2 Round 6,497 to the nearest thousand.
a 5,000
(b) 6,000
c 7,000
d NG

3 What is the value of the 7 in 279,654?
a tens
b thousands
(c) ten thousands
d NG

4
  6,372
+ 2,348
a 8,610
(b) 8,720
c 9,720
d NG

5
  62,438
+ 33,675
(a) 96,113
b 95,014
c 95,003
d NG

6
  907
− 329
(a) 578
b 622
c 588
d NG

7
  37,243
− 18,376
a 18,877
(b) 18,867
c 28,867
d NG

8 48 ÷ 6 =
a 5
b 7
c 9
(d) NG

9 4)96
a 2 R1
b 21 R2
(c) 24
d NG

10 3)85
a 20 R5
(b) 28 R1
c 28
d NG

11 What would you use to measure a candle?
a mm
(b) cm
c m
d NG

12 What would you use to measure a person's weight?
a oz
(b) lbs
c tons
d NG

13
  5,344
×   2
a 7,566
(b) 10,688
c 11,688
d NG

14
  36
× 46
a 320
b 2,656
(c) 1,656
d NG

3-digit Quotients

pages 191-192

Objective

To divide 3-digit numbers by 1-digit numbers, no remainders

Materials

place value materials

Mental Math

Dictate the following:

1. 20×12 (240)
2. 13×30 (390)
3. 100×50 (5,000)
4. 70×8 (560)
5. 30×40 (1,200)
6. $\$4 \times 50$ ($200)
7. $\$10 \times 40$ ($400)
8. 16×60 (960)

Skill Review

Review 2-digit division by writing the following problems on the board:

$$\overset{(12)}{3\overline{)36}} \quad \overset{(24)}{4\overline{)96}} \quad \overset{(12)}{6\overline{)72}} \quad \overset{(11)}{5\overline{)55}} \quad \overset{(12)}{7\overline{)84}}$$

Have students work the problems and then check by multiplying the quotient by the divisor.

Dividing, 3-digit Quotients

Sidney and Alicia have 254 invitations to address for a PTA social. How many invitations will each girl have to do?

We want to know the number of invitations each girl will write.

All together, there are __254__ invitations.

There are __2__ girls doing the addressing. To find the number of invitations each girl will write, we divide __254__ by __2__.

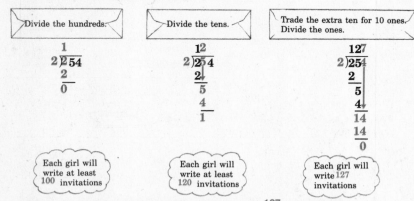

Divide the hundreds.	Divide the tens.	Trade the extra ten for 10 ones. Divide the ones.

Each girl will write at least 100 invitations.

Each girl will write at least 120 invitations.

Each girl will write 127 invitations.

Sidney and Alicia will each have to write __127__ invitations.

Getting Started

Divide. Show your work.

1. $\overset{212}{3\overline{)636}}$
2. $\overset{114}{4\overline{)456}}$

Copy and divide.

3. $392 \div 2$ 196
4. $714 \div 6$ 119

191

Teaching the Lesson

Introducing the Problem Have students describe the picture. Have a student read the problem and tell what they are to find. (number of invitations each girl will need to do) Ask students what information is given. (There are 254 invitations and 2 girls.) Ask what operation should be used. (division) Have students read and complete the sentences through each stage of the plan. Guide students through the division steps in the model. Tell students to multiply their solution by 2 to check their work.

Developing the Skill Write **426** on the board. Ask students how many hundreds, tens and ones are in the number. (4,2,6) Put out place-value materials to represent 426 and ask students to help you put the materials in 2 equal groups by starting with the hundreds, then the tens and the ones last. Tell students they are dividing 426 by 2. Ask students to name the number in each group. (213) Repeat for $316 \div 2$ where the extra hundred is traded for 10 tens and the extra ten is traded for 120 ones. Continue for $816 \div 6$ and $645 \div 5$.

191

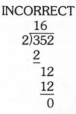

Practice

Divide. Show your work.

1. $\begin{array}{r} 122 \\ 3\overline{)366} \end{array}$

2. $\begin{array}{r} 324 \\ 2\overline{)648} \end{array}$

3. $\begin{array}{r} 111 \\ 5\overline{)555} \end{array}$

4. $\begin{array}{r} 211 \\ 4\overline{)844} \end{array}$

5. $\begin{array}{r} 217 \\ 3\overline{)651} \end{array}$

6. $\begin{array}{r} 214 \\ 4\overline{)856} \end{array}$

7. $\begin{array}{r} 329 \\ 2\overline{)658} \end{array}$

8. $\begin{array}{r} 112 \\ 6\overline{)672} \end{array}$

9. $\begin{array}{r} 242 \\ 4\overline{)968} \end{array}$

10. $\begin{array}{r} 113 \\ 7\overline{)791} \end{array}$

11. $\begin{array}{r} 253 \\ 3\overline{)759} \end{array}$

12. $\begin{array}{r} 469 \\ 2\overline{)938} \end{array}$

Copy and Do

13. $575 \div 5$
115

14. $736 \div 4$
184

15. $832 \div 2$
416

16. $864 \div 3$
288

17. $996 \div 2$
498

18. $579 \div 3$
193

19. $912 \div 4$
228

20. $597 \div 3$
199

Apply

Solve these problems.

21. A computer printer printed 924 lines in 4 minutes. How many lines were printed each minute?
231 lines

22. Ivan helped his grandfather box up his collection of 462 books that he will donate to 3 school libraries. How many books will each library receive?
154 books

23. There are 3 feet in 1 yard. Change 486 feet to yards.
162 yards

24. The pond is 165 yards wide. How wide is the pond in feet?
495 feet

192

3-digit Quotient With Remainder

Objective

To divide 3-digit numbers by 1-digit numbers, with remainder

Materials

place value materials

Mental Math

Dictate the following:

1. $2/6 + 2/6 + 1/6$ ($5/6$)
2. $.1 + .4 + .3$ (.8)
3. 20×400 (8,000)
4. $11 + 22 + 33$ (66)
5. $(8 \times 6) + (10 \times 6)$ (108)
6. $7 \times (48 \div 6)$ (56)
7. 5 lb 2 oz $\times$ 8 (41 lb)
8. 25¢ $\times$ 6 ($1.50)

Skill Review

Review 2-digit division with a reminder by writing the following problems on the board:

(13R1) (13R2) (14R1) (11R6)
2)37   7)93   6)85   7)83

Have students work the problems and then check by multiplying the quotient by the divisor and adding the remainder.

Dividing, Remainders

The Quick Computer Company manufactures computer disks. In one day, the company made 748 disks. How many packages did the company produce? How many disks were left over?

We want to find the number of packages of disks produced and the number of disks left over.

Quick Computer made ___748___ disks in one day.

There are ___6___ disks in a package. To find the number of packages, we divide

___748___ by ___6___. The remainder is the number of disks left over.

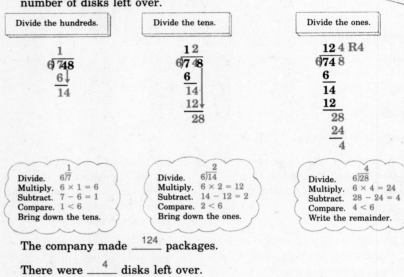

Divide the hundreds.	Divide the tens.	Divide the ones.
1 6)748 6↓ 14	12 6)748 6↓ 14 12↓ 28	124 R4 6)748 6 14 12 28 24 4

Divide. 6)7 — 1 Multiply. $6 \times 1 = 6$ Subtract. $7 - 6 = 1$ Compare. $1 < 6$ Bring down the tens.	Divide. 6)14 — 2 Multiply. $6 \times 2 = 12$ Subtract. $14 - 12 = 2$ Compare. $2 < 6$ Bring down the ones.	Divide. 6)28 — 4 Multiply. $6 \times 4 = 24$ Subtract. $28 - 24 = 4$ Compare. $4 < 6$ Write the remainder.

The company made ___124___ packages.

There were ___4___ disks left over.

Getting Started

Divide. Show your work.

1. 212 R1
4)849

2. 134 R3
5)673

Copy and divide.

3. $793 \div 2$
396 R1

4. $796 \div 6$
132 R4

193

Teaching the Lesson

Introducing the Problem Tell students the picture shows a machine which sorts floppy computer discs. Have a student read the problem. Ask students what 2 questions are to be answered. (number of packages of discs and number of discs left over) Ask what information is needed. (number of disks in all and number per package) Ask what information is given. (748 disks in all and 6 per package) Have students read and complete the sentences. Guide students through the division in the model, pointing out the remainder. Have students check by multiplying and adding the remainder.

Developing the Skill Write 6)732 on the board and have a student work the problem, talking through each step. (122) Ask students if there is a remainder. (no) Now write 6)735 on the board. Have students talk through each step as you work the problem. (122) Ask students if there are ones left-over. (yes) Ask how many ones. (3) Remind students that left-over ones in division are called the remainder as you write **R3** after the 122. Tell students that the left-over number must be less than the divisor. Repeat the procedure for 2)268, 2)269, 8)992 and 8)999.

Practice

Divide. Show your work.

1. 4)449   112 R1
2. 6)682   113 R4
3. 3)635   211 R2
4. 2)869   434 R1

5. 5)594   118 R4
6. 4)925   231 R1
7. 7)784   112
8. 6)826   137 R4

9. 8)916   114 R4
10. 5)727   145 R2
11. 2)916   458
12. 7)919   131 R2

13. 4)857   214 R1
14. 6)885   147 R3
15. 3)558   186
16. 5)962   192 R2

Copy and Do

17. 437 ÷ 2   218 R1
18. 789 ÷ 4   197 R1
19. 896 ÷ 8   112
20. 416 ÷ 3   138 R2

21. 775 ÷ 6   129 R1
22. 815 ÷ 5   163
23. 593 ÷ 2   296 R1
24. 779 ÷ 7   111 R2

25. 956 ÷ 7   136 R4
26. 651 ÷ 3   217
27. 912 ÷ 5   182 R2
28. 852 ÷ 4   213

Apply

Solve these problems.

29. Li is packing tomato plants into boxes that hold 5 plants each. Li has 598 plants. How many boxes will she need? How many plants will be left over?
119 boxes   3 left over plants

30. Mr. Hawthorne is putting 4 chairs at each table in his cafe. If Mr. Hawthorne has 462 chairs, how many tables can he supply?
115 tables

194

Zeros in Quotients

pages 195-196

Objective

To divide 3-digit numbers by 1-digit numbers for quotients with zeros

Materials

place value materials

Mental Math

Ask if each is divisible by 3:

1. 102 (yes)
2. 391 (no)
3. 475 (no)
4. 261 (yes)
5. 592 (no)
6. 375 (yes)
7. 87 (yes)
8. 799 (no)

Skill Review

Review dividing a 3-digit number by a 1-digit number by having students work the following problems on the board:

(212)	(239R1)	(114R3)
4)848	3)718	5)573

(122R5)	(118R3)	(326)
6)737	7)829	2)652

Dividing, Zeros in Quotients

How long will it take for 424 quarts of blood to be pumped through the human heart?

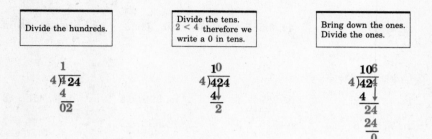

4 QUARTS EACH MINUTE

We want to know the number of minutes it will take for the heart to pump ___424___ quarts of blood.

We know that ___4___ quarts are pumped each ___minute___.

To find the length of time it takes to pump the blood, we divide ___424___ by ___4___.

Divide the hundreds.	Divide the tens. 2 < 4 therefore we write a 0 in tens.	Bring down the ones. Divide the ones.
1 4)424 4 02	10 4)424 4↓ 2	106 4)424 4↓ 24 24 0

It will take ___106___ minutes to pump 424 quarts of blood through the heart.

Getting Started

Divide. Show your work.

1. 7)735 — 105
2. 5)534 — 106 R4
3. 3)309 — 103
4. 2)803 — 401 R1

Copy and divide.

5. 826 ÷ 8 — 103 R2
6. 816 ÷ 4 — 204
7. 600 ÷ 6 — 100
8. 948 ÷ 9 — 105 R3

195

Teaching the Lesson

Introducing the Problem Have a student read the problem aloud and tell what is to be solved. (number of minutes it will take for 242 quarts of blood to be pumped through the human heart) Ask students if the information in the picture is needed to solve the problem. (yes) Have a student read the sign. Have students complete the sentences to solve the problem. Guide students through each step of the division. Tell students to multiply their answer by the number of quarts per minute to see if they get the total number of quarts.

Developing the Skill Write 3)627 on the board. Have students divide the hundreds and then tell if there are enough tens to be divided by 3. (no) Tell students we then place a zero in tens place in the quotient to show that each of the 3 groups gets no tens. Write the zero in the quotient. Tell students we must trade the 2 tens for 20 ones and then see how many ones in all. (27) Ask students to divide the 27 ones by 3 and tell the number. (9) Now have students multiply the quotient by the divisor to check their answer. Repeat for 9)997 and 4)830 where the quotients have zero and a remainder.

Practice

Divide. Show your work.

1. 6)618 — 103
2. 4)832 — 208
3. 7)749 — 107
4. 2)816 — 408

5. 9)927 — 103
6. 5)545 — 109
7. 8)854 — 106 R6
8. 9)906 — 100 R6

Copy and Do

9. 436 ÷ 4
   109
10. 200 ÷ 2
    100
11. 709 ÷ 7
    101 R2
12. 651 ÷ 6
    108 R3

13. 837 ÷ 8
    104 R5
14. 320 ÷ 3
    106 R2
15. 529 ÷ 5
    105 R4
16. 973 ÷ 9
    108 R1

Apply

Solve these problems.

17. How many gallons of water are there in 408 quarts?
    102 gallons

18. Change 318 feet to yards.
    106 yards

19. Change 36 feet to inches.
    432 inches

20. How many pints are there in 216 quarts?
    432 pints

EXCURSION

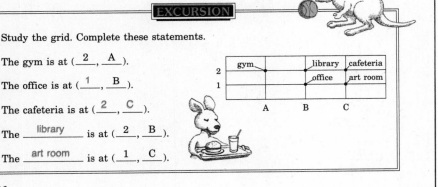

Study the grid. Complete these statements.

The gym is at (2 , A).

The office is at (1 , B).

The cafeteria is at (2 , C).

The library is at (2 , B).

The art room is at (1 , C).

196

More Dividing

pages 197-198

Objective

To divide 3-digit numbers by 1-digit numbers for 2-digit quotients

Materials

place value materials

Mental Math

Have students name the decade number that is closest to the answer:

1. $82 \div 2$ (40)
2. 27×3 (80)
3. $(6 \times 9) + 12$ (70)
4. $69 \div 3$ (20)
5. 25×3 (80)
6. $(32 \div 8) + 7$ (10)
7. ½ of 96 (50)
8. $10 + 60 - 43$ (30)

Skill Review

Ask students to tell the numbers from 1 through 9 that can be divided by 3 evenly and those having a remainder. Continue for numbers from 1 through 9 when divided by 2, 4, 6, 7, 9.

Dividing Larger Dividends

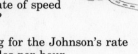

It took the Johnson family 6 hours to drive from Argus to Chester, through Lincoln. What was their rate of speed in miles per hour?

We are looking for the Johnson's rate of speed in miles per hour.

We need to add ___136___ and ___188___ to get the total distance traveled.

The Johnsons traveled ___324___ miles from Argus to Chester.

It took them ___6___ hours to make this trip.

To find the rate of speed, we divide the total miles by the number of hours. We divide ___324___ by ___6___.

Divide the hundreds. 3 < 6 Trade the 3 hundreds for 30 tens.	Divide the tens. Write the quotient digit above the tens.	Trade the 2 tens for 20 ones. Divide the ones.

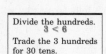

$$6\overline{)324}$$

$$\begin{array}{r} 5 \\ 6\overline{)324} \\ 30 \\ \hline 2 \end{array}$$

$$\begin{array}{r} 54 \\ 6\overline{)324} \\ 30\downarrow \\ \hline 24 \\ 24 \\ \hline 0 \end{array}$$

The Johnson family drove at a rate of speed of ___54___ miles per hour.

Getting Started

Divide. Show your work.   Copy and divide.

1. $\begin{array}{r}69\\4\overline{)276}\end{array}$

2. $\begin{array}{r}54\text{ R1}\\8\overline{)433}\end{array}$

3. $850 \div 9$ 94 R4

4. $249 \div 3$ 83

197

Teaching the Lesson

Introducing the Problem Have students tell what can be learned from the pictured map. (Argus to Lincoln is 136 miles and Lincoln to Chester is 188 miles.) Ask students how the map could be used to find the distance from Argus to Chester. (add the distances from Argus to Lincoln and from Lincoln to Chester) Have a student read the problem and tell what is to be solved. (speed at which the Johnsons travelled) Ask what information would be used to find the speed. (distance travelled and hours they drove) Have students complete the sentences and work the division in the model with you. Have them complete the solution sentence.

Developing the Skill Write $4\overline{)424}$ on the board. Ask students if the 4 hundreds can be divided by 4. (yes) Erase the 4 hundreds and replace with a 1. Ask students if the 1 hundred can be divided by 4. (no) Tell students that since the 1 hundred cannot be divided by 4, they must trade the 1 hundred for 10 tens. Have students tell how many tens in all. (12) Ask students to divide the 12 tens by 4. (3) Write 3 in tens place in the quotient and ask students if any tens are leftover and need to be traded for ones. (no) Have students divide the ones (1) and tell if there is a remainder. (no) Have students check the solution by multiplying. Repeat the procedure for $9\overline{)549}$ and $8\overline{)709}$.

Practice

Divide. Show your work.

1. 7)511 → 73
2. 4)268 → 67
3. 2)108 → 54
4. 6)354 → 59

5. 9)468 → 52
6. 3)159 → 53
7. 5)262 → 52 R2
8. 8)290 → 36 R2

9. 5)476 → 95 R1
10. 2)137 → 68 R1
11. 9)755 → 83 R8
12. 4)317 → 79 R1

13. 8)363 → 45 R3
14. 3)209 → 69 R2
15. 7)539 → 77
16. 6)540 → 90

Copy and Do

17. 627 ÷ 8
    78 R3
18. 248 ÷ 5
    49 R3
19. 615 ÷ 9
    68 R3
20. 137 ÷ 2
    68 R1

21. 316 ÷ 6
    52 R4
22. 209 ÷ 7
    29 R6
23. 312 ÷ 4
    78
24. 115 ÷ 3
    38 R1

25. 423 ÷ 7
    60 R3
26. 196 ÷ 5
    39 R1
27. 120 ÷ 3
    40
28. 517 ÷ 6
    86 R1

Apply

Solve these problems.

29. It is 385 miles from Al's house to his grandparents. It takes Al's family 7 hours to drive there. What is their rate of speed in miles per hour?
    55 miles per hour

30. The cafeteria serves 6-ounce glasses of milk. How many full glasses of milk can be poured from 225 ounces of milk?
    37 glasses

31. How many pints are there in 10 gallons?
    80 pints

32. How much farther is it from Lincoln to Chester than from Lincoln to Argus? Use the map on page 197.
    52 miles

198

Correcting Common Errors

When the first digit in the dividend cannot be divided by the divisor, some students simply ignore it.

INCORRECT
23
3)169

Have students use place-value materials to model the problem. The first step is to trade 1 hundred for 10 tens, giving a total of 16 tens. With such experiences, students soon will "see" the 16 tens right away without having to regroup or work with manipulatives.

Enrichment

Have students tell how they know, without working the problem, that the quotient of 6)542 will be 2 digits. Tell them to write 4 such problems for a friend to solve and check.

Practice

Remind students they will need to provide additional information in the word problems, and that some may require operations other than division. Have students complete the page independently.

Extra Credit *Biography*

Galileo Galilei, an Italian scientist of the 16th and 17th centuries, is best remembered for his work in astronomy. He invented the telescope, and within one year had observed and written about sunspots, mountains on the moon, Jupiter's moons, the phases of Venus and the rings of Saturn. But Galileo also went against many popular views of the time. He theorized that objects fall at the same rate, no matter how much they weigh. Galileo also supported Copernicus' view that the sun was the center of the solar system. The church officials of the time said this view went against the Holy Scriptures, and they threatened Galileo with torture if he refused to take back what he had said. Today, we know that Galileo was right about our solar system.

Dividing Money

pages 199-200

Objective

To divide money through $9.99 by 1-digit numbers

Materials

Mental Math

Have students complete each comparison: a century is to 25 as:

1. 12 is to (3).
2. 60 is to (15).
3. 40 is to (10).
4. 16 oz is to (4 oz).
5. 36 is to (9).
6. $1 is to (25¢).
7. 48 is to (12).
8. 56 is to (14).

Skill Review

Write the following problems on the board to have students review dividing a 3-digit number by a 1-digit number for a 2-digit quotient:

(78R2) (69R4) (93R8) (55)
8)626   6)418   9)845   5)275

Dividing Money

At the print shop, Fred had 5 copies made of his science report. What was the cost of each copy?

We want to know the cost of making one copy of Fred's report.

Fred paid $2.45 to get 5 copies made. To find the cost of one copy, we divide the cost of all the copies by the number of copies made.

We divide $2.45 by 5.

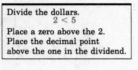

Divide the dollars. 2 < 5 Place a zero above the 2. Place the decimal point above the one in the dividend.	Trade the 2 dollars for 20 dimes. Divide the dimes.	Trade the 4 left-over dimes for pennies. Divide the pennies. Place the dollar sign.

$$\begin{array}{r} 0. \\ 5)\overline{\$2.45} \end{array}$$

$$\begin{array}{r} 0.4 \\ 5)\overline{\$2.45} \\ \underline{2\,0} \\ 4 \end{array}$$

$$\begin{array}{r} \$0.49 \\ 5)\overline{\$2.45} \\ \underline{2\,0} \\ 45 \\ \underline{45} \\ 0 \end{array}$$

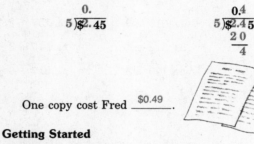

One copy cost Fred $0.49.

Getting Started

Divide. Show your work.

1. $0.48
   7)$3.36

2. $2.27
   3)$6.81

3. $0.45
   4)$1.80

Copy and divide.

4. $7.45 ÷ 5
   $1.49

5. $4.74 ÷ 6
   $0.79

6. $9.09 ÷ 9
   $1.01

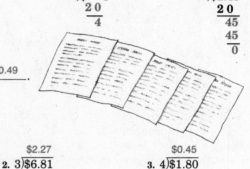

199

Teaching the Lesson

Introducing the Problem Have a student read the problem aloud and tell what is to be found. (the cost of 1 copy) Ask students what information is given in the picture. (The bill reads $2.45.) Ask how many reports Fred had printed. (5) Have students read and complete the sentences to solve the problem. Guide them through the division steps in the model. Have students multiply to check their answer.

Developing the Skill Write 4)$3.24 on the board. Tell students this problem is like those they have been doing but there is a dollar sign and decimal point. Ask if there are enough dollars to divide by 4. (no) Tell students that in dollar amounts, we place a zero in the dollar place to show no dollars. Tell students we then trade the 3 dollars for dimes. Ask how many dimes in all. (32) Have students divide the dimes by 4, record the 8 in tens place and subtract the 32 from 32. (0) Ask students to divide the ones (1) and subtract the product of 4 from 4. (0) Tell students that the quotient of a money problem must have a dollar sign and decimal point. Have students read the quotient of $.82. Ask a student to multiply to check the answer. Repeat the procedure for 7)$3.55 and 9)$8.01.

199

Practice

Divide. Show your work.

1. $\dfrac{\$1.29}{4\overline{)\$5.16}}$ 2. $\dfrac{\$0.42}{8\overline{)\$3.36}}$ 3. $\dfrac{\$1.18}{7\overline{)\$8.26}}$

4. $\dfrac{\$1.03}{3\overline{)\$3.09}}$ 5. $\dfrac{\$0.64}{9\overline{)\$5.76}}$ 6. $\dfrac{\$3.79}{2\overline{)\$7.58}}$

7. $\dfrac{\$1.87}{5\overline{)\$9.35}}$ 8. $\dfrac{\$0.48}{6\overline{)\$2.88}}$ 9. $\dfrac{\$1.02}{7\overline{)\$7.14}}$

10. $\dfrac{\$0.77}{6\overline{)\$4.62}}$ 11. $\dfrac{\$1.18}{8\overline{)\$9.44}}$ 12. $\dfrac{\$3.06}{3\overline{)\$9.18}}$

Copy and Do

13. $\$6.25 \div 5$
$1.25

14. $\$4.24 \div 8$
$0.53

15. $\$1.38 \div 2$
$0.69

16. $\$8.13 \div 3$
$2.71

17. $\$9.52 \div 7$
$1.36

18. $\$1.56 \div 6$
$0.26

19. $\$9.20 \div 4$
$2.30

20. $\$7.65 \div 9$
$0.85

Apply

Solve these problems.

21. Rob bought 5 paperback books for $7.95. What was the cost of each book?
$1.59

22. Two hundred eight quarts of milk are poured into gallon jugs. How many jugs will be needed?
52 jugs

23. Vince and Don bought 3 birthday cards, each costing $1.20. They decided to split the cost evenly. How much did each boy pay?
$1.80

24. A 6-pound package of hamburger costs $8.34. How much would a 4-pound package cost?
$5.56

200

Dividing, Multiples of 10

pages 201-202

Objective

To divide 2- or 3-digit multiples of 10 by 2-digit multiples of 10

Materials

Mental Math

Ask students is there a remainder?

1. $25 \div 3$ (yes)
2. $84 \div 7$ (no)
3. $666 \div 2$ (no)
4. $35 \div 6$ (yes)
5. $100 \div 4$ (no)
6. $68 \div 6$ (yes)
7. $38 \div 3$ (yes)
8. $72 \div 2$ (no)

Skill Review

Remind students that when we multiply a number by a multiple of 10, we multiply the numbers that are not zero and follow that product with the number of zeros in the factors. Ask students to write and solve the following problems as you dictate them: $26 \times 20 = (520)$, $2 \times 400 = (800)$, $40 \times 50 = (2,000)$, $70 \times 9 = (630)$, $75 \times 20 = (1,500)$

Dividing by Multiples of 10

Madalana is making bows that use 20 centimeters of ribbon each. How many bows can Madalana make with one spool of ribbon?

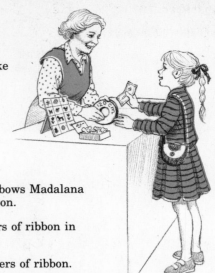

We want to find the number of bows Madalana can make from one spool of ribbon.

Madalana uses __20__ centimeters of ribbon in one bow.

A spool contains __120__ centimeters of ribbon.

To find the number of bows, we divide __120__

by __20__.

We know that: We use this fact $\begin{array}{r} 6 \\ 20\overline{)120} \\ \underline{120} \\ 0 \end{array}$
$12 \div 2 = 6.$ to find
                      $120 \div 20.$

Madalana can make __6__ bows from one spool of ribbon.

Getting Started

Divide. Show your work.

1. $\overset{2}{30\overline{)60}}$
2. $\overset{4}{20\overline{)80}}$
3. $\overset{8}{80\overline{)640}}$
4. $\overset{6}{40\overline{)240}}$

Copy and divide.

5. $60 \div 20$
   $\overset{}{3}$
6. $90 \div 30$
   $\overset{}{3}$
7. $320 \div 40$
   $\overset{}{8}$
8. $360 \div 60$
   $\overset{}{6}$

201

Teaching the Lesson

Introducing the Problem   Have a student read the problem aloud and describe the picture. Ask students what question is to be answered. (number of bows that can be made from 1 spool of ribbon) Ask what information is given. (1 spool holds 120 centimeters of ribbon and 1 bow uses 20 centimeters of ribbon.) Have students complete the sentences, estimate, and work the division with you. Tell students to check their solution by multiplying, after filling in the solution.

Developing the Skill   Write $80\overline{)240}$ on the board. Tell students we know 2 is too small to divide by 80, and 24 is also too small, so we need to find how many 80's are in 240. Tell students we look at the 8 and think of how many 8's are in 24. (3) Remind students that they learned that 8×3 is 24 and therefore 80×3 is 240. Therefore, there are 3 eights in 24 and there are 3 eighties in 240. Write 3 in the quotient in ones place. Ask students to multiply 80 by 3 to see if the answer is correct. Write $90\overline{)810}$ on the board and remind students we ask ourselves how many 9's are in 81 to estimate the number of 90's in 810. Help students complete and check the problem. Continue having students use partial divisors to work $80\overline{)720}$ and $90\overline{)180}$.

Practice

Divide. Show your work.

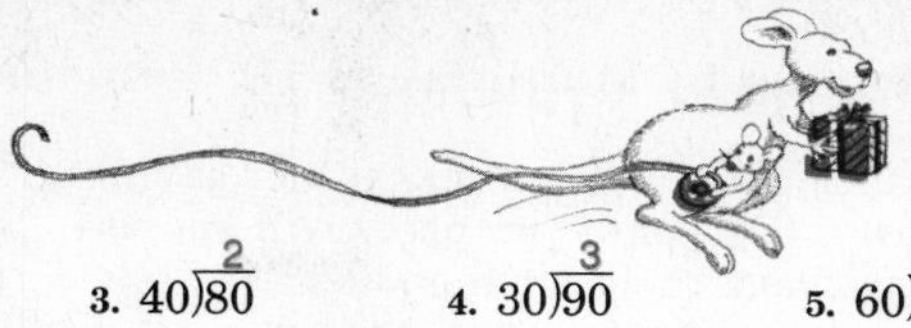

1. $20\overline{)60}$ = 3
2. $30\overline{)60}$ = 2
3. $40\overline{)80}$ = 2
4. $30\overline{)90}$ = 3
5. $60\overline{)60}$ = 1

6. $50\overline{)150}$ = 3
7. $70\overline{)490}$ = 7
8. $60\overline{)300}$ = 5
9. $80\overline{)240}$ = 3
10. $90\overline{)270}$ = 3

11. $40\overline{)360}$ = 9
12. $20\overline{)160}$ = 8
13. $70\overline{)490}$ = 7
14. $90\overline{)270}$ = 3
15. $80\overline{)400}$ = 5

16. $30\overline{)150}$ = 5
17. $60\overline{)480}$ = 8
18. $50\overline{)200}$ = 4
19. $60\overline{)240}$ = 4
20. $90\overline{)540}$ = 6

21. $20\overline{)100}$ = 5
22. $40\overline{)320}$ = 8
23. $30\overline{)210}$ = 7
24. $70\overline{)140}$ = 2
25. $80\overline{)240}$ = 3

Copy and Do

26. $270 \div 30$
9
27. $450 \div 50$
9
28. $420 \div 60$
7
29. $560 \div 80$
7
30. $200 \div 40$
5
31. $810 \div 90$
9

Apply

Solve these problems.

32. There are 30 pounds of potatoes in each sack. How many sacks of potatoes will be needed to fill an order of 180 pounds?
6 sacks

33. There are 20 small glasses of juice in each quart. How many glasses can be made from 80 quarts?
1,600 glasses

202

More Dividing Multiples of 10

pages 203-204

Objective

To divide 2- or 3-digit numbers by 2-digit multiples of 10

Materials

Mental Math

Have students name a unit of measure for:

1. metric length. (cm, dm, m, km)
2. customary capacity. (c, pt, qt, gal)
3. metric capacity. (L, mL)
4. customary length. (in., ft, yd, mi)
5. customary weight. (oz, lb, ton)
6. metric weight. (g, kg)
7. metric length less than dm. (cm)

Skill Review

Ask students to write and solve the division problem which tells the number of 40's in 280. (280 ÷ 40 = 7) Dictate the following for more practice in writing and solving problems: 50's in 850, 70's in 840, 20's in 900, 60's in 720, 40's in 960.

Dividing by Multiples of 10, Remainders

Pam counted 625 pennies in her savings. How many penny wrappers will she use to package them for the bank? How many pennies will she have left over?

We want to find the number of penny wrappers that Pam needs.

She has ___625___ pennies in her savings.

A penny wrapper will hold ___50___ pennies. To find the number of penny wrappers needed, we divide the total number of pennies by the number of pennies in one wrapper.

We divide ___625___ by ___50___.

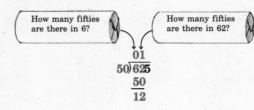

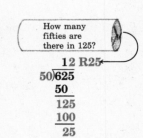

Pam needs ___12___ wrappers. She will have

___25___ pennies left over.

Getting Started

Divide. Show your work.

1. 3 R8
   10)38

2. 18 R5
   40)725

3. 1 R20
   70)90

4. 36 R18
   20)738

Copy and divide.

5. 87 ÷ 40
   2 R7

6. 906 ÷ 70
   12 R66

7. 67 ÷ 50
   1 R17

8. 959 ÷ 90
   10 R59

203

Teaching the Lesson

Introducing the Problem Have students tell about the picture. Discuss coin wrappers and reasons for using them. (ease in handling, counting and shipping quantities) Have a student read the problem and tell what is to be solved. (number of penny wrappers Pam will use and number of pennies left over) Ask students what information is given. (Pam has 625 pennies and 1 wrapper holds 50 pennies.) Have students read and complete the sentences to solve the problem. Guide them through the division in the model. Tell students to multiply and add the remainder to check their solution.

Developing the Skill Write 40)715 on the board. Ask students how many 40's are in 7. (none) Tell students we can think of zero in hundreds place in the quotient. Ask how many 40's are in 71. (1) Write 1 in tens place and tell students to subtract the product of 1 × 40 from 71 to see how many tens are left. (31) Tell students to think of 31 tens as 310 ones and tell how many ones in all. (315) Ask how many 40's in 315. (7) Have students tell the product of 7 × 40 as you write a 7 in ones place. (280) Ask the number of ones left-over. (35) Have a student write the remainder in the answer. Have a student check by multiplying and then adding the remainder. Repeat for 50)726 and 30)504.

Practice

Divide. Show your work.

1. $2\ R8$ $30\overline{)68}$
2. $1\ R35$ $40\overline{)75}$
3. $8\ R3$ $10\overline{)83}$
4. $4\ R16$ $20\overline{)96}$

5. 12 $50\overline{)600}$
6. 12 $80\overline{)960}$
7. $11\ R54$ $70\overline{)824}$
8. $12\ R10$ $60\overline{)730}$

9. $4\ R5$ $80\overline{)325}$
10. $19\ R5$ $40\overline{)765}$
11. $35\ R8$ $20\overline{)708}$
12. $8\ R36$ $50\overline{)436}$

13. $12\ R2$ $70\overline{)842}$
14. $4\ R65$ $90\overline{)425}$
15. $6\ R15$ $60\overline{)375}$
16. $28\ R12$ $30\overline{)852}$

Copy and Do

17. $800 \div 40$ — 20
18. $735 \div 60$ — $12\ R15$
19. $867 \div 30$ — $28\ R27$
20. $428 \div 70$ — $6\ R8$
21. $650 \div 50$ — 13
22. $329 \div 80$ — $4\ R9$

Apply

Solve these problems.

23. There are 40 paper clips to a box. How many boxes will you buy if you need 360 paper clips for your project?
    9 boxes
24. A container holds 80 milliliters of juice. How many containers are needed to hold 480 milliliters of juice?
    6 containers
25. Robbie worked 40 hours a week on his summer job. His paycheck for one week amounted to $240. How much did Robbie make per hour?
    $6
26. There are 50 clothespins to a box. Each box costs $3.75. How much will 350 clothespins cost?
    $26.25

204

2-digit Divisors

pages 205-206

Objective

To divide 3-digit numbers by 2-digit numbers for 1-digit quotients

Materials

Mental Math

Ask students which is more:

1. 40×40 or $180 \div 1$? (40×40)
2. 8×700 or 9×600? (8×700)
3. 1/2 of 320 or $244 \div 2$? (1/2 of 320)
4. 8 lb or 9×16 oz? (9×16 oz)
5. 42 dimes or 16 quarters? (42 dimes)
6. 8 gal or 33 qt? (33 qt)
7. $(6 \times 30) + 18$ or 3×70? (3×70)

Skill Review

To help students see more complicated division problems as multiples of 10 divided by multiples of 10, have students rewrite each of the following problems by rounding both numbers to the nearest 10: $94\overline{)789}$ (90, 790), $66\overline{)118}$ (70, 120), $56\overline{)478}$ (60, 480), $61\overline{)127}$ (60, 130), $78\overline{)326}$ (80, 330).

Dividing, 2-digit Divisors

It averages 42 miles per gallon; it has a 16 gallon gas tank; it needs unleaded gas...

The Valdez family is going to visit relatives 379 miles from their home. About how many gallons of gas will the Valdez family use on the trip?

We want to find the number of gallons of gas the Valdez family will use on their trip.

The Valdez family is driving ___379___ miles.

Their car gets ___42___ miles on each gallon of gas. To find the number of gallons needed we divide the distance by the number of miles per gallon. We divide ___379___ by ___42___.

Round the divisor to its nearest 10. Estimate how many 40's are in 379.

Since $379 \div 40$ is about 9, use 9 as the digit in the quotient.

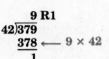

✔ We check a division problem by using multiplication, and addition if there is a remainder.

```
    42      divisor
  ×  9      × quotient
  ------
   378
  +  1      + remainder
  ------
   379      dividend
```

The Valdez family will use about ___9___ gallons of gas on their trip.

Getting Started

Divide. Check your work.

Copy and divide.

1. $22\overline{)176}$
   8

2. $49\overline{)325}$
   6 R31

3. $225 \div 31$
   7 R8

4. $142 \div 18$
   7 R16

205

Teaching the Lesson

Introducing the Problem Have students tell about the picture. Have students tell what information is given in the operator's manual. Ask a student to read the problem and tell what is to be solved. (approximate number of gallons of gas to be used on a trip) Help students decide which information will be used. (The trip is 379 miles and the car gets about 42 miles per gallon.) Have students read and complete the sentences to solve the problem. Guide them through the rounding and division in the model.

Developing the Skill Write $59\overline{)486}$ on the board. Tell students we want to see how many 59's are in 486 and we know neither 4 nor 48 is divisible by 59. Tell students we then must find a number which multiplied by 59 will be about 486. Tell students we can think of 59 as rounded to 60, and think of how many 60's are in 486. Remind students they know that 8 sixties are 480 and 480 is almost 486, so 8 would be a good estimate. Write **8** in the quotient of $59\overline{)486}$ and have a student multiply 8×59 to see if the estimate is about right. (472) Have a student subtract to see how many are left. (14) Tell students that since 14 is less than 59, then the remainder is 14. Repeat for $92\overline{)786}$ and $46\overline{)187}$.

205

Practice

Divide. Check your work.

1. $\overset{5\ R7}{31\overline{)162}}$

2. $\overset{6\ R22}{38\overline{)250}}$

3. $\overset{7\ R4}{42\overline{)298}}$

4. $\overset{8\ R22}{72\overline{)598}}$

5. $\overset{7\ R26}{37\overline{)285}}$

6. $\overset{6\ R18}{28\overline{)186}}$

7. $\overset{5\ R13}{51\overline{)268}}$

8. $\overset{8\ R13}{19\overline{)165}}$

9. $\overset{9\ R11}{89\overline{)812}}$

Copy and Do

10. 268 ÷ 43
    6 R10
11. 208 ÷ 45
    4 R28
12. 846 ÷ 92
    9 R18
13. 198 ÷ 32
    6 R6
14. 658 ÷ 73
    9 R1
15. 575 ÷ 67
    8 R39
16. 322 ÷ 46
    7
17. 585 ÷ 68
    8 R41
18. 148 ÷ 21
    7 R1
19. 561 ÷ 58
    9 R39
20. 389 ÷ 53
    7 R18
21. 233 ÷ 32
    7 R9
22. 248 ÷ 28
    8 R24
23. 616 ÷ 83
    7 R35
24. 321 ÷ 47
    6 R39

Apply

Solve these problems.

25. Miss Douglas worked 71 hours during one pay period. She made $568. How much does Miss Douglas earn in one hour?
    $8

26. Mr. Kowalski put 14 gallons of gas into his car. His car averages 37 miles to a gallon. About how far can Mr. Kowalski drive his car on this tank of gas?
    518 miles

206

Trial Quotients

Objective

To estimate quotients

Materials

Mental Math

Tell students to round each to the nearest ten to estimate:

1. 683 + 124 (800)
2. 16 × 31 (600)
3. 89 ÷ 13 (9)
4. 67 × 22 (1,400)
5. 443 ÷ 16 (22)
6. 78 ÷ 19 (4)
7. 1,251 − 703 (550)
8. 346 × 21 (7,000)

Skill Review

Review multiplying 2-digit numbers by 1-digit numbers.

Trial Quotients

Raoul is keeping a diary on the development of the puppy his parents gave him. Help him convert the puppy's weight to pounds and left-over ounces.

We want to know the puppy's weight in pounds and ounces.

The scale reads that the puppy weighs __121__ ounces.

There are __16__ ounces in a pound.

To convert from ounces to pounds, we divide the total number of ounces by the number of ounces in one pound.

We divide __121__ by __16__.

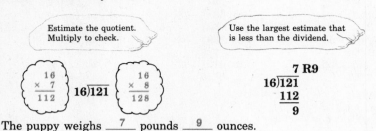

Estimate the quotient. Multiply to check.

Use the largest estimate that is less than the dividend.

$$\begin{array}{r} 16 \\ \times\ 7 \\ \hline 112 \end{array} \quad 16\overline{)121} \quad \begin{array}{r} 16 \\ \times\ 8 \\ \hline 128 \end{array}$$

$$\begin{array}{r} 7\ R9 \\ 16\overline{)121} \\ \underline{112} \\ 9 \end{array}$$

The puppy weighs __7__ pounds __9__ ounces.

Getting Started

Divide.

1. $12\overline{)108}$ → 9

2. $53\overline{)348}$ → 6 R30

3. $31\overline{)248}$ → 8

4. 268 ÷ 27
   9 R25

5. 144 ÷ 24
   6

6. 340 ÷ 84
   4 R4

207

Teaching the Lesson

Introducing the Problem Ask a student to read the problem aloud and tell what it asks. (the puppy's weight in pounds and ounces) Ask students the weight of the puppy. (121 ounces) Ask what information is still needed. (16 ounces equal 1 pound) Have students read and complete the sentences. Guide them through the conversion in the model. Ask students how they can check their answer. (multiply 7 by 16 and add 9)

Developing the Skill Write $47\overline{)368}$ on the board. Tell students to look for a number that, when multiplied by 47, will be slightly less than 368. Tell students another way to estimate the quotient is to guess and check. Tell students we think of the number of 47's in 368 and know it well be less than 10, since 10 forty-sevens would be 470 and way too much. Tell students we then think of how much less than 10 and could guess there may be 8 forty-sevens in 368. Have a student find the product of 47 × 8 to check this guess. (376) Ask if 8 was a good guess. (no, too high) Tell students to check to see if 7 forty-sevens would be a better estimate by multiplying 47 × 7. (329) Tell students the 7 is the better estimate because it is close to but less than 368. Have students find the remainder. (39) Repeat to estimate quotients for $54\overline{)476}$ and $39\overline{)263}$.

Practice

Divide.

1. 43)344
   8

2. 56)180
   3 R12

3. 72)608
   8 R32

4. 37)163
   4 R15

5. 94)715
   7 R57

6. 86)816
   9 R42

7. 16)100
   6 R4

8. 48)312
   6 R24

9. 55)476
   8 R36

Copy and Do

10. 250 ÷ 39
    6 R16

11. 512 ÷ 62
    8 R16

12. 273 ÷ 57
    4 R45

13. 288 ÷ 48
    6

14. 743 ÷ 85
    8 R63

15. 214 ÷ 96
    2 R22

16. 536 ÷ 75
    7 R11

17. 160 ÷ 23
    6 R22

18. 308 ÷ 68
    4 R36

Apply

Solve these problems.

19. How many feet and inches are there in 105 inches?
    8 feet 9 inches

20. How many days are there in 168 hours?
    7 days

EXCURSION

These nine numbers have one thing in common. Can you find what it is?

222	315	2,124
405	999	6,111
210	1,905	1,158

They are all multiples of 3.

208

Partial Dividends

pages 209-210

Objective

To divide 3-digit numbers by 2-digit numbers using a partial dividend

Materials

Mental Math

Ask students to name a multiple of 3:

1. >100 (102, 105, etc.)
2. equal to 174 ÷ 2 (87)
3. with 4 digits (1,002, etc.)
4. with 9 hundreds (900, 903, etc.)
5. < 212 (3, 6, 9, . . . 210)
6. whose digits have a sum of 27 (999, 9,891, etc.)
7. with 1 zero (102, 120, etc.)
8. which is divisible by 27 (81, 108, etc.)

Skill Review

Review problems such as 80)480 and 30)540 for students to practice dividing multiples of 10 using partial divisors and dividends.

Partial Dividends

Quan's job at the dairy is to run the machine that attaches the labels to the bottles. Quan figures that 330 bottles pass through the machine in one hour. How many cases of milk are labeled in one hour? How many bottles are left over?

We want to know the number of cases of milk Quan labels in one hour.

Quan's machine labels __330__ bottles in an hour.

There are __12__ bottles in a case.
To find the number of cases, we divide the total number of bottles by the number in one case. We divide __330__ by __12__.

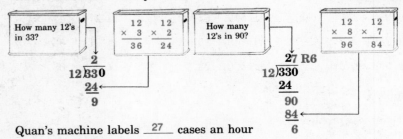

Quan's machine labels __27__ cases an hour

with __6__ bottles labeled for the next case.

Getting Started

Divide.

1. 18 R7
   26)475

2. 23 R1
   36)829

3. 3 R56
   63)245

Copy and divide.

4. 930 ÷ 24
   38 R18

5. 856 ÷ 35
   24 R16

6. 758 ÷ 48
   15 R38

209

Teaching the Lesson

Introducing the Problem Have a student read the problem aloud and describe the picture. Ask students what question is to be answered. (number of cases of milk labeled in 1 hour) Ask students what information is given in the problem. (330 bottles are labeled in 1 hour.) Ask students what information is needed from the picture. (12 bottles are in each case.) Have students complete the sentences and solve the problem. Guide students through the steps in the model. Have students multiply and add the remainder to check their work.

Developing the Skill Write 26)436 on the board. Tell students we want to know the number of 26's in 436. Since we know that 4 is too small, we look at the number of 26's in 43. Write **1** above the 3 and tell students we subtract the product of 1 × 26 from 43 to see how many tens are left. (17) Tell students we then trade the 17 tens for 170 ones and add the 6 ones for 176 ones. Tell students we want to find the number of 26's in 176 and can estimate that there are 6 because 6 × 26 is 156. Record the 6, subtract to find the remainder and record the remainder of 20. Have students multiply and add the remainder to check the problem. Repeat for 8)732 and 21)265.

Practice

Divide.

1. 24)536 22 R8

2. 18)627 34 R15

3. 39)846 21 R27

4. 21)756 36

5. 58)427 7 R21

6. 63)915 14 R33

7. 37)854 23 R3

8. 16)898 56 R2

9. 84)963 11 R39

Copy and Do

10. 945 ÷ 45
21

11. 763 ÷ 17
44 R15

12. 812 ÷ 26
31 R6

13. 518 ÷ 63
8 R14

14. 838 ÷ 28
29 R26

15. 946 ÷ 37
25 R21

16. 683 ÷ 12
56 R11

17. 915 ÷ 48
19 R3

18. 709 ÷ 29
24 R13

Apply

Solve these problems.

19. Mr. Wallace drove 385 miles at 55 miles per hour. How many hours did Mr. Wallace drive?
7 hours

20. How many pounds are there in 950 ounces?
59 pounds 6 ounces

21. Sandy ran 720 meters running wind sprints. She ran 45 meters each time. How many sprints did Sandy run?
16 sprints

22. Richard used 864 grams of batter to make cookies for the bake sale. Each cookie weighed 24 grams and was sold for 15¢. How much did the bake sale make from Richard's cookies?
$5.40

210

Correcting Common Errors

Watch for students who fail to notice that their first estimated partial quotient is not enough.

INCORRECT	CORRECT
310 R14	40 R14
23)934	23)934
69	92
244	14
230	0
14	14

Each time they subtract, students should check to make sure the difference is less than the divisor. If it is not, then the quotient figure should be increased.

Enrichment

Tell students to show their work and tell why 31 is not a good quotient estimate to find how many 26's are in 832.

Practice

Remind students that they know several ways to divide by a 2-digit number. Remind students that a word problem may require more than one operation.

Extra Credit *Logic*

Read the following story to students and tell them to use logic to answer the question. A salesman was talking to a customer in a pet store. "This parrot is a fantastic bird," said the pet store salesman. "He can give the correct answer to absolutely every mathematical problem he hears." The customer was so impressed he bought the parrot. But when he got it home, the bird wouldn't answer a single problem. If what the salesman said was true, why wouldn't the bird answer the problems? (The parrot was deaf.)

Finding Averages

Objective

To find averages

Materials

Mental Math

Have students name the remainder:

1. $73 \div 9$ (1)
2. $67 \div 7$ (4)
3. $50 \div 8$ (2)
4. $70 \div 9$ (7)
5. $84 \div 9$ (3)
6. $49 \div 5$ (4)
7. $29 \div 3$ (2)
8. $53 \div 6$ (5)

Skill Review

Review column addition through 4-digit numbers. Have students read the sum of each problem and tell how many numbers were added to reach that sum.

Finding Averages

Andy, Bart and Larry are the top collectors of autographs in their class. Andy has 14 autographs. Bart has 9 and Larry has 13. What is the average number of autographs each boy has?

We want to find the average number of autographs each of the boys has.

Andy has __14__ autographs, Bart has __9__ and

Larry has __13__.

To find the average, we add all the individual numbers of autographs, and divide that sum by

the number of boys. We add __14__, __9__ and

__13__ and divide by __3__.

$$\begin{array}{r} 14 \\ 9 \\ +13 \\ \hline 36 \end{array} \qquad 3\overline{)36} \;\; ^{12}$$

The three boys have an average of __12__ autographs each.

Getting Started

Find the average of each set of numbers.

1. 16, 54

    __35__

2. 34, 27, 48, 51, 35

    __39__

3. 158, 196, 243, 203

    __200__

211

Teaching the Lesson

Introducing the Problem Have a student read the problem aloud and tell what is to be solved. (the average number of the 3 boys' autographs) Ask what useful information is given in the problem. (Andy has 14, Bart has 9 and Larry has 13 autographs.) Have students complete the sentences with you. Guide them through the addition and division in the model to find the average number of autographs.

Developing the Skill Tell students an **average** gives us an idea of how a group of numbers can be represented by 1 number. Write on the board: **10 13 8 7 12.** Tell students we want to find the average of these numbers, so we add them together and divide by 5. Have students add the numbers. (50) Ask students why we would divide by 5. (There are 5 numbers in all.) Have students divide 50 by 5. (10) Tell students that 10 is the average of these 5 numbers. Tell students that if the group of numbers told us how many students were absent with the flu each day for a week, then we would say an average of 10 students were absent each day that week. Have students find the average of 14, 18 and 100 and then discuss why the average is much larger than the 2 smaller numbers and much smaller than 100. (The smaller numbers pull the average down and the larger number pulls it up.) Give students more groups of numbers which average to a whole number.

211

Practice

Find the average of each set of numbers.

1. 17, 28, 36 _27_
2. 146, 254 _200_
3. 86, 58, 37, 49, 45 _55_
4. 624, 534, 810, 712 _670_
5. 1,496, 4,868 _3,182_
6. 3,467, 2,948, 4,511 _3,642_

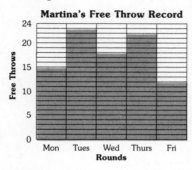

Apply

Solve these problems.

7. Maria bowled games of 125, 136, and 138. What was Maria's average score?
   133

8. Mrs. Jordan drove 243 miles on Monday and 485 miles on Tuesday. What was her average driving distance for one day?
   364 miles

9. Art earned $243 in May, $316 in June, $375 in July and $286 in August. What was Art's average monthly earnings?
   $305

10. Arnold took part in a free-throw contest. His scores were 13, 19, 17, 18 and 18. Find his average score and how many times he scored below it.
    Average 17, scored once below average

Use this graph of Martina's free throws to solve problems 11 through 16.

11. In which round did Martina get her highest score? _Tues_

12. What was her highest score? _23_

13. In which round did Martina get her lowest score? _Fri_

14. What was her lowest score? _12_

Martina's Free Throw Record

15. What was Martina's average? _18_

212

Correcting Common Errors

Some students may always divide by 2 or some other incorrect number when they are finding averages. Have them work in small cooperative groups to discuss why the divisor must be the number of addends. They should recognize that, if the average represents each addend, then the number of addends times the average should have the same total as the sum of all the addends. Then have students make up sets of 3, 4, and 5 addends and give them to other groups to find the averages.

Enrichment

Tell students to use the World Almanac to find the average population of the 3 largest cities in your state.

Practice

Remind students that when finding an average the divisor is the number of examples in all. Tell students to check to be sure their answers are reasonable. Ask students some questions about the graph to review graph skills. Have students complete the page independently.

Extra Credit   *Measurement*

Have students make a floor plan of their classroom. Review with them the meaning of **scale.** Have them make a list of all the measurements they will need, such as the length of each wall; width of doors and windows, and where they are placed in the walls, etc. For this project, also have them measure the furniture. Discuss why they will not need to make height measurements. Showing them a sample floor plan will help. Complete the measurements with students, using a meter stick. Give each a sheet of graph paper, a metric ruler, and tell them to use the scale: 1 cm = 1 meter. Have students draw the floor plan to scale on the graph paper. On another sheet of graph paper, have them draw the furniture to scale. Tell them to cut the furniture out and place it in the room where it belongs. Let them have fun arranging their classroom in many different ways.

212

Problem Solving, Make a Graph

pages 213-214

Objective

To make a bar graph to solve problems

Materials

*simple bar graphs for display

Mental Math

Ask students how many:

1. (¼ of 20) × 60 (300)
2. 7's in 81 × 7 (81)
3. hours in 2 ¼ days (54)
4. inches in 4 yards (144)
5. zeros in product of 100 × 6 × 10 (3)
6. digits in quotient of 720 ÷ 9 (2)
7. 100 ÷ (½ of 40) (5)
8. hours and minutes in your school day

Making a Graph

On each bounce, Amy's playground ball bounces exactly half the height from which it is dropped. She drops the ball from a height of 48 inches. How high will it bounce on the fourth bounce?

★ SEE

We want to know how high Amy's ball will bounce on the fourth bounce.

We know that the ball bounces $\frac{1}{2}$ of the height of the previous bounce. Amy starts by dropping the ball from a height of __48__ inches.

★ PLAN

We can make a bar graph of the ball's height after each bounce. Each bar will be one half the height of the one before it.

★ DO

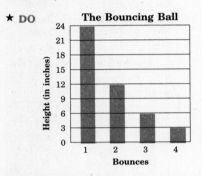

The Bouncing Ball

The ball will bounce __3__ inches on the fourth bounce.

★ CHECK

48 ÷ 2 = __24__

__24__ ÷ __2__ = __12__   __12__ ÷ __2__ = __6__   __6__ ÷ __2__ = __3__

213

Teaching the Lesson

Display some simple bar graphs and tell students that a **bar graph** organizes data and provides a picture for easy comparison of facts. Tell students that making a graph is another way to solve a problem. Help students tell about each graph as you stress that a graph has a vertical axis and a horizontal axis. Ask students some questions that can be answered by the displayed graphs. Have a student read the problem. Tell students there are several ways to solve it and making a graph is the way they will use in this lesson. Have a students read the SEE and PLAN stages. Ask students what increments they should use for each axis. Have students complete the graph and the solution sentence. Tell students to complete the CHECK stage to verify their solution. Have students read the sentences aloud.

Apply

Make a graph to help solve each problem.

1. The length of blue whales reaches up to 30 meters, gray whales up to 15 meters, humpback whales up to 15 meters, killer whales up to 9 meters and bowhead whales up to 18 meters. Create a graph to illustrate and compare the lengths of these whales.
See Solution Notes.

2. A plain pizza costs $5.00. Each additional item, such as mushrooms, onions and peppers, costs $0.50 each. Provide a graph to show the costs and the number of extra items on pizzas less than $8.00.
See Solution Notes.

3. It costs 25 cents to mail a package that weighs one ounce or less, and 20 cents for each additional ounce or fraction thereof. Construct a graph to illustrate the weight and cost of packages that can be mailed for under one dollar.
See Solution Notes.

4. Gwen babysits for $2.00 per hour before midnight, and $3.00 per hour after midnight. Make a graph to show Gwen's total earnings if she babysits from 9 PM one evening until 3 AM the next morning.
See Solution Notes.

5. The Fighting Flamingos, a football team, played 10 games during their regular season. A graph kept by their coach showed that their average score per game was 15 points. How many points in all did they score during their season?
150 points

6. A graph shows that, in their first game, the six players on the Beastly Bulls Basketball Team made these scores: 18, 14, 12, 20, 16, and 10. If they want to have no less than the same total score in their next game, what average number of points must each player score?
15 points

7. When computing to find data for a graph, how do you know that you can find $350 \div 70$ by finding $35 \div 7$?
See Solution Notes.

8. When dividing to find data, how can you use a calculator to check a quotient with a remainder?
See Solution Notes.

214

Extra Credit   *Creative Practice Logic*

Here are more toothpick puzzles for students to solve:

1. Here are 9 toothpicks to form 3 triangles in a row. Now move just 3 toothpicks and form 5 new triangles.

2. Start with 8 toothpicks. Now devise a way to take away 3, and still have 8 left. (The toothpicks form the Roman numeral, VIII.)
Ask students to create 2 more toothpick puzzles of their own, to challenge another student.

Solution Notes

Remind students that the bars on a bar graph can run vertically or horizontally, depending on which information is placed on which axis.

1. Students' graphs will show the length of each kind of whale.
2. Help students read the problem critically to find what the highest priced pizza on their graph will be. ($7.50)

Plain	: $5.00
1 item	: $5.50
2 items	: $6.00
3 items	: $6.50
4 items	: $7.00
5 items	: $7.50

3. Students should extend their graphs to the first price over $1.

1 oz	: $.25
2 oz	: .45
3 oz	: .65
4 oz	: .85
5 oz	: $1.05

4. Help students establish the range of hours from the completion of the first hour, through 3 AM.

10:00	$2
11:00	$4
12:00	$6
1:00	$9
2:00	$12
3:00	$15

Higher-Order Thinking Skills

5. [Analysis] Students should recognize that, should the team score the average number of points each game, the total would equal the number of points they actually did score.
6. [Comprehension] The answer is the same as the average in the first game.
7. [Synthesis] When both the dividend and the divisor are multiplied by the same number, the quotient will be the same.
8. [Analysis] Multiply the divisor and the quotient, add the remainder, and then compare to see that the result equals the dividend.

Calculators, Comparing Costs

pages 215-216

Objective

To use a calculator to do comparison shopping

Materials

calculators
priced product containers in 2 sizes

Mental Math

Ask students is each evenly divisible by 5?

1. 1,986 (no)
2. 2,000,000 (yes)
3. 4,695 (yes)
4. 10 × 46 (yes)
5. 10 + 15 + 4 (no)
6. 5 × 45 (yes)
7. 38 ÷ 2 (no)
8. ½ of 460 (yes)

Skill Review

To review calculator codes for money amounts, dictate amounts of money for students to enter into their calculators. Now have students write calculator codes on the board for problems of 1- and 2-digit numbers times money amounts. Have students complete the codes.

Calculators, Comparison Shopping

Mrs. Anderson is shopping for the best buy in cereal. Should Mrs. Anderson buy the 8-ounce or the 14-ounce box of cereal?

We are looking for the best buy in cereal.

The 8-ounce box is sold for __$2.40__ .

The 14-ounce box costs __$3.92__ .

To decide which is the better buy, we must find the **unit cost** of each container. To find the unit cost, we divide the total cost of the package by the number of units in it. We divide

$2.40 by 8 and __$3.92__ by __14__ .

Complete these codes.

2 [·] 4 [÷] 8 [=] (__0.3__)

0.3 means $0.30 or 30¢.

3 [·] 92 [÷] 14 [=] (__0.28__)

0.28 means $__0.28__ or __28__ ¢.

30¢ > 28¢

The __14__-ounce box is the better buy.

Find the cost per ounce. Circle the best buy.

1. 8 ounces $4.48 __$0.56__ per oz
    (16 ounces) $8.32 __$0.52__ per oz

2. 6 ounces $1.74 __$0.29__ per oz
    (8 ounces) $2.24 __$0.28__ per oz
    12 ounces $3.72 __$0.31__ per oz

215

Teaching the Lesson

Introducing the Problem Have a student read the problem aloud and tell what is to be solved. (which of the 2 boxes of cereal is the best buy) Ask students what information is given. (8-ounce box costs $2.40 and the 14-ounce box costs $3.92.) Have students read and complete the sentences. Help them complete the calculator codes in the model to find the best buy.

Developing the Skill Tell students we often need to know which packages in a store give us the most for our money. Tell students we have to find the **unit cost,** or the cost per cookie in each bag of cookies, in order to know which package is the best buy. Tell students to suppose there are 2 bags of cookies, one costing $2.80 for 40 cookies and the other costing $3.00 for 50 cookies. Tell students that to find the cost of 1 cookie in each bag, we divide $2.80 by 40 and $3.00 by 50. Have students write and solve the calculator codes. Remind them to enter the deci-mal point, and that no dollar sign will appear on the screen. Ask students which bag offers a cookie at a lower price. ($3.00 bag) Repeat for 22 tea bags for $2.64 and 35 bags for $3.85.

Practice

Use your calculator to find each answer.

1. $\frac{\$1.23}{7)\$8.61}$

2. $\frac{\$1.54}{6)\$9.24}$

3. $\frac{\$0.20}{15)\$3.00}$

4. $\frac{\$0.26}{26)\$6.76}$

5. $\frac{\$0.24}{34)\$8.16}$

6. $\frac{\$0.06}{59)\$3.54}$

Find the cost per unit. Circle the best buy.

7. (6 ounces $1.38) _$0.23_ per oz
   8 ounces $1.92 _$0.24_ per oz

8. 24 ounces $7.68 _$0.32_ per oz
   (36 ounces $8.28) _$0.23_ per oz

9. (7 pounds $3.01) _$0.43_ per lb
   9 pounds $3.96 _$0.44_ per lb
   15 pounds $6.75 _$0.45_ per lb

10. 6 pounds $2.40 _$0.40_ per lb
    15 pounds $6.00 _$0.40_ per lb
    (25 pounds $9.75) _$0.39_ per lb

Apply

Solve these problems.

11. A store sells one brand of hot chocolate at 3 pounds for $2.43. They sell another brand at 5 pounds for $3.95. Which is the better buy? How much do you save?

5 pounds for $3.95   2¢ per pound

12. The school store sells binders for $1.35 each. If you buy 6 binders of the same color, the cost is $7.74. How much is saved if you buy 6 red binders?

$0.36

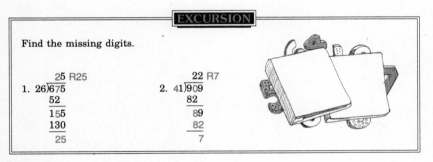

EXCURSION

Find the missing digits.

1.
```
       25 R25
   26)675
       52
      155
      130
       25
```

2.
```
       22 R7
   41)909
       82
       89
       82
        7
```

216

Practice

Remind students to place a dollar sign and decimal point in each answer of a money amount. Tell them the word problems involve more than one operation. Tell students to complete the page independently.

Excursion

In any missing-digit problem, it is usually the inverse operation that helps find the missing numbers. Students can find the missing digits in the first problem by thinking 26 times 2 is 52; 7 minus 2 is 5 and 5 minus 5 is 0.

Correcting Common Errors

Some students may try to determine the best buy without finding the unit cost. Have them work in cooperative groups to discuss why just because they are paying less for a group of items, they are not necessarily getting a better buy. Students should recognize that the cost of 1 unit in one grouping may be more than it would be if you bought another grouping of a like item. Students should recognize also that "better buy" assumes two offerings of equal quality. Have students visit grocery stores as a group project to find examples.

Enrichment

Have students make a table of unit costs for 2 different-sized containers, for each of 4 different products used often in their homes.

Extra Credit *Numeration*

Tell students they will be playing Tic Tac Toe with numbers, following some very special rules. Have students draw the standard game grid of 3 rows of 3 squares. Working in pairs, tell one student to use only odd numbers, 1 through 9. The other player will use only even numbers, 2 through 10. To play, each player writes any of his allowed numbers in any square, on each turn. The winning player completes a row, either horizontal, diagonal or vertical, whose sum is 15. Students will see plans of strategy after they play a few times. Each time they must reach a sum of 15 to win. Have students reverse roles so they have a chance to work with both even and odd numbers.

Chapter Test

page 217

Item	Objective
1-2	Divide 3-digit number by 1-digit number to get 3-digit quotient without remainder (See pages 191-192)
3-4	Divide 3-digit number by 1-digit number to get 3-digit quotient with remainder (See pages 193-194)
5-8	Divide 3-digit number by 1-digit number to get 3-digit quotient with zero (See pages 195-196)
9-12	Divide 3-digit number by 1-digit number to get 2-digit quotient with or without remainder (See pages 197-198)
13-16	Divide money by 1-digit number (See pages 199-200)
17-20	Divide 3-digit number by 2-digit number to get 1-digit quotient with or without remainder (See pages 201-203)
21-24	Divide 3-digit number by 2-digit number to get 2-digit quotient with or without remainder (See pages 207-209)
25-28	Find the average of a set of numbers (See pages 211-212)

Divide. Show your work.

1. $3\overline{)696}$ = 232

2. $5\overline{)555}$ = 111

3. $2\overline{)843}$ = 421 R1

4. $4\overline{)725}$ = 181 R1

5. $2\overline{)408}$ = 204

6. $3\overline{)900}$ = 300

7. $4\overline{)813}$ = 203 R1

8. $3\overline{)625}$ = 208 R1

9. $6\overline{)276}$ = 46

10. $7\overline{)399}$ = 57

11. $8\overline{)623}$ = 77 R7

12. $9\overline{)738}$ = 82

13. $4\overline{)\$8.24}$ = $2.06

14. $8\overline{)\$5.76}$ = $0.72

15. $2\overline{)\$9.38}$ = $4.69

16. $7\overline{)\$4.97}$ = $0.71

17. $27\overline{)162}$ = 6

18. $39\overline{)247}$ = 6 R13

19. $53\overline{)325}$ = 6 R7

20. $44\overline{)428}$ = 9 R32

21. $18\overline{)468}$ = 26

22. $25\overline{)\$9.00}$ = $0.36

23. $39\overline{)756}$ = 19 R15

24. $57\overline{)912}$ = 16

Find the average of each set of numbers.

25. 6, 9, 4, 5, 1

   5

26. 92, 46, 75, 55

   67

27. 47, 69, 101, 45, 48

   62

28. 155, 245, 125

   175

217

CUMULATIVE REVIEW

Circle the letter of the correct answer.

1 Round 639 to the nearest hundred.
- (a) 600
- b 700
- c NG

2 Round 7,350 to the nearest thousand.
- (a) 7,000
- b 8,000
- c NG

3 What is the value of the 1 in 517,296?
- a thousands
- (b) ten thousands
- c hundred thousands
- d NG

4
$$3,249 + 1,816$$
- a 4,055
- b 4,065
- (c) 5,065
- d NG

5
$$26,795 + 27,586$$
- a 43,271
- b 44,381
- (c) 54,381
- d NG

6
$$829 - 136$$
- (a) 693
- b 713
- c 793
- d NG

7
$$13,053 - 12,875$$
- (a) 178
- b 1,178
- c 1,822
- d NG

8 Choose the better estimate of weight.
- (a) 1.5 g
- b 1.5 kg

9 Choose the better estimate of volume.
- (a) 350 mL
- b 350 L

10
$$\$2.79 \times 6$$
- a $12.24
- (b) $16.74
- c $16.76
- d NG

11 39 × 23
- a 117
- (b) 897
- c 7,107
- d NG

12
$$607 \times 58$$
- a 34,204
- b 35,786
- c 36,206
- (d) NG

13 624 ÷ 3
- a 28
- (b) 208
- c 280
- d NG

14 42)226
- a 5
- b 5 R26
- c 526
- (d) NG

□ score

218

Cumulative Review

page 218

Item	Objective
1	Round numbers to nearest 100 (See pages 31-34)
2	Round numbers to nearest 1,000 (See pages 33-34)
3	Identify place value through hundred thousands (See pages 27-28)
4	Add two 4-digit numbers (See pages 47-48)
5	Add two 5-digit numbers (See pages 53-54)
6	Subtract two 3-digit numbers (See pages 63-64)
7	Subtract two 5-digit numbers with zero in minuend (See pages 67-68)
8	Determine appropriate metric unit of weight (See pages 159-160)
9	Determine appropriate metric unit of capacity (See pages 157-158)
10	Multiply money by 1-digit number (See pages 169-170)
11	Multiply two 2-digit numbers trading ones (See pages 179-180)
12	Multiply 3-digit number by 2-digit number (See pages 181-182)
13	Divide 3-digit number by 1-digit number to get 3-digit quotient with zero (See pages 195-196)
14	Divide 3-digit number by 2-digit number to get 1-digit quotient with remainder (See pages 201-203)

Alternate Cumulative Review

Circle the letter of the correct answer.

1 Round 756 to the nearest hundred.
- a 700
- (b) 800
- c NG

2 Round 9,227 to the nearest thousand.
- a 8,000
- (b) 9,000
- c 10,000
- d NG

3 What is the value of the 2 in 213,654?
- a thousands
- b ten thousands
- (c) hundred thousands
- d NG

4
$$6,157 + 2,937$$
- a 8,084
- (b) 9,094
- c 9,194
- d NG

5
$$16,975 + 37,145$$
- (a) 54,120
- b 53,010
- c 43,121
- d NG

6
$$348 - 162$$
- a 286
- b 226
- (c) 186
- d NG

7
$$18,062 - 16,473$$
- a 2,411
- (b) 1,589
- c 2,699
- d NG

8 What unit would you use to weigh a loaf of bread?
- a mg
- b kg
- (c) g
- d NG

9 What unit of capacity would you use to measure a small pond?
- a mL
- (b) L

10
$$\$4.86 \times 4$$
- a $13.24
- b $17.64
- (c) $19.44
- d NG

11
$$27 \times 34$$
- a 189
- b 698
- (c) 918
- d NG

12
$$409 \times 36$$
- a 3,681
- b 14,728
- c 14,984
- (d) NG

13 6)636
- a 16
- b 160
- (c) 106
- d NG

14 67)423
- a 60 R21
- (b) 6 R21
- c 621
- d NG

218

Points, Lines, Segments

pages 219-220

Objective

To identify and draw points, lines and line segments

Materials

*straight-edge
rulers marked in centimeters
cards measuring less than 15 cm
graph paper

Mental Math

Have students give the perimeter of a:

1. square with 1 side of 14. (56)
2. hexagon whose sides are each 22. (132)
3. triangle whose sides are each 38. (114)
4. 60 × 150 rectangle. (420)
5. octagon whose sides are each 110. (880)
6. triangle with sides of 28, 72 and 14. (114)
7. 48 × 400 rectangle. (896)

Skill Review

Have students work in pairs to find the length and width of various-sized cards. Have students write the measurements on the cards, trade cards with another pair of students and check each others' measurements.

Points, Lines and Segments

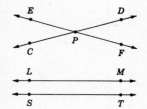

"As the crow flies" is an expression which means the shortest distance between two points. Draw a line that shows this distance between Becky's and Allen's homes.

The shortest path from Becky's house to Allen's house is a straight path through the park.

A straight path from point A to point B is called a **line segment.** Points A and B are called **endpoints.**

   We say: **segment** AB or **segment** BA.
   We write: $\overline{AB}$ or $\overline{BA}$.

A line segment is part of a **line.** A line extends indefinitely in both directions.

   We say: **line** XY or **line** YX.
   We write: $\overleftrightarrow{XY}$ or $\overleftrightarrow{YX}$.

✔ Notice that a line does not have endpoints.

Some lines **intersect.** Line CD intersects line EF at point P.

Some lines do not intersect. Line LM is **parallel** to line ST.

Getting Started

Write the name for this figure.

1. • C

   Point C

Draw and label this figure.

2. $\overline{TV}$

T •————————————• V

219

Teaching the Lesson

Introducing the Problem Have a student read the problem and describe the picture. Ask what is to be done. (draw the shortest path from Becky's to Allen's house) Have students read with you and find the points of each figure on the right as they are described in the model. Have students draw the line in the picture to show the shortest distance. Have students check by measuring their line segment and comparing that measurement with the total distance travelled along the streets.

Developing the Skill Use the straight-edge to draw line segment XY on the board and have students measure it. Tell students the figure is called **segment** XY or segment YX because it is the same. Tell students **the shortest distance from endpoint X to endpoint Y is a straight path.** Extend the segment beyond both endpoints and draw an arrow at each end. Tell students that the figure is now called **line** XY or line YX because its ends do not have definite points. Draw **lines BC** and **DE** on the board so they intersect and tell students these lines cross each other and are called **intersecting lines.** Draw **lines MN** and **OP** on the board as parallel lines and tell students **parallel lines** would never meet. Have students use graph paper to practice drawing and labeling parallel and intersecting lines.

Practice

Write the name for each figure.

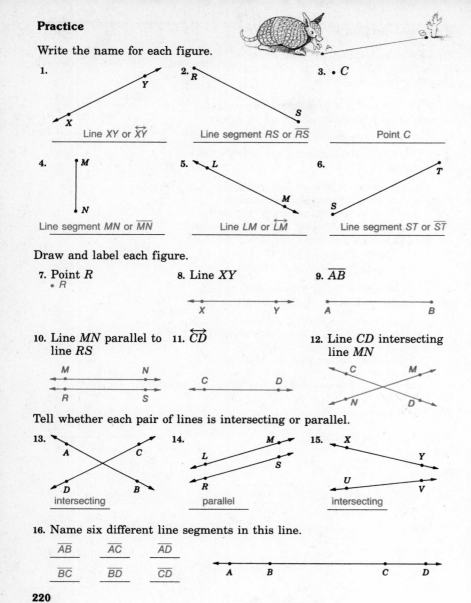

1. Line *XY* or $\overleftrightarrow{XY}$

2. Line segment *RS* or $\overline{RS}$

3. Point *C*

4. Line segment *MN* or $\overline{MN}$

5. Line *LM* or $\overrightarrow{LM}$

6. Line segment *ST* or $\overline{ST}$

Draw and label each figure.

7. Point *R*
   • *R*

8. Line *XY*

9. $\overline{AB}$

10. Line *MN* parallel to line *RS*

11. $\overleftrightarrow{CD}$

12. Line *CD* intersecting line *MN*

Tell whether each pair of lines is intersecting or parallel.

13. intersecting

14. parallel

15. intersecting

16. Name six different line segments in this line.

$\overline{AB}$ $\overline{AC}$ $\overline{AD}$

$\overline{BC}$ $\overline{BD}$ $\overline{CD}$

220

Rays and Angles

pages 221-222

Objective

To identify and draw rays and angles

Materials

*flashlight
*straight-edge
rulers

Mental Math

Have students divide each by 30, then subtract 7:

1. 300 (3)
2. 1,200 (33)
3. 660 (15)
4. 900 (23)
5. 1,800 (53)
6. 33,000 (1,093)
7. 450 (8)
8. 750 (18)

Skill Review

Write the following on the board in a column: $\overline{LM}$, $\overrightarrow{PQ}$, **line TU parallel to line CD, point 2, line segment MN, line segment DE intersecting line segment RS.** Have students draw and label the identifying qualities of each figure. (See definitions on page 219.)

Rays and Angles

Turn on a flashlight. The beam of light is like a special kind of line called a ray. How is a beam like a ray?

A **ray** is part of a line. It has one endpoint and extends indefinitely in one direction.
We write: ray AB or $\overrightarrow{AB}$.

✔ Notice that we name a ray starting with its endpoint.

Two rays which intersect at a common endpoint form an **angle.** The common endpoint is called the **vertex** of the angle. We name an angle using 1 or 3 of its points.
This is angle CDE or $\angle CDE$
angle EDC or $\angle EDC$
angle D or $\angle D$.

✔ Notice when using three points to name an angle, the vertex is always in the middle.

An angle that forms a square corner is called a **right angle.** $\angle XYZ$ is a right angle. The ⌐ shows the angle is a right angle.
The flashlight beam, like the ray, has __one__

endpoint and extends indefinitely in __one__ direction.

Getting Started

Write the name of each figure.

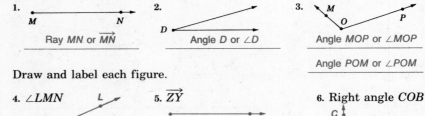

1. Ray MN or $\overrightarrow{MN}$
2. Angle D or $\angle D$
3. Angle MOP or $\angle MOP$

Angle POM or $\angle POM$

Draw and label each figure.

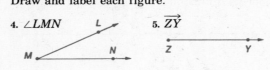

4. $\angle LMN$
5. $\overrightarrow{ZY}$

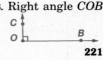

6. Right angle COB

221

Teaching the Lesson

Introducing the Problem Have a student read the problem and describe the picture. Ask what question is to be answered. (Tell how the beam of a flashlight differs from a line segment or a line.) Have students read with you to learn about a **ray,** an **angle,** a **vertex** and a **right angle.** Have students explain why a flashlight beam is a ray, and have them complete the solution statement.

Developing the Skill Have a student use a straight-edge to draw a line from point A. Tell the student to stop at some point and draw an arrow at that place. Tell students this is an example of a ray. Have several students draw rays on the board. Tell students to label the endpoint and another point along their rays and then tell the names of their rays. Now draw and label 2 rays from a common endpoint and tell students the 2 rays form an angle and their common endpoint is called the vertex. Tell students that angles are named by the vertex letter or 3 letters with the vertex in the middle. Tell students the symbol $\angle$ can replace the word angle. Have students label and name the angle. Now draw a square or rectangle on the board and highlight each corner. Tell students the angle in each corner is called a right angle. Have students identify right angles in the classroom.

Write the name of each figure.

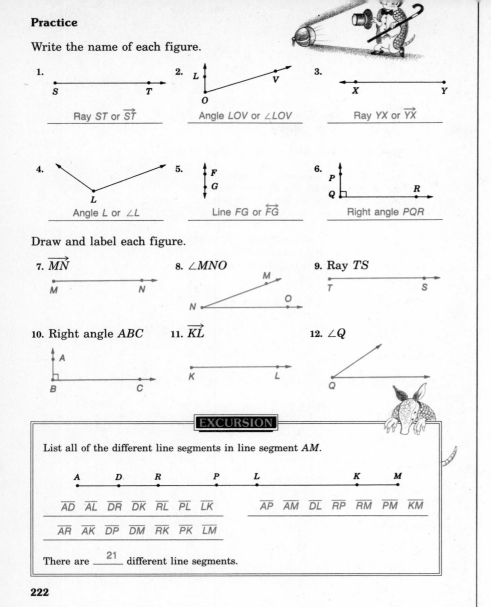

1.

S   T

Ray *ST* or $\overrightarrow{ST}$

2.

L

O   V

Angle *LOV* or ∠*LOV*

3.

X   Y

Ray *YX* or $\overrightarrow{YX}$

4.

L

Angle *L* or ∠*L*

5.

F

G

Line *FG* or $\overleftrightarrow{FG}$

6.

P

Q   R

Right angle *PQR*

Draw and label each figure.

7. $\overrightarrow{MN}$

M   N

8. ∠*MNO*

M

N   O

9. Ray *TS*

T   S

10. Right angle *ABC*

A

B   C

11. $\overleftrightarrow{KL}$

K   L

12. ∠*Q*

Q

EXCURSION

List all of the different line segments in line segment *AM*.

A   D   R   P   L   K   M

$\overline{AD}$ $\overline{AL}$ $\overline{DR}$ $\overline{DK}$ $\overline{RL}$ $\overline{PL}$ $\overline{LK}$ $\overline{AP}$ $\overline{AM}$ $\overline{DL}$ $\overline{RP}$ $\overline{RM}$ $\overline{PM}$ $\overline{KM}$

$\overline{AR}$ $\overline{AK}$ $\overline{DP}$ $\overline{DM}$ $\overline{RK}$ $\overline{PK}$ $\overline{LM}$

There are ___21___ different line segments.

Correcting Common Errors

Watch for students who do not put the letter for the vertex in the middle when they are naming an angle. Have students work in pairs, drawing two rays with a common endpoint, labeling the endpoint X. Discuss how point X is the endpoint of both rays and the vertex of the angle. Then have students name a point, other than the endpoint, on each ray and use these points to name the angle.

Enrichment

Have students draw and label an open or closed figure having 6 angles, 1 of which is a right angle. Then tell them to use a square corner to draw the right angle, and the 6 ways to name each angle in their figure.

Practice

Remind students that rays and angles can be named by using the word and identifying initials, or by the symbols and initials. Remind students to always use a straight-edge to draw lines and line segments. Have students complete the page independently.

Excursion

Review the definition of line segment. Have students list the segments as they find them. Remind them to use the symbol for line segment. Have students determine how many of the segments do not contain $\overline{RP}$.

Extra Credit   *Biography*

Grace Chisholm Young was a woman pioneer in the field of mathematics. She was born in England in 1868, at a time when industry and modern technology were developing at an amazing rate. Grace was fortunate to be accepted into a girls' boarding school. Her education there prepared her to take university entrance exams. Encouraged by her father, Grace enrolled at Girton College, a women's college. Later, she met resistance at home when she applied to graduate school. She was able to enter the University at Göttingen, in Germany, and was granted a doctorate, the first official doctorate granted a woman in Germany in any subject area. Besides her research work in mathematics, Grace wrote the ***First Book of Geometry.*** She felt that to help students learn solid geometry concepts, they should make and handle three-dimensional figures. Grace provided diagrams in her book that students could cut, fold and refold into different shapes. This idea made learning geometry easier and more fun.

Plane Figures

Objective

To identify and draw plane figures

Materials

*cube

Mental Math

Tell students to give their total cost for 1 year if:

1. $600/month ($7,200)
2. $20 bi-weekly ($520)
3. $150/quarter year ($600)
4. $2,600 semi-annually ($5,200)
5. $760 bi-annually ($380)
6. $4/week ($208)
7. $89/6 months ($178)
8. $200 bi-monthly ($1,200)

Skill Review

Have students draw and label figures on the board such as: 2 rays with C as their vertex, a figure having 2 square angles, a closed figure having 3 line segments, etc.

Plane Figures

How can the images on a movie screen help you to understand plane figures?

A **plane** figure is a "flat" shape. It has length and width, but no height. We think of plane figures as being on the surface of things, like the figures drawn on this page, or images we see in a movie. Some plane figures are called **polygons**.

Polygons have only straight sides. **Triangles** have 3 sides; **quadrilaterals** have 4; **pentagons** have 5; **hexagons** have 6 and **octagons** have 8 sides.

Plane figures, like movie images, have ___length___

and ___width___, but no height.

Getting Started

Name the kind of polygon. List its line segments and angles.

1. Kind of polygon	2. Line segments		3. Angles	
Hexagon	$\overline{AB}$	$\overline{DE}$	$\angle A$	$\angle D$
	$\overline{BC}$	$\overline{EF}$	$\angle B$	$\angle E$
	$\overline{CD}$	$\overline{FA}$	$\angle C$	$\angle F$

Draw and label each polygon.

4. Triangle XYZ

5. Quadrilateral $ABCD$

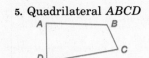

223

Teaching the Lesson

Introducing the Problem Have a student describe the picture and then read the problem. Ask students what kind of figures will be discussed in this lesson. (plane figures) Have students read with you through the first 2 sentences about polygons. Ask what figure has no straight sides. (circle) Have students complete the sentences as they continue to read with you and complete the solution statement.

Developing the Skill Ask students if they have ever seen a movie in 3-D. Tell students that the term, 3-D, means 3-dimensional and that length and width are 2 of the dimensions. Tell students the images on a 3-D movie also have height or depth and seem to reach out to us. Show a cube and tell students that it has the 3 dimensions of length, width and height. Point to each surface as you tell students that each side has only length and width so we call it a plane figure. Point out some plane figures in the room and tell students they are called **polygons** because they are plane figures with straight sides. Show a circular surface as you tell students the circle is a plane figure. Ask why a circle is not a polygon. (has no straight sides) Have students find more plane figures in the room.

Practice

Name each kind of polygon.
List its line segments and angles.

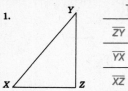

1.
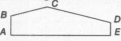

Triangle

$\overline{ZY}$	$\angle Z$
$\overline{YX}$	$\angle Y$
$\overline{XZ}$	$\angle X$

2.

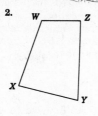

Quadrilateral

$\overline{WZ}$	$\angle W$
$\overline{ZY}$	$\angle Z$
$\overline{YX}$	$\angle Y$
$\overline{XW}$	$\angle X$

3.

Pentagon

$\overline{AB}$	$\angle A$
$\overline{BC}$	$\angle B$
$\overline{CD}$	$\angle C$
$\overline{DE}$	$\angle D$
$\overline{EA}$	$\angle E$

4.

Hexagon

$\overline{XW}$	$\angle W$
$\overline{WV}$	$\angle V$
$\overline{VU}$	$\angle U$
$\overline{UT}$	$\angle T$
$\overline{TS}$	$\angle S$
$\overline{SX}$	$\angle X$

Draw and label each polygon.

5. Pentagon *ABCDE*

6. Quadrilateral *RSTU*

7. Octagon *ABCDEFGH*

Complete the chart.

Polygon	Number of Sides	Number of Angles
8. Triangle	3	3
9. Quadrilateral	4	4
10. Pentagon	5	5
11. Hexagon	6	6
12. Octagon	8	8

224

Congruent and Similar Figures

pages 225-226

Objective
To identify figures that are congruent or similar

Materials
*cardboard square
*2 × 3 and 4 × 6 unit cardboard rectangles
*cardboard triangle
*congruent and similar figures
*figures not similar or congruent
transparent graph paper
rulers

Mental Math
Have students give the total cost if gasoline is $1.10/gal:

1. 10 gal ($11)
2. 11 gal ($12.10)
3. 7 gal ($7.70)
4. 10½ gal ($11.55)
5. 2¹⁄₁₀ gal ($2.31)
6. 20 gal ($22)

Skill Review
Dictate the following angles for students to draw and label on the board: ∠ C, ∠ XYZ, right angle STU, ∠ R, ∠ KLM. Have students use rulers and trace square corners for the right angle.

Congruent and Similar Figures

Plane figures that have the same shape are called **similar** figures. Plane figures that have the same shape and size are called **congruent** figures. Which triangle is similar to triangle *ABC*? Which triangle is congruent to triangle *ABC*?

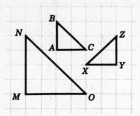

Graph paper can be used to show that two plane figures are similar.

Triangle *ABC* is similar to triangles <u>MNO</u> and <u>YZX</u>. Tracing paper can be used to show that two plane figures are congruent.

Triangle *ABC* is congruent to triangle <u>YZX</u>.

Getting Started

Tell if each pair of figures is similar. Write **yes** or **no.**

1. _____ yes _____

2. _____ no _____

Tell if each pair of figures is congruent. Write **yes** or **no.**

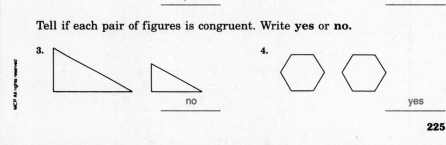

3. _____ no _____

4. _____ yes _____

225

Teaching the Lesson

Introducing the Problem Have a student read the problem aloud and tell what is to be answered. (which triangle is **similar** to and which is **congruent** to triangle ABC) Ask students what information is known about similar plane figures. (They have the same shape.) Ask what is known about congruent figures. (They have the same shape and size.) Have students read with you as they complete the sentences.

Developing the Skill Before class, trace the triangle on the board and then rotate the cardboard and trace a second triangle. Have students use graph paper to duplicate the figures and cut them out. Tell students to lay one on top of the other to see if they are the same size and shape. (yes) Tell students the 2 triangles are congruent because they have the same size and shape. Make 2 tracings of the cardboard square on the board. Tell students the 2 squares are also congruent. Make tracings of the 2 rectangles on the board and tell students the rectangles are called similar figures because they have the same shape but not the same size. Have students identify congruent and similar plane figures in the classroom. If possible, have students trace the figures on the board to check their identification.

Practice

Tell if each pair of figures is similar. Write **yes** or **no**.

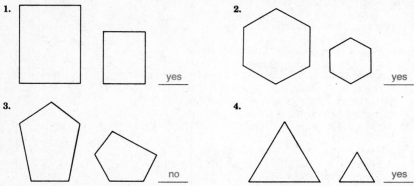

1. _yes_

2. _yes_

3. _no_

4. _yes_

Tell if each pair of figures is congruent. Write **yes** or **no**.

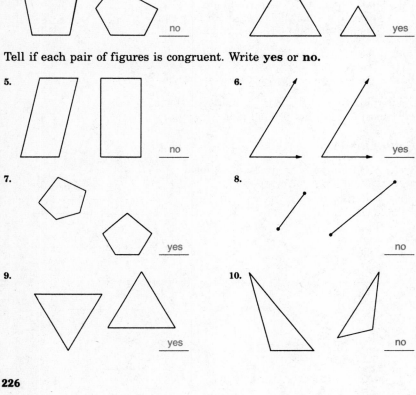

5. _no_

6. _yes_

7. _yes_

8. _no_

9. _yes_

10. _no_

226

Practice

Allow students to use tracing paper to check their work if needed. Have students complete the page independently.

Mixed Practice

1. $7.98 ÷ 3 ($2.66)
2. 600 × 7 (4,200)
3. 2,050 − 923 (1,127)
4. 27 × 436 (11,772)
5. 58,176 + 4,950 (63,126)
6. 738 + 297 + 30 (1,065)
7. 21,753 − 17,247 (4,506)
8. 123 ÷ 6 (20 R3)
9. 608 × 9 (5,472)
10. 57 ÷ 6 (9 R3)

Extra Credit *Numeration*

Have students create their own pictorial number system, for numbers one through ten, by using common objects to stand for each number. Tell students the object chosen to represent a number must have a reference to that number. For example: sun = one; eyes = two; triangle = three, etc. Tell them to write a story problem using their number system, and give it to a partner to solve. Extend the activity by having students write their phone numbers and addresses, using their systems.

Parallelograms

pages 227-228

Objective

To identify parallelograms

Materials

congruent and similar figures
parallelograms, rectangles and
 squares
rulers

Mental Math

Ask students if six 3-ft boards could be
cut from:

1. two 10-ft boards. (yes)
2. one 12-ft board. (no)
3. four 5-ft boards. (no)
4. two 12-ft boards. (yes)
5. five 6-ft boards. (yes)
6. one 12-ft and one 4 ft board. (no)
7. six 5-ft boards. (yes)
8. six 40-in. boards. (yes)

Skill Review

Have a student select 2 congruent figures from a display. Repeat for 2 similar figures. Continue for more practice
in identifying congruent and similar figures.

Parallelograms

Which of these plane figures have
sides made of parallel lines?

Figure __C__ has no straight lines.

Figure __A__ has no sets of parallel lines.

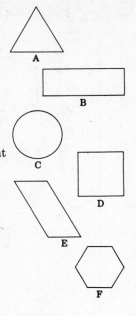

A polygon in which the opposite sides are
parallel and congruent is called a
parallelogram. Congruent sides are line
segments which have the same length.
A **rectangle** is a parallelogram with four right

angles. Figure __B__ is a rectangle.
A **square** is a parallelogram with four right
angles and all four sides congruent.

Figure __D__ is a square.

✔ A square is a special kind of rectangle.

Figures __B__, __D__, __E__ and __F__ have
opposite sides that are parallel.

Getting Started

Write a **P** for a parallelogram, an **R** for a
rectangle and an **S** for a square.

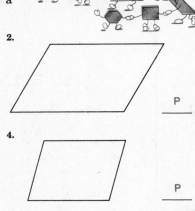

1.   R

2.   P

3.   S

4.   P

227

Teaching the Lesson

Introducing the Problem Have a student read the
problem question aloud. Ask students what they will look
for in each pictured figure. (parallel lines) Remind students
that all the figures except C are polygons. Have students
read with you as they complete the sentences. Have students read the definition and complete the solution sentence.

Developing the Skill Trace a parallelogram on the
board. Tell students the figure is called a **parallelogram**
because its opposite sides are parallel and congruent. Tell
students that **congruent sides** are line segments which
have the same length. Show students a rectangle and ask if
it is a parallelogram. (yes) Ask why. (Opposite sides are parallel and congruent.) Box all 4 angles in the rectangle to
show they are right angles. Tell students the parallelogram is
called a **rectangle** because it has 4 right angles. Mix rectangles with other parallelograms and have students find all the

rectangles and tell how they differ from the other parallelograms. (The rectangle angles are all right angles.) Show students a square as you tell them that a **square** is a parallelogram that is a special kind of rectangle because all 4 of its
sides are congruent and it has 4 right angles. Now show
squares, rectangles and other parallelograms and have students tell the name of each and the qualities which give the
figure its name.

227

Practice

Write a **P** for a parallelogram, an **R** for a rectangle and an **S** for a square.

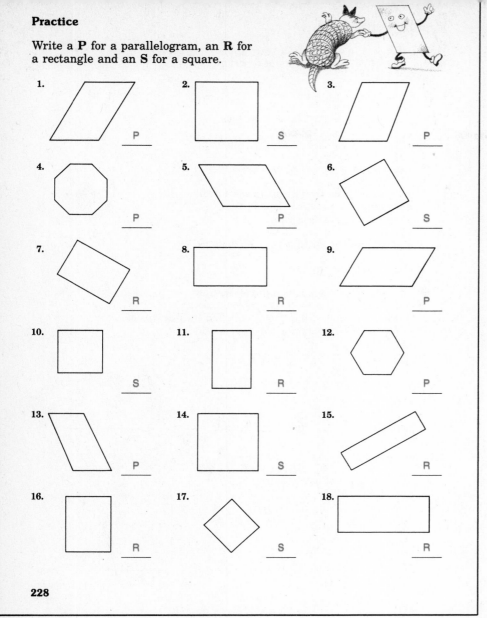

1. _P_

2. _S_

3. _P_

4. _P_

5. _P_

6. _S_

7. _R_

8. _R_

9. _P_

10. _S_

11. _R_

12. _P_

13. _P_

14. _S_

15. _R_

16. _R_

17. _S_

18. _R_

228

Perimeter

Objective

To find the perimeter of a figure

Materials

*rectangle and square
rulers

Mental Math

Tell students to name the figure or
part of a figure:

1. has no height or depth. (plane)
2. 2 lines which cross is. (intersecting
   lines)
3. has 2 endpoints. (line segment)
4. is a point of an angle. (vertex)
5. is formed by 2 intersecting rays.
   (angle)
6. is part of a line. (line segment)
7. is a square corner. (right angle)
8. has no endpoints. (line)

Skill Review

Review column addition through
2-digit addends. Have students write
and solve the problem 14 + 14 +
14 + 14 using an operation other than
simple addition. (4 × 14 = 56) Repeat
for 16 + 9 + 9 + 16. ((2 × 16) +
(2 × 9) = 50) Provide additional problems.

Finding Perimeter

Dino is fencing in a section of the yard
for his dog. How many meters of fencing
will Dino need?

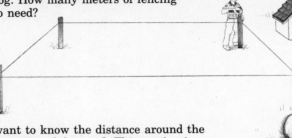

We want to know the distance around the
dog's section of the yard. This section is
shaped like a parallelogram.

The length of the section is _____5_____ meters.

Its width is _____3_____ meters.

✔ Because the section is a parallelogram,
we know that opposite sides are congruent.
The distance around a polygon is called its **perimeter.**
To find the perimeter of a polygon, we add the

lengths of all the sides. We add ___3___, ___5___,

___3___ and ___5___.

3 + 5 + 3 + 5 = ___16___

Dino will need ___16___ meters of fencing.

Getting Started

Find the perimeter of each figure.

1.

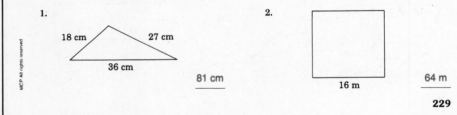

18 cm   27 cm

36 cm

81 cm

2.

16 m

64 m

229

Teaching the Lesson

Introducing the Problem Ask a student to describe the
picture, read the problem aloud and tell what information is
given. (The space to be fenced is 3 meters on one side and
5 meters on another.) Ask what information is needed. (the
length of the other 2 sides) Have students read aloud with
you as they complete the sentences to solve the problem.

Developing the Skill Trace a rectangle on the board.
Have students identify the figure. (rectangle) Ask students
what length the opposite side would be as you write **7 ft**
along the length of the rectangle. (7 ft) Ask how we know
this. (The figure is a rectangle and opposite sides are congruent.) Repeat for a width of **3 ft.** Tell students they can
find the **perimeter** or the distance around this figure by
adding 7 ft + 7 ft + 3 ft + 3 ft. Have a student find the perimeter. (20 ft) Change the dimensions to 13 cm by 28 cm
and have students find the perimeter. (82 cm) Show students how to find the perimeter by adding 2 times the
length and 2 times the width. Have students find the perimeter of a square when only 1 side is known. Show students
that side + side + side + side is the same as 4 times the
measurement of 1 side.

Practice

Find the perimeter of each figure.

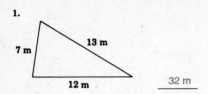

1.

7 m / 13 m / 12 m
<u>32 m</u>

2.

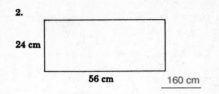

24 cm / 56 cm
<u>160 cm</u>

3.

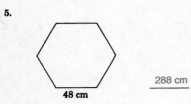

15 m / 50 m / 40 m / 45 m / 24 m
<u>174 m</u>

4.

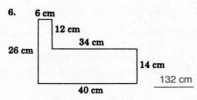

10 cm / 34 cm
<u>88 cm</u>

5.

hexagon 48 cm
<u>288 cm</u>

6.

6 cm / 12 cm / 34 cm / 26 cm / 14 cm / 40 cm
<u>132 cm</u>

Apply

Solve these problems.

7. Find the perimeter of a rectangular parking lot that is 28 meters wide and 47 meters long.
   150 meters

8. Allison drew a pentagon with all the sides equal in length. One side was 14 centimeters. What was the perimeter of the pentagon?
   70 centimeters

9. Complete the table.

Squares

Side	1 cm	2 cm	3 cm	4 cm	5 cm
Perimeter	4 cm	8 cm	12 cm	16 cm	20 cm

230

Area

pages 231-232

Objective

To find the area of a figure

Materials

various-sized rectangular pieces of
graph paper
rulers

Mental Math

Tell students to solve for R:

1. $20 \times R = 100$ (5)
2. $160 \div R = 8$ (20)
3. $8 \times R = 888$ (111)
4. $(3 \times R) + (6 \times R) = 108$ (12)
5. $4 \times R = 240$ (60)
6. $(2 \times R) + (2 \times 10) = 28$ (4)
7. ½ of $R = 720$ (1440)
8. $R \div 5 = 150$ (750)

Skill Review

Draw an array of X's on the board
that has 3 rows of 6 X's each. Ask
students how many rows of X's there
are. (3) Ask how many X's in each
row. (6) Ask students how to find the
total number of X's. (multiply 3×6 or
6×3) Repeat for more problems to
find the total number when objects are
arranged in rows.

Finding Area

Dino's father told him the dog had
to have at least 10 square meters
to play in. Will the fenced-in section
be big enough?

We need to find the number of square meters
in the dog's yard.

There are ___3___ rows of square meters with

___5___ square meters in each row.
The number of square units is called the **area**.
To find the area of the section, we multiply
the number of rows by the number of square

units in each row. We multiply ___3___ by ___5___.

___3___ × ___5___ = ___15___

There are ___15___ square meters in the dog's yard.

It ___is___ big enough for the dog.
We can use a **formula** to find the area of a
rectangle or a square.

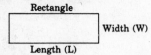

Rectangle   Width (W)
Length (L)

Square
Side (S)

✔ The area of a rectangle is
equal to the length multiplied
by the width.
$A = L \times W$

✔ The area of a square is
equal to the side multiplied by
itself.
$A = S \times S$

Getting Started

Find the area of each figure.

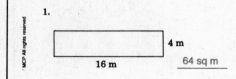

1.
4 m
16 m
64 sq m

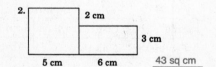

2.
2 cm
3 cm
5 cm   6 cm
43 sq cm

231

Teaching the Lesson

Introducing the Problem Have a student read the
problem and tell what needs to be answered. (if the fenced
space is 10 or more square meters) Ask what information is
known. (length is 5 meters and width is 3 meters) Ask what
is known about the shape of the figure. (It is a rectangle.)
Have students read the rules in the model aloud and study
the figure. Have them fill in the sentences and solve the
problem.

Developing the Skill Show students a piece of graph
paper. Tell students to find the number of units on the
whole sheet. Help students see that each unit is a square.
Tell students we find the number of square units in all by
multiplying the length by the width. Have a student find the
number of square units. (Answers will vary.) Write the number on the board as ____ **square units.** Tell students we
call this the **area** and area is always given in square units
whether it is square miles, square feet or square centimeters.
Have students find the areas of more rectangular sheets of
graph paper. Write **A = L × W** on the board and tell students this is the **formula** to find the area of a rectangle.
Repeat the activity for a square sheet of graph paper and
have students discover that the formula for the area of a
square is $A = S \times S$.

Practice

Find the area of each figure.

1.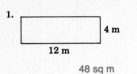

4 m
12 m
48 sq m

2.

9 m
81 sq m

3.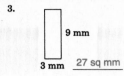

9 mm
3 mm   27 sq mm

Find the area of the shaded part of each figure.

4.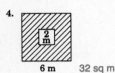

2 m
6 m   32 sq m

5.

3 cm
4 cm
3 cm
6 cm   48 sq cm

6.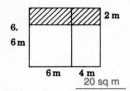

2 m
6 m
6 m   4 m
20 sq m

Apply

Solve these problems.

7. Fritz is tiling a floor. The length of the floor is 3 meters. The width is 4 meters. How many square meters of floor does Fritz have to tile?
12 square meters

8. Wayne has a rectangle that is 5 centimeters by 12 centimeters. Rosa has a square that is 8 centimeters on a side. Who has the larger polygon? How much larger is it?
Rosa   4 square centimeters

Complete these tables.

9.

Rectangles					
Length	6 cm	5 cm	2 cm	5 cm	9 cm
Width	4 cm	4 cm	8 cm	7 cm	2 cm
Area	24 sq cm	20 sq cm	16 sq cm	35 sq cm	18 sq cm

10.

Squares					
Side	2 cm	10 cm	6 cm	11 cm	7 cm
Area	4 sq cm	100 sq cm	36 sq cm	121 sq cm	49 sq cm

232

Correcting Common Errors

Some students may have difficulty relating multiplication to finding the area of a square or rectangle. Have them work with partners and use grid paper to draw a variety of rectangles and squares that are a whole number of units long and wide. One member of the pair should count the small squares to find the area of each figure and the other write the multiplication sentence that also gives the area.

Enrichment

Tell students to write the formulas for finding the perimeter of a rectangle and a square. Tell them to use these formulas to find the perimeter of a square whose area is 36 square miles and the perimeter of a rectangle whose area is 90 square yards.

Practice

Write the formulas **A = L × W** for a rectangle and **A = S × S** for a square on the board. Ask students how they will find the area of just the shaded part in problem 7. (Find the area of the whole square and subtract the area of the small white square.) Tell students they are to find the area of each rectangle and square in the tables. Have students complete the page independently.

Extra Credit   *Statistics*

Use one-inch graph paper and give each student a sheet that is 10 inches on a side. Ask them to look at the pattern you have begun on the board:

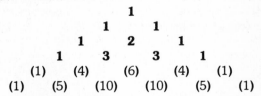

```
              1
          1       1
      1       2       1
  1       3       3       1
(1)   (4)   (6)   (4)   (1)
(1)  (5)  (10)  (10)  (5)  (1)
```

Have students copy the section on the board and continue with the pattern. This arrangement is called "Pascal's Triangle." Point out that each number in the pattern is the sum of the two numbers diagonally above it. They should also note that each number represents the number of different ways there are to get to that square from the top-most square.

Solid Figures

pages 233-234

Objective

To identify solid figures

Materials

*wood cube
*rectangular prism
*pyramid

Mental Math

Tell students to identify:

1. 16 days from now.
2. 62 years from today.
3. 15 hours from now.
4. 2 hours 15 min ago.
5. 7 months ago.
6. hours and minutes yet till 3:45 PM.
7. 36 years and 1 one month from today.
8. 3 weeks ago.

Skill Review

Show students the wood cube. Have a student name any length and width for 1 of its surfaces. Have students find the area and perimeter of that surface. Repeat for the rectangular prism.

Solid Figures

Many of the objects used in our daily lives are shaped like these solid figures.

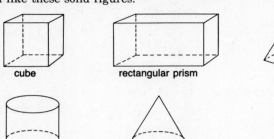

cube   rectangular prism   pyramid

cylinder   cone   sphere

Solid figures have length and width as well as height. Some figures such as cubes, prisms and pyramids have **faces, vertices** and **edges.**

edge
face
vertex

Getting Started

Write the name of the solid figure you see in each object.

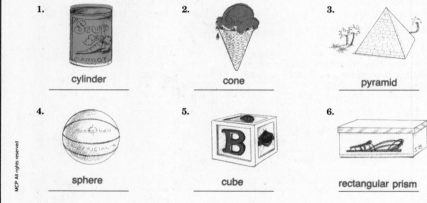

1. cylinder
2. cone
3. pyramid
4. sphere
5. cube
6. rectangular prism

233

Teaching the Lesson

Introducing the Problem Have a student read the statement about **solid** figures. Have students point to each pictured figure as you say its name. Have students read the information about solid figures and some properties of cubes, prisms and pyramids in the model.

Developing the Skill Remind students that they have been learning about plane or flat figures. Hold up a wood cube and point to one surface at a time as you tell students that each surface is a plane figure with length and width. Tell students that if we look at the cube as a whole object, it has length, width and height. Tell students we call any figure which has all 3 dimensions a **solid figure.** Point to one of the surfaces and tell students the plane figure or surface is called a **face** of the cube. Ask how many faces the cube has. (6) Run your finger along the edge of the cube as you tell students the edge of a solid figure is called an **edge.** Have students count the edges on the cube. (12) Show a corner on the cube as you tell students the line segments meet at the corners to form a **vertex.** Have students count the number of vertices on the cube. (8) Repeat the activity for a rectangular prism and a pyramid.

233

Practice

Write the name of the solid figure you see in each object.

1.

cylinder

2.

cone

3.

cube

4.

rectangular prism

5.

sphere

6.

pyramid

Complete the table.

7.

Figure	Number of Faces	Number of Edges	Number of Vertices
	6	12	8
	6	12	8
	5	8	5

EXCURSION

These polygons are **symmetrical.** They can be divided into two identical figures by at least one line segment, called an **axis of symmetry.** Draw all axes of symmetry for these polygons. Complete the chart.

Symmetrical Polygons

Sides	4	4	6	8
Angles	4	4	6	8
Axes of Symmetry	2	2	3	4

234

Practice

Have students complete the page independently.

Excursion

Demonstrate the property of symmetry for the triangle. Point out that regular polygons have more than one axis of symmetry. Provide cutouts of the other polygons on a larger scale for students to trace on scratch paper.

Extra Credit   *Statistics*

Have students use the Pascal's triangle they have already drawn in the preceding extra credit or ask them to compose a new one. Ask them to complete 9 rows. Have each student graph the numbers appearing in the last line on the graph paper.

```
                1
              1   1
            1   2   1
          1   3   3   1
        1   4   6   4   1
      1   5   10   10   5   1
    1   6   15   20   15   6   1
  1   7   21   35   35   21   7   1
1 8  28  56  70  56  28 8 1
1 2  3   4    5   6    7 8 9
              column
```

```
              70
              60
Pascal   50
number  40
              30
              20
              10
column     1 2 3 4 5 6 7 8 9
```

Volume

pages 235-236

Objective

To find the volume of a rectangular
prism

Materials

*rectangular prism
*building cubes

Mental Math

Tell students to find the value of P if:

1. $P \times P \times P = 27$ (3)
2. $P \times P \times P \times P = 16$ (2)
3. $P \div P = 1$ (any number)
4. $P \times 30 = 0$ (0)
5. $4 \times P = 320$ sq mi (80 mi)
6. $P \times P \times P = 8$ (2)
7. $8 \times (P + 6) = 64$ (2)
8. 62 ft $\times P = 124$ sq ft (2 ft)

Skill Review

Have students work the following
problems from the board: **3 × 6 × 2**
(36), **8 × 10 × 4** (320), **6 × 7 × 9**
(378), **3 × 3 × 3** (27), **20 × 20 × 26**
(10,400)

Finding Volume

One way to measure the volume of a solid figure is in
cubic units. How many cubic units are in this prism?

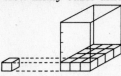

We want to find the volume of the prism in
cubic units.

The prism is ___3___ cubic units long, ___4___

cubic units wide and ___5___ cubic units high.
To find the volume, we multiply the number
of cubic units in a row, by the number of rows,
and again by the number of units high. We

multiply ___3___ by ___4___ by ___5___.

There are ___60___ cubic units in the prism.
We can use a formula to find the volume of
a prism.

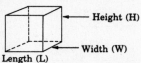

✔ The volume of a prism is equal to the
length multiplied by the width multiplied by
the height.
$$V = L \times W \times H$$

Getting Started

Find the volume of each prism.

1.
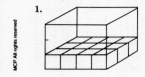
36 cubic units

2.
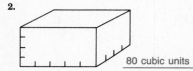
80 cubic units

235

Teaching the Lesson

Introducing the Problem Have a student read the
problem aloud and tell what is to be solved. (How many
cubic units are in the prism?) Have students tell what information is known. (There are 4 rows of 3 units each and
there are 5 layers.) Have students read and complete the
sentences to solve the problem. Tell students to check their
solution by finding the number of cubes on each layer and
multiplying by 5. Have students read the formula for finding
the volume of a prism as you explain it.

Developing the Skill Show students a rectangular
prism. Remind students that a solid figure has length, width
and height. Have students count the solid's faces, vertices
and edges. Tell students they will now learn how to find the
volume or the number of **cubic units** in the whole prism.
Write on the board: **Length = 7 units, Width = 6 units**
and **Height = 4 units.** Tell students they are to find the
number of cubic units in 1 layer by multiplying the length
times the width (42) and then multiply that number by the
number of layers or the height of the prism. (168) Write
V = L × W × H on the board as you tell students that the
volume of a solid figure is length × width × height. Tell students that volume is reported in cubic units such as cubic
yards, cubic meters, etc.

235

Practice

Find the volume of each prism.

1.

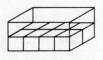

16 cubic units

2.

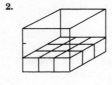

27 cubic units

3.

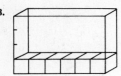

24 cubic units

4.

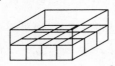

24 cubic units

5.

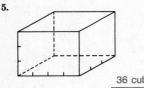

36 cubic units

6.

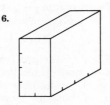

32 cubic units

7.

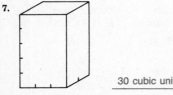

30 cubic units

8.

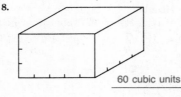

60 cubic units

9.

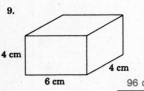

4 cm

6 cm

4 cm

96 cubic cm

10.

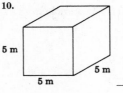

5 m

5 m

5 m

125 cubic m

236

Problem Solving
Restate Problem

pages 237-238

Objective

To solve a problem by restating it

Materials

Mental Math

Have students identify each stage of problem solving:

1. carry out the plan (Do)
2. look for careless errors (Check)
3. ask if answer is reasonable (Check)
4. decide on operations to be used (Plan)
5. find a solution (Do)
6. decide what is to be solved (See)
7. last stage (Check)
8. state facts known (See)

Restating the Problem

Jose has $1.20 in coins. Half of the coins are nickels and half are dimes. What is the value of his nickels?

★ SEE

We want to find the value of Jose's nickels.

We know that the value of all his coins is ___$1.20___.

Half of the coins are ___nickels___ and half are ___dimes___.

★ PLAN

We want to express this problem in our own way, so that we really understand it.

★ DO

Any way we reword the problem is good, as long as it helps us to understand it better. We know that the value of any number of nickels is one half the value of the same number of dimes.

Number of Nickels	Number of Dimes	Value of Nickels	Value of Dimes	Total Value of the Coins
6	6	$0.30	$0.60	$0.90
7	7	$0.35	$0.70	$1.05
8	8	$0.40	$0.80	$1.20

The value of Jose's nickels is ___$0.40___

★ CHECK

$$5¢ \times 8 = 40¢ \qquad 10¢ \times 8 = 80¢ \qquad 80¢ + 40¢ = \underline{\$1.20}$$

237

Teaching the Lesson

Ask students to think about the last time they became confused when someone told them something. Tell students that in order to understand what we hear, we often reword it in our own words, either aloud or silently. Ask students if they have ever read something that was confusing. Tell students that again we often find ourselves **restating** in our own words, what we have read. We often need to reread something several times before we can even restate it. Have a student read the problem aloud and tell what is to be found. (the value of the nickels) Tell students that a first reading of this problem leads us to assume that, since the number of nickels and dimes is the same, then their values would also be the same. So we would take half of $1.20 for a 60¢ value of the nickels. Tell students that we have to keep in mind, though, that a nickel's value is 1/2 that of a dime so we cannot just take half of $1.20. Have students read aloud with you through the See, Plan and Do stages and complete the table to solve the problem. Have students complete the sentences in the Check stage to verify their solution.

Apply

Restate the problem in your own words to help you understand each problem.

1. Suppose there are 20 pinchworms in the bottom of a jar. Pinchworms multiply so fast that they double their number every minute. If it takes forty minutes for the pinchworms to fill the jar, how long will it take them to fill half of the jar?
39 minutes

2. It is 34 miles from Bay City to Hamilton. It is 28 miles from Hamilton to Glenville. Glenville is on the road between Bay City and Hamilton. How far is it from Bay City to Glenville?
6 miles

3. Kyle bought a bat for $5.00. Then he bought a ball costing $3.00 more than the bat. How much did he spend?
$13.00

4. If you can buy thirteen stamps for a cent and a quarter, what is the cost of one stamp?
$0.02

5. If five boys can paint 5 garages in five days, how many boys would it take to paint 25 garages in 25 days?
5 boys

6. I went to the store and bought several items that were the same price. I bought as many items as the number of total cents in the cost of each item. My bill was $1.44. How many items did I buy?
12 items

7. Read Exercise 3 again. What if the ball costs half as much as the bat. Then how much would Kyle spend for both items?
$7.50

8. Explain how you would find the perimeter of a rectangle if the length were given in feet and the width in yards?
See Solution Notes.

9. Read Exercise 2 again. Write a problem about 3 towns where the distance between the nearest two towns is 5 miles and between the two towns farthest apart is 15 miles.
Answers will vary.

10. Suppose you know the length, width, and area of a rectangle. How would you affect the area if you double just the length? if you double both the length and width?
See Solution Notes.

238

Solution Notes

Have students restate each problem before beginning to solve it.

1. Careful reading will help students see that the number of pinchworms is unnecessary information. The important information is that pinchworms double their number *every* minute. If the jar is full in 40 minutes, then it is half-full in 1 minute less than 40 minutes.

2. Students should see that they should draw a map to locate the three cities in relation to each other.

3. At first glance the students are likely to respond that Kyle spent $8. Upon closer reading they should see that the ball cost $8 since it was $3 more than the bat.

4. Help students restate the cost of 13 stamps as 26¢.

5. Help students restate the problem to see that the 5 boys paint 1 garage each day and, therefore, the 5 boys can paint 25 garages in 25 days.

6. Students need to restate the problem to realize that the cost of each item equals the number of items bought. Thus they are looking for a number that, multiplied by itself, is 144.

Higher-Order Thinking Skills

7. [Analysis] The ball costs half as much as the bat, or $2.50; so, together the items cost $5.00 + $2.50, or $7.50.

8. [Synthesis] Either change the length to a number of yards or the width to a number of feet.

9. [Synthesis] A sample answer is "The distance between A and B is 5 miles. Peter Rabbit runs from A to C, a distance of 15 miles. How far is it from B to C?"

10. [Synthesis] If the student doubles the length, the area is doubled; if the student doubles both dimensions, the area is multiplied by 4.

Chapter Test

page 239

Item	Objective
1	Identify points, line segments, lines, intersecting lines and parallel lines (See pages 219-220)
2-3	Identify rays, angles and right angles (See pages 221-222)
4-6	Identify polygons by number of sides (See pages 223-224)
7-8	Find perimeter of a polygon (See pages 229-230)
9-10	Find area of rectangles and squares (See pages 231-232)
11-12	Find volume of a box (See pages 235-236)

Write the name for each figure.

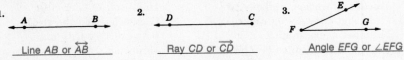

1. Line AB or $\overleftrightarrow{AB}$

2. Ray CD or $\overrightarrow{CD}$

3. Angle EFG or ∠EFG

Name each kind of polygon.

4. Hexagon

5. Square

6. Parallelogram

Find the perimeter.

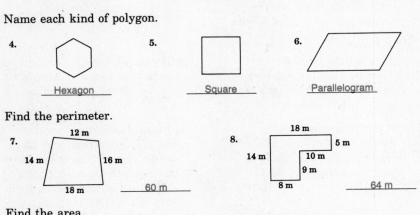

7. 12 m, 14 m, 16 m, 18 m — 60 m

8. 18 m, 5 m, 14 m, 10 m, 9 m, 8 m — 64 m

Find the area.

9. 6 cm, 9 cm — 54 sq cm

10. 5 cm — 25 sq cm

Find the volume.

11. 48 cubic units

12. 30 cubic units

239

Circle the letter of the correct answer.

1 Round 4,628 to the nearest thousand.
- a 4,000
- (b) 5,000
- c NG

2 What is the value of the 9 in 629,206?
- a tens
- (b) thousands
- c hundred thousands
- d NG

3   4,268
    + 3,659
- a 7,817
- (b) 7,927
- c 8,927
- d NG

4   39,475
    + 26,628
- (a) 66,103
- b 56,003
- c 56,093
- d NG

5 707 − 388
- a 389
- b 419
- c 481
- (d) NG

6   36,239
    − 14,856
- a 22,383
- b 22,623
- (c) 21,383
- d NG

7 Choose the better estimate of volume.
- (a) 1 pint
- b 1 quart

8 $3.25 × 7
- a $21.45
- b $21.75
- (c) $22.75
- d NG

9   16
    × 45
- a 144
- b 490
- (c) 720
- d NG

10   516
    × 25
- a 3,612
- b 12,700
- (c) 12,900
- d NG

11 27)162
- a 5
- (b) 6
- c 7
- d NG

12 629 ÷ 34
- a 18
- (b) 18 R17
- c 180 R17
- d NG

13 Name the figure.
- a square
- b rectangle
- (c) parallelogram
- d NG

☐ score

240

Item	Objective
1	Round numbers to nearest 100 (See pages 33-34)
2	Identify place value through hundred thousands (See pages 27-28)
3	Add two 4-digit numbers (See pages 47-48)
4	Add two 5-digit numbers (See pages 51-52)
5	Subtract two 3-digit numbers with zero in minuend (See pages 65-66)
6	Subtract two 5-digit numbers (See pages 67-68)
7	Determine appropriate customary unit of capacity (See pages 149-150)
8	Multiply money by 1-digit number (See pages 169-170)
9	Multiply two 2-digit numbers trading ones (See pages 179-180)
10	Multiply 3-digit number by 2-digit number (See pages 181-182)
11	Divide 3-digit number by 2-digit number to get 1-digit quotient without remainder (See pages 205-208)
12	Divide 3-digit number by 2-digit number to get 2-digit quotient with remainder (See pages 209-210)
13	Identify polygons by number of sides (See pages 223-224)

Alternate Cumulative Review

Circle the letter of the correct answer.

1 Round 6,318 to the nearest thousand
- (a) 6,000
- b 7,000
- c NG

2 What is the value of the 4 in 429,631?
- a ten thousands
- (b) hundred thousands
- c millions
- d NG

3   6,422
    + 3,398
- a 9,710
- (b) 9,820
- c 9,830
- d NG

4   44,378
    + 18,664
- a 62,942
- b 62,932
- (c) 63,042
- d NG

5 604 − 276 =
- (a) 328
- b 338
- c 428
- d NG

6   46,328
    − 12,974
- a 33,454
- b 33,554
- c 34,354
- (d) NG

7 What would you use to measure gas in a car?
- a pints
- b quarts
- (c) gallons
- d NG

8 $4.64
    ×8
- a $32.14
- b $36.22
- (c) $37.12
- d NG

9   27
    ×36
- a 732
- b 243
- (c) 972
- d NG

10   623
    ×34
- a 20,182
- b 21,172
- c 21,082
- (d) NG

11 32)224
- (a) 7
- b 7 R1
- c 7 R10
- d NG

12 756 ÷ 25 =
- a 3 R6
- (b) 30 R6
- c 36
- d NG

13 What is the name for a six-sided figure?
- a pentagon
- (b) hexagon
- c octagon
- d NG

240

Fractions

pages 241-242

Objective

To write a fraction for part of a whole or part of a set

Materials

*rectangle marked in thirds

Mental Math

Tell students to find the average of:

1. 3, 7, and 8. (6)
2. 40, 50 and 120. (70)
3. 0 and 100. (50)
4. 17 and 7. (12)
5. 20, 30 and 40. (30)
6. 70 and 100. (85)
7. 600 and 900. (750)
8. 1, 2, and 21. (8)

Skill Review

Review division of 3- and 4-digit dividends by 2-digit divisors, by having a student work the problem at the board. Have another student check each problem by multiplying and adding any remainder.

Fractional Parts

Fractions can help you talk about a part of a figure, or some of a set of things.
What part of this rectangle is red?
What part of this set of cars is not red?

The rectangle is divided into __5__ equal parts.

__1__ part is red.

We write: $\frac{1}{5}$ ← numerator
 ← denominator   We say: **one-fifth**.

$\frac{1}{5}$ __ of the rectangle is red.

There are __3__ cars.

__2__ cars are not red.

We write: $\frac{2}{3}$.   We say: **two-thirds**.

$\frac{2}{3}$ __ of the cars are not red.

Getting Started

Write a fraction to show what part of each figure is red.

1.
$\frac{2}{6}$

2.
$\frac{7}{8}$

3.
$\frac{3}{8}$

Write a fraction to show what part of each set is *not* red.

4.
$\frac{2}{10}$

5.
$\frac{3}{5}$

6.
$\frac{2}{9}$

241

Teaching the Lesson

Introducing the Problem Have a student describe the picture and tell what is to be answered. (what part of the rectangle is red and what part of the set of cars is red) Ask what is known about the rectangle. (It has 5 parts the same size.) Ask what is known about the set of cars. (There are 3 in all.) Have students complete the sentences and answer the questions as they read with you.

Developing the Skill Have 6 students stand in a group. Tell 1 of the 6 to step aside and ask students how many of the 6 moved away. (1) Write ⅙ on the board and tell students we read this fraction as one-sixth. Tell students the top number is called the **numerator** and it tells what part of the whole group we are talking about. Tell students the bottom number is called the **denominator** and it tells how many are in the whole group. Have another student step aside. Ask how many students moved away from the group of 6 as you write ⅖ on the board. Continue for ³⁄₆, ⁴⁄₆, ⁵⁄₆ and ⁶⁄₆. Trace a rectangle on the board and show its 3 equal parts. Have a student shade 1 of the 3 parts and write the fraction to show the shaded part. (⅓) Continue for ⅔ and ³⁄₃. Repeat for other parts of whole sets of objects.

241

Practice

Write a fraction to show what part of each figure is red.

1. $\frac{3}{4}$

2. $\frac{4}{6}$

3. $\frac{1}{5}$

4. $\frac{5}{10}$

5. $\frac{6}{12}$

6. $\frac{3}{8}$

Write a fraction to show what part of each set is *not* red.

7. $\frac{1}{3}$

8.  $\frac{2}{5}$

9.  $\frac{3}{4}$

10.  $\frac{3}{8}$

Apply

Solve these problems.

11. What part of the triangle is red?
 $\frac{1}{3}$

12. What part of the octagon is *not* red?
 $\frac{2}{8}$

13. What part of the set of coins are pennies?
 $\frac{4}{7}$

14. Write a fraction to show what part of the set of figures are squares.
 $\frac{6}{10}$

242

Practice

Tell students to read the directions for each section carefully. Have students complete the page independently.

Mixed Practice

1. 8.95×6 ($53.70)
2. $130 \div 63$ (2 R4)
3. $2,075 - 927$ (1,148)
4. $65,237 + 30,207$ (95,444)
5. $7 + 5 \times 7$ (42)
6. $320 \div 40$ (8)
7. $376 + 408 + 27$ (811)
8. $56.95 - 28.38$ ($28.57)
9. $562 \div 3$ (187 R1)
10. 52×34 (1,768)

Extra Credit *Numeration*

Put the following example on the board and explain to students that it is possible to multiply with Roman numerals.

XXVII	27
× XVI	×16
CCCCXXXII	432

Remind students to use the whole value of the Roman numeral when multiplying, rather than each digit. Ask students to try and solve these problems. Check each problem with Arabic numerals to make sure the answers are correct.

XIV	14	XXV	25	LV	55
×III	× 3	× IV	× 4	× V	× 5
(XLII)	(42)	(C)	(100)	(CCLXXV)	(275)

Challenge students to write their own multiplication problems using Roman numerals.

242

Unit Fraction of a Number

pages 243-244

Objective

To find the number of objects in 1 part of a set

Materials

counters
division fact cards
graph paper
scissors

Mental Math

Ask students how much more or less than $10 is:

1. $1.68 ($8.32 less)
2. 4 × $2.52 (8¢ more)
3. $5.75 ÷ 2 ($2.25 less)
4. $10 ÷ 4 ($7.50 less)
5. $8.15 + 2.50 (65¢ more)
6. $3.50 × 2 ($3 less)
7. $4.96 + 7 ($1.96 more)
8. $7.50 × 2 ($5 more)

Skill Review

Have students work in pairs with fact cards to review the basic division facts. Have the first student give the quotient of a fact with the partner stating the related multiplication fact.

Finding a Fraction of a Number, Unit Fractions

Mark and Nadia are trading seashells for a science project. Mark gives Nadia $\frac{1}{5}$ of his seashells. How many seashells does Nadia get from Mark?

We want to know how many seashells Mark gives Nadia.

Mark has ___15___ seashells.

He gives Nadia ___$\frac{1}{5}$___ of them.

To find the number of seashells, we need

to multiply ___15___ by $\frac{1}{5}$.

✔ $\frac{1}{5}$ is called a **unit fraction** because its numerator is one. To multiply by a unit fraction, we simply divide by the denominator.

We divide ___15___ by ___5___.

Mark gave Nadia ___3___ seashells.

Getting Started

Find the part.

1. $\frac{1}{3}$ of 9 = ___3___
2. $\frac{1}{6}$ of 36 = ___6___
3. $\frac{1}{2}$ of 32 = ___16___
4. $\frac{1}{4}$ of 48 = ___12___
5. $\frac{1}{7}$ of 84 = ___12___
6. $\frac{1}{12}$ of 180 = ___15___

243

Teaching the Lesson

Introducing the Problem Have a student describe the picture. Have another student read the problem aloud and tell what is to be solved. (find the number of butterflies Mark gives Nadia) Ask what information is known. (Mark has 15 and gives ⅕ of them to Nadia.) Have students complete the sentences as they read with you to solve the problem. Tell them to do the division mentally. Have students see if 5 times their solution equals the total Mark had.

Developing the Skill Have students work in pairs and 12 counters. Have students move ½ of the counters in each group. (6) Tell students that the 12 counters were divided into equal parts and 1 of those 2 parts has 6 in it. Tell students that ⅙ is called a **unit fraction** because its numerator is 1. Tell students that to find ⅙ of a number we divide the number by the denominator. Repeat for ⅓ of 12, ¼ of 12, ⅙ of 12 and ¹⁄₁₂ of 12. Have students put out 18 counters and find ½, ⅓, ⅙, ⅑ and ¹⁄₁₈ of them.

243

Practice

Find the part.

1. $\frac{1}{5}$ of 15 = ___3___   2. $\frac{1}{8}$ of 64 = ___8___   3. $\frac{1}{9}$ of 63 = ___7___

4. $\frac{1}{6}$ of 36 = ___6___   5. $\frac{1}{3}$ of 36 = ___12___   6. $\frac{1}{2}$ of 56 = ___28___

7. $\frac{1}{7}$ of 84 = ___12___   8. $\frac{1}{4}$ of 96 = ___24___   9. $\frac{1}{3}$ of 87 = ___29___

10. $\frac{1}{6}$ of 96 = ___16___   11. $\frac{1}{12}$ of 72 = ___6___   12. $\frac{1}{15}$ of 750 = ___50___

Apply

Solve these problems.

13. Betty bought 36 pencils. She gave $\frac{1}{2}$ to her best friend. How many pencils did Betty give away?
18 pencils

14. Peter's baseball team played 21 games. They won $\frac{1}{3}$ of the games. How many games did the team win?
7 games

15. There are 32 children in Mr. Chan's class. One day, $\frac{1}{4}$ of the class was absent. How many children were absent?
8 children

16. There are 65 children on the school swimming team. $\frac{1}{5}$ of the team are boys. How many boys are on the team?
13 boys

17. Mr. James planted 56 trees. $\frac{1}{4}$ of the trees were maples. How many trees were maples?
14 trees

18. Mrs. Spencer bought 156 apples. $\frac{1}{6}$ of the apples were spoiled. How many apples were not spoiled?
130 apples

19. Ed lives 924 meters from school. He jogs $\frac{1}{3}$ of the way there and walks the rest. How far does Ed walk?
616 meters

20. A book has 216 pages. Lucia has read $\frac{1}{12}$ of the book. How many pages does Lucia have left to read?
198 pages

244

Fraction of a Number

Objective

To find a fraction of a number

Materials

$5 and $1 bills

Mental Math

Tell students to answer true or false:

1. A square is a rectangle. (T)
2. Parallel lines do not meet. (T)
3. A line has endpoints. (F)
4. A pyramid has 5 vertices. (T)
5. Congruent means same size and shape. (T)
6. A ray has 1 endpoint. (T)
7. A square has only 2 right angles. (F)
8. The volume of a solid is L x W. (F)

Skill Review

Remind students that to multiply by a unit fraction, we divide by the denominator. Have students illustrate and solve the following on the board: ⅙ of 18 (3), ¼ of 16 (4), ⅕ of 10 (2), ⅓ of 6 (2), ¼ of 12 (3).

Finding a Fraction of a Number

Pat looks for sales when she shops for clothes. How much will Pat pay for a blouse?

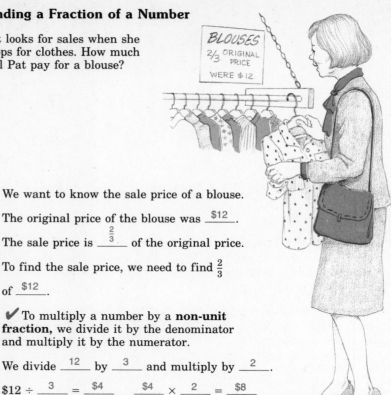

We want to know the sale price of a blouse.

The original price of the blouse was $12 .

The sale price is $\frac{2}{3}$ of the original price.

To find the sale price, we need to find $\frac{2}{3}$ of $12 .

✔ To multiply a number by a **non-unit fraction,** we divide it by the denominator and multiply it by the numerator.

We divide 12 by 3 and multiply by 2 .

$12 ÷ 3 = $4 $4 × 2 = $8

Pat will pay $8 for a blouse.

Getting Started

Find the part.

1. $\frac{3}{4}$ of 16 = 12
2. $\frac{5}{8}$ of $72 = $45
3. $\frac{3}{16}$ of 224 = 42

Find the sale price.

4. $\frac{5}{6}$ of a price of $36
$30

5. $\frac{3}{5}$ of a price of $365
$219

6. $\frac{2}{3}$ of a price of $672
$448

245

Teaching the Lesson

Introducing the Problem   Have a student describe the picture and read the sign. Have a student read the problem and tell what is to be found. (the sale price of the blouse) Ask what information is known. (The original cost was $12 and the sale price is ⅔ of the original price.) Have students read and complete the sentences to solve the problem. Guide them through the division and multiplication in the model. Students can check their solution with play money.

Developing the Skill   Write ⅓ of 9 on the board and have students illustrate it by drawing a set of nine, and circling three and solve the problem. Remind students that ⅓ is a unit fraction and 9 ÷ 3 = 3. Now write **⅔ of 9** and have students circle 2 of the 3 groups and tell how many. (6) Tell students ⅔ is a **non-unit fraction** and to find ⅔ of 9, we divide the 9 by 3 (3) and multiply by the numerator. (3 × 2 = 6) Have students work ⅗ of 45 (27), ¾ of 16 (12), ⅞ of 56 (49), and 2/9 of 81 (18).

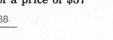

Practice

Find the part.

1. $\frac{2}{5}$ of $25 = \underline{\$10}$
2. $\frac{3}{4}$ of $36 = \underline{\$27}$
3. $\frac{3}{8}$ of 72 = \underline{27}

4. $\frac{5}{6}$ of 54 = \underline{45}
5. $\frac{7}{9}$ of 90 = \underline{70}
6. $\frac{2}{3}$ of $30 = \underline{\$20}$

7. $\frac{7}{8}$ of 64 = \underline{56}
8. $\frac{3}{4}$ of $84 = \underline{\$63}$
9. $\frac{5}{11}$ of $121 = \underline{\$55}$

Find the sale price.

10. $\frac{3}{8}$ of a price of $56
\underline{\$21}

11. $\frac{3}{4}$ of a price of $28
\underline{\$21}

12. $\frac{3}{5}$ of a price of $25
\underline{\$15}

13. $\frac{5}{7}$ of a price of $49
\underline{\$35}

14. $\frac{5}{8}$ of a price of $80
\underline{\$50}

15. $\frac{2}{3}$ of a price of $33
\underline{\$22}

16. $\frac{2}{3}$ of a price of $57
\underline{\$38}

17. $\frac{5}{6}$ of a price of $126
\underline{\$105}

18. $\frac{1}{2}$ of a price of $96
\underline{\$48}

Apply

Solve these problems.

19. Sweaters are on sale for $\frac{2}{3}$ of the original price. Before the sale, the sweaters were $42 each. What is the sale price?
$28

20. The original price of a jogging suit was $80. It is on sale for $\frac{3}{4}$ of the price. How much can be saved if you buy the suit on sale?
$20

Find the sale price.

21.

Sale $\frac{2}{5}$ off			
Original Price	$65	$140	$585
Sale Price	$39	$84	$351

246

246

Equivalent Fractions

pages 247-248

Objective

To understand equivalent fractions

Materials

strips of paper
crayons

Mental Math

Ask students how many 4-foot lengths of fence are needed to fence a:

1. perimeter of 720 ft. (180)
2. 12-ft square. (12)
3. 16-ft × 56-ft rectangle. (36)
4. 28-ft × 24-ft rectangle. (26)
5. perimeter of 364 ft. (91)
6. a 32-ft and 28-ft side. (15)
7. hexagon with each side 24 ft. (36)
8. octagon with each side 8 ft. (16)

Skill Review

Have a student write any unit fraction on the board and then find that part of a random number you give. Repeat for other unit fractions and then continue the activity for non-unit fractions. Be sure each number you give is a multiple of the fraction's denominator.

Understanding Equivalent Fractions

Fractions that name the same amount are called **equivalent fractions.** Name the two fractions that are equivalent to $\frac{2}{3}$.

We want to find two fractions that are equivalent to $\frac{2}{3}$.

Each large rectangle below is the same size. We can shade in the same amount of space in each rectangle to find equivalent **thirds, sixths** and **twelfths.**

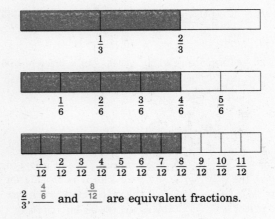

$\frac{1}{3}$ $\frac{2}{3}$

$\frac{1}{6}$ $\frac{2}{6}$ $\frac{3}{6}$ $\frac{4}{6}$ $\frac{5}{6}$

$\frac{1}{12}$ $\frac{2}{12}$ $\frac{3}{12}$ $\frac{4}{12}$ $\frac{5}{12}$ $\frac{6}{12}$ $\frac{7}{12}$ $\frac{8}{12}$ $\frac{9}{12}$ $\frac{10}{12}$ $\frac{11}{12}$

$\frac{2}{3}$, $\frac{4}{6}$ and $\frac{8}{12}$ are equivalent fractions.

Getting Started

Write the missing numerators.

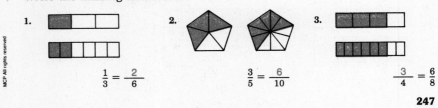

1. $\frac{1}{3} = \frac{2}{6}$

2. $\frac{3}{5} = \frac{6}{10}$

3. $\frac{3}{4} = \frac{6}{8}$

Teaching the Lesson

Introducing the Problem Ask a student to read the problem aloud and tell what is to be found. (2 fractions that name the same amount as ⅔) Have students read the model as they shade the parts of the rectangles and complete the solution statement. Have students tell what part is shaded in each pictured circle. (⅔, ⁴⁄₆, ⁸⁄₁₂)

Developing the Skill Have students fold paper in half and color 1 of the halves. Tell students to fold the paper in half again and tell how many parts in all. (4) Ask students how many parts are colored. (2) Ask what part of the whole is colored. (²⁄₄) Tell students to fold again and tell how many parts. (8) Ask what part of the whole is colored. (⁴⁄₈) Write ½ = ²⁄₄ = ⁴⁄₈ on the board and tell students that the fractions ½, ²⁄₄ and ⁴⁄₈ are **equivalent** fractions because each describes the same part of the whole paper.

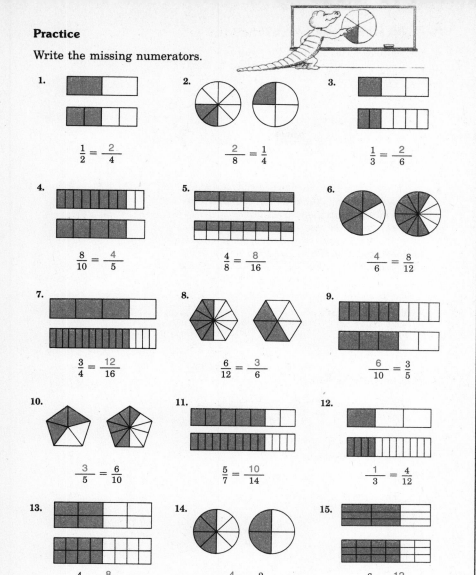

Practice

Write the missing numerators.

1.
$$\frac{1}{2} = \frac{2}{4}$$

2.
$$\frac{2}{8} = \frac{1}{4}$$

3.
$$\frac{1}{3} = \frac{2}{6}$$

4.
$$\frac{8}{10} = \frac{4}{5}$$

5.
$$\frac{4}{8} = \frac{8}{16}$$

6.
$$\frac{4}{6} = \frac{8}{12}$$

7.
$$\frac{3}{4} = \frac{12}{16}$$

8.
$$\frac{6}{12} = \frac{3}{6}$$

9.
$$\frac{6}{10} = \frac{3}{5}$$

10.
$$\frac{3}{5} = \frac{6}{10}$$

11.
$$\frac{5}{7} = \frac{10}{14}$$

12.
$$\frac{1}{3} = \frac{4}{12}$$

13.
$$\frac{4}{8} = \frac{8}{16}$$

14.
$$\frac{4}{8} = \frac{2}{4}$$

15.
$$\frac{6}{9} = \frac{12}{18}$$

248

Correcting Common Errors

Students may have difficulty finding a missing numerator when given diagrams for equivalent fractions. Have them work with partners with two rectangles, one separated into 4 equal parts, and the other into 16 equal parts. Have them color half of each rectangle. Ask: What fraction names half of the first rectangle? (2⁄4) What fraction names half of the second rectangle? (8⁄16) Write the following on the chalkboard: ½ = 2⁄4 = 8⁄16.

Enrichment

Tell students to find the equivalent fractions for the non-shaded portions in each problem on page 248.

Practice

Remind the students that the denominator tells the number of parts in all and the numerator tells the number of parts we are talking about. Have students complete the page independently.

Extra Credit *Sets*

Have students use the letters of their names as a set. Tell them to write a fraction to answer these questions:

1. What fraction of the letters in your first name are vowels?
2. What fractions of the letters are consonants?
3. What fraction of the letters in your first and last name are vowels?
4. What fraction are consonants?

For the next set of questions, tell students to use the names of all the students in the class.

5. What fraction of the first names in the class begin with the letter J? (or any other letter)
6. What fraction of the first names have 4 or less letters?
7. What fraction of the girl's names have more than 5 letters?

248

Find Equivalent Fractions

pages 249-250

Objective

To find equivalent fractions by multi-plying the numerator and denomi-nator by the same number

Materials

*rectangle

Mental Math

Ask students who is taller:

1. Joe is 49 in., Uri is 5 ft. (Uri)
2. Beth is 4 ft, Chun is 52 in. (Chun)
3. Zita is 1 m, Amit is 1,200 cm. (Amit)
4. Mick is 4½ ft, Don is 50 in. (Mick)
5. Mia is 1½ m, Tim is 1,600 cm. (Tim)
6. Al is 2 yd, Julio is 68 in. (Al)
7. Renee is 3½ ft, Cam is 1½ yd. (Cam)

Skill Review

Write **36** on the board. Ask students how many 2's are in 36. (18) Ask how many 3's (12), 4's (9), 6's (6), 9's (4) and 12's (3) are in 36. Repeat for 2's, 3's, 4's, 6's, 8's, 12's and 24's in 48.

Finding Equivalent Fractions

Bobbi discovered a shortcut for finding equivalent fractions. Use her shortcut to find the missing numbers.

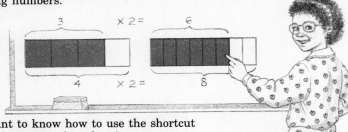

We want to know how to use the shortcut for finding equivalent fractions.
We can compare the shaded areas of the rectangles, to see what has happened to the numerator and denominator in equivalent fractions.

$$\frac{3}{4} = \frac{6}{8}$$

What number times 4 equals 8? What number times 3 equals 6?

$$\frac{3}{4} \times \frac{2}{2} = \frac{6}{8}$$

The numerator and the denominator are multiplied by the same number.

Usually, we know the denominator of an equivalent fraction.

What number times 3 equals 12?

$$\frac{2}{3} = \frac{8}{12}$$

Multiply the numerator by the same number.

Getting Started

Write the missing numerators.

1. $\frac{1}{4} = \frac{2}{8}$

2. $\frac{1}{3} = \frac{3}{9}$

3. $\frac{1}{6} = \frac{4}{24}$

4. $\frac{2}{3} = \frac{8}{12}$

5. $\frac{3}{4} = \frac{12}{16}$

6. $\frac{4}{5} = \frac{24}{30}$

7. $\frac{1}{2} = \frac{2}{4} = \frac{3}{6} = \frac{4}{8} = \frac{5}{10} = \frac{6}{12} = \frac{7}{14} = \frac{8}{16} = \frac{9}{18}$

249

Teaching the Lesson

Introducing the Problem Have a student read the problem aloud and tell what is to be done. (use the shortcut to find equivalent fractions) Ask students to tell what is known from the picture. (¾ is equivalent to ⁶⁄₈.) Have students read with you as they complete the fractions.

Developing the Skill Trace the rectangle to show 3 congruent rectangles on the board. Shade ½ of the 1st one and ²⁄₄ of the 2nd one. Tell students that the same amount of each rectangle is shaded, but there are 2 times as many parts in the 2nd one and therefore 2 times as many parts are shaded. Ask students how much of the 1st rectangle is shaded. (½) Ask what part of the 2nd rectangle is shaded. (²⁄₄) Write ½ = ²⁄₄ on the board. Tell students that the de-nominator of 4 is 2 times the denominator of 2 in the 1st fraction, and the numerator 2 is 2 times the numerator of 1. Now shade ⁴⁄₈ of the 3rd rectangle and ask students to tell the fraction. (⁴⁄₈) Write ½ = ²⁄₄ = ⁴⁄₈ on the board. Show

students that 8 is 2 × 4 and 4 is 1 × 4. Tell students that when we multiply both the numerator and the denominator by the same number, we are multiplying by 1 and multiply-ing any number by 1 does not change the number's value. Have students now use the multiplication shortcut to find an equivalent fraction for ⅞ in sixteenths and for ⁶⁄₇ in four-teenths.

249

Practice

Write the missing numerators.

1. $\frac{1}{4} = \frac{3}{12}$ 2. $\frac{5}{6} = \frac{15}{18}$ 3. $\frac{3}{5} = \frac{9}{15}$ 4. $\frac{4}{7} = \frac{12}{21}$

5. $\frac{2}{3} = \frac{8}{12}$ 6. $\frac{5}{6} = \frac{20}{24}$ 7. $\frac{5}{8} = \frac{15}{24}$ 8. $\frac{3}{10} = \frac{6}{20}$

9. $\frac{5}{9} = \frac{10}{18}$ 10. $\frac{7}{8} = \frac{56}{64}$ 11. $\frac{3}{9} = \frac{9}{27}$ 12. $\frac{3}{4} = \frac{12}{16}$

13. $\frac{4}{5} = \frac{16}{20}$ 14. $\frac{3}{7} = \frac{12}{28}$ 15. $\frac{5}{8} = \frac{10}{16}$ 16. $\frac{1}{9} = \frac{5}{45}$

17. $\frac{4}{7} = \frac{12}{21}$ 18. $\frac{5}{9} = \frac{30}{54}$ 19. $\frac{3}{4} = \frac{24}{32}$ 20. $\frac{4}{5} = \frac{24}{30}$

21. $\frac{6}{11} = \frac{18}{33}$ 22. $\frac{5}{12} = \frac{30}{72}$ 23. $\frac{3}{16} = \frac{18}{96}$ 24. $\frac{5}{24} = \frac{15}{72}$

25. $\frac{1}{3} = \frac{2}{6} = \frac{3}{9} = \frac{4}{12} = \frac{5}{15} = \frac{6}{18} = \frac{7}{21} = \frac{8}{24} = \frac{9}{27}$

26. $\frac{2}{3} = \frac{4}{6} = \frac{6}{9} = \frac{8}{12} = \frac{10}{15} = \frac{12}{18} = \frac{14}{21} = \frac{16}{24} = \frac{18}{27}$

27. $\frac{1}{4} = \frac{2}{8} = \frac{3}{12} = \frac{4}{16} = \frac{5}{20} = \frac{6}{24} = \frac{7}{28} = \frac{8}{32} = \frac{9}{36}$

28. $\frac{3}{4} = \frac{6}{8} = \frac{9}{12} = \frac{12}{16} = \frac{15}{20} = \frac{18}{24} = \frac{21}{28} = \frac{24}{32} = \frac{27}{36}$

29. $\frac{1}{5} = \frac{2}{10} = \frac{3}{15} = \frac{4}{20} = \frac{5}{25} = \frac{6}{30} = \frac{7}{35} = \frac{8}{40} = \frac{9}{45}$

30. $\frac{1}{6} = \frac{2}{12} = \frac{3}{18} = \frac{4}{24} = \frac{5}{30} = \frac{6}{36} = \frac{7}{42} = \frac{8}{48} = \frac{9}{54}$

250

Correcting Common Errors

Some students may not multiply the numerator by the same number used for the denominator. When students are working with a problem such as $\frac{2}{9} = \frac{\square}{18}$, ask them to write the factors so they multiply the numerator of the first fraction by the correct number.
$$\frac{2 \times 2}{9 \times 2} = \frac{4}{18}$$

Enrichment

Tell students to write 10 equivalent fractions for each of the following: $\frac{4}{10}$, $\frac{7}{12}$, $\frac{3}{11}$.

Practice

Remind students that they will be finding equivalent fractions. Have students complete the page independently.

Mixed Practice

1. $259 \div 35$ (7 R14)
2. $356 + 4,279$ (4,635)
3. $4,000 - 1,356$ (2,644)
4. $\$15.29 + 3.82 + 9.47$ (\$28.58)
5. 26×401 (10,426)
6. $107 \div 9$ (11 R8)
7. $\$2.53 \times 8$ (\$20.24)
8. $25,376 + 7,208$ (32,584)
9. 78×36 (2,808)
10. $14 + 7 \times 2$ (28)

Extra Credit   *Numeration*

Have students research the Babylonian number system, one through ten. Help students make Babylonian numerals on clay. Have them use a pencil to carve these shapes:

      ᴠ for ones   ⟨ for tens

Explain that the Babylonians grouped their symbols to make them easier to read. For example: ⟨⟨ ᴠᴠ

Have each member of the class write the following numbers on their clay "tablets":

    95   34   64   81

This activity could be completed more quickly, if necessary, using paper and pencils or crayons.

Comparing Fractions

pages 251-252

Objective

To find the greater of 2 fractions

Materials

Mental Math

Have students find:

1. ⅖ of 100. (40)
2. ³⁄₁₀ of 20. (6)
3. 2/7 of 91. (26)
4. ⅝ of 96. (60)
5. 6/9 of 108. (72)
6. 4/9 of 54. (24)
7. 7/12 of 144. (84)
8. ¹⁄₁₀₀ of 1,800. (18)

Skill Review

Draw a number line from 0 to 1 on the board. Divide the line into twelfths and write ³⁄₁₂ in place. Have students write the additional twelfths as you dictate the fractions in random order. Repeat for a number line divided into forty-eighths.

Comparing Fractions

Elaina and Kurt are trying to find out which fraction is greater, $\frac{3}{4}$ or $\frac{2}{3}$. Help them compare the fractions.

We want to know if $\frac{3}{4}$ is greater or less than $\frac{2}{3}$.

We can draw two number lines.

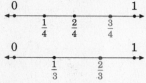

$\frac{3}{4}$ is closer to 1. Therefore $\frac{3}{4}$ is _____greater_____ than $\frac{2}{3}$.

We write: $\frac{3}{4} > \frac{2}{3}$.

We can also say $\frac{2}{3}$ **is less than** $\frac{3}{4}$. We write: $\frac{2}{3} < \frac{3}{4}$.

✔ Numbers to the right on number lines are always greater. Numbers to the left are always less.

We can also find equivalent fractions that have the same denominator. Then we can compare the numerators.

$$\frac{3 \times 3}{4 \times 3} = \frac{9}{12} \qquad \frac{2 \times 4}{3 \times 4} = \frac{8}{12} \qquad \frac{9}{12} > \frac{8}{12}$$

Getting Started

Use the number lines to compare the fractions.
Write < or > in the circle.

1. $\frac{1}{2} \bigcirc \frac{1}{4}$

2. $\frac{3}{5} \bigcirc \frac{2}{3}$

Use equivalent fractions to compare the fractions.
Write < or > in the circle.

3. $\frac{3}{4} \bigcirc \frac{3}{8}$

4. $\frac{3}{8} \bigcirc \frac{1}{2}$

5. $\frac{2}{3} \bigcirc \frac{5}{6}$

6. $\frac{5}{6} \bigcirc \frac{4}{5}$

251

Teaching the Lesson

Introducing the Problem Have a student read the problem aloud and tell what is to be solved. (find the greater fraction: ¾ or ⅔) Have a student tell about the picture. Have students tell how many pieces would be ¾ (3) and how many would be ⅔. (2) Have students complete the number lines, sentence and fractions with you. Ask students what conclusion Elaina and Kurt should come to.

Developing the Skill Remind students that numbers on a number line are always greater to the right and less to the left. Draw two number lines from 0 to 1 on the board. Divide the top number line into fifths and show ⅗. Divide the bottom number line into tenths and show ⁴⁄₁₀. Tell students that ⅗ is greater because it is closer to 1. Tell students we can also find out which number is greater by finding a multiple of both 5 and 10 to show ⅗ and ⁴⁄₁₀ as fractions with the same denominator. Ask students what number is a multiple of 5 and 10. (Answers will vary.) Show students how ⁶⁄₁₀ and ⁴⁄₁₀, ⁸⁄₂₀ and ¹²⁄₂₀, etc. can be used to compare the numbers to find the greater. Repeat to compare ⅖ and 3/7.

Practice

Use the number lines to compare the fractions.
Write < or > in the circle.

1. $\frac{1}{3}$ ⊘ $\frac{1}{5}$ (>)

2. $\frac{3}{5}$ ⊘ $\frac{2}{8}$ (>)

3. $\frac{4}{6}$ ⊘ $\frac{3}{4}$ (<)

4. $\frac{1}{2}$ ⊘ $\frac{6}{10}$ (<)

5. $\frac{5}{8}$ ⊘ $\frac{2}{3}$ (<)

6. $\frac{7}{10}$ ⊘ $\frac{3}{5}$ (>)

Use equivalent fractions to compare the fractions.
Write < or > in the circle.

7. $\frac{5}{6}$ (>) $\frac{1}{2}$

8. $\frac{5}{8}$ (<) $\frac{3}{4}$

9. $\frac{2}{3}$ (>) $\frac{3}{12}$

10. $\frac{1}{2}$ (>) $\frac{4}{16}$

11. $\frac{1}{4}$ (>) $\frac{1}{12}$

12. $\frac{1}{4}$ (<) $\frac{3}{8}$

13. $\frac{5}{9}$ (>) $\frac{1}{2}$

14. $\frac{3}{5}$ (<) $\frac{5}{6}$

15. $\frac{5}{12}$ (<) $\frac{1}{2}$

16. $\frac{7}{9}$ (>) $\frac{2}{3}$

17. $\frac{5}{6}$ (>) $\frac{7}{12}$

18. $\frac{3}{4}$ (<) $\frac{4}{5}$

19. $\frac{1}{2}$ (>) $\frac{1}{3}$

20. $\frac{3}{7}$ (>) $\frac{3}{8}$

21. $\frac{2}{3}$ (<) $\frac{3}{4}$

22. $\frac{3}{8}$ (<) $\frac{5}{6}$

Apply

Solve these problems.

23. Cleve ran $\frac{2}{5}$ of a mile. Gary ran $\frac{5}{10}$ of a mile. Who ran farther?
    Gary

24. Jennie read for $\frac{3}{4}$ of an hour. Myra read for $\frac{5}{6}$ of an hour. Who read longer?
    Myra

252

252

Fractions to Lowest Terms

pages 253-254

Objective

To simplify a fraction to its lowest terms

Materials

Mental Math

Ask students how many years from:

1. present to 2100
2. 1776 to present
3. 1492 to present
4. your birth year to 2000
5. present to 2001
6. 1865 to present
7. 1945 to present
8. 1929 to present

Skill Review

Have students write equivalent fractions for ½ and ⅓ until they find fractions which have the same denominator. (³⁄₆, ²⁄₆) Have students tell if ½ is >, <, or = to ⅓. (>) Repeat for equivalent fractions for ⅑ and ⅙ (²⁄₁₈, ³⁄₁₈) and ⅓ and ⅕. (⁵⁄₁₅, ³⁄₁₅)

Finding Simplest Terms

Mr. Granger bought 18 cans of motor oil on sale. What fraction of the case of oil did he buy?

We want to know what part of a case Mr. Granger bought.

Mr. Granger bought $\frac{18}{}$ cans of oil.

There are $\frac{24}{}$ cans in a case.

✔ Remember that the denominator names the total number of parts, and the numerator names the number of parts you are counting.

$$\frac{18}{24} = \frac{\text{the number of cans bought}}{\text{the number of cans in a case}}$$

This fraction can be **simplified**. The numerator and the denominator of a fraction are called the **terms** of a fraction. To simplify a fraction, we name it in its lowest terms. To simplify a fraction, we divide the numerator and the denominator by the same number.

$$\frac{18 \div 6}{24 \div 6} = \frac{3}{4}$$

✔ A fraction is in its lowest terms when the terms cannot be divided by any common factor other than 1.

Mr. Granger bought $\frac{3}{4}$ of a case of oil.

Getting Started

Simplify each fraction.

1. $\frac{6}{8} = \frac{3}{4}$ 2. $\frac{5}{15} = \frac{1}{3}$ 3. $\frac{3}{9} = \frac{1}{3}$ 4. $\frac{10}{20} = \frac{1}{2}$ 5. $\frac{16}{24} = \frac{2}{3}$

253

Teaching the Lesson

Introducing the Problem Have a student read the problem and tell what is to be found. (what part of the case of oil Mr. Granger bought) Ask what data is known. (He bought 18 cans and a case contains 24 cans.) Have students read aloud with you as they complete the sentences to solve the problem. Reinforce the explanation of simplified terms.

Developing the Skill Remind students that multiplying a fraction by any form of 1, does not change the value of the fraction. Tell students that dividing a fraction by any form of 1, also does not change the value of the fraction. Tell students that the numerator and denominator of a fraction are called its **terms**. Write **24/40** on the board and tell students that if there is a number other than 1 that can be divided evenly into both the numerator and denominator, then the fraction is not **simplified** or in **lowest terms**. Ask students if 24 and 40 can both be divided by 2. (yes) Ask

students if there is any larger number which can be divided into both 24 and 40. (4,8) Simplify the fraction by dividing by 2 repeatedly and then show students how much quicker it is to bring a fraction to lowest terms if we find the largest divisor possible. Have students find the lowest terms for the fractions, ⁶⁄₉ (⅔), ¹⁶⁄₂₀ (⅘), ³⁰⁄₄₅ (⅔) and ¹⁷⁄₅₁ (⅓). Then write various fractions on the board and have students simplify them.

Practice

Simplify each fraction.

1. $\frac{10}{15} = \frac{2}{3}$
2. $\frac{6}{9} = \frac{2}{3}$
3. $\frac{4}{12} = \frac{1}{3}$
4. $\frac{5}{10} = \frac{1}{2}$
5. $\frac{6}{18} = \frac{1}{3}$

6. $\frac{4}{20} = \frac{1}{5}$
7. $\frac{4}{24} = \frac{1}{6}$
8. $\frac{4}{16} = \frac{1}{4}$
9. $\frac{8}{12} = \frac{2}{3}$
10. $\frac{9}{18} = \frac{1}{2}$

11. $\frac{6}{12} = \frac{1}{2}$
12. $\frac{14}{16} = \frac{7}{8}$
13. $\frac{15}{25} = \frac{3}{5}$
14. $\frac{6}{10} = \frac{3}{5}$
15. $\frac{3}{12} = \frac{1}{4}$

16. $\frac{16}{20} = \frac{4}{5}$
17. $\frac{9}{12} = \frac{3}{4}$
18. $\frac{24}{48} = \frac{1}{2}$
19. $\frac{8}{16} = \frac{1}{2}$
20. $\frac{8}{32} = \frac{1}{4}$

21. $\frac{10}{12} = \frac{5}{6}$
22. $\frac{16}{24} = \frac{2}{3}$
23. $\frac{4}{8} = \frac{1}{2}$
24. $\frac{27}{36} = \frac{3}{4}$
25. $\frac{12}{16} = \frac{3}{4}$

EXCURSION

Find the missing digits.

1.
```
    7 , 0  4  2
 -  3 , 5  6  5
    3 , 4  7  7
```

2.
```
    2  4 , 0  6  8
       4  9 , 4  7  3
  +  1  5 , 7  3  4
       8  9 , 2  7  5
```

3.
```
    3  6 , 4  7  3
    1  7 , 2  8  7
  +  1  9 , 9  6  4
    7  3 , 7  2  4
```

254

Practice

Remind students to divide both terms of the fraction by the same number. Have students complete the page independently.

Excursion

Guide students through the first problem to see how they must use inverse operation to find the missing digit. They subtract 2 from 7 to find what must be added to 2 to equal 7.

Extra Credit   *Geometry*

Ask students to collect pictures of bridges and bring them to class. If there is a bridge near the school, suggest that students visit the bridge and draw it as best they can. Display all the photos, copies, and drawings. Have students study the display and write a description of the shapes they see repeated in many of the bridges. (As students may already have discovered, a triangle is a very rigid shape. Bridge struts often form patterns of triangles. Arches are a common shape in masonry bridges.)

Mixed Numbers

pages 255-256

Objective

To write a whole or mixed number for a fraction greater than or equal to 1

Materials

*congruent circles in halves, thirds, etc.

Mental Math

Ask students is there a remainder?

1. $553 \div 5$ (yes)
2. $78 \div 3$ (no)
3. $686 \div 2$ (no)
4. $1,003 \div 5$ (yes)
5. $703 \div 3$ (yes)
6. $211 \div 2$ (yes)
7. $985 \div 5$ (no)
8. $72,403 \div 3$ (yes)

Skill Review

Have students simplify the following dictated fractions: $^{40}/_{60}$ ($^2/_3$), $^{35}/_{50}$ ($^7/_{10}$), $^{12}/_{18}$ ($^2/_3$), $^{12}/_{24}$ ($^1/_2$), $^{15}/_{45}$ ($^1/_3$), $^{18}/_{72}$ ($^1/_4$), $^{25}/_{55}$ ($^5/_{11}$), $^{93}/_{99}$ ($^{31}/_{33}$), $^{54}/_{72}$ ($^3/_4$), $^{24}/_{40}$. ($^3/_5$)

Understanding Mixed Numbers

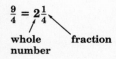

Miguel is on the track team. His coach told him to run 9 laps during practice. How many miles did Miguel run?

We want to know how far Miguel ran.

Miguel ran $\underline{9}$ laps.

Each lap is $\underline{\frac{1}{4}}$ mile long.
We can use a number line to help us understand this distance.

$$0 \quad\quad\quad 1 \; 1\tfrac{1}{4} \; 1\tfrac{2}{4} \; 1\tfrac{3}{4} \; 2 \; 2\tfrac{1}{4} \; 2\tfrac{2}{4} \; 2\tfrac{3}{4} \; 3$$

$$\tfrac{1}{4} \; \tfrac{2}{4} \; \tfrac{3}{4} \; \tfrac{4}{4} \; \tfrac{5}{4} \; \tfrac{6}{4} \; \tfrac{7}{4} \; \tfrac{8}{4} \; \tfrac{9}{4} \; \tfrac{10}{4} \; \tfrac{11}{4} \; \tfrac{12}{4}$$

Miguel ran a total of $\frac{9}{4}$ miles.

✔ A fraction whose numerator is larger than its denominator can be renamed as a **mixed number** by using division. We divide the numerator by the denominator.

We divide $\underline{9}$ by $\underline{4}$.

$$\frac{9}{4} = 4\overline{)9} \quad \leftarrow \text{whole number}$$
$$\underline{8}$$
$$1 \leftarrow \text{number of fourths left}$$

$$\frac{9}{4} = 2\tfrac{1}{4}$$
whole number   fraction

$2\tfrac{1}{4}$ is called a **mixed number.**

Miguel ran $\underline{2\tfrac{1}{4}}$ miles.

✔ Remember to simplify fractions.

$$\frac{16}{6} = 2\tfrac{4}{6} = 2\tfrac{2}{3} \quad\quad \frac{15}{3} = 5$$

Getting Started

Rename each fraction as a whole or mixed number, and simplify.

1. $\frac{7}{3} = 2\tfrac{1}{3}$
2. $\frac{5}{2} = 2\tfrac{1}{2}$
3. $\frac{10}{8} = 1\tfrac{1}{4}$
4. $\frac{8}{4} = 2$
5. $\frac{16}{10} = 1\tfrac{3}{5}$

255

Teaching the Lesson

Introducing the Problem   Have a student read the problem aloud and tell what is being asked. (how many miles Miguel ran) Ask what information is known. (He ran 9 laps and each lap is ¼ mile.) Have students study the number line and complete the sentences with you. Have students read each checked statement with you and complete the solution sentence.

Developing the Skill   Review the number line on page 255. Write ⁷⁄₂ on the board. Tell students this fraction is more than the whole number 1. Tell students we read this fraction as seven halves and, since we know there are 2 halves in 1 whole, we divide the 7 by 2 to see how many wholes are in 7 halves. Show the division to find there are 3 wholes and 1 left-over half. Tell students that 3 ½ is called a **mixed number.** Write ¹⁸⁄₄ on the board. Have a student divide 18 by 4. (4 ²⁄₄) Ask students if the ²⁄₄ is in simplified terms. (no) Have a student simplify the fraction and then tell the mixed number. (4 ½) Have students rename, and simplify if necessary, each of the following fractions as whole numbers or mixed numbers: ¹⁴⁄₄ (3 ½), ²⁶⁄₁₂ (2 ⅙), ¹⁶⁄₄ (4) and ²⁸⁄₁₆. (1 ¾)

255

Practice

Rename each fraction as a whole or mixed number, and simplify.

1. $\frac{5}{4} = 1\frac{1}{4}$ 2. $\frac{8}{3} = 2\frac{2}{3}$ 3. $\frac{6}{4} = 1\frac{1}{2}$ 4. $\frac{7}{2} = 3\frac{1}{2}$ 5. $\frac{9}{3} = 3$

6. $\frac{8}{5} = 1\frac{3}{5}$ 7. $\frac{12}{8} = 1\frac{1}{2}$ 8. $\frac{14}{6} = 2\frac{1}{3}$ 9. $\frac{16}{3} = 5\frac{1}{3}$ 10. $\frac{18}{4} = 4\frac{1}{2}$

11. $\frac{25}{10} = 2\frac{1}{2}$ 12. $\frac{30}{12} = 2\frac{1}{2}$ 13. $\frac{30}{9} = 3\frac{1}{3}$ 14. $\frac{24}{16} = 1\frac{1}{2}$ 15. $\frac{21}{7} = 3$

16. $\frac{44}{8} = 5\frac{1}{2}$ 17. $\frac{16}{10} = 1\frac{3}{5}$ 18. $\frac{63}{12} = 5\frac{1}{4}$ 19. $\frac{40}{6} = 6\frac{2}{3}$ 20. $\frac{86}{20} = 4\frac{3}{10}$

21. $\frac{16}{5} = 3\frac{1}{5}$ 22. $\frac{14}{10} = 1\frac{2}{5}$ 23. $\frac{18}{8} = 2\frac{1}{4}$ 24. $\frac{36}{6} = 6$ 25. $\frac{42}{9} = 4\frac{2}{3}$

26. $\frac{16}{7} = 2\frac{2}{7}$ 27. $\frac{72}{9} = 8$ 28. $\frac{9}{5} = 1\frac{4}{5}$ 29. $\frac{29}{10} = 2\frac{9}{10}$ 30. $\frac{49}{7} = 7$

Correcting Common Errors

When students write a fraction as a mixed number, they may write the whole-number part and forget the fraction part. Have students work in pairs rewriting a problem such as $\frac{5}{4}$ as $\frac{4}{4} + \frac{1}{4} = 1 + \frac{1}{4}$, or $1\frac{1}{4}$.

Enrichment

Tell students to rename the following fractions and simplify if necessary: $\frac{216}{42}$ ($5\frac{1}{7}$), $\frac{378}{189}$ (2), $\frac{500}{200}$ ($2\frac{1}{2}$), $\frac{1000}{300}$ ($3\frac{1}{3}$), $\frac{896}{7}$. (128)

EXCURSION

Fill in the blanks and circles with numbers and signs that correctly complete the pattern.

$9 \times 9 = \underline{81}$ $8 + 1 = \underline{9}$

$9 \times 8 = \underline{72}$ $7 + 2 = \underline{9}$

$9 \times 7 = \underline{63}$ $6 \oplus 3 = \underline{9}$

$9 \times 6 = \underline{54}$ $5 \oplus 4 = \underline{9}$

$9 \times 5 = \underline{45}$ $4 \oplus 5 = \underline{9}$

Write a sentence that describes the pattern.

If we add the digits of the product of 9 and a

single-digit number, the sum should be 9.

Practice

Remind students to simplify any fractions which are part of mixed numbers. Have students complete the problems independently.

Excursion

When any number has been multiplied by 9, the sum of the digits in the answer is always 9. Have the students extend the pattern through 9×2.

Extra Credit *Biography*

Duplicate the following for students: Emmy Noether was a German mathematician of the twentieth century. She taught mathematics both in Europe and in the United States. Some famous mathematicians she worked with included: Weyl, Hasse, Klein, Brauer, Hilbert and Einstein. She taught at Bryn Mawr and Princeton Universities in the United States, and at Moscow and Gottingen Universities in Russia and Europe. Using this information about Noether, fill in missing

words below, using the clues given. When you finish, the boxed letters will spell the part of mathematics in which Emmy Noether did her best work.

(B r [a] u e r)   a friend
(W e y [l])   a friend
([G] o t t i n g e n) a university in Europe
(E i n s t [e] i n)   a friend
(H i l [b] e r t)   a friend
(P [r] i n c e t o n)   a university in the U.S.
(B r y n M [a] w r)   a university in the U.S.

Fractions in Measurement

pages 257-258

Objective

To measure lengths to the nearest ¼-inch

Materials

*large paper ruler marked in ¼ inches
rulers marked in ¼ inches

Mental Math

Have students give an equivalent fraction for:

1. ½ (³⁄₆, etc.)
2. ⅓ (²⁄₆, etc.)
3. ¼ (²⁄₈, etc.)
4. ⅕ (²⁄₁₀, etc.)
5. ⅔ (⁴⁄₆, etc.)
6. ¾ (⁹⁄₁₂, etc.)
7. ⅝ (¹⁰⁄₁₆, etc.)
8. ¹⁄₁₀ (¹⁰⁄₁₀₀, etc.)

Skill Review

Have students write a whole or simplified mixed number for the following dictated fractions: ⁹⁄₄ (2 ¼), ¹⁷⁄₄ (4 ¼), ²²⁄₄ (5 ½), ³⁶⁄₄ (9), ⁴²⁄₄. (10 ½)

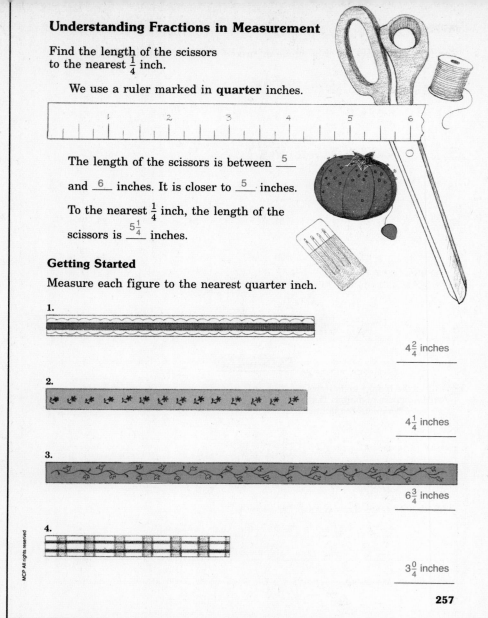

Understanding Fractions in Measurement

Find the length of the scissors to the nearest $\frac{1}{4}$ inch.

We use a ruler marked in **quarter** inches.

The length of the scissors is between __5__

and __6__ inches. It is closer to __5__ inches.

To the nearest $\frac{1}{4}$ inch, the length of the

scissors is __$5\frac{1}{4}$__ inches.

Getting Started

Measure each figure to the nearest quarter inch.

1. $4\frac{2}{4}$ inches

2. $4\frac{1}{4}$ inches

3. $6\frac{3}{4}$ inches

4. $3\frac{0}{4}$ inches

257

Teaching the Lesson

Introducing the Problem Have a student read the problem and tell what is to be done. (Measure the scissors to the nearest ¼-inch.) Have students complete the sentences as they read aloud with you to solve the problem.

Developing the Skill Post the large paper ruler on the board. Ask students how many fourth-inch spaces are between each 2 numbers on the ruler. (4) Tell students that each of those spaces measures a quarter inch. Write **¼ = 1** on the board to remind students that 4 fourths equals 1 whole. Tell students that 4 quarter inches equals 1 whole inch. Have students find various points such as 2 ½, 3 ¾ and 5 ¼ inches on the large ruler. Use a pointer to locate a point on the ruler that is approximately 5 ⁵⁄₁₆ inches and have students tell if the point is closer to 5 ¼ inches or 5 ½ inches. (5 ¼) Repeat for other measurements to be found to the nearest quarter-inch. Include some points closer to the whole inch for students to name the measurement closest to a whole number and zero fourths.

Practice

Measure each figure to the nearest quarter inch.

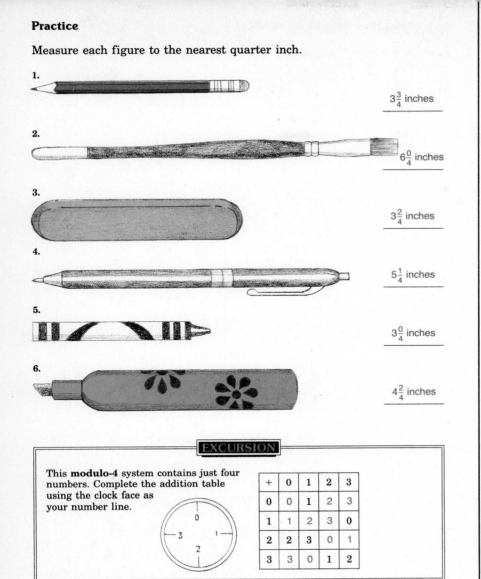

1. $3\frac{3}{4}$ inches

2. $6\frac{0}{4}$ inches

3. $3\frac{2}{4}$ inches

4. $5\frac{1}{4}$ inches

5. $3\frac{0}{4}$ inches

6. $4\frac{2}{4}$ inches

EXCURSION

This **modulo-4** system contains just four numbers. Complete the addition table using the clock face as your number line.

+	0	1	2	3
0	0	1	2	3
1	1	2	3	0
2	2	3	0	1
3	3	0	1	2

258

Practice

Remind students to record each measurement to the nearest fourth-inch. Have students complete the measurements independently.

Excursion

Define a modulus as a closed system of numbers. To demonstrate the use of the clock face as a closed number line, perform addition by starting at the first addend and moving clockwise as many places as the second addend. Students can construct a modulo-5 clock face and complete its addition table.

Extra Credit   *Statistics*

Present the class with a list of anonymous scores for a test, perhaps this week's spelling test. List the scores in random order. Have them group the scores in five point categories: 0–5, 6–10, 11–15, 16–20, 21–25, 26–30, 31–35, 36–40, and so on. Now have them tally the number of scores within each five point range. Explain that the number of scores in each category is called the "frequency." Discuss with students and have them list on the board when this kind of frequency tally might be made. (monitoring traffic on a roadway, tracking season totals for a basketball team, etc.)

Ratio

pages 259-260

Objective

To use a ratio to compare 2 quantities

Materials

counters in 2 colors

Mental Math

Have students solve for N:

1. $5/8 = {}^n/16$ (10)
2. $2/3 = {}^n/18$ (12)
3. $^n/5 = 4/10$ (2)
4. $1/2 = {}^n/100$ (50)
5. $4/n = 8/20$ (10)
6. $7/6 = 1 - {}^n/6$ (1)
7. $2-5/6 = {}^{17}/n$ (6)
8. $1/n = 4/8$ (2)

Skill Review

Have students compare $7/8$ and $3/4$ and tell which fraction is greater. Repeat for more comparisons of simple fractions.

Understanding Ratios

A **ratio** is a comparison of two quantities. The ratio of black pins to all pins on the bulletin board is 3 to 9. This can also be written as the fraction $\frac{3}{9}$. The ratio of black pins to white pins is 3 to 4, or $\frac{3}{4}$. What is the ratio of red pins to black pins?

We want to compare the number of red pins to the number of black pins.

There are ___2___ red pins.

There are ___3___ black pins.

The ratio of red pins to black pins is ___2___

to ___3___, or the fraction ___$\frac{2}{3}$___.

Getting Started

Write each comparison as a fraction.

1. white pins to black pins
   $\frac{4}{3}$

2. white pins to all pins
   $\frac{4}{9}$

3. red pins to black pins
   $\frac{2}{3}$

4. black pins to red pins
   $\frac{3}{2}$

Write each comparison as a ratio.

5. red pins to all pins
   2 to 9

6. red pins to white pins
   2 to 4

259

Teaching the Lesson

Introducing the Problem Have a student read the problem aloud and tell what is to be found. (Write a **ratio** to compare the number of red pins to black pins.) Ask what information is given. (There are 3 black pins and 2 red pins.) Ask students to tell what is known about a ratio. (It is a comparison of 2 quantities and can be written as a fraction.) Have students read with you as they complete the sentences.

Developing the Skill Draw on the board:

XXX   RR

XX   R

Tell students we want to compare the number of X's to the number of R's. Write on the board:

X   R

5   to   3   or $5/3$

Tell students that the ratio of 5 to 3 or $5/3$ compares the X's to the R's. Now tell students that to compare the number of R's to X's we give the number of R's first and then the X's. Write on the board:

R   X

3   to   5   or $3/5$

Have students compare the number of R's to all the letters. (3 to 8 or $3/8$) Repeat for number of X's to all letters. (5 to 8 or $5/8$) Add more X's and/or R's and have students write ratios to compare the new numbers.

259

Practice

Write each comparison as a fraction.

1. white marbles to all marbles $\frac{6}{12}$

2. white marbles to red marbles $\frac{6}{4}$

3. black marbles to red marbles $\frac{2}{4}$

4. red marbles to white marbles $\frac{4}{6}$

5. black marbles to all marbles $\frac{2}{12}$

6. red marbles to black marbles $\frac{4}{2}$

7. white marbles to black marbles $\frac{6}{2}$

8. red marbles to all marbles $\frac{4}{12}$

Write each comparison as a ratio.

9. truck to cars

_3 to 2_

10. cars to engines

_2 to 4_

11. engines to all vehicles

_4 to 9_

12. engines to trucks

_4 to 3_

13. trucks to all vehicles

_3 to 9_

14. trucks to engines

_3 to 4_

15. trucks and cars to all vehicles

_5 to 9_

16. cars to trucks and engines

_2 to 7_

260

Practice

Have students complete the page independently.

Mixed Practice

1. $2{,}056 \div 81$ (25 R31)
2. 320×42 (13,440)
3. $\frac{3}{5}$ of 25 (15)
4. $6{,}021 - 4{,}317$ (1,704)
5. $\frac{4}{5} = \frac{}{35}$ (28)
6. $3{,}658 + 23 + 176$ (3,857)
7. $253 \div 5$ (50R3)
8. $\frac{}{4} = \frac{12}{16}$ (3)
9. 8×700 (5,600)
10. $23{,}156 - 6{,}080$ (17,076)

260

Problem Solving
Select Notation

pages 261-262

Objective

To select a suitable notation to record
and solve a problem

Materials

Mental Math

Have students simplify:

1. 2/6 (1/3)
2. 4/8 (1/2)
3. 4/10 (2/5)
4. 3/9 (1/3)
5. 6/9 (2/3)
6. 5/10 (1/2)
7. 4/24 (1/6)
8. 8/20 (2/5)

Selecting Notation

Jack's mother is making pudding which calls for
4 cups of milk. Show how she gets this amount
using her 7-cup and 5-cup measures.

★ **SEE**

We want to show how Jack's mother measures 4
cups of milk.

★ **PLAN**

We use ordered pairs to represent the amount of
milk in the measuring cups. The first number
represents the amount in the 7-cup measure.
The second number represents the amount in
the 5-cup measure.

★ **DO**

(7, 0)   She fills the ___7___-cup measure.

(2, 5)   She pours ___5___ cups from the 7-cup into the ___5___-cup.

(2, 0)   She empties the ___5___-cup into the carton.

(0, 2)   She pours ___2___ cups from the ___7___-cup to the ___5___-cup.

(7, 2)   She refills the ___7___-cup.

(4, 5)   She pours ___3___ cups from the ___7___-cup into the ___5___-cup.

Jack's mother now has exactly ___4___ cups of milk
in the first measuring cup.

★ **CHECK**

We can check our work by reviewing each step
in the solution and by asking ourselves these questions:

1. Did we use more than 7 cups
   for the first measuring cup
   in any step? __No__

2. Did we use more than 5 cups
   for the second measuring cup
   in any step? __No__

261

Teaching the Lesson

Tell students that some problems require us to use symbols
to represent data and that symbols can be numbers, letters
or abstract diagrams. Tell students that 2 people solving the
same problem may decide to use different but appropriate
symbols to represent the data. Give students some examples
such as using the symbols P for penny, D for dime, when
dealing with a money problem, etc. or using a 1 to desig-
nate a favorite song, and a 2 for the second favorite. Help
students come up with other examples of symbols. Tell students that the use of symbols helps us record information as
we work toward solving a problem. Have a student read the
problem aloud and tell what is to be done. (Show how to
measure out exactly 4 cups of milk using only a 7- and
5-cup measure.) Have students read aloud with you through
the Plan stage. Write **7 cup 5 cup** on the board and tell
students that we will write a number in each column to
show the amount of milk in that container. Tell students the
2 numbers will always be in the same order, with the first

number meaning the contents of the 7-cup measure and the
second number meaning the contents of the 5-cup measure.
Have a student read the first step of the Do stage aloud and
pause for restating of the meaning of the ordered pair of
numbers. Have students complete the sentence. Continue
through the Do stage to solve the problem. Have students
read the Check stage aloud with you as they complete the
problems and answer the questions.

Apply

Select notation to help you solve these problems.

1. Fran and Beth gathered blackberries and filled an 8-quart pail. They want to divide the berries equally. Besides the 8-quart pail, they have a 5-quart pail and a 3-quart pail. Describe how they can divide the berries equally.
See Solution Notes.

2. Arthur Ant wants to crawl from A to G on this cube. How many different 3-sided trips can he make? He must stay on an edge of the cube at all times.
6 trips

3. Terry has a 3-ounce and a 7-ounce container. Her recipe calls for exactly 5 ounces of water. Using only these two containers, show how Terry can measure exactly 5 ounces of water.
See Solution Notes.

4. The Chess Club is having a tournament. Each of the 7 members will play every other member one time. The champion will be the member with the most wins. How many games will be played?
21 games

5. Jodi's favorite clothes include four sweatshirts, three pairs of jeans and two pairs of tennis shoes. How many days in a row can she wear a different outfit using her favorite clothes?
24 days

6. At Washington School, for every 7 students who walked to school, 25 rode a school bus. At Lincoln School, the ratio was 9 to 30. At which school is the ratio of walkers to riders greater? Prove your answer.
At Washington School

7. In a math contest, the first 10 students to give an answer to the following challenge were given a puzzle book for a prize. "Show 3 fractions in order from least to greatest where the numerators are arranged from greatest to least." What answer would you give?
Answers will vary.

8. Tom and Juan were playing a game making up special numbers. To make up the first number, Tom said 3 and Juan said 5. To make up another number, Tom said 6 and Juan said 10. Then Tom said 9 and Juan said 15. What kind of numbers are they making up?
See Solution Notes.

262

Extra Credit   *Creative Drill*

Have each student bring an old calendar to class. Have students complete any or all of the following activities, using their calendars. Tell students to:

1. Draw a ring around the multiples of 7. What pattern do you see? (The multiples of 7 are in a row.) Will this pattern always be the same? (yes)

2. Complete the following column addition: the sum of all Tuesdays; the sum of all Fridays; the sum of the last full week of the month.

3. Box in a 3 × 3 square on the calendar and find the average of the numbers in the box. Repeat for another 3 × 3 box. What pattern do you see? (The average is the number in the center square.) Test the pattern in another 3 × 3 box.

Challenge students to come up with their own calendar activities.

Solution Notes

1. In this solution the first number represents the amount of berries in the 8 qt pail, the second number represents the 5 qt pail and the third number represents the 3 qt pail: (3,5,0), (3,2,3), (6,2,0), (6,0,2), (1,5,2) (1,4,3), (4,4,0).

2. Students should use the corner letters to denote possible trips: ADHG, ADCG, ABFG, ABCG, AEFG, AEHG

3. Students will recognize this problem as being similar to the problem in the lesson. One solution is: (7,0), (4,3), (4,0), (1,3), (1,0), (0,1), (7,1), (5,3).

4. Students may use a more abstract notation to solve this problem: Discuss the pattern that is evident in the solution. (6 + 5 + 4 + 3 + 2 + 1 = 21)

5. After solving the problem, talk about the number of each of the items and the relationship they have with the solution. (4 × 3 × 2 = 24) Shirts, A,B,C,D,; Jeans 1,2,3; Shoes X,Y: Day 1-A,1,X; Day 2-A,1,Y: Day 3-A,2,X; Day 4-A,2,Y; Day 5-A,3,X; Day 6-A,3,Y. Repeat for each of the 4 sweatshirts: 6 × 4 = 24 days.

Higher-Order Thinking Skills

6. [Evaluation] Proofs will vary but might include answers such as the fraction $7/25$ is greater than the fraction $9/30$, or the quotient of 25 ÷ 7 is greater than the quotient of 30 ÷ 9.

7. [Synthesis] A possible answer is $14/28$, $13/24$, $12/22$.

8. [Analysis] They are making up equivalent fractions; Tom is naming numerators, and Juan is naming denominators.

Chapter Test

page 263

Item	Objective
1-2	Write a fraction for part of whole or part of set (See pages 241-242)
3	Find unit fraction of a number (See pages 243-244)
4-5	Find a fraction of number by multiplying and dividing (See pages 245-246)
6-9	Find equivalent fractions by multiplying (See pages 249-250)
10-13	Compare and order fractions (See pages 251-252)
14-17	Reduce a fraction to lowest terms (See pages 253-254)
18-21	Write an improper fraction as mixed number (See pages 255-256)
22-23	Write mixed numbers for lengths measured to nearest ¼ inch (See pages 257-258)
24-25	Use ratios to compare two quantities (See pages 259-260)

Write the fraction.

1. What part of the square is red? $\frac{1}{4}$

2. What part of the coins are *not* pennies? $\frac{2}{5}$

Find the part.

3. $\frac{1}{5}$ of 25 = 5

4. $\frac{3}{4}$ of 24 = 18

5. $\frac{7}{8}$ of a price of \$48
$\underline{\$42}$

Write the missing numerators.

6. $\frac{3}{5} = \frac{6}{10}$ 7. $\frac{3}{8} = \frac{6}{16}$ 8. $\frac{2}{3} = \frac{6}{9}$ 9. $\frac{5}{6} = \frac{20}{24}$

Write < or > in the circle.

10. $\frac{2}{3} \bigcirc \frac{5}{9}$ 11. $\frac{1}{2} \bigcirc \frac{3}{8}$ 12. $\frac{7}{12} \bigcirc \frac{3}{4}$ 13. $\frac{2}{3} \bigcirc \frac{1}{4}$

Simplify each fraction.

14. $\frac{6}{10} = \frac{3}{5}$ 15. $\frac{4}{8} = \frac{1}{2}$ 16. $\frac{9}{12} = \frac{3}{4}$ 17. $\frac{14}{16} = \frac{7}{8}$

Write each fraction as a mixed number.

18. $\frac{7}{2} = 3\frac{1}{2}$ 19. $\frac{8}{3} = 2\frac{2}{3}$ 20. $\frac{16}{5} = 3\frac{1}{5}$ 21. $\frac{15}{12} = 1\frac{1}{4}$

Measure the figures to the nearest quarter inch.

22.
$\underline{1\frac{3}{4} \text{ inches}}$

23.
$\underline{2\frac{2}{4} \text{ inches}}$

Write the comparisons as fractions.

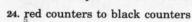

24. red counters to black counters
$\frac{5}{9}$

25. black counters to all counters
$\frac{9}{14}$

Circle the letter of the correct answer.

1 Round 925 to the nearest hundred.
a 800
b 900
c NG

2 What is the value of the 6 in 473,160?
a tens
b hundreds
c thousands
d NG

3 3,416
+ 7,396
a 10,702
b 10,712
c 10,812
d NG

4 23,089
+ 29,716
a 42,805
b 52,705
c 52,795
d NG

5 1,836
− 929
a 907
b 913
c 917
d NG

6 28,016
− 19,951
a 11,945
b 17,065
c 8,065
d NG

7 Choose the better estimate of weight.
a i gram
b 1 kilogram

8 $4.26 × 8
a $31.08
b $32.68
c $34.08
d NG

9 424
× 28
a 5,040
b 5,172
c 11,772
d NG

10 144 ÷ 9
a 1 R5
b 1 R6
c 1 R7
d NG

11 890 ÷ 26
a 3 R11
b 34
c 34 R6
d NG

12 Find the perimeter.
6 cm
9 cm
a 12 cm
b 18 cm
c 30 cm
d NG

13 Find the area.
a 14 units
b 20 sq units
c 24 sq units
d NG

☐ score

264

Cumulative Review

page 264

Item	Objective
1	Round numbers to nearest 100 (See pages 31-34)
2	Identify place value through hundred thousands (See pages 27-28)
3	Add two 4-digit numbers (See pages 47-48)
4	Add two 5-digit numbers (See pages 51-52)
5	Subtract from a 4-digit number (See pages 67-68)
6	Subtract two 5-digit numbers with zero in minuend (See pages 67-68)
7	Determine appropriate metric unit of weight (See pages 159-160)
8	Multiply money by 1-digit number (See pages 169-170)
9	Multiply 3-digit number by 2-digit number (See pages 181-182)
10	Divide 3-digit number by 1-digit number to get 2-digit quotient without remainder (See pages 197-198)
11	Divide 3-digit number by 2-digit number to get 2-digit quotient with remainder (See pages 209-210)
12	Find perimeter of a rectangle (See pages 229-230)
13	Find area of a rectangle (See pages 231-232)

Alternate Cumulative Review

Circle the letter of the correct answer.

1 Round 693 to the nearest hundred.
a 600
b 700

2 What is the value of the 9 in 629,873?
a hundreds
b thousands
c ten thousands
d NG

3 2,327
+ 6,486
a 8,713
b 8,803
c 8,813
d NG

4 25,217
+ 38,887
a 63,104
b 64,004
c 64,094
d NG

5 3,742
− 1,816
a 1,926
b 1,936
c 2,126
d NG

6 46,025
− 28,864
a 17,161
b 17,261
c 2,126
d NG

7 What unit would you use to measure a sack of flour?
a g
b kg

8 $6.47
×7
a $42.29
b $44.89
c $45.29
d NG

9 643
×26
a 15,618
b 16,618
c 16,718
d NG

10 8)136
a 17
b 17 R2
c 18
d NG

11 649 ÷ 23 =
a 28
b 28 R5
c 28 R8
d NG

12 Find the perimeter of a rectangle 5 cm by 8 cm.
a 20 cm
b 40 cm
c 91 cm
d NG

13 Find the area of a 4 cm square.
a 8 sq cm
b 16 sq cm
c 32 sq cm
d NG

264

Adding Fractions

pages 265-266

Objective

To add fractions having common denominators

Materials

fraction circles

Mental Math

Have students tell the larger fraction:

1. ½ or ⅓ (½)
2. ⅗ or ⁴⁄₁₀ (⅗)
3. ⁵⁄₁₀ or ⅗ (⅗)
4. ⁴⁄₁₅ or ⁶⁄₃₀ (⁴⁄₁₅)
5. ¼ or ²⁄₁₂ (¼)
6. ⅔ or ⅘ (⅘)
7. ½ or ¹⁷⁄₃₀ (¹⁷⁄₃₀)
8. ²⁄₇ or ¹⁄₁₄ (²⁄₇)

Skill Review

Have students identify fractions as you show 2 of 8 parts of a circle, 3 of 5 parts, etc. Then dictate fractions and have students show the correct number of parts of the appropriate fraction circle.

Adding Fractions with Common Denominators

Ellen spent the afternoon at the zoo. She walked from the elephant's cage to the monkey house, and on to the lion's den. How far did Ellen walk?

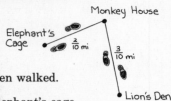

We want to find how many miles Ellen walked.

She walked $\frac{2}{10}$ of a mile from the elephant's cage

to the monkey house, and $\frac{3}{10}$ of a mile more to the lion's den.

To find the total distance Ellen walked, we add $\frac{2}{10}$ and $\frac{3}{10}$. We can use the number line.

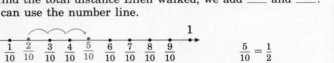

$$\frac{5}{10} = \frac{1}{2}$$

✔ To add fractions with common denominators, we can also add the numerators and put the sum over the denominator. We simplify the answer if necessary.

$$\frac{2}{10} + \frac{3}{10} = \frac{2+3}{10} = \frac{5}{10} = \frac{1}{2}$$

$$\begin{array}{r} \frac{2}{10} \\ + \frac{3}{10} \\ \hline \frac{5}{10} = \frac{1}{2} \end{array}$$

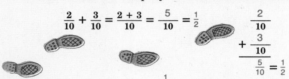

Ellen walked a total of $\frac{1}{2}$ of a mile to reach the lion's den.

Getting Started

Add. Simplify if necessary.

1. $\frac{5}{12} + \frac{6}{12} = \frac{11}{12}$

2. $\frac{3}{8} + \frac{3}{8} = \frac{3}{4}$

3. $\begin{array}{r} \frac{3}{5} \\ + \frac{1}{5} \\ \hline \frac{4}{5} \end{array}$

4. $\begin{array}{r} \frac{5}{16} \\ + \frac{7}{16} \\ \hline \frac{3}{4} \end{array}$

265

Teaching the Lesson

Introducing the Problem Have students describe the picture. Ask a student to read the problem and tell what is to be answered. (how far Ellen walked at the zoo) Ask students what information is known. (It is ²⁄₁₀ miles from the elephant's cage to the monkey house and ³⁄₁₀ miles from the monkey house to the lion's den.) Have students read with you as they complete the sentences and study the number line. Have students read the checked information aloud and guide them through the fraction addition in the model. Have students use the fraction circles to check their work, and complete the solution sentence.

Developing the Skill Write ⅖ + ⅕ on the board. Ask students to name the denominator in each addend. (5 or fifths) Draw a number line from 0 through 1 on the board and write ⁰⁄₅ through ⁵⁄₅ along it. Have a student locate ⅖ on the number line, go ⅕ more to the right and tell the

stopping place. (⅗) Tell students that when fractions have the same denominator, we can add the numerators and write that sum over the denominator. Write ⅖ + ⅕ = $\frac{2+1}{5}$ = ⅗ on the board. Write ⅖ + ⅕ = ⅗ vertically on the board. Repeat the procedure for ⁴⁄₇ + ²⁄₇, ³⁄₉ + ³⁄₉, etc. and have students simplify answers when necessary.

265

Practice

Add. Simplify if necessary.

1. $\frac{3}{8} + \frac{2}{8} = \frac{5}{8}$ 2. $\frac{5}{7} + \frac{1}{7} = \frac{6}{7}$ 3. $\frac{3}{9} + \frac{3}{9} = \frac{2}{3}$ 4. $\frac{7}{12} + \frac{4}{12} = \frac{11}{12}$

5. $\frac{3}{10} + \frac{5}{10} = \frac{4}{5}$ 6. $\frac{2}{6} + \frac{1}{6} = \frac{1}{2}$ 7. $\frac{5}{12} + \frac{3}{12} = \frac{2}{3}$ 8. $\frac{3}{16} + \frac{1}{16} = \frac{1}{4}$

9. $\frac{1}{6}$
$+\frac{3}{6}$
$\overline{\frac{2}{3}}$

10. $\frac{5}{9}$
$+\frac{2}{9}$
$\overline{\frac{7}{9}}$

11. $\frac{2}{5}$
$+\frac{2}{5}$
$\overline{\frac{4}{5}}$

12. $\frac{5}{11}$
$+\frac{3}{11}$
$\overline{\frac{8}{11}}$

13. $\frac{1}{4}$
$+\frac{2}{4}$
$\overline{\frac{3}{4}}$

14. $\frac{5}{8}$
$+\frac{1}{8}$
$\overline{\frac{3}{4}}$

15. $\frac{5}{10}$
$+\frac{1}{10}$
$\overline{\frac{3}{5}}$

16. $\frac{4}{12}$
$+\frac{5}{12}$
$\overline{\frac{3}{4}}$

17. $\frac{1}{6}$
$+\frac{3}{6}$
$\overline{\frac{2}{3}}$

18. $\frac{3}{16}$
$+\frac{5}{16}$
$\overline{\frac{1}{2}}$

19. $\frac{7}{12}$
$+\frac{3}{12}$
$\overline{\frac{5}{6}}$

20. $\frac{4}{8}$
$+\frac{1}{8}$
$\overline{\frac{5}{8}}$

21. $\frac{7}{9}$
$+\frac{1}{9}$
$\overline{\frac{8}{9}}$

22. $\frac{5}{12}$
$+\frac{3}{12}$
$\overline{\frac{2}{3}}$

23. $\frac{1}{3}$
$+\frac{1}{3}$
$\overline{\frac{2}{3}}$

24. $\frac{3}{7}$
$+\frac{2}{7}$
$\overline{\frac{5}{7}}$

25. $\frac{4}{10}$
$+\frac{4}{10}$
$\overline{\frac{4}{5}}$

26. $\frac{5}{8}$
$+\frac{1}{8}$
$\overline{\frac{3}{4}}$

27. $\frac{3}{16}$
$+\frac{9}{16}$
$\overline{\frac{3}{4}}$

28. $\frac{1}{5}$
$+\frac{3}{5}$
$\overline{\frac{4}{5}}$

Apply

Solve these problems.

29. Wally lives $\frac{2}{5}$ of a mile from school. It is another $\frac{1}{5}$ of a mile from school to the store. How far is it from Wally's house to the store? $\frac{3}{5}$ of a mile

30. Rochelle painted $\frac{3}{8}$ of a mural. Martha painted $\frac{1}{8}$ of it. How much of the mural did the girls paint?
$\frac{1}{2}$ of the mural

266

266

Subtracting Fractions

pages 267-268

Objective

To subtract fractions having common denominators

Materials

fraction circles

Mental Math

Ask students how many in:

1. ⅖ of 15. (6)
2. ⅕ + ⅗. (⅘)
3. 60 × ½ of 2. (60)
4. 16 × (6 ÷ 3). (32)
5. ⅝ + ⅝. (1¼)
6. months in ⅓ of 1 year. (4)
7. lowest terms for ¹²⁄₁₆. (¾)
8. lowest terms for ³⁵⁄₉₀. (⁷⁄₁₈)

Skill Review

Review addition of fractions with common denominators by having students work the following problems at the board:

⁴⁄₁₅ + ¹⁰⁄₁₅ (¹⁴⁄₁₅), ³⁄₁₂ + ⁶⁄₁₂ (¾), ⁹⁄₁₃ + ¹⁄₁₃ (¹⁰⁄₁₃), ⁴⁄₉ + ²⁄₉ (⅔), ⁷⁄₁₆ + ⁵⁄₁₆ (¾).

Subtracting Fractions with Common Denominators

Barry, Cal and Craig challenged each other to an 8-minute run around the school track. How much farther did Barry run than Cal?

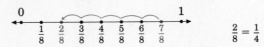

We want to know how much farther Barry ran than Cal.

Barry ran $\frac{7}{8}$ of a mile, while Cal ran $\frac{5}{8}$ of a mile in the same period of time. To find the difference in the distance Barry and Cal ran, we subtract $\frac{5}{8}$ from $\frac{7}{8}$. We can use a number line.

$$\frac{2}{8} = \frac{1}{4}$$

✔ To subtract fractions with common denominators, we can also subtract the numerators and write the difference over the denominator. We simplify the answer if necessary.

$$\frac{7}{8} - \frac{5}{8} = \frac{7-5}{8} = \frac{2}{8} = \frac{1}{4}$$

$$\begin{array}{r} \frac{7}{8} \\ -\frac{5}{8} \\ \hline \frac{2}{8} = \frac{1}{4} \end{array}$$

Barry ran $\frac{1}{4}$ of a mile farther than Cal.

Getting Started

Subtract. Simplify if necessary.

1. $\frac{4}{5} - \frac{1}{5} = \frac{3}{5}$

2. $\frac{9}{10} - \frac{3}{10} = \frac{3}{5}$

3. $\begin{array}{r} \frac{7}{8} \\ -\frac{7}{8} \\ \hline 0 \end{array}$

4. $\begin{array}{r} \frac{8}{12} \\ -\frac{5}{12} \\ \hline \frac{1}{4} \end{array}$

267

Teaching the Lesson

Introducing the Problem Ask a student to read the problem and tell what is to be solved. (how much farther Barry ran than Cal) Ask students what information is given. (In 8 minutes Barry ran ⅞ miles, Cal ran ⅝ miles, Craig ran ⁶⁄₈ miles.) Ask what, if any, of this information is unnecessary to solve the problem. (8 minutes and Craig ran ⁶⁄₈ miles) Have students read with you as they complete the sentences. Have them study the number line and rule, and do the subtraction in the model. Have students use the fraction circles to check their work.

Developing the Skill Draw a number line from 0 through 1 on the board and write ⁰⁄₉ through ⁹⁄₉ along it. Write ⁸⁄₉ − ³⁄₉ on the board and ask students to name the denominator in each fraction. (9 or ninths) Have a student locate ⁸⁄₉ on the number line, go backward ³⁄₉ and tell the

stopping place. (⁵⁄₉) Tell students that when fractions have the same denominator, we can subtract the numerator and write that difference over the denominator. Write ⁸⁄₉ − ³⁄₉ = $\frac{8-3}{9}$ = ⁵⁄₉ on the board. Write ⁸⁄₉ − ³⁄₉ vertically on the board. Then repeat the procedure for ⁷⁄₁₂ − ¹⁄₁₂, ⁹⁄₁₀ − ⁴⁄₁₀, etc. and have students simplify their answers when necessary.

267

Practice

Subtract. Simplify if necessary.

1. $\dfrac{4}{5} - \dfrac{1}{5} = \dfrac{3}{5}$ 2. $\dfrac{5}{6} - \dfrac{2}{6} = \dfrac{1}{2}$ 3. $\dfrac{3}{11} - \dfrac{2}{11} = \dfrac{1}{11}$ 4. $\dfrac{12}{16} - \dfrac{8}{16} = \dfrac{1}{4}$

5. $\dfrac{5}{8} - \dfrac{2}{8} = \dfrac{3}{8}$ 6. $\dfrac{7}{12} - \dfrac{4}{12} = \dfrac{1}{4}$ 7. $\dfrac{6}{7} - \dfrac{3}{7} = \dfrac{3}{7}$ 8. $\dfrac{5}{6} - \dfrac{1}{6} = \dfrac{2}{3}$

9. $\dfrac{7}{9}$ $-\dfrac{4}{9}$ $\dfrac{1}{3}$ 10. $\dfrac{4}{5}$ $-\dfrac{3}{5}$ $\dfrac{1}{5}$ 11. $\dfrac{6}{8}$ $-\dfrac{1}{8}$ $\dfrac{5}{8}$ 12. $\dfrac{11}{12}$ $-\dfrac{5}{12}$ $\dfrac{1}{2}$ 13. $\dfrac{13}{16}$ $-\dfrac{5}{16}$ $\dfrac{1}{2}$

14. $\dfrac{9}{10}$ $-\dfrac{1}{10}$ $\dfrac{4}{5}$ 15. $\dfrac{5}{7}$ $-\dfrac{2}{7}$ $\dfrac{3}{7}$ 16. $\dfrac{7}{9}$ $-\dfrac{3}{9}$ $\dfrac{4}{9}$ 17. $\dfrac{5}{8}$ $-\dfrac{1}{8}$ $\dfrac{1}{2}$ 18. $\dfrac{7}{11}$ $-\dfrac{3}{11}$ $\dfrac{4}{11}$

19. $\dfrac{11}{16}$ $-\dfrac{9}{16}$ $\dfrac{1}{8}$ 20. $\dfrac{7}{8}$ $-\dfrac{3}{8}$ $\dfrac{1}{2}$ 21. $\dfrac{7}{10}$ $-\dfrac{2}{10}$ $\dfrac{1}{2}$ 22. $\dfrac{12}{15}$ $-\dfrac{9}{15}$ $\dfrac{1}{5}$ 23. $\dfrac{11}{12}$ $-\dfrac{2}{12}$ $\dfrac{3}{4}$

24. $\dfrac{2}{3}$ $-\dfrac{1}{3}$ $\dfrac{1}{3}$ 25. $\dfrac{7}{16}$ $-\dfrac{3}{16}$ $\dfrac{1}{4}$ 26. $\dfrac{8}{15}$ $-\dfrac{3}{15}$ $\dfrac{1}{3}$ 27. $\dfrac{5}{10}$ $-\dfrac{2}{10}$ $\dfrac{3}{10}$ 28. $\dfrac{9}{16}$ $-\dfrac{7}{16}$ $\dfrac{1}{8}$

Apply

Solve these problems.

29. Cindy had $\dfrac{5}{8}$ of a liter of milk left in the bottle. She drinks $\dfrac{3}{8}$ of a liter with her lunch. How much milk does Cindy have left? $\dfrac{1}{4}$ of a liter

30. Clay did $\dfrac{9}{10}$ of his homework correctly. David did $\dfrac{7}{10}$ of his correctly. How much more homework did Clay do correctly?
$\dfrac{1}{5}$ more

268

Mixed Practice

1. $\dfrac{3}{8} = \dfrac{}{24}$ (9)
2. 368×29 (10,672)
3. Simplify: $\dfrac{16}{20}$ $\left(\dfrac{4}{5}\right)$
4. $\$6.93 \times 4$ (\$27.72)
5. $508 - 39$ (469)
6. $25,176 + 18,927$ (44,103)
7. 723×47 (33,981)
8. Simplify: $\dfrac{24}{9}$ $\left(2\dfrac{2}{3}\right)$
9. $31,205 - 16,571$ (14,634)
10. $638 \div 27$ (23 R17)

Extra Credit *Numeration*

Explain that computers use a special system of numbers and letters because they understand only two signals: electric current on and electric current off. Ask students to look up the binary number system in the library. Give each student a place value chart and ask them to fill in the place values for binary numbers through 512.

place value

512	256	128	64	32	16	8	4	2	1

Explain that in each place combinations of only two digits, 0 and 1, are possible. Ask them to write the binary numbers for these decimal numbers:

7 (111)	3 (11)	20 (10100)
36 (100100)	14 (1110)	73 (1001001)

Simplifying Fractions

pages 269-270

Objective

To add or subtract fractions and simplify the answer

Materials

Mental Math

Have students complete each comparison: 6 is to 2 as 12 is to 4 as:

1. 18 is to (6)
2. 120 is to (40)
3. 72 is to (24)
4. 3 is to (1)
5. 24 is to (8)
6. 30 is to (10)
7. 99 is to (33)
8. 63 is to (21)

Skill Review

Have students write a whole or mixed number for each of the following dictated fractions: $^{16}/_4$ (4), $^{10}/_3$ (3⅓), $^{16}/_5$ (3⅕), $^{22}/_{10}$ (2⅕), $^{17}/_{13}$ (1⁴⁄₁₃)

Adding and Subtracting Fractions, Simplifying

Leah is going skiing on Tuesday, if fresh snow falls in the mountains. She checks with the Weather Bureau on Monday before she packs her gear. How much new snow fell over the weekend?

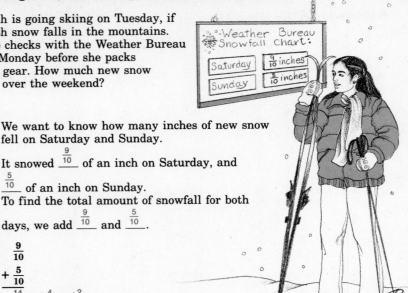

We want to know how many inches of new snow fell on Saturday and Sunday.

It snowed $\frac{9}{10}$ of an inch on Saturday, and $\frac{5}{10}$ of an inch on Sunday.

To find the total amount of snowfall for both days, we add $\frac{9}{10}$ and $\frac{5}{10}$.

$$\frac{9}{10}$$
$$+\frac{5}{10}$$
$$\frac{14}{10} = 1\frac{4}{10} = 1\frac{2}{5}$$

✔ Remember to simplify fractions. This includes changing fractions to whole or mixed numbers, if necessary.

It snowed $\underline{1\frac{2}{5}}$ inches over the weekend.

Getting Started

Add or subtract. Simplify if necessary.

1. $\frac{3}{5} + \frac{4}{5} = 1\frac{2}{5}$

2. $\frac{7}{8} - \frac{1}{8} = \frac{3}{4}$

3. $\frac{3}{4} + \frac{3}{4} = 1\frac{1}{2}$

4. $\frac{19}{12} - \frac{13}{12} = \frac{1}{2}$

5. $\frac{5}{6}$
$+\frac{4}{6}$
$1\frac{1}{2}$

6. $\frac{9}{16}$
$-\frac{1}{16}$
$\frac{1}{2}$

7. $\frac{1}{2}$
$-\frac{1}{2}$
0

8. $\frac{17}{10}$
$-\frac{3}{10}$
$1\frac{2}{5}$

9. $\frac{3}{10}$
$+\frac{7}{10}$
1

269

Teaching the Lesson

Introducing the Problem Have a student read the problem aloud and tell what is to be found. (the amount of new snow over the weekend) Have students tell what information is given in the problem or picture. (⁹⁄₁₀ inch of snow fell on Saturday and ⁵⁄₁₀ inch fell on Sunday.) Have students complete the sentences. Guide them through the simplification in the model, and solve the problem. Have students draw a number line in tenths to check their answer.

Developing the Skill Tell students that fraction answers must always be given in the simplest form possible. Tell students this may mean simplifying a fraction to its lowest terms or changing a fraction to a whole or mixed number. Write ¹⁴⁄₅ on the board and ask students how this fraction would be simplified. (change to mixed number) Have a student change the fraction to 2⅘. Ask if 2⅘ can be further simplified. (no) Have students talk through the work to solve ⅘ + ⅗. (⁷⁄₅ = 1⅖) Repeat for ¹⁹⁄₁₀ − ⁸⁄₁₀, 1⁴⁄₂ + ½, ¹³⁄₈ − ⁵⁄₈, etc.

269

Practice

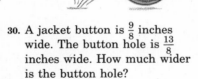

Add or subtract. Simplify if necessary.

1. $\frac{5}{8} + \frac{6}{8} = 1\frac{3}{8}$ 2. $\frac{11}{9} - \frac{1}{9} = 1\frac{1}{9}$ 3. $\frac{11}{12} - \frac{3}{12} = \frac{2}{3}$ 4. $\frac{7}{5} + \frac{3}{5} = 2$

5. $\frac{19}{8} - \frac{9}{8} = 1\frac{1}{4}$ 6. $\frac{3}{4} + \frac{6}{4} = 2\frac{1}{4}$ 7. $\frac{5}{6} + \frac{5}{6} = 1\frac{2}{3}$ 8. $\frac{11}{15} - \frac{6}{15} = \frac{1}{3}$

9. $\begin{array}{r} \frac{9}{10} \\ +\frac{6}{10} \\ \hline 1\frac{1}{2} \end{array}$
10. $\begin{array}{r} \frac{4}{7} \\ +\frac{6}{7} \\ \hline 1\frac{3}{7} \end{array}$
11. $\begin{array}{r} \frac{18}{9} \\ -\frac{12}{9} \\ \hline \frac{2}{3} \end{array}$
12. $\begin{array}{r} \frac{2}{3} \\ +\frac{2}{3} \\ \hline 1\frac{1}{3} \end{array}$
13. $\begin{array}{r} \frac{18}{10} \\ -\frac{2}{10} \\ \hline 1\frac{3}{5} \end{array}$

14. $\begin{array}{r} \frac{14}{15} \\ +\frac{4}{15} \\ \hline 1\frac{1}{5} \end{array}$
15. $\begin{array}{r} \frac{10}{12} \\ +\frac{8}{12} \\ \hline 1\frac{1}{2} \end{array}$
16. $\begin{array}{r} \frac{7}{10} \\ +\frac{9}{10} \\ \hline 1\frac{3}{5} \end{array}$
17. $\begin{array}{r} \frac{9}{16} \\ -\frac{5}{16} \\ \hline \frac{1}{4} \end{array}$
18. $\begin{array}{r} \frac{9}{5} \\ +\frac{3}{5} \\ \hline 2\frac{2}{5} \end{array}$

19. $\begin{array}{r} \frac{7}{12} \\ +\frac{8}{12} \\ \hline 1\frac{1}{4} \end{array}$
20. $\begin{array}{r} \frac{17}{12} \\ -\frac{3}{12} \\ \hline 1\frac{1}{6} \end{array}$
21. $\begin{array}{r} \frac{7}{8} \\ -\frac{1}{8} \\ \hline \frac{3}{4} \end{array}$
22. $\begin{array}{r} \frac{9}{12} \\ +\frac{7}{12} \\ \hline 1\frac{1}{3} \end{array}$
23. $\begin{array}{r} \frac{10}{6} \\ +\frac{4}{6} \\ \hline 2\frac{1}{3} \end{array}$

24. $\begin{array}{r} \frac{19}{10} \\ +\frac{11}{10} \\ \hline 3 \end{array}$
25. $\begin{array}{r} \frac{17}{12} \\ -\frac{13}{12} \\ \hline \frac{1}{3} \end{array}$
26. $\begin{array}{r} \frac{13}{4} \\ +\frac{13}{4} \\ \hline 6\frac{1}{2} \end{array}$
27. $\begin{array}{r} \frac{15}{16} \\ +\frac{11}{16} \\ \hline 1\frac{5}{8} \end{array}$
28. $\begin{array}{r} \frac{13}{6} \\ -\frac{3}{6} \\ \hline 1\frac{2}{3} \end{array}$

Apply

Solve these problems.

29. Bonnie drank $\frac{5}{8}$ of a quart of water before the basketball game and $\frac{7}{8}$ of a quart after. How much water did Bonnie drink? $1\frac{1}{2}$ quarts

30. A jacket button is $\frac{9}{8}$ inches wide. The button hole is $\frac{13}{8}$ inches wide. How much wider is the button hole? $\frac{1}{2}$ of an inch

270

270

Adding Uncommon Denominators

pages 271-272

Objective

To add 2 fractions when 1 denominator is a multiple of the other denominator

Materials

*fraction circles

Mental Math

Tell students to name a mixed number for:

1. 2 hrs 15 min. (2¼)
2. 3 hrs 12 min. (3⅕)
3. 11 hrs 5 min. (11 1/12)
4. 4 hrs 2 min. (4 1/30)
5. 10 hrs 20 min. (10⅓)
6. 70 min. (1⅙)
7. 375 min. (6¼)
8. 200 min. (3⅓)

Skill Review

Have students complete the following problems on the board to review equivalent fractions:

½ = ⁽²⁾⁄₄ = ⁽³⁾⁄₆ = ⁽⁴⁾⁄₈ = ⁵⁄₍₁₀₎
⅓ = ²⁄₍₆₎ = ⁽³⁾⁄₉ = ⁴⁄₍₁₂₎
¼ = ⁽²⁾⁄₈ = ³⁄₍₁₂₎ = ⁽⁴⁾⁄₁₆ = ⁵⁄₍₂₀₎

Adding Fractions, Uncommon Denominators

Following the veterinarian's advice, Tony is keeping track of his new kitten's weight gain. How many pounds has his kitten gained after 2 weeks?

weight gain:
week 1 ½ pound
week 2 ¾ pound
week 3 1 pound
week 4 ¼ pound

We want to find the kitten's weight gain for the first two weeks.

It gained $\frac{1}{2}$ of a pound the first week

and $\frac{3}{4}$ of a pound the second week. To find how many pounds the kitten

gained in 2 weeks, we add $\frac{1}{2}$ and $\frac{3}{4}$.

The fractions do not have common denominators.	Find equivalent fractions with common denominators.	Add the fractions. Simplify.
$\frac{1}{2}$ $+\frac{3}{4}$	$\frac{1 \times 2}{2 \times 2} = \frac{2}{4}$ $+\frac{3}{4} = \frac{3}{4}$	$\frac{2}{4}$ $+\frac{3}{4}$ $\frac{5}{4} = 1\frac{1}{4}$

Tony's kitten has gained $1\frac{1}{4}$ pounds in the first 2 weeks.

Getting Started

Add. Simplify if necessary.

1. $\frac{3}{5}$ $+\frac{3}{10}$ = $\frac{9}{10}$

2. $\frac{2}{3}$ $+\frac{5}{6}$ = $1\frac{1}{2}$

3. $\frac{3}{8}$ $+\frac{1}{4}$ = $\frac{5}{8}$

4. $\frac{7}{12}$ $+\frac{2}{3}$ = $1\frac{1}{4}$

Copy and add.

5. $\frac{1}{2} + \frac{3}{4}$ $1\frac{1}{4}$

6. $\frac{2}{5} + \frac{3}{20}$ $\frac{11}{20}$

7. $\frac{1}{6} + \frac{3}{12}$ $\frac{5}{12}$

271

Teaching the Lesson

Introducing the Problem Have students describe the picture and tell what is to be found. (the weight gain of the kitten after 2 weeks) Ask what information will be used from the chart. (½ lb gain in week 1 and ¾ lb gain in week 2) Have students complete the sentences. Guide them through the problems in the model to find the kitten's weight gain in the first 2 weeks. Have students use a number line marked in halves and fourths to check their answer.

Developing the Skill Tell students that when we add 2 fractions that have different denominators, we see if the larger denominator is a multiple of the other. Write ⅔ + 3/6 on the board and ask students if 6 is a multiple of 3. (yes) Tell students we then find an equivalent fraction for ⅔ which has 6 as its denominator. (4/6) Write 4/6 + 3/6 on the board and have students add and simplify the answer. (1⅙) Repeat the procedure to solve 6/12 + 4/6 (1⅙), 5/8 + 13/16 (1 7/16), 7/9 + ⅔ (1 4/9) and 5/20 + 9/10 (1 3/20).

271

Practice

Add. Simplify if necessary.

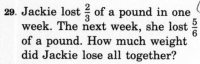

1. $\dfrac{1}{3}$ $+\dfrac{1}{6}$ $\dfrac{1}{2}$

2. $\dfrac{4}{2}$ $+\dfrac{3}{8}$ $2\dfrac{3}{8}$

3. $\dfrac{7}{12}$ $+\dfrac{3}{4}$ $1\dfrac{1}{3}$

4. $\dfrac{2}{3}$ $+\dfrac{5}{9}$ $1\dfrac{2}{9}$

5. $\dfrac{7}{8}$ $+\dfrac{3}{4}$ $1\dfrac{5}{8}$

6. $\dfrac{5}{12}$ $+\dfrac{5}{6}$ $1\dfrac{1}{4}$

7. $\dfrac{3}{5}$ $+\dfrac{7}{10}$ $1\dfrac{3}{10}$

8. $\dfrac{3}{14}$ $+\dfrac{2}{7}$ $\dfrac{1}{2}$

9. $\dfrac{7}{15}$ $+\dfrac{3}{5}$ $1\dfrac{1}{15}$

10. $\dfrac{9}{10}$ $+\dfrac{1}{2}$ $1\dfrac{2}{5}$

11. $\dfrac{12}{5}$ $+\dfrac{7}{20}$ $2\dfrac{3}{4}$

12. $\dfrac{9}{16}$ $+\dfrac{8}{4}$ $2\dfrac{9}{16}$

Copy and Do

13. $\dfrac{3}{4} + \dfrac{7}{2}$ $4\dfrac{1}{4}$

14. $\dfrac{2}{3} + \dfrac{7}{9}$ $1\dfrac{4}{9}$

15. $\dfrac{5}{8} + \dfrac{1}{4}$ $\dfrac{7}{8}$

16. $\dfrac{7}{10} + \dfrac{3}{5}$ $1\dfrac{3}{10}$

17. $\dfrac{11}{12} + \dfrac{3}{4}$ $1\dfrac{2}{3}$

18. $\dfrac{4}{3} + \dfrac{5}{12}$ $1\dfrac{3}{4}$

19. $\dfrac{2}{3} + \dfrac{9}{6}$ $2\dfrac{1}{6}$

20. $\dfrac{7}{15} + \dfrac{4}{5}$ $1\dfrac{4}{15}$

21. $\dfrac{7}{9} + \dfrac{7}{18}$ $1\dfrac{1}{6}$

22. $\dfrac{9}{16} + \dfrac{7}{8}$ $1\dfrac{7}{16}$

23. $\dfrac{9}{20} + \dfrac{9}{10}$ $1\dfrac{7}{20}$

24. $\dfrac{4}{9} + \dfrac{2}{3}$ $1\dfrac{1}{9}$

25. $\dfrac{9}{2} + \dfrac{3}{16}$ $4\dfrac{11}{16}$

26. $\dfrac{8}{10} + \dfrac{4}{5}$ $1\dfrac{3}{5}$

27. $\dfrac{13}{12} + \dfrac{1}{6}$ $1\dfrac{1}{4}$

28. $\dfrac{5}{7} + \dfrac{11}{21}$ $1\dfrac{5}{21}$

Apply

Solve these problems.

29. Jackie lost $\dfrac{2}{3}$ of a pound in one week. The next week, she lost $\dfrac{5}{6}$ of a pound. How much weight did Jackie lose all together?

$1\dfrac{1}{2}$ pounds

30. Lanny used $\dfrac{1}{2}$ of a roll of red paper and $\dfrac{5}{8}$ of a roll of green paper to wrap gifts. How much paper did Lanny use?

$1\dfrac{1}{8}$ rolls

272

Subtracting Uncommon Denominators

pages 273-274

Objective

To subtract fractions when 1 denominator is a multiple of the other denominator

Materials

*fraction circles

Mental Math

Ask what fraction names this part of a year:

1. 1 month (1/12)
2. 3 months (1/4)
3. 5 days (5/365 or 1/73)
4. 10 months (5/6)
5. 9 months (3/4)

Ask what mixed number names these months in years:

6. 18 months (1½)
7. 25 months (2 1/12)
8. 40 months (3⅓)

Skill Review

Have students solve the following problems to review subtraction of fractions: 2/3 − 1/3 (1/3), 5/4 − 3/4 (1/2), 9/5 − 4/5 (1), 11/2 − 7/2 (2), 20/12 − 2/12 (1½)

Subtracting Fractions, Uncommon Denominators

Dina is making cookies to give to the new neighbors. The recipe calls for $\frac{7}{8}$ of a cup of peanut butter. How much more peanut butter does Dina need?

We want to find how much more peanut butter Dina needs to measure.

She needs $\frac{7}{8}$ of a cup and has

already measured $\frac{1}{2}$ of a cup.

To find how much more peanut butter

is needed, we subtract $\frac{1}{2}$ from $\frac{7}{8}$.

The fractions do not have common denominators.

$$\frac{7}{8}$$
$$-\frac{1}{2}$$

Find equivalent fractions with common denominators.

$$\frac{7}{8} = \frac{7}{8}$$
$$-\frac{1}{2} \times \frac{4}{4} = \frac{4}{8}$$

Subtract the fractions.

$$\frac{7}{8}$$
$$-\frac{4}{8}$$
$$\overline{\frac{3}{8}}$$

Dina needs $\frac{3}{8}$ of a cup more peanut butter.

Getting Started

Subtract.

1. $\frac{5}{6}$ $-\frac{1}{3}$ $\frac{1}{2}$

2. $\frac{7}{12}$ $-\frac{1}{2}$ $\frac{1}{12}$

3. $\frac{15}{16}$ $-\frac{3}{4}$ $\frac{3}{16}$

4. $\frac{7}{9}$ $-\frac{2}{3}$ $\frac{1}{9}$

Copy and subtract.

5. $\frac{5}{8} - \frac{1}{2}$ $\frac{1}{8}$

6. $\frac{11}{12} - \frac{2}{3}$ $\frac{1}{4}$

7. $\frac{17}{15} - \frac{4}{5}$ $\frac{1}{3}$

8. $\frac{6}{3} - \frac{7}{12}$ $1\frac{5}{12}$

273

Teaching the Lesson

Introducing the Problem Have a student read the problem aloud and tell what is to be found. (how much more peanut butter Dina needs) Ask what information is given in the problem and the picture. (Recipe calls for 7/8 of a cup and Dina has already measured ½ of a cup.) Have students complete the sentences. Guide them through the subtraction in the model to solve the problem. Tell students their solution plus ½ of a cup must equal 7/8 of a cup.

Developing the Skill Tell students that if fractions have uncommon denominators, we look to see if the larger denominator is a multiple of the other. Write 7/9 − 1/3 on the board. Ask if 9 is a multiple of 3. (yes) Have a student find the number of ninths in 1/3. (3/9) Remind students that when the denominators are the same, we merely subtract the numerators. Have a student subtract 3/9 from 7/9. (4/9) Repeat for more problems where 1 fraction's denominator is a multiple of the other denominator.

Practice

Subtract.

1. $\dfrac{3}{4}$
$-\dfrac{1}{2}$
$\dfrac{1}{4}$

2. $\dfrac{5}{8}$
$-\dfrac{1}{4}$
$\dfrac{3}{8}$

3. $\dfrac{7}{2}$
$-\dfrac{9}{10}$
$2\dfrac{3}{5}$

4. $\dfrac{7}{9}$
$-\dfrac{2}{3}$
$\dfrac{1}{9}$

5. $\dfrac{9}{6}$
$-\dfrac{7}{12}$
$\dfrac{11}{12}$

6. $\dfrac{8}{5}$
$-\dfrac{5}{10}$
$1\dfrac{1}{10}$

7. $\dfrac{9}{12}$
$-\dfrac{2}{3}$
$\dfrac{1}{12}$

8. $\dfrac{11}{4}$
$-\dfrac{7}{8}$
$1\dfrac{7}{8}$

9. $\dfrac{15}{6}$
$-\dfrac{7}{12}$
$1\dfrac{11}{12}$

10. $\dfrac{17}{10}$
$-\dfrac{4}{5}$
$\dfrac{9}{10}$

11. $\dfrac{13}{5}$
$-\dfrac{9}{15}$
2

12. $\dfrac{7}{6}$
$-\dfrac{2}{3}$
$\dfrac{1}{2}$

Copy and Do

13. $\dfrac{7}{9} - \dfrac{1}{3}$ $\dfrac{4}{9}$

14. $\dfrac{8}{3} - \dfrac{5}{6}$ $1\dfrac{5}{6}$

15. $\dfrac{7}{5} - \dfrac{8}{10}$ $\dfrac{3}{5}$

16. $\dfrac{15}{16} - \dfrac{7}{8}$ $\dfrac{1}{16}$

17. $\dfrac{11}{12} - \dfrac{1}{4}$ $\dfrac{2}{3}$

18. $\dfrac{9}{4} - \dfrac{3}{8}$ $1\dfrac{7}{8}$

19. $\dfrac{7}{3} - \dfrac{5}{9}$ $1\dfrac{7}{9}$

20. $\dfrac{9}{2} - \dfrac{7}{4}$ $2\dfrac{3}{4}$

21. $\dfrac{8}{5} - \dfrac{13}{10}$ $\dfrac{3}{10}$

22. $\dfrac{13}{16} - \dfrac{1}{4}$ $\dfrac{9}{16}$

23. $\dfrac{16}{12} - \dfrac{5}{4}$ $\dfrac{1}{12}$

24. $\dfrac{11}{6} - \dfrac{4}{3}$ $\dfrac{1}{2}$

25. $\dfrac{24}{15} - \dfrac{7}{5}$ $\dfrac{1}{5}$

26. $\dfrac{16}{3} - \dfrac{26}{9}$ $2\dfrac{4}{9}$

27. $\dfrac{7}{4} - \dfrac{7}{8}$ $\dfrac{7}{8}$

28. $\dfrac{25}{12} - \dfrac{5}{4}$ $\dfrac{5}{6}$

Apply

Solve these problems.

29. A recipe for custard calls for $\dfrac{3}{4}$ of a cup of milk. Joel has only $\dfrac{1}{2}$ of a cup of milk. How much milk does Joel need to borrow? $\dfrac{1}{4}$ of a cup

30. Naomi lives $\dfrac{3}{5}$ of a mile from Deven. She has walked $\dfrac{5}{10}$ of a mile so far. How much further must she walk to get to Deven's?
$\dfrac{1}{10}$ of a mile

274

Finding Common Multiples

pages 275-276

Objective

To find a common multiple to add or subtract fractions

Materials

Mental Math

Dictate the following:

1. 78×2 (156)
2. $6,000 - 28$ (5,972)
3. $800 \div 40$ (20)
4. $(10 \times 60) \div 30$ (20)
5. $(54 \div 6)+(72 \div 8)$ (18)
6. $(\frac{1}{12}$ of $72) \times 6$ (36)
7. $(48 \times 2) \div 6$ (16)
8. $\frac{2}{3}$ of 24 (16)

Skill Review

Have a student write the first 6 multiples of each number on the board as students count by 2's, 3's, 4's, 5's, 6's, 7's, 8's and 9's. Have students find the least common multiple of 4 and 5, 2 and 3, etc.

Adding and Subtracting Fractions, Finding Common Multiples

Marty spends her allowance on school supplies, food and entertainment. What part of her allowance does she spend on school supplies and entertainment?

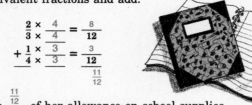

Marty's Allowance

We want to know what fractional part of her money Marty spends on school supplies and entertainment. Marty spends $\frac{2}{3}$ of her allowance on school supplies and $\frac{1}{4}$ of her allowance on entertainment. To find the total portion spent on these two things, we add $\frac{2}{3}$ and $\frac{1}{4}$.

We need to find the smallest number that is a multiple of both 3 and 4. This is called the **least common multiple.**

Multiples of 3: 3   6   9   12   15
Multiples of 4: 4   8   12   16

The least common multiple of 3 and 4 is 12.
Use the least common multiple as the common denominator. Find the equivalent fractions and add.

$$\frac{2}{3} \times \frac{4}{4} = \frac{8}{12}$$
$$+ \frac{1}{4} \times \frac{3}{3} = \frac{3}{12}$$
$$\frac{11}{12}$$

Marty spends $\frac{11}{12}$ of her allowance on school supplies and entertainment.

Getting Started

Add or subtract.

1. $\frac{3}{5}$
   $+ \frac{1}{3}$
   $\overline{\frac{14}{15}}$

2. $\frac{5}{6}$
   $- \frac{1}{4}$
   $\overline{\frac{7}{12}}$

3. $\frac{2}{3}$
   $+ \frac{1}{2}$
   $\overline{1\frac{1}{6}}$

Copy and add or subtract.

4. $\frac{21}{3} - \frac{11}{9}$ $5\frac{7}{9}$

5. $\frac{1}{7} + \frac{1}{3}$ $\frac{10}{21}$

6. $\frac{3}{4} + \frac{5}{6}$ $1\frac{7}{12}$

7. $\frac{8}{6} - \frac{2}{5}$ $\frac{14}{15}$

275

Teaching the Lesson

Introducing the Problem Have a student read the problem and tell what is to be solved. (total spent on school supplies and entertainment) Ask what information is known. (She spends $\frac{2}{3}$ on school supplies, $\frac{1}{12}$ on food and $\frac{1}{4}$ on entertainment.) Ask what information is not needed to solve the problem. ($\frac{1}{12}$ on food) Have students read with you as they complete the sentences and read the list of least common multiples in the model. Tell students their solution plus $\frac{1}{12}$ must equal 1 whole. Have them complete the addition and solution sentence.

Developing the Skill Write $\frac{1}{12} + \frac{1}{5}$ on the board. Ask students if 1 denominator is a multiple of the other. (no) Tell students that they must find a denominator which is a multiple of both 2 and 5. Have students write the multiples of 2 and 5 through 30 on the board. Ask a student to circle the common multiples (10, 20, 30) and underline the least common multiple. (10) Tell students we call 10 the **least common denominator** because it is the lowest common multiple of 2 and 5. Write $\frac{1}{2} = \frac{(5)}{10}$ and $\frac{1}{5} = \frac{(2)}{10}$ on the board. Have a student find equivalent fractions and then add. ($\frac{7}{10}$) Repeat for $\frac{7}{9}+\frac{2}{6}$ ($1\frac{1}{9}$), $\frac{2}{3} - \frac{1}{4}$ ($\frac{5}{12}$), $\frac{20}{4} - \frac{2}{5}$ ($4\frac{3}{5}$), $\frac{4}{7}+\frac{5}{3}$ ($2\frac{5}{21}$) and $\frac{11}{12} - \frac{4}{6}$ ($\frac{1}{4}$).

275

Practice

Add or subtract.

1. $\dfrac{1}{5}$
    $+\dfrac{1}{3}$
    $\overline{\dfrac{8}{15}}$

2. $\dfrac{1}{4}$
    $+\dfrac{5}{6}$
    $\overline{1\dfrac{1}{12}}$

3. $\dfrac{2}{3}$
    $-\dfrac{1}{2}$
    $\overline{\dfrac{1}{6}}$

4. $\dfrac{3}{8}$
    $-\dfrac{1}{6}$
    $\overline{\dfrac{5}{24}}$

5. $\dfrac{3}{4}$
    $+\dfrac{4}{5}$
    $\overline{1\dfrac{11}{20}}$

6. $\dfrac{7}{3}$
    $-\dfrac{3}{4}$
    $\overline{1\dfrac{7}{12}}$

7. $\dfrac{3}{4}$
    $+\dfrac{5}{8}$
    $\overline{1\dfrac{3}{8}}$

8. $\dfrac{11}{12}$
    $-\dfrac{2}{3}$
    $\overline{\dfrac{1}{4}}$

9. $\dfrac{7}{8}$
    $-\dfrac{3}{16}$
    $\overline{\dfrac{11}{16}}$

10. $\dfrac{3}{2}$
    $+\dfrac{2}{3}$
    $\overline{2\dfrac{1}{6}}$

11. $\dfrac{7}{10}$
    $+\dfrac{4}{5}$
    $\overline{1\dfrac{1}{2}}$

12. $\dfrac{18}{5}$
    $-\dfrac{9}{4}$
    $\overline{1\dfrac{7}{20}}$

Copy and Do

13. $\dfrac{1}{2} + \dfrac{1}{9}$ $1\dfrac{11}{18}$

14. $\dfrac{7}{4} - \dfrac{5}{3}$ $\dfrac{1}{12}$

15. $\dfrac{5}{6} - \dfrac{4}{9}$ $\dfrac{7}{18}$

16. $\dfrac{7}{8} + \dfrac{5}{3}$ $2\dfrac{13}{24}$

17. $\dfrac{2}{3} - \dfrac{1}{5}$ $\dfrac{7}{15}$

18. $\dfrac{11}{12} - \dfrac{5}{6}$ $\dfrac{1}{12}$

19. $\dfrac{13}{9} - \dfrac{25}{36}$ $\dfrac{3}{4}$

20. $\dfrac{5}{9} + \dfrac{7}{6}$ $1\dfrac{13}{18}$

21. $\dfrac{1}{2} - \dfrac{2}{5}$ $\dfrac{1}{10}$

22. $\dfrac{3}{5} + \dfrac{1}{4}$ $\dfrac{17}{20}$

23. $\dfrac{11}{6} - \dfrac{8}{9}$ $\dfrac{17}{18}$

24. $\dfrac{9}{8} + \dfrac{4}{3}$ $2\dfrac{11}{24}$

25. $\dfrac{7}{8} + \dfrac{5}{6}$ $1\dfrac{17}{24}$

26. $\dfrac{7}{10} + \dfrac{4}{5}$ $1\dfrac{1}{2}$

27. $\dfrac{5}{8} + \dfrac{1}{12}$ $\dfrac{17}{24}$

28. $\dfrac{9}{4} - \dfrac{6}{7}$ $1\dfrac{11}{28}$

Apply

Use the chart to help solve these problems.

29. How much rain fell on Monday and Wednesday? $1\dfrac{7}{20}$ of an inch

30. How much more rain fell on Monday than on Tuesday? $\dfrac{1}{12}$ of an inch

Rainfall in Inches	
Monday	$\dfrac{3}{4}$ inch
Tuesday	$\dfrac{2}{3}$ inch
Wednesday	$\dfrac{3}{5}$ inch

276

276

Adding Mixed Numbers

pages 277-278

Objective

To add 2 mixed numbers

Materials

Mental Math

Tell students to name 2 fractions which equal:

1. 1 (⅟₁, ²⁄₂, etc.)
2. 2 (²⁄₁, ⁴⁄₂, etc.)
3. 13
4. 4
5. 15
6. 9
7. 11
8. 8

Skill Review

Review addition of fractions by having students solve the following problems at the board: ⅔+½ (1⅙), ⅞+¾ (1⅝), ⅚+⅞ (1¹⁷⁄₂₄), ¹¹⁄₁₂+⅝ (1¹³⁄₂₄)

Adding Mixed Numbers

Betsy answers the telephone afternoons, at the library. How many hours does Betsy's time card show she has worked this week?

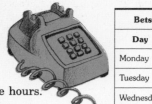

Betsy Kline	
Day	**Hours**
Monday	$3\frac{3}{4}$
Tuesday	$2\frac{1}{2}$
Wednesday	
Thursday	

We want to know Betsy's part-time hours.

Betsy worked $\underline{\ \ 3\frac{3}{4}\ \ }$ hours on Monday and

$\underline{\ \ 2\frac{1}{2}\ \ }$ hours on Tuesday.

To find her total hours, we add $\underline{\ \ 3\frac{3}{4}\ \ }$ and $\underline{\ \ 2\frac{1}{2}\ \ }$.

> **Add the fractions.**
> **Add the whole numbers.**

$$3\frac{3}{4} = 3\frac{3}{4}$$
$$+\ 2\frac{1}{2} = 2\frac{2}{4}$$
$$\overline{\phantom{+\ 2\frac{1}{2} =}\ 5\frac{5}{4}}$$

> **Simplify by trading.**

$$5\frac{5}{4} = 5 + \frac{4}{4} + \frac{1}{4}$$
$$5 + 1 + \frac{1}{4} = 6\frac{1}{4}$$

✔ Remember that a fraction with the same numerator and denominator, is equal to one.

Betsy has worked $\underline{\ \ 6\frac{1}{4}\ \ }$ hours this week.

Getting Started

Add. Simplify your answers if necessary.

1. $3\frac{1}{2}$
$+\ 7\frac{7}{8}$
$\overline{11\frac{3}{8}}$

2. $5\frac{2}{3}$
$+\ 2\frac{1}{6}$
$\overline{7\frac{5}{6}}$

3. $9\frac{1}{2}$
$+\ 5\frac{2}{3}$
$\overline{15\frac{1}{6}}$

4. $16\frac{1}{6}$
$+\ 25\frac{5}{9}$
$\overline{41\frac{13}{18}}$

Copy and add.

5. $10\frac{2}{3} + 4\frac{1}{6}$
$14\frac{5}{6}$

6. $5\frac{1}{9} + 5\frac{3}{18}$
$10\frac{5}{18}$

7. $8\frac{2}{5} + 9\frac{8}{10}$
$18\frac{1}{5}$

8. $11\frac{1}{4} + 2\frac{1}{3}$
$13\frac{7}{12}$ **277**

Teaching the Lesson

Introducing the Problem Have a student read the problem aloud and tell what is to be found. (number of hours Betsy worked this week) Ask students what information is known. (She worked 3¾ hours on Monday and 2½ hours on Tuesday.) Have students complete the sentences and the addition in the model with you. Have students use a number line to check their addition.

Developing the Skill Write **4+6** vertically on the board and have students give the sum. (10) Write **⅔+¾** vertically on the board and have a student work the problem. (1⁵⁄₁₂) Now write **4⅔+6¾** vertically on the board and tell students we add mixed numbers just like we add whole numbers and fractions. Tell students we add the fractions first because their sum may be 1 whole or more. Ask students the sum of the fractions in twelfths. (¹⁷⁄₁₂) Tell students we know that ¹⁷⁄₁₂ is ¹²⁄₁₂ plus ⁵⁄₁₂. Remind students that ¹²⁄₁₂ equals 1. Tell students we record the ⁵⁄₁₂ and trade the ¹²⁄₁₂ for 1 whole which is added to 4+6 for 11. Record the 11 and ask students to read the mixed number. (11⁵⁄₁₂) Repeat the procedure for 6²⁄₇+2³⁄₇ (8⁵⁄₇) 5⅜+4⅞ (10¼) and 2½+⅝. (3⅛)

Practice

Add. Simplify your answers if necessary.

1. $5\frac{1}{6}$
   $+ 6\frac{1}{2}$
   $\overline{11\frac{2}{3}}$

2. $4\frac{2}{3}$
   $+ 7\frac{5}{8}$
   $\overline{12\frac{7}{24}}$

3. $6\frac{3}{4}$
   $+ 5\frac{1}{3}$
   $\overline{12\frac{1}{12}}$

4. $9\frac{1}{2}$
   $+ 7\frac{1}{8}$
   $\overline{16\frac{5}{8}}$

5. $16\frac{3}{5}$
   $+ 9\frac{3}{4}$
   $\overline{26\frac{7}{20}}$

6. $6\frac{7}{8}$
   $+ 15\frac{5}{6}$
   $\overline{22\frac{17}{24}}$

7. $1\frac{2}{3}$
   $+ 6\frac{5}{8}$
   $\overline{8\frac{7}{24}}$

8. $6\frac{3}{5}$
   $+ 7\frac{3}{4}$
   $\overline{4\frac{7}{20}}$

Copy and Do

9. $5\frac{3}{8} + 2\frac{1}{2}$
   $7\frac{7}{8}$

10. $9\frac{7}{16} + 3\frac{7}{8}$
    $13\frac{5}{16}$

11. $4\frac{2}{3} + 16\frac{3}{4}$
    $21\frac{5}{12}$

12. $7\frac{3}{5} + 15\frac{1}{4}$
    $22\frac{17}{20}$

13. $6\frac{7}{10} + 16\frac{3}{4}$
    $23\frac{9}{20}$

14. $14\frac{7}{10} + 8\frac{5}{6}$
    $23\frac{8}{15}$

15. $9\frac{7}{8} + 7\frac{11}{12}$
    $17\frac{19}{24}$

16. $7\frac{2}{3} + 9\frac{5}{9}$
    $17\frac{2}{9}$

17. $6\frac{5}{6} + 18\frac{3}{7}$
    $25\frac{11}{42}$

18. $2\frac{4}{5} + 6\frac{3}{8}$
    $9\frac{7}{40}$

19. $5\frac{7}{8} + 7\frac{5}{7}$
    $13\frac{33}{56}$

20. $3\frac{2}{3} + 6\frac{3}{5}$
    $10\frac{4}{15}$

EXCURSION

Use the diagram to answer **true** or **false** for each statement.

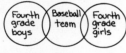

1. All of the fourth grade boys play on the baseball team. _False_

2. Only fourth graders play on the baseball team. _False_

3. There are boys and girls on the team. _True_

4. Some fourth graders do not play on the team. _True_

278

Practice

Remind students to add the fractions first and trade any fractions that equal a whole number. Have students work the problems independently.

Excursion

Discuss venn diagrams. Have students write one more true statement and one more false statement from the diagram after they have completed the excursion.

Correcting Common Errors

Some students may add the fractions but forget to add the whole numbers. Have students work in pairs to solve the problem ⅙ + ⅞, writing the answer in simplified form (2⅗). Next, ask them to rewrite ⅙ and ⅞ as mixed numbers and add (1⅕ + 1⅖ = 2⅗) and compare their answers to see that they are the same. Have them repeat the procedure for 9/4 + ⅜ (2⅝).

Enrichment

Tell students to write a problem of 2 mixed numbers whose sum is 25⅘, if 1 of the addends is 4 1/20.

Extra Credit   *Logic*

Dictate the following situation for students to solve: One day Jane White, Amy Black and Beth Brown were walking to school. "Look at our skirts," said one girl. "One is white, one is black and one is brown, just like our names."

"Yes," said the girl with the black skirt, "but none of us is wearing the skirt color to match our own name."

"I noticed that too," said Jane White. Can you identify what color skirt each girl is wearing? (Jane White, brown skirt; Amy Black, white skirt; Beth Brown, black skirt. Jane White isn't wearing white, because that would match her name. And she isn't wearing black, because she was talking to the girl wearing black. So her skirt has to be brown. Amy Black can't be wearing her name color and she can't be wearing brown, because Jane's skirt is brown. So Amy must be wearing white. That leaves the black skirt for Beth Brown.)

278

Adding Three Mixed Numbers

pages 279-280

Objective

To add 3 mixed numbers

Materials

Mental Math

Have students name 2 common multiples for:

1. 2 + 3 (6, 12, etc.)
2. 2 + 5 (10, 20, etc.)
3. 2 + 4 (4, 8, etc.)
4. 2 + 6 (6, 12, etc.)
5. 2 + 7 (14, 28, etc.)
6. 3 + 4 (12, 24, etc.)
7. 3 + 5 (15, 30, etc.)
8. 3 + 6 (6, 12, etc.)

Skill Review

Write on the board: **2⅔ + 4⅚** (7½), **9⅗ + 11⅓** (20¹⁴⁄₁₅), **24⅚ + 43⅞** (68¹⁷⁄₂₄). Have a student talk through each problem as it is worked. Continue for similar problems.

Adding Three Mixed Numbers

Marianne built a model trading fort as part of her project on fur traders in Colonial America. Find the perimeter of her triangular fort.

We want to know the perimeter of Marianne's fort.

The sides of the triangle measure $\underline{4\frac{3}{4}}$ inches, $\underline{3\frac{1}{2}}$ inches and $\underline{6\frac{3}{5}}$ inches.

To find the perimeter, we add $\underline{4\frac{3}{4}}$, $\underline{3\frac{1}{2}}$, and $\underline{6\frac{3}{5}}$.

$$4\frac{3}{4} = 4\frac{15}{20}$$
$$3\frac{1}{2} = 3\frac{10}{20}$$
$$+\ 6\frac{3}{5} = 6\frac{12}{20}$$
$$13\frac{37}{20} = 14\frac{17}{20}$$

The perimeter of Marianne's fort is $\underline{14\frac{17}{20}}$ inches.

Getting Started

Add.

1. $2\frac{1}{2}$
   $3\frac{5}{8}$
   $+\ 2\frac{1}{4}$
   $\overline{8\frac{3}{8}}$

2. $3\frac{1}{3}$
   $6\frac{5}{9}$
   $+\ 4\frac{5}{6}$
   $\overline{14\frac{13}{18}}$

3. $9\frac{1}{2}$
   $3\frac{4}{5}$
   $+\ 7\frac{5}{6}$
   $\overline{21\frac{2}{15}}$

4. $7\frac{5}{12}$
   $9\frac{3}{8}$
   $+\ 4\frac{5}{6}$
   $\overline{21\frac{5}{8}}$

Copy and add.

5. $10\frac{3}{4} + 2\frac{1}{2} + 7\frac{1}{4}$ $20\frac{1}{2}$

6. $5\frac{1}{3} + 11\frac{5}{6} + \frac{1}{12}$ $17\frac{1}{4}$

7. $11\frac{1}{4} + 3\frac{2}{3} + 2\frac{1}{2}$ $17\frac{5}{12}$

279

Teaching the Lesson

Introducing the Problem Have a student read the problem aloud and tell what is to be found. (the perimeter of the fort) Ask what information is given. (The triangle has sides of 4¾, 3½ and 6⅗ inches.) Ask students how to find the perimeter of a triangle. (side + side + side) Have students complete the sentences and addition in the model with you. Have students use a number line to check their solution.

Developing the Skill Write **2 + 3 + 5** vertically on the board. Have a student solve the problem. (10) Tell students that just as we can add several whole numbers, we can add several fractions. Write ½ + ⅓ + ¼ vertically on the board and tell students that we must find the least common denominator for these 3 fractions. Have a student list the multiples of 2, 3 and 4 on the board to find the least common multiple. (12) Have a student find the number of twelfths in ½, ⅓ and ¼ and write the equivalent fractions. (⁶⁄₁₂, ⁴⁄₁₂, ³⁄₁₂) Have students tell the sum in simplest form. (1¹⁄₁₂) Remind students that the answer is a mixed number because it is more than 1. Have students solve more addition problems of 3 simple fractions or mixed numbers with uncommon denominators.

279

Practice

Add.

1. $5\frac{1}{5}$
    $7\frac{3}{10}$
    $+ 4\frac{9}{20}$
    ___
    $16\frac{19}{20}$

2. $6\frac{2}{3}$
    $5\frac{1}{6}$
    $+ 7\frac{5}{9}$
    ___
    $19\frac{7}{18}$

3. $3\frac{2}{5}$
    $4\frac{1}{2}$
    $+ 9\frac{5}{10}$
    ___
    $17\frac{2}{5}$

4. $2\frac{5}{8}$
    $3\frac{5}{12}$
    $+ 4\frac{1}{4}$
    ___
    $10\frac{7}{24}$

5. $7\frac{5}{9}$
    $6\frac{1}{3}$
    $+ 4\frac{5}{6}$
    ___
    $18\frac{13}{18}$

6. $7\frac{3}{4}$
    $2\frac{1}{3}$
    $+ 11\frac{5}{6}$
    ___
    $21\frac{11}{12}$

7. $8\frac{5}{8}$
    $7\frac{1}{2}$
    $+ 9\frac{2}{3}$
    ___
    $25\frac{19}{24}$

8. $16\frac{3}{4}$
    $2\frac{1}{2}$
    $+ 5\frac{2}{3}$
    ___
    $24\frac{11}{12}$

Copy and Do

9. $5\frac{1}{8} + 6\frac{2}{3} + 9\frac{1}{2}$ $21\frac{7}{24}$

10. $7\frac{3}{4} + 5\frac{3}{8} + 7\frac{1}{2}$ $20\frac{5}{8}$

11. $9\frac{2}{3} + 4\frac{3}{5} + 6\frac{7}{10}$ $20\frac{29}{30}$

12. $4\frac{1}{3} + 8\frac{5}{6} + 12\frac{5}{9}$ $25\frac{13}{18}$

13. $16\frac{1}{2} + 5\frac{2}{3} + 8\frac{1}{4}$ $30\frac{5}{12}$

14. $7\frac{5}{6} + 9\frac{1}{2} + 8\frac{2}{3}$ 26

EXCURSION

Unscramble each word, writing one letter in a box. To find the mystery word, unscramble the circled letters.

ATIRO	Ⓡ	A	T	I	O				
DERECU	R	E	D	U	Ⓒ	E			
MYSIPFLI	S	Ⓘ	M	P	L	I	Ⓕ	Y	
MERUNATOR	N	U	M	E	Ⓐ	T	O	R	
EVTELQUAIN	E	Q	U	I	V	A	L	E	Ⓝ	Ⓣ
WLETOS MREST	L	Ⓞ	W	E	S	T	T	E	R	M	S

Mystery Word: | F | R | A | C | T | I | O | N |

280

Practice

Remind students to find a common denominator for all 3 fractions. Have students complete the problems independently.

Excursion

Tell students they are looking for words from their vocabulary list for fractions.

Extra Credit

This activity will help students sharpen their estimation and measurement skills. Collect a variety of objects that are measured by varying units, such as those suggested in the following questions. Tell students to estimate to answer the following, as you hold up each item; (Answers will vary.)

1. How much does this full box of salt weigh?
2. How many pages in this volume of the encyclopedia? How many pages in the entire set?
3. How many inches long is the baseball bat?
4. What is the circumference of this baseball, in centimeters?
5. How many of these crayons, put end to end, would equal 2 feet?

Have volunteers measure each object to find the exact answer and write these on the board. Have students compare their guesses to the measurements.

Problem Solving
Open Sentence

pages 281-82

Objective

To write an open sentence to solve a problem

Materials

Mental Math

Have students name a unit fraction whose value is less than:

1. $\frac{1}{10}$ ($\frac{1}{11}$, $\frac{1}{12}$, etc.)
2. $\frac{1}{2}$
3. $\frac{1}{6}$
4. $\frac{1}{3}$
5. $\frac{1}{4}$
6. $\frac{1}{15}$
7. $\frac{1}{20}$
8. $\frac{1}{5}$

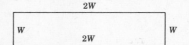
Writing an Open Sentence

A mouse walks around the outside of a rectangular-shaped piece of cheese whose length is twice its width. If the mouse walks a total of 12 inches, what are the dimensions of the piece of cheese?

★ **SEE**

We want to find the length and the width of the piece of cheese.
The perimeter of the cheese is __12__ inches.
The length of cheese is double that of its width.
If the width were 1 inch, the length would be __2__ inches.

★ **PLAN**

We make a sketch of the rectangle of cheese with notation. If we label the width with the letter W, then the length is $2 \times W$. This is written: **2W.**

```
              2W
   ┌──────────────────────┐
 W │          2W          │ W
   └──────────────────────┘
```

The mouse walks:
$2W + W + 2W + W$.

★ **DO**

The **open sentence** that represents the distance the mouse walks is:

$2W + W + 2W + W = 12$
$6W = 12$ (We add all the W's. $2 + 1 + 2 + 1 = 6$)
$W = 2$ (We reason if 6 times $W = 12$, the value of W must be 2 because $6 \times 2 = 12$.)

If $W = 2$, then $2W =$ __4__ .

The width of the cheese is __2__ inches and the length is __4__ inches.

★ **CHECK**

We can check by adding the lengths and widths together.

$4 + 2 + 4 + 2 =$ __12__

281

Teaching the Lesson

Tell students that 2 plus another number equals 30. Tell students we want to know the other number. Write on the board: **2 + some number = 30.** Tell students it would be easier if we could use a symbol to talk about the unknown number. For example, write **2 + Y = 30** on the board. Tell students this is called an open sentence because there is an unknown quantity. Ask students what number plus 2 will equal 30. (28) Now tell students that 20 plus 2 times another number equals 30. Tell students we write an open sentence using a letter for the unknown number. Write **20 + 2 × R = 30** on the board. Tell students we can simplify $2 \times R$ by writing it 2R as you write **20 + 2R = 30** on the board. Tell students that if 2R + 20 = 30 then 2R would equal 30 − 20 or 10 as you write **2R = 10** on the board. Ask what number times 2 equals 10. (5) Write **R = 5** on the board. Tell students we must check the work as you write **20 + 2 × 5 = 30** on the board. Remind students we work multiplication first as you write **20 + 10 = 30** on the board.

Have a student read the problem aloud and tell what is to be found. (the dimensions of the cheese) Ask what information is known. (The cheese is a rectangular shape, its length is 2 times its width and total distance around it is 12 inches.) Remind students that 2 times an unknown number can be written as 2R. Have students read with you and complete the SEE, PLAN and DO sections to solve the problem. Have them CHECK their solution.

Apply

Write an open sentence to solve each problem.

1. What is the perimeter of a square checkerboard 12 inches on a side?
48 inches

2. Find the distance around a square pasture 45 meters on each side.
180 meters

3. There are 12 months in a year. Write an open sentence to find how many months are in 10 years. How many months are in 100 years?
120 months   1,200 months

4. A basketball game is played where a team scored 55 points. Twenty baskets were made and each is worth two points. A free throw is worth one point. Write an open sentence to help find how many free throws were made.
15 free throws

5. The sum of two numbers is 90. The first number is twice the second number. What are the two numbers?
60, 30

6. Jason has saved 3 nickels, 8 dimes and an unknown number of quarters. The total amount he saved is $2.20. Write an open sentence that will help determine the number of quarters Jason has.
5 quarters

7. Read Exercise 4 again and the open sentence that you wrote. What if the team scored 60 points? How would you change your open sentence to fit this new situation?
Answers will vary.

8. Read Exercise 5 again. Rewrite the exercise so that the second number is twice the first number. Then explain how this changes the answer.
See Solution Notes.

9. Write an addition sentence to show two like fractions that have a sum with a different denominator.
Answers will vary.

10. Explain how you could write many different number sentences where 1 is the sum of two like fractions.
Answers will vary.

282

Solution Notes

1. Let P = perimeter
   Let S = length of side
   P = 4S
   P = 4 × 12 inches
   P = 48 inches

2. Let P = perimeter of pasture
   Let S = length of side
   P = 4S
   P = 4 × 45
   P = 180 meters

3. Let M = number of months
   Let Y = number of years
   M = 12Y
   M = 12 × 10
   M = 120 months
   M = 12 × 100
   M = 1,200 months

4. Let F = number of free throws
   Let B = number of baskets
   2B + F = 55
   2 × 20 + F = 55
   40 + F = 55
   F = 15 free throws

5. Let A = 2nd number
   2A + A = 90
   3A = 90
   A = 30

6. Let N = 5
   Let D = 10
   Let Q = 25
   3N + 8D + ?Q = 220
   (3 × 5) + (8 × 10) + (?Q) = 220
   15 + 80 + (?Q) = 220
   95 + (?Q) = 220
   ?Q = 125
   ? = 5

Higher-Order Thinking Skills

7. [Analysis] Any correct answer will, in some way, indicate changing 55 in the open sentence to 60.

8. [Synthesis] Change the second sentence to say, "The second number is twice the first number." This really does not change the answer; the two numbers are still 30 and 60.

9. [Synthesis] The answer will be any two fractions with like denominators with a sum that can be simplified, such as $1/8 + 3/8 = 1/2$.

10. [Analysis] Show fractions where the sum of the numerators equals the denominator; for example, $1/4 + 3/4 = 1$.

Extra Credit *Logic*

Dictate the following and have students answer the questions:

Matthew shares his bedroom with his younger brother, who gets up after Matthew goes to school. This presents a problem for Matthew because he has to pick out his school clothes in the dark. One of his drawers is filled with an even number of unpaired black socks and white socks. What is the smallest number of socks he must pull from the drawer to make sure he will have a matched pair? How many socks must he pull out to make sure he has two pairs of matching socks? If Matthew is going camping and needs 9 pairs of socks, what is the least number he must pull out? Ask students to identify the pattern they see. (3, 5, 19. If Matthew needs X number of pairs then double X plus 1 is the number of socks he needs.)

282

Calculators and Fractions

pages 283-284

Objective

To use a calculator to find a sale price

Materials

calculators

Mental Math

Have students name the least common multiple for:

1. 3 and 4. (12)
2. 5 and 4. (20)
3. 6 and 9. (18)
4. 12 and 5. (60)
5. 2 and 8. (8)
6. 4 and 6. (12)
7. 7 and 3. (21)
8. 5 and 3. (15)

Skill Review

Review finding unit and non-unit parts of numbers by having students work the following dictated problems at the board: $\frac{2}{3}$ of 468 (312), $\frac{4}{5}$ of 750 (600), $\frac{3}{10}$ of 100 (30), $\frac{7}{9}$ of 846 (658).

Calculators and Fractions

Mr. Smalley is writing a newspaper advertisement for computers and stereos in his store. What are the sale prices of the computers and the stereos?

We want to know the sale prices of the computers and the stereos.

The original price of a computer was ___$599___.

The original price of a stereo was ___$750___.

Mr. Smalley is offering $\frac{3}{5}$ off each original price.
To find sale prices, we first must find the money that the customer saves. This is called the **discount**. To find the discount, we find $\frac{3}{5}$

of the original price. We multiply ___$599___ by

___3___ and divide by ___5___.

599 ⊠ 3 ÷ 5 ⊟ (359.4)

The discount on the computer is ___$359.40___.
To find the sale price, we subtract the discount

from the original price. We subtract ___$359.40___

from ___$599___.

The sale price of the computer is ___$239.60___.
Complete the codes to find the sale price of the stereo.

___750___ ⊠ ___3___ ÷ ___5___ ⊟ (450)

The discount on the stereo is ___$450___.

___750___ ⊟ ___450___ ⊟ (300)

The sale price of the stereo is ___$300___.

Clearance
Big Discounts
$\frac{3}{5}$ off all stock!
→ Computers
were: $599. now:
→ Stereos
were: $750 now:

283

Teaching the Lesson

Introducing the Problem Have students describe the picture, read the problem aloud and tell what is to be found. (the sale prices of computers and stereos) Ask students what information is known. (Computers were $599 and stereos were $750 before the sale of $\frac{3}{5}$ off.) Have students complete the sentences and the calculator codes to find the sale prices. Have students use their calculators to find each original price to check their solutions.

Developing the Skill Write **$1.00** on the board. Tell students they are to find the sale price of a $1.00 item if it is now $\frac{3}{4}$ off. Have a student write the calculator code of **1 × 3 ÷ 4 =** on the board. Remind them that the order of operations does not matter, for example 1 × 3 ÷ 4 or 1 ÷ 4 × 3. Tell students we multiply the 1 by the numerator and divide by the denominator to find the amount off or the **discount**. Have students work the code. (0.75) Remind students that 0.75 means 75¢ off. Tell students that we now want to find the new cost after the discount, so we subtract the discount from the original price. Have a student write the calculator code on the board (1 − .75 =) Have students check their work by adding .75 and .25. Have students find the sale price of an $879 item if it is $\frac{2}{3}$ off. ($293)

283

Practice

Use your calculator to find these discounts.

1. $\frac{2}{3}$ of $186 $124

2. $\frac{5}{8}$ of $924 $577.50

3. $\frac{4}{5}$ of $625 $500

Complete the table.

4.
Original Price	$1,260	$2,124	$1,125	$5,275
Fraction Off	$\frac{3}{4}$	$\frac{5}{6}$	$\frac{2}{9}$	$\frac{1}{2}$
Discount	$945	$1,770	$250	$2,637.50
Sale Price	$315	$354	$875	$2,637.50

Apply

Solve these problems.

5. Miss Ortiz received a bonus equal to $\frac{2}{5}$ of her yearly salary. Her yearly salary is $37,500. How much is her bonus?
$15,000

6. Mr. Roberts bought a car for $7,950. He paid $\frac{3}{10}$ of the amount as a down payment. How much does Mr. Roberts still owe on the car?
$5,565

EXCURSION

Steve figured out a shortcut to help Mr. Smalley find the sale prices faster. He figured that if the customer was going to **save** $\frac{3}{5}$ of the original price, he would **pay** $\frac{2}{5}$ of it. Steve used this calculator code to find the sales price of the stereo.

750 $\times$ 2 $\div$ 5 $=$ 300

Complete this table to show that Steve's shortcut works everytime.

Original Price	Fraction Off	Sale Price
$870	$\frac{2}{3}$	$290
$850	$\frac{1}{4}$	$637.50
$648	$\frac{5}{9}$	$288
$1,026	$\frac{5}{6}$	$171
$460	$\frac{3}{10}$	$322

284

Practice

Help students plan how to solve the first problem in the table. Then have students complete the page independently.

Excursion

Encourage students to think of the original price as the "one whole thing or 100%". Remind them that one can be expressed as a fraction where the numerator and denominator are the same. Thus, the original price of $870 in the first problem is 100%, or 1 or ⅔. If we do not pay ⅔ of the $870, the ⅓ that remains is the cost.

Correcting Common Errors

Some students may confuse the discount and the sale price. Have them work with partners to find the sale price of a sweater marked $48 with ⅔ off. One student should write the calculator code for the discount,

48 $\times$ 2 $\div$ 3 $=$,

and use a calculator to find it ($32); the other students should write the calculator code for the sale price,

48 $-$ 32 $=$,

and use the calculator to find it ($16). Then have them reverse their roles.

Enrichment

Tell students to cut an ad from the newspaper that shows an item on sale at a fraction of its original price. Tell them to find the sale price.

Extra Credit *Numeration*

Prepare two sets of cards with the numerals 0 through 9 on them, and one set showing +, − and = signs. Divide students into groups of at least six each. Shuffle the cards, let each student in a group draw a card. Have each group, within a fifteen second period, arrange their cards to form the largest number possible. Then give each group the operation sign cards and tell them to use any or all of the cards, to form a true number sentence. Repeat by having groups draw different cards. Extend the activity by changing the time limit, or adding other criteria to the numbers they are to form.

Chapter Test

Item	Objective
1, 3, 6-8	Add fractions with common denominators (See pages 265-266)
2, 4, 5	Subtract fractions with common denominators (See pages 267-268)
10, 14	Subtract fractions when one denominator is a multiple of other denominator (See pages 273-274)
9, 11-13 15-16, 21-22	Add or subtract two fractions with uncommon denominators (See pages 275-276)
17-20, 23	Add two mixed numbers (See pages 277-278)
24	Subtract two mixed numbers (See pages 273-274)

CHAPTER TEST

Add or subtract. Simplify your answers.

1. $\frac{3}{5}$
  $+\frac{1}{5}$
  $\overline{\frac{4}{5}}$

2. $\frac{7}{8}$
  $-\frac{4}{8}$
  $\overline{\frac{3}{8}}$

3. $\frac{2}{7}$
  $+\frac{4}{7}$
  $\overline{\frac{6}{7}}$

4. $\frac{5}{9}$
  $-\frac{4}{9}$
  $\overline{\frac{1}{9}}$

5. $\frac{11}{12}$
  $-\frac{5}{12}$
  $\overline{\frac{1}{2}}$

6. $\frac{5}{10}$
  $+\frac{3}{10}$
  $\overline{\frac{4}{5}}$

7. $\frac{1}{8}$
  $+\frac{5}{8}$
  $\overline{\frac{3}{4}}$

8. $\frac{5}{6}$
  $+\frac{1}{6}$
  $\overline{1}$

9. $\frac{2}{3}$
  $+\frac{3}{4}$
  $\overline{1\frac{5}{12}}$

10. $\frac{7}{8}$
  $-\frac{1}{4}$
  $\overline{\frac{5}{8}}$

11. $\frac{3}{5}$
  $+\frac{1}{3}$
  $\overline{\frac{14}{15}}$

12. $\frac{7}{4}$
  $+\frac{5}{6}$
  $\overline{2\frac{7}{12}}$

13. $\frac{9}{8}$
  $-\frac{5}{6}$
  $\overline{\frac{7}{24}}$

14. $\frac{11}{10}$
  $-\frac{4}{5}$
  $\overline{\frac{3}{10}}$

15. $\frac{9}{6}$
  $+\frac{7}{9}$
  $\overline{2\frac{5}{18}}$

16. $\frac{15}{12}$
  $-\frac{9}{8}$
  $\overline{\frac{1}{8}}$

17. $3\frac{5}{8}$
  $+2\frac{3}{4}$
  $\overline{6\frac{3}{8}}$

18. $7\frac{2}{3}$
  $+8\frac{1}{4}$
  $\overline{15\frac{11}{12}}$

19. $6\frac{3}{4}$
  $+6\frac{1}{6}$
  $\overline{12\frac{11}{12}}$

20. $3\frac{2}{3}$
  $+9\frac{1}{8}$
  $\overline{12\frac{19}{24}}$

21. $\frac{1}{2} + \frac{3}{7}$
  $\frac{13}{14}$

22. $\frac{2}{3} - \frac{2}{7}$
  $\frac{8}{21}$

23. $4\frac{3}{5} + 2\frac{1}{2}$
  $7\frac{1}{10}$

24. $7\frac{3}{10} - 1\frac{1}{4}$
  $6\frac{1}{20}$

Circle the letter of the correct answer.

1 Round 6,296 to the nearest thousand.
- (a) 6,000
- b 7,000
- c NG

2 What is the place value of the 7 in 736,092?
- a hundreds
- b thousands
- (c) hundred thousands
- d NG

3
$$5,629 + 24,751$$
- a 29,380
- b 30,370
- (c) 30,380
- d NG

4
$$8,096 - 1,998$$
- a 7,008
- b 7,098
- c 7,902
- (d) NG

5 Choose the better estimate of weight.
- (a) 1 gram
- b 1 kilogram

6
$$\$3.26 \times 6$$
- a $18.26
- b $18.56
- (c) $19.56
- d NG

7 67 × 53
- a 536
- b 3,451
- (c) 3,551
- d NG

8 8)265
- a 3 R1
- b 30 R1
- (c) 33 R1
- d NG

9 760 ÷ 42
- a 1 R4
- (b) 18 R4
- c 180 R4
- d NG

10 Find the perimeter.

12 in.   12 in.
15 in.
- a 24 in.
- b 27 in.
- (c) 39 in.
- d NG

11 Find the area.
- (a) 24 sq units
- b 48 sq units
- c 48 units
- d NG

12 $\frac{2}{3}$ of 48 = n
   n = ?
- a 8
- b 16
- (c) 32
- d NG

☐ score

286

Cumulative Review

page 286

Item	Objective
1	Round numbers to nearest 1,000 (See pages 33-34)
2	Identify place value through hundred thousands (See pages 27-28)
3	Add two numbers up to 5-digits (See pages 51-52)
4	Subtract two 4-digit numbers with zero in minuend (See pages 67-68)
5	Determine appropriate metric unit of weight (See pages 159-160)
6	Multiply money by 1-digit number (See pages 169-170)
7	Multiply two 2-digit numbers with trading (See pages 179-180)
8	Divide 3-digit number by 1-digit number to get 2-digit quotient with remainder (See pages 197-198)
9	Divide 3-digit number by 2-digit number to get 2-digit quotient with remainder (See pages 209-210)
10	Find perimeter of a parallelogram (See pages 229-230)
11	Find area of a rectangle (See pages 231-232)
12	Find fraction of a number by multiplying and dividing (See pages 245-246)

Alternate Cumulative Review

Choose the letter of the correct answer.

1 Round 4,500 to the nearest thousand.
- a 4,000
- (b) 5,000
- c NG

2 What is the value of the 2 in 326,985?
- a hundreds
- b thousands
- (c) ten thousands
- d NG

3
$$4,765 + 34,527$$
- a 38,292
- b 39,282
- (c) 39,292
- d NG

4
$$9,084 - 2,276$$
- (a) 6,808
- b 6,818
- c 7,818
- d NG

5 What unit would you use to weigh a pencil?
- a ounces
- b pounds
- c tons
- (d) NG

6
$$\$7.35 \times 7$$
- a $49.45
- b $51.15
- c $51.55
- (d) NG

7 32 × 74 =
- (a) 2,368
- b 2,378
- c 2,468
- d NG

8 9)380
- a 4 R2
- b 40 R2
- (c) 42 R2
- d NG

9 496 ÷ 32 =
- (a) 15 R16
- b 105 R16
- c 150 R16
- d NG

10 Find the perimeter of a triangle 9 cm by 9 cm by 10 cm.
- a 81 cm
- (b) 28 cm
- c 90 cm
- d NG

11 Find the area of a 6 m square.
- a 13 sq m
- b 26 sq m
- (c) 36 sq m
- d NG

12 $\frac{3}{4}$ of 84 =
- a 21
- b 28
- (c) 63
- d NG

286

Tenths

pages 287-288

Objective

To understand and write tenths

Materials

Mental Math

Ask students how many in:

1. ⅔ of 81 (18)
2. 4,261 + 9 tens − 6 ones (4,345)
3. (280 ÷ 14) × 2 (40)
4. ½ + ⅙ (⅔)
5. ⁵⁄₇ − ¹⁄₁₄ (⁹⁄₁₄)
6. 2½ + 6⅓ (8⅚)
7. LCM of 4, 3 and 6 (12)
8. LCM of 9, 3 and 6 (18)

Skill Review

Dictate the following fractions for students to write on the board: 3⁵⁄₁₀, 4⁶⁄₁₀, 13⁹⁄₁₀, 27⁴⁄₁₀, 9³⁄₁₀. Now have students dictate fractions for other students to write on the board.

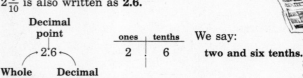

DECIMALS 14

Understanding Tenths

Rita is painting scenery panels for the class play. What decimal represents the number of panels Rita has painted so far?

We want to know how many panels are painted.

Each panel has __10__ equal parts.

Rita has painted __2__ complete panels, and __6__ parts of another panel.

We can write this as the mixed number $2\frac{6}{10}$.

$2\frac{6}{10}$ is also written as **2.6**.

Decimal point

⌒2.6⌒

ones	tenths
2	6

We say:
two and six tenths.

Whole number   Decimal number

Rita has painted __2.6__ panels so far.

✔ Decimal numbers can name parts less than 1. Rita still has 0.4 of a panel to paint.

⌒0.4⌒

0 ones   4 tenths

We say: **four tenths.**

✔ Whole numbers can be written as decimal numbers. When she completes her job, Rita will have painted 3.0 panels.

⌒3.0⌒

3 ones   0 tenths

We say: **three.**

Getting Started

Write the decimal numbers to show how many panels are painted.

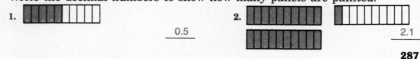

1. _____

0.5

2. _____

2.1

287

Teaching the Lesson

Introducing the Problem Have a student read the problem and tell what is to be found. (The decimal number that tells the number of panels painted so far.) Ask what information is given in the picture. (2 panels have 10 parts painted and 1 has 6 of 10 painted.) Have students complete the sentences. Read the rules and definitions in the model with them, as they discuss the situation beyond the solution sentence.

Developing the Skill Write 1²⁄₁₀ on the board. Have a student read the mixed number. Tell students that any fraction which has 10 as the denominator can be written as a **decimal**. Write **1.2** on the board and tell students that 1²⁄₁₀ can be written as 1 and a decimal point and then the numerator of the fraction. Write **4⁵⁄₁₀** and **4.5** on the board and tell students that the decimal point separates the whole number from the fraction part and we say the word, **and**, for the decimal point when reading a decimal. Write ¹⁄₁₀ and **0.1** on the board and tell students that less than 1 whole is written with a zero as a place holder in the whole number place. Write **4.0** on the board and tell students a whole number can be written as a decimal number with zero as a place holder in tenths place. Write on the board: **three and seven tenths, 3.7, 3⁷⁄₁₀.** Dictate more decimals for students to tell the number words, the decimal and the fraction.

287

Practice

Write the decimal number to show how many panels are painted.

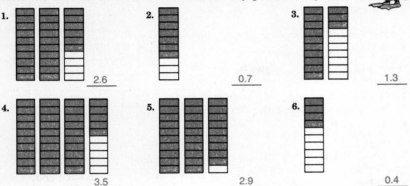

1. _____ 2.6

2. _____ 0.7

3. _____ 1.3

4. _____ 3.5

5. _____ 2.9

6. _____ 0.4

Write the decimal numbers.

7. six and four tenths _6.4_

8. three and seven tenths _3.7_

9. eight tenths _0.8_

10. $\frac{7}{10}$ _0.7_

11. eight _8.0_

12. four and one tenth _4.1_

13. $5\frac{9}{10}$ _5.9_

14. one and nine tenths _1.9_

15. seven and two tenths _7.2_

Write the decimal numbers in words.

16. 4.3
four and three tenths

17. 5.9
five and nine tenths

18. 6.0
six

19. 2.7
two and seven tenths

20. 7.0
seven

21. 6.9
six and nine tenths

22. 9.9
nine and nine tenths

23. 5.3
five and three tenths

24. 1.5
one and five tenths

288

Correcting Common Errors

Some students may forget that the decimal point separates the whole-number part from tenths and write six and seven tenths as 0.67. Have these students work with partners taking turns writing the decimal for each of the following on a place-value chart.

six tenths	(0.6)
four and six tenths	(4.6)
three tenths	(0.3)
five and three tenths	(5.3)
nine tenths	(0.9)
one and nine tenths	(1.9)

After writing each decimal number, partners should discuss whether the notation is correct and why or why not.

Enrichment

Tell students to draw a picture to show each of the following decimals: 0.4, 2.9, 4.3, 6.0.

Practice

Read the directions for each problem section with students, and have them complete the page independently.

Extra Credit *Probability*

One way of expressing the probability of two events occurring is to create an **ordered pair**, symbols that express the different ways in which some event could take place. Give students the example of throwing a 6 on one die and then drawing a heart from a deck of cards. Have them use the symbols 1, 2, 3, 4, 5, 6 as the first number in the ordered pair, representing all the possible outcomes of the throw of one die. Ask them to use the letters H, S, D, C (heart, spade, diamond, club) for the four ways of drawing one suit from a pack containing four suits. Ask one student to write the ordered pair describing a throw of 3 and a draw of a diamond. (3, D) Have students list all the ordered pairs for throwing a die and drawing a card. [(1, H) (1, S) (1, D) (1, C) (2, H) (2, S) (2, D) (2, C) (3, H) (3, S) (3, D) (3, C) (4, H) (4, S) (4, D) (4, C) (5, H) (5, S) (5, D) (5, C) (6, H) (6, S) (6, D) (6, C)]

Hundredths

pages 289-290

Objective

To understand and write hundredths

Materials

Mental Math

Ask students if all the factors of 8 are 1, 2, 4 and 8, what are all the factors of:

1. 6 (1, 2, 3, 6)
2. 3 (1, 3)
3. 4 (1, 2, 4)
4. 5 (1, 5)
5. 7 (1, 7)
6. 9 (1, 3, 9)
7. 10 (1, 2, 5, 10)
8. 12 (1, 2, 3, 4, 6, 12)

Skill Review

Review decimals in tenths by having students write each of the following in words, as a fraction and as a decimal: $2\frac{5}{10}$, $\frac{9}{10}$, $14\frac{1}{10}$, $5\frac{3}{10}$, $6\frac{2}{10}$, $7\frac{8}{10}$, $261\frac{4}{10}$.

Understanding Hundredths

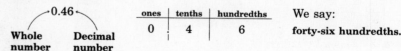

Dick is making a mosaic tile picture to enter in the art contest. What decimal represents the part of the picture that Dick has tiled so far?

We want to know what part of the picture is finished.

The mosaic will be made up of __100__ tiles.

Dick has used __46__ of the tiles so far. We can write this as the fraction $\frac{46}{100}$.

$\frac{46}{100}$ is also written as **0.46**.

0.46

Whole number **Decimal number**

ones	tenths	hundredths
0	4	6

We say: **forty-six hundredths.**

Dick has finished __0.46__ of the mosaic tile picture so far.

Getting Started

Write the decimal number for the part that has been tiled.

1. _____ 2. _____ 3. _____

1.09   3.50   2.85

Write the decimal numbers.

4. six and twelve hundredths __6.12__

5. $3\frac{36}{100}$ __3.36__

Write the decimal numbers in words.

6. 0.08 __eight hundredths__

289

Teaching the Lesson

Introducing the Problem Have a student read the problem aloud and tell what is to be found. (the decimal which tells how much of the picture is tiled) Ask students what information is given in the picture. (There are 10 rows of 10 squares or 100 squares in all. 46 squares are tiled.) Have students read with you as they complete the sentences. Help them put the number correctly in the place value chart and complete the solution sentence.

Developing the Skill Write $\frac{42}{100}$ on the board. Tell students that just as we can write $\frac{4}{10}$ as 0.4, we can also write any fraction which has a denominator of 100 as a decimal. Write **.42** on the board and tell students we read this decimal just like the fraction of $\frac{42}{100}$. Write **forty-two hundredths** on the board and have a student read it. Write on the board:

Ones	.	Tenths	Hundredths
0	.	**4**	**2**
(1)	.	(3)	(9)
(1)	.	(5)	(0)
(10)	.	(0)	(1)

Dictate the following decimals for students to write on the grid: 1.39, 1.50, 10.01. Have students write each decimal as a fraction and in number words.

289

Practice

Write the decimal number for the part that has been tiled.

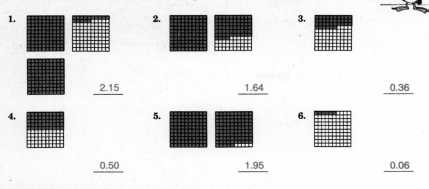

1. _2.15_

2. _1.64_

3. _0.36_

4. _0.50_

5. _1.95_

6. _0.06_

Write the decimal numbers.

7. four and twenty-five hundredths _4.25_

8. six and nine hundredths _6.09_

9. $14\frac{12}{100}$ _14.12_

10. five hundredths _0.05_

11. seven and thirty-seven hundredths _7.37_

12. seventy-eight hundredths _0.78_

Write the decimal numbers in words.

13. 3.29

three and twenty-nine _hundredths_

14. 6.01

six and one _hundredths_

15. 3.09

three and nine _hundredths_

16. 7.50

seven and fifty _hundredths_

17. 0.16

sixteen hundredths

18. 0.03

three hundredths

five hundredths

290

Thousandths

Objective

To understand and write thousandths

Materials

Mental Math

Tell students to find the average of:

1. 2½ and 3½ (3)
2. 4¼, 7¼ and ²⁄₄ (4)
3. 6, 9 and 3 (6)
4. 4⅔ and 15⅓ (10)
5. 200, 300 and 850 (450)
6. ⁷⁄₁₂ and 1⁵⁄₁₂ (1)
7. 3¼ and ¾ (2)
8. 68 and 100 (84)

Skill Review

Review decimals in tenths and hundredths by having students write each of the following decimals on a place value grid showing ones, tenths and hundredths: 1.4, 2.66, .38 8.2, 5.7 and 0.16.

Understanding Thousandths

Mr. Half Moon buys supplies for the school cafeteria. He is purchasing a quantity of meat. How much meat is he buying?

We want to understand the reading on the scale.

The scale reads ___36.258___ pounds.
We can use a place value chart to help understand the weight of the meat.

tens	ones	tenths	hundredths	thousandths
3	6	2	5	8

We say: **thirty-six and two hundred fifty-eight thousandths.**

Mr. Half Moon bought ___36.258___ pounds of meat.

Getting Started

Name the place value of the red digit.

1. 76.2⁵9 ___hundredths___   2. 116.³26 ___tenths___   3. ⁴.405 ___ones___

Name the place value of 4 in each number.

4. 390.421 ___tenths___   5. 26.045 ___hundredths___   6. 5.324 ___thousandths___

Write the decimal numbers.

7. two and sixteenth thousandths   8. seven thousandths

___2.016___   ___0.007___

Write the decimal numbers in words.

9. 29.752 ___twenty-nine and seven hundred fifty-two thousandths___

10. 5.308 ___five and three hundred eight thousandths___

Teaching the Lesson

Introducing the Problem   Have a student read the problem and tell what is to be found. (the amount of meat being bought) Have students complete the sentences and write the number correctly in the place value chart. Have them read the number words and complete the solution sentence.

Developing the Skill   Remind students that we write ⁴⁸⁄₁₀₀ as **0.48** and ⁴⁄₁₀ as **0.4** as you write the fractions and decimals on the board. Write ²³⁸⁄₁₀₀₀ on the board and tell students we write thousandths as decimals also. Write **0.238** on the board and tell students we read this decimal just like the fraction of ²³⁸⁄₁₀₀₀. Write **two hundred thirty-eight thousandths** on the board and have a student read it. Now write **0.860** on the board and tell students the 8 is in tenths place, the 6 is in hundredths place and the zero is in thousandths place. Have a student read the number. Write **6.016** on the board and have students tell the place value of each number. (6 ones, 0 tenths, 1 hundredth, 6 thousandths) Repeat for the following decimal numbers: 12.001, 8.206, 46.200, 0.103, 4.555. Help students write and read each decimal number in number words.

Practice

Name the place value of the red digit.

1. 13.26 ___tenths___

2. 67.154 ___thousandths___

3. 7.041 ___tenths___

4. 216.107 ___hundredths___

5. 2.478 ___hundredths___

6. 23.840 ___thousandths___

Name the value of the 6 in each number.

7. 21.684 ___tenths___

8. 16.273 ___ones___

9. 143.86 ___hundredths___

10. 4.156 ___thousandths___

11. 29.465 ___hundredths___

12. 3.006 ___thousandths___

Write the decimal numbers.

13. five and fifty-seven thousandths ___5.057___

14. twelve and one hundred thirty-nine thousandths ___12.139___

15. two hundred and four hundred eight thousandths ___200.408___

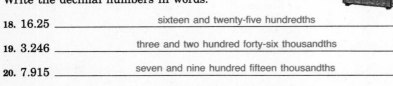

16. eighty-one and seven thousandths ___81.007___

17. fifty-three and seventy-five thousandths ___53.075___

Write the decimal numbers in words.

18. 16.25 ___sixteen and twenty-five hundredths___

19. 3.246 ___three and two hundred forty-six thousandths___

20. 7.915 ___seven and nine hundred fifteen thousandths___

21. 12.021 ___twelve and twenty-one thousandths___

22. 14.6 ___fourteen and six tenths___

23. 29.003 ___twenty-nine and three thousandths___

24. 114.218 ___one hundred fourteen and two hundred eighteen thousandths___

25. 16.255 ___sixteen and two hundred fifty-five thousandths___

26. 39.608 ___thirty-nine and six hundred eight thousandths___

292

Practice

Remind students to be careful about spelling, and have them complete the page independently.

Mixed Practice

1. $\frac{2}{3} + \frac{5}{8}$ $\left(1\frac{7}{24}\right)$
2. 957×24 (22,968)
3. Simplify: $\frac{28}{36}$ $\left(\frac{7}{9}\right)$
4. $1,300 \div 48$ (27 R4)
5. $\frac{7}{9} - \frac{1}{6}$ $\left(\frac{11}{18}\right)$
6. $295 \div 7$ (42 R1)
7. $\frac{7}{8}$ of 40 (35)
8. $3\frac{3}{5} + 2\frac{3}{4}$ $\left(6\frac{7}{20}\right)$
9. $250.00 - 176.32$ ($73.68)
10. $65,176 + 84 + 2,174$ (67,434)

Extra Credit *Logic*

Read or duplicate the following for students to solve: A girl saw two boys in the park who looked exactly alike. She asked, "Are you boys twins?"

"No, we are not twins," one boy replied.

"That's right," said the other. "We have the same parents, we were born on the same day of the same year, but we're not twins." How can this be so? (The boys are 2 members of a set of triplets.)

Comparing Decimals

Objective

To compare decimal numbers

Materials

Mental Math

Have students name the amount of money:

1. 1.02 ($1.02)
2. 15.10 ($15.10)
3. .76 (76¢)
4. 4.10 ($4.10)
5. 19.36 ($19.36)
6. 7.70 ($7.70)
7. 3.90 ($3.90)
8. .15 (15¢)

Skill Review

Review place value by writing numbers such as **405.06** on the board. Ask students to name the value of each digit. Now write **405 ◯ 406** on the board and have a student write >, < or = in the circle to show the relationship of the numbers. (<) Repeat for more whole number comparisons.

Comparing Decimals

Three friends competed in the Special Olympics. One of their best events was the 100-meter dash. In what order did they finish the race?

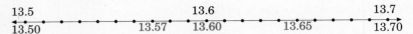

100 meter dash		
Bert	42	13.6
Dennis	12	13.65
Daryle	8	13.57

We want to know in what order the boys finished the race.

We can arrange __13.6__, __13.65__ and __13.57__ in order on a number line.

```
13.5                    13.6                              13.7
13.50        13.57    13.60          13.65          13.70
```

✔ Notice that 13.5 and 13.50 are the same value. Zeros to the far right of a decimal number do not change its value.

We can also compare them one place value at a time.

13.60	13.60	13.60	13.65
13.65	13.65	13.65	
13.57	13.57		
Same number of ones	13.57 has fewer tenths, 13.57 is the least.	13.60 has fewer hundredths. 13.60 is less than 13.65.	13.65 is the greatest.

__Daryle__ finished the race first. __Bert__ was second and __Dennis__ finished third.

Getting Started

Write <, = or > in the circle.

1. 9.35 (>) 9.27
2. 3.09 (<) 3.90
3. 7.5 (=) 7.50
4. 0.3 (>) 0.2
5. 4.24 (<) 4.28
6. 8.6 (<) 8.66

Write the numbers in order from least to greatest.

7. 2.3, 1.67, 1.95
   _1.67, 1.95, 2.3_
8. 7.1, 6.9, 7
   _6.9, 7, 7.1_
9. 3.5, 3.49, 3.2
   _3.2, 3.49, 3.5_

293

Teaching the Lesson

Introducing the Problem Have a student read the problem aloud and tell what is to be found. (who finished 1st, 2nd and 3rd in the race) Talk with students about the Special Olympics for handicapped persons. Ask what information is given about the boys' times in the 100 meter dash. (Bert ran in 3.6 seconds, Dennis in 3.65 seconds and Daryle in 3.57 seconds.) Have students complete the sentences. Guide them through the comparisons in the model. Have students write each decimal as a mixed number to check their solution.

Developing the Skill Write **2.6** and **2.60** on the board and tell students the values of these numbers are the same since zeros to the far right of a decimal number do not change its value. Write **8.1** and **8.12** on the board and tell students we want to know which number is larger. Tell students we must write 8.1 in hundredths in order to compare its value to 8.12. Have a student write 8.1 in hundredths. (8.10) Ask which is larger. (8.12) Ask why. (Both numbers have 8 ones and 1 tenth but 8.12 has 2 hundredths while 8.10 has no hundredths.) Write **8.1 ◯ 8.12** on the board and have a student write the sign to compare the numbers. (<) Repeat for comparing 6.72 and 6.6, 3.02 and 3.1. Now have students compare 1.26, 1.1 and 1.2.

Practice

Write <, = or > in the circle.

1. 5.64 ⊖ 5.78
2. 3.21 ⊖ 3.30
3. 5.71 ⊙ 5.17

4. 9.03 ⊖ 9.30
5. 7.50 ⊜ 7.5
6. 2.39 ⊙ 2.35

7. 8.6 ⊖ 8.68
8. 9.23 ⊖ 9.32
9. 0.6 ⊜ 0.60

10. 4.75 ⊙ 4.6
11. 8.25 ⊖ 8.3
12. 0.16 ⊖ 0.17

Write the numbers in order from least to greatest.

13. 4.6, 3.5, 3.9
    3.5, 3.9, 4.6

14. 6.26, 6.38, 6.16
    6.16, 6.26, 6.38

15. 8.15, 8.1, 8
    8, 8.1, 8.15

16. 7.1, 7.15, 7.09
    7.09, 7.1, 7.15

17. 4.36, 4.16, 5.03
    4.16, 4.36, 5.03

18. 9.2, 9.14, 9.27
    9.14, 9.2, 9.27

19. 18.21, 18.12, 18.22
    18.12, 18.21, 18.22

20. 37.08, 38.7, 37.8
    37.08, 37.8, 38.7

EXCURSION

Complete the magic square and then make one of your own.

16	2	12
6	10	14
8	18	4

Answers will vary.

The sum of each row, column, and diagonal in the first square is ___30___.

294

294

Adding Decimals

pages 295-296

Objective

To add decimals through thousandths

Materials

Mental Math

Have students name the number that is:

1. ½ less than 1. (½)
2. ⅔ of 12. (4)
3. ⅞ less than ¹⁵⁄₈. (1)
4. ¹⁄₁₅ more than ¹⁄₃₀. (¹⁄₁₀)
5. ¼ + ⅓. (⁷⁄₁₂)
6. ½ + ¼ + ¼. (1)
7. ⅖ of 400. (160)
8. 3¹⁄₁₀ more than 6¹⁄₁₀. (9¹⁄₅)

Skill Review

Review addition of whole numbers and money amounts by having students write and solve the following dictated problems at the board: 38 + 27 (65), 235 + 579 (814), 45 + 83 + 72 (200), $56.12 + 25.49 ($81.61), $18.02 + 46.58 + 7.89 ($72.49)

Adding Decimals

Iron is an important mineral in each person's diet. It is recommended that young people, 11 to 14 years old, have at least 18 milligrams of iron daily. How much iron is found in a meal of sirloin steak and spinach?

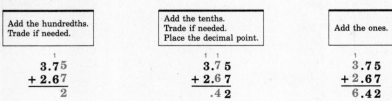

We want to find the amount of iron in a meal of steak and spinach.

The steak has __3.75__ milligrams of iron.

The spinach has __2.67__ milligrams of iron. To find the total milligrams of iron, we add

__3.75__ and __2.67__ .

Add the hundredths. Trade if needed.	Add the tenths. Trade if needed. Place the decimal point.	Add the ones.
$$\begin{array}{r} 3.7\overset{1}{5} \\ + 2.67 \\ \hline 2 \end{array}$$	$$\begin{array}{r} \overset{1}{3}.\overset{1}{7}5 \\ + 2.67 \\ \hline .42 \end{array}$$	$$\begin{array}{r} \overset{1}{3}.75 \\ + 2.67 \\ \hline 6.42 \end{array}$$

The steak and the spinach contain __6.42__ milligrams of iron.

✔ When adding decimals, alignment of the decimal point and place values is important.

7.4 + 8.62 should be set up:

$$\begin{array}{r} 7.4 \\ + 8.62 \\ \hline 16.02 \end{array}$$

Getting Started

Add.

1. $$\begin{array}{r} 4.9 \\ + 2.6 \\ \hline 7.5 \end{array}$$
2. $$\begin{array}{r} 5.83 \\ + 2.16 \\ \hline 7.99 \end{array}$$
3. $$\begin{array}{r} 17.59 \\ + \ 8.6 \\ \hline 26.19 \end{array}$$
4. $$\begin{array}{r} 37.25 \\ + 18.77 \\ \hline 56.02 \end{array}$$

Copy and add.

5. 39.2 + 18.5
   57.7
6. 47.3 + 21.19
   68.49
7. 13.6 + 92.5 + 53.8
   159.9

295

Teaching the Lesson

Introducing the Problem Have a student read the problem aloud and tell what is to be solved. (the number of milligrams of iron in the sirloin steak and the spinach) Ask what information is given. (4½ oz of steak has 3.75 mg of iron, ⅔ c of spinach has 2.67 mg of iron.) Have students read with you as they complete the sentences. Guide them through the addition in the model, emphasizing correct placement of the decimal point. Tell students to write the decimals as mixed numbers and then add to check their work.

Developing the Skill Write **2.8 + 3.67** on the board and tell students that to add decimals, we write the numbers vertically so that the decimal points are aligned and tenths are under tenths, etc. Remind students that we add money that way also. Write **2.8 + 3.67** vertically on the board and tell students there are only 7 hundredths so 7 is recorded in hundredths place. Tell students that 8 tenths plus 6 tenths equal 14 tenths or 1 whole and 4 tenths, so 4 is recorded in tenths place. Tell students the decimal point is placed in line under those in the addends, and the whole numbers are added plus the 1 that was traded. Have students read the answer. (6.47) Have students work the following problems for more practice: 5.67 + 7.2 (12.87), 9.65 + 24.58 (34.23), 6.35 + 1.261 + 4.376 (11.987), 1.203 + 2.02 + 4.5 (7.723).

295

Practice

Add.

1. 5.3 + 2.4 7.7	2. 9.2 + 3.5 12.7	3. 7.72 + 6.13 13.85	4. 3.64 + 8.3 11.94
5. 7.8 + 3.9 11.7	6. 4.86 + 2.71 7.57	7. 39.5 + 8.64 48.14	8. 17.763 + 26.389 44.152
9. 7.8 13.2 + 15.7 36.7	10. 26.4 19.7 + 38.6 84.7	11. 9.87 11.586 + 27.197 48.653	12. 52.174 48.8 + 19.78 120.754

Copy and Do

13. 39.7 + 18.58
   58.28

14. 72.69 + 28.36
   101.05

15. 78.15 + 87.85
   166

16. 38.09 + 27.284
   65.374

17. 46.9 + 52.48
   99.38

18. 37.75 + 39.28
   77.03

19. 58.164 + 9.283
   67.447

20. 15.96 + 83.48
   99.44

21. 10.36 + 15.4 + 12.75
   38.51

22. 39.75 + 48.16 + 58.03
   145.94

23. 87.51 + 93.78 + 48.62
   229.91

Apply

Solve these problems.

24. Lee swam the first lap of the butterfly-stroke race in 25.48 seconds. She swam the second lap in 27.59 seconds. What was Lee's total time?
   53.07 seconds

25. A container holds 3.25 liters of liquid. Another container holds 4.65 liters. How much liquid can both containers hold when they are filled?
   7.9 liters

296

Correcting Common Errors

When students are adding decimal numbers, they may add the decimal-parts to the right of the decimal point and the whole-number parts separately.

INCORRECT	CORRECT
7.8 + 4.5 11.13	7.8 + 4.5 12.3

Have students work the problems on a place-value chart, renaming from right to left, just as they do when adding whole numbers.

Enrichment

Tell students to find the average of 3.26, 5.686 and 3.054. (4)

Practice

Remind students to trade when necessary. Have students complete the page independently.

Extra Credit *Statistics*

Bring in an example of a public opinion poll reported in the newspaper. Read the results of the poll and explain that there are companies, like the Harris Polls, that do nothing but ask people their opinions. When they have sampled enough people, they report on the results of the poll and this tells us what people think about some issue. Ask the class to work in groups to write an opinion poll about a topic important in their school. Ask them to write a number of questions related to the topic they select. Give them time to conduct their survey at lunch or before school. Have each group report on the results of their survey. They should include in their report the frequency of each answer to the questions they asked.

Subtracting Decimals

pages 297-298

Objective

To subtract decimals through thousandths

Materials

Mental Math

Ask students if 80 minutes is 1⅓ hours, how many hours or what part of an hour is:

1. 4 minutes (¹⁄₁₅)
2. 30 minutes (½)
3. 90 minutes (1½)
4. 190 minutes (3⅙)
5. 40 minutes (⅔)
6. 15 minutes (¼)
7. 45 minutes (¾)
8. 127 minutes (2⁷⁄₆₀)

Skill Review

Review subtraction of whole numbers and money amounts by having students solve the following problems at the board: 68 − 35 (33), 75 − 48 (27), $36.71 − 17.95 ($18.76), 508 − 289 (219), $6.25 − 1.97 ($4.28)

Subtracting Decimals

A barometer is used to help forecast the weather. As the barometer rises, the weather becomes clear and dry. How much did the barometer change from noon to 9:00 PM?

Noon	76.28 cm
3:00	75.38 cm
6:00	74.75 cm
9:00	74.57 cm

We want to know the change in the barometer between noon and 9 PM.

At noon the barometer read __76.28__ centimeters.

At 9:00 PM the barometer read __74.57__ centimeters.

To find the total change, we subtract __74.57__

from __76.28__ .

Subtract the hundredths. Trade if needed.	Subtract the tenths. Trade if needed. Place the decimal point.	Subtract the ones.
$$\begin{array}{r} 76.2\overset{8}{} \\ -\ 74.5\overset{7}{} \\ \hline 1 \end{array}$$	$$\begin{array}{r} 7\overset{5\ 12}{6.2}8 \\ -\ 74.57 \\ \hline .71 \end{array}$$	$$\begin{array}{r} 7\overset{5}{6}.28 \\ -\ 74.57 \\ \hline 1.71 \end{array}$$

The barometer dropped __1.71__ centimeters.

✔ Remember to align the decimal points and place values before you subtract.

Getting Started

Add.

1. $$\begin{array}{r} 8.9 \\ -\ 3.6 \\ \hline 5.3 \end{array}$$

2. $$\begin{array}{r} 14.25 \\ -\ 11.21 \\ \hline 3.04 \end{array}$$

3. $$\begin{array}{r} 89.145 \\ -\ 17.964 \\ \hline 71.181 \end{array}$$

4. $$\begin{array}{r} 57.04 \\ -\ 29.17 \\ \hline 27.87 \end{array}$$

Copy and subtract.

5. 82.7 − 39.9
   42.8

6. 485.18 − 210.16
   275.02

7. 280.36 − 198.75
   81.61

297

Teaching the Lesson

Introducing the Problem   Have a student read the problem and tell what is to be found. (the barometer change from noon to 9:00 PM) Ask what information is needed from the picture. (Noon reading was 76.28 cm and 9:00 reading was 74.57 cm.) Have students complete the sentences as they read aloud with you. Guide them through the subtraction in the model. Tell students their answer plus the 9:00 reading should equal the noon reading.

Developing the Skill   Tell students that the decimal point must be aligned in subtraction of decimals just as in addition of decimals. Write **12.078 − 4.269** vertically on the board and talk through the subtraction, trading and decimal point placement as you work the problem. Have a student read the answer. (7.809) Dictate more problems for students to write and solve at the board. Have students read their answers.

Practice

Subtract.

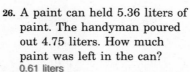

1. 4.7
    − 3.2
    1.5

2. 8.9
    − 2.5
    6.4

3. 7.84
    − 6.31
    1.53

4. 9.79
    − 5.16
    4.63

5. 41.56
    − 14.81
    26.75

6. 59.27
    − 28.93
    30.34

7. 38.28
    − 26.57
    11.71

8. 67.35
    − 48.72
    18.63

9. 74.117
    − 29.585
    44.532

10. 87.68
    − 52.99
    34.69

11. 38.046
    − 19.787
    18.259

12. 51.168
    − 38.274
    12.894

Copy and Do

13. 8.36 − 4.51
    3.85

14. 12.97 − 7.83
    5.14

15. 19.21 − 11.75
    7.46

16. 27.36 − 18.58
    8.78

17. 73.09 − 5.78
    67.31

18. 35.59 − 14.96
    20.63

19. 68.25 − 19.68
    48.57

20. 40.65 − 29.76
    10.89

21. 92.17 − 37.82
    54.35

22. 80.034 − 27.389
    52.645

23. 73.17 − 29.98
    43.19

24. 53.706 − 19.851
    33.855

Apply

Solve these problems.

25. Jim's long jump measured
    4.26 meters. Dave's was 5.03
    meters long. How much
    longer was Dave's jump?
    0.77 meters

26. A paint can held 5.36 liters of
    paint. The handyman poured
    out 4.75 liters. How much
    paint was left in the can?
    0.61 liters

27. A pair of slacks costs $29.79.
    A shirt costs $7.95. How
    much will one pair of slacks
    and two shirts cost?
    $45.69

28. How much change will be left
    if a suit costing $89.50, is
    paid for with a one hundred
    dollar bill?
    $10.50

298

298

Adding and Subtracting Decimals

pages 299-300

Objective

To add and subtract decimals through thousandths

Materials

Mental Math

Dictate the following:

1. $^2/_{10} + ^4/_{10} + ^6/_{10}$ ($1^1/_5$)
2. $^7/_{15} + ^8/_{15}$ (1)
3. $^9/_2 + ^3/_2$ (6)
4. 3.2 + 6.6 (9.8)
5. 2.10 + 3.36 (5.46)
6. $1^1/_5 + ^2/_{10}$ (2)
7. $3^2/_{10} + 15^1/_{10} + ^9/_{10}$ ($19^1/_5$)
8. $^{12}/_{11} - ^1/_{11}$ (1)

Skill Review

Review addition and subtraction of money by having students work the following problems at the board:

$48.01 − 19.73 ($28.28)
$32.56 + 15.95 ($48.51)
$23.50 − 19.70 ($3.80)
$54.90 + 26.19 ($81.09)
$46.00 − 19.26 ($26.74)

Adding and Subtracting Decimals

Two animals known for their speed are the quarterhorse and the greyhound. Both were timed over a distance of a quarter mile. How much faster is the quarterhorse?

We want to know how much faster the quarterhorse ran the quarter mile.

The quarterhorse ran __47.5__ miles per hour.

The greyhound ran __39.35__ miles per hour.

To compare the two speeds, we subtract __39.35__

from __47.5__.

✔ Remember that writing a zero to the far right of a decimal number does not change its value.

$$
\begin{array}{r}
47.5 \longrightarrow 47.50 \\
- 39.35 \longrightarrow - 39.35 \\
\hline
8.15
\end{array}
$$

The quarterhorse ran __8.15__ miles per hour faster than the greyhound.

	Miles Per Hour
Quarterhorse	47.5
Greyhound	39.35

Getting Started

Add or subtract.

1.
$$
\begin{array}{r} 9.7 \\ + 11.39 \\ \hline 21.09 \end{array}
$$

2.
$$
\begin{array}{r} 27.56 \\ - 13.9 \\ \hline 13.66 \end{array}
$$

3.
$$
\begin{array}{r} 54.275 \\ + 13.3 \\ \hline 67.575 \end{array}
$$

4.
$$
\begin{array}{r} 96 \\ - 12.85 \\ \hline 83.15 \end{array}
$$

Copy and add or subtract.

5. 14.68 − 7.32
   7.36

6. 9.63 + 2.7 + 3
   15.33

7. (12 − 7.5) + 6.2
   10.7

299

Teaching the Lesson

Introducing the Problem Have a student read the problem aloud and tell what is to be solved. (how much faster a quarterhorse ran than a greyhound) Ask what information is given in the picture and problem. (The quarterhorse ran the quarter mile at a speed of 47.5 miles per hour and the greyhound ran at 39.35 miles per hour.) Have students read with you as they complete the sentences. Guide them through the placement of the zero in the model subtraction to solve the problem.

Developing the Skill Remind students that adding a zero to the far right of the decimal point does not change its value. Write **67.1** on the board and ask a student to write the number in hundredths. (67.10) Have a student subtract 17.26 from 67.10 and tell the difference. (49.84) Have a student write **80.2 − 17.261** on the board and tell what must be done before the subtraction can be done. (add zero to write 80.2 in thousandths as 80.200) Have a student talk through the subtraction and tell the difference. (62.939) Now have a student write 40.21 + 36.9 on the board. Remind students that 36.9 has no hundredths so the total hundredths is 1. Have a student complete the problem. (77.11) Have students work the following problems at the board for additional practice: 25.4 − 7.781 (17.619), 29 − 4.6 (24.4), (15 − 2.9) + 6.19 (18.29), (9.24 + 6.596) − 8 (7.836)

Add or subtract.

1. $\begin{array}{r} 5.6 \\ + 7.15 \\ \hline 12.75 \end{array}$
2. $\begin{array}{r} 6.23 \\ + 5.9 \\ \hline 12.13 \end{array}$
3. $\begin{array}{r} 9.27 \\ - 3.1 \\ \hline 6.17 \end{array}$
4. $\begin{array}{r} 8.5 \\ - 6.15 \\ \hline 2.35 \end{array}$

5. $\begin{array}{r} 7.24 \\ + 11.6 \\ \hline 18.84 \end{array}$
6. $\begin{array}{r} 5.28 \\ - 1.965 \\ \hline 3.315 \end{array}$
7. $\begin{array}{r} 8 \\ + 3.5 \\ \hline 11.5 \end{array}$
8. $\begin{array}{r} 9.27 \\ - 6.9 \\ \hline 2.37 \end{array}$

9. $\begin{array}{r} 17.21 \\ - \quad 9 \\ \hline 8.21 \end{array}$
10. $\begin{array}{r} 25.3 \\ + 48.68 \\ \hline 73.98 \end{array}$
11. $\begin{array}{r} 76.467 \\ + 80.984 \\ \hline 157.451 \end{array}$
12. $\begin{array}{r} 14 \\ - \quad 9.52 \\ \hline 4.48 \end{array}$

Copy and Do

13. 39 + 6.58
    45.58
14. 27.39 − 19.8
    7.59
15. 4.26 + 8.2 + 9.5
    21.96
16. 17.9 − 11.764
    6.136
17. 15.2 + 36.48 + 9
    60.68
18. (28.39 + 14.6) − 9.8
    33.19
19. (48 − 16.3) + 8.25
    39.95
20. (47.8 − 15.27) − 7.3
    25.23
21. (43 + 18.12) − 6.9
    54.22

Apply

Solve these problems.

22. How much more does the sugar packet weigh than the salt?

SUGAR 14.25 grams
4.85 grams

23. Find the total snowfall.

Monday	3.6 cm
Tuesday	2.48 cm
Wednesday	2 cm

8.08 cm

24. How much longer is the knife than the fork?

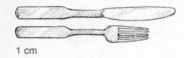

1 cm

25. Find the perimeter.

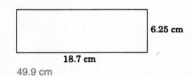

6.25 cm
18.7 cm
49.9 cm

300

Correcting Common Errors

Some students may not align decimal numbers correctly in an addition or subtraction problem. Have them align the decimal points and then annex zeros so that there are the same number of places to the right of the decimal point in both of the numbers in the problem; e.g.

$\begin{array}{r} 8.5 \\ - 6.15 \\ \hline \end{array}$ $\begin{array}{r} 8.50 \\ - 6.15 \\ \hline 2.35 \end{array}$

Enrichment

Tell students to consult the World Almanac to find out how much less yearly precipitation Mobile, Alabama has than Hawaii's Mount Waialeale. They should discover Mt. Waialeale is the rainiest place in the world, receiving 460 inches per year.

Practice

Remind students to add zero to the right of the minuend if the subtrahend is extended to more places. Have students complete the page independently.

Extra Credit *Creative Drill*

Put the following on the board, or duplicate for students. Directions: Fill in the blanks below. The first letters, reading down, spells the name of a famous mathematician and astronomer of the third century B.C. He knew the earth was round and made a measurement of the distance around it.

(E L E V E N)

(R O M A N)

(A R E A)

(T H O U S A N D S)

(O N)

(S E C O N D)

(T W E L V E)

(H O U R S)

(E V E N)

(N I N E)

(E Q U A L S)

(S E Q U E N C E)

What number divides both 22 and 121 exactly?
V is the _?_ numeral for 5.
6 square centimeters is the _?_ of this rectangle.
In 6,092, the 6 is in the _?_ place.
In the picture , P is _?_ both lines.
B is the _?_ letter of the alphabet.
A dozen is _?_ .
There are 24 _?_ in a day.
2, 4, 6 and 8 are all _?_ numbers.
54 ÷ 6 equals _?_ .
3 + 4 _?_ 7. (What word fills the blank?)
1, 3, 5, 7 are numbers in a _?_ .

Multiplying By a Decimal

pages 301-302

Objective

To multiply a whole number by a decimal through hundredths

Materials

Mental Math

Have students name the fraction to compare:

1. 4 boys to 9 girls. (4/9)
2. 9 boys to 4 girls. (9/4)
3. 6 moms to 8 dads. (6/8)
4. 22 students to 1 teacher. (22/1)
5. 2 teachers to 39 students. (2/39)
6. 5 houses to 9 cars. (5/9)
7. boys to girls in your class.
8. girls to boys in your class.

Skill Review

Review multiplication through 3-digit numbers by 2-digit numbers.

Multiplying a Whole Number and a Decimal

Joe and Melissa are using a square table to display their fresh vegetables at the town centennial. They want to hang bunting along 3 sides of the table. How many meters of bunting do they need?

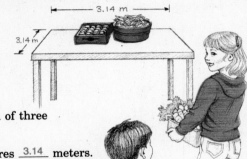

We want to know the length of three sides of a square table.

One side of the table measures __3.14__ meters. To find the length of three sides, we multiply the length of one side by three.

We multiply __3.14__ by __3__.

> Multiply as whole numbers.

$$
\begin{array}{r}
3.14 \leftarrow \text{2 decimal places} \\
\times \quad 3 \leftarrow \text{0 decimal places} \\
\hline
9.42 \leftarrow \text{2 decimal places}
\end{array}
$$

✔ Remember the product has the same number of decimal places as the factors.

Joe and Melissa need __9.42__ meters of bunting.

Getting Started

Place the decimal points in these products.

1. $\begin{array}{r} 8.12 \\ \times \quad 5 \\ \hline 40.60 \end{array}$
2. $\begin{array}{r} 12.9 \\ \times \quad 3 \\ \hline 38.7 \end{array}$
3. $\begin{array}{r} 79 \\ \times 2.5 \\ \hline 197.5 \end{array}$
4. $\begin{array}{r} 7.08 \\ \times \quad 56 \\ \hline 396.48 \end{array}$

Copy and multiply.

5. 9.43×7
   66.01
6. 14.3×12
   171.6
7. 17×0.39
   6.63
8. 2.06×37
   76.22

301

Teaching the Lesson

Introducing the Problem Have a student read the problem aloud and tell what is to be found. (the length, in meters, of 3 sides of a table) Ask what information is known. (1 side of the table is 3.14 meters.) Have students read with you and complete the sentences. Guide them through the multiplication in the model, to solve the problem. Tell students to add to check their solution.

Developing the Skill Write **2.64** on the board and ask students to tell how many decimal places there are. (2) Repeat the question for 16.1 (1), 29.26 (2), 19 (0), 95.06 (2). Now write **16.24 × 6** vertically on the board and have students do the multiplication. (9744) Tell students that there must be a decimal point in this product because we are multiplying hundredths. Ask students how many decimal places are in the problem. (2) Tell students that when multiplying decimals we count the number of decimal places in the problem and have that same number of decimal places in the product. Place the decimal point to show 97.44 as the answer. Have students read the product. Repeat the procedure for 62 × 2.4 (148.8), 7.6 × 8 (60.8), 40.02 × 58 (2321.16), 761 × 0.12 (91.32) and 1.26 × 18 (22.68).

Practice

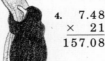

Place the decimal points in these products.

1. $\begin{array}{r} 7.29 \\ \times\ \ \ 8 \\ \hline 58.32 \end{array}$

2. $\begin{array}{r} 14.6 \\ \times\ \ \ 6 \\ \hline 87.6 \end{array}$

3. $\begin{array}{r} 85 \\ \times\ 4.1 \\ \hline 348.5 \end{array}$

4. $\begin{array}{r} 7.48 \\ \times\ \ \ 21 \\ \hline 157.08 \end{array}$

5. $\begin{array}{r} 15.2 \\ \times\ \ \ 7 \\ \hline 106.4 \end{array}$

6. $\begin{array}{r} 3.25 \\ \times\ \ \ 9 \\ \hline 29.25 \end{array}$

7. $\begin{array}{r} 9.6 \\ \times\ \ 9 \\ \hline 86.4 \end{array}$

8. $\begin{array}{r} 7.05 \\ \times\ \ \ 8 \\ \hline 56.40 \end{array}$

Multiply.

9. $\begin{array}{r} 8.8 \\ \times\ \ 6 \\ \hline 52.8 \end{array}$

10. $\begin{array}{r} 15 \\ \times\ 2.9 \\ \hline 43.5 \end{array}$

11. $\begin{array}{r} 281 \\ \times\ 5.3 \\ \hline 1489.3 \end{array}$

12. $\begin{array}{r} 8.65 \\ \times\ \ \ 9 \\ \hline 77.85 \end{array}$

13. $\begin{array}{r} 3.75 \\ \times\ \ \ 17 \\ \hline 63.75 \end{array}$

14. $\begin{array}{r} 6.85 \\ \times\ \ \ 60 \\ \hline 411 \end{array}$

15. $\begin{array}{r} 325 \\ \times\ 0.28 \\ \hline 91 \end{array}$

16. $\begin{array}{r} 5.9 \\ \times\ 36 \\ \hline 212.4 \end{array}$

Copy and Do

17. 96 × 3.5
    336
18. 8.7 × 27
    234.9
19. 3.25 × 9
    29.25
20. 12 × 7.36
    88.32

21. 18 × 0.29
    5.22
22. 4.8 × 37
    177.6
23. 0.73 × 9
    6.57
24. 85 × 7.83
    665.55

25. 5.9 × 372
    2194.8
26. 28.9 × 6
    173.4
27. 1.75 × 8
    14
28. 85 × 0.16
    13.6

Apply

Solve these problems.

29. A dictionary weighs 2.15 pounds. How much does a stack of four dictionaries weigh?
    8.6 pounds

30. Find the area of a rectangle that has a length of 6.75 inches and a width of 9 inches.
    60.75 square inches

302

Correcting Common Errors

Some students may count the decimal places from the left in the numeral for the product instead of from the right to place the decimal point. Have them count the number of decimal places in the factors before they multiply and write this number in a circle beside the place for the product with an arrow to show the direction to count.

$\begin{array}{r} 7.29 \\ \times\ \ \ \ 8 \\ \hline \leftarrow ②\end{array}$

Enrichment

Tell students to match each problem to its product:

0.4 × 62 — 2,480
0.62 × 4 — 2,480
620 × 4 — 2.48
6.2 × 40 — 24.8
62 × 40 — 248
0.4 × 62 — 2,480
6.2 × 400 — 24.8

Practice

Remind students that the number of decimal places in the answer must be the same as in the problem. Have students complete the page independently.

Mixed Practice

1. 2,153 × 72 (155,016)
2. $\frac{3}{8} + \frac{1}{6}\ \left(\frac{13}{24}\right)$
3. (46 − 13.8) + 7.23 (39.43)
4. Simplify: $\frac{21}{49}\ \left(\frac{3}{7}\right)$
5. 457 ÷ 16 (28 R9)
6. 16 + 19.354 (35.354)
7. $\frac{7}{9} - \frac{1}{3}\ \left(\frac{4}{9}\right)$
8. 1,953 + 15,828 (17,781)
9. 11.235 − 6.47 (4.765)
10. $90.00 − 65.22 ($24.78)

Extra Credit *Applications*

Have students research the names of the cities where the last 5 summer Olympics were held. Have them find the distance each city is from their home town. Divide students into groups and have them complete the following activities using their data:

1. Order the cities from closest to farthest away.
2. Find the total distance travelled by someone from their city attending all five Olympics.
3. Choose one of the cities and prepare a numeration report about it, that might include the following data: population, annual rainfall, denomination of currency, total attendance at Olympic events, etc.

Multiplying Decimals

Objective

To multiply 2 decimal numbers through hundredths

Materials

Mental Math

Have students find the equivalent fraction for:

1. 1/2 in tenths. (⁵/₁₀)
2. 2/3 in twelfths. (⁸/₁₂)
3. 9/12 in fourths. (³/₄)
4. 2/7 in fourteenths. (⁴/₁₄)
5. 4/6 in thirds. (²/₃)
6. 75/100 in fourths. (³/₄)
7. 1/10 in hundredths. (¹⁰/₁₀₀)
8. 1/10 in thousandths. (¹⁰⁰/₁₀₀₀)

Skill Review

Review multiplying a whole number and a decimal by having students solve the following problems at the board: 6.4×7 (44.8), 25.7×15 (385.5), 68×1.7 (115.6), 2.09×53 (110.77). Now dictate the same digits but change the decimal point to show that the product remains the same but the decimal point may change.

Multiplying Decimals

A barrel of maple syrup holds 1.5 times as much syrup as a pail. How much syrup does a barrel hold?

PAIL 4.5 L

BARREL ? L

We want to know how much maple syrup a barrel holds.

A pail holds __4.5__ liters of syrup.

A barrel holds __1.5__ times as much syrup. To find how much a barrel can hold, we multiply __4.5__ by __1.5__.

Multiply as whole numbers.

$$\begin{array}{r} 4.5 \\ \times\ 1.5 \\ \hline 225 \\ 45 \\ \hline 675 \end{array}$$

The product has the same number of decimal places as both factors together.

$$\begin{array}{r} 4.5 \leftarrow \text{1 place} \\ \times\ 1.5 \leftarrow \text{1 place} \\ \hline 225 \\ 45 \\ \hline 6.75 \leftarrow \text{2 places} \end{array}$$

A barrel holds __6.75__ liters of syrup.

Getting Started

Multiply.

1. $\begin{array}{r} 3.7 \\ \times\ 2.4 \\ \hline 8.88 \end{array}$

2. $\begin{array}{r} 8.9 \\ \times\ 5 \\ \hline 44.5 \end{array}$

3. $\begin{array}{r} 3.8 \\ \times\ 0.7 \\ \hline 2.66 \end{array}$

4. $\begin{array}{r} 0.9 \\ \times\ 0.6 \\ \hline 0.54 \end{array}$

Copy and multiply.

5. 18×3.5 63
6. 0.24×3 0.72
7. 12.6×1.9 23.94
8. 8.5×6.8 57.8

303

Teaching the Lesson

Introducing the Problem Have a student read the problem aloud and tell what is to be found. (the amount of syrup 1 barrel holds) Ask what information is given in the problem or the picture. (A pail holds 4.5 L and a barrel holds 1.5 times that amount.) Have students read with you as they complete the sentences. Guide them through the multiplication in the model to solve the problem.

Developing the Skill Remind students that in multiplying decimals, we find the product of the factors and then count the decimal places in the factors. Remind students the product must have the same number of decimal places as are in the problem. Write **2.6 × 2.1** on the board and ask students how many decimal places will be in the product. (2) Repeat for $9 \times .62$ (2), 605×2.6 (1), 26.9×8.2 (2), 4.6×20 (1) and 6.4×29.9 (2) Now have students work the problems and place the decimal points in their products.

Practice

Multiply.

1. 4.6
    × 2.7
    ——
    12.42

2. 3.26
    × 8
    ——
    26.08

3. 12.7
    × 6
    ——
    76.2

4. 14.3
    × 2.5
    ——
    35.75

5. 7.09
    × 5
    ——
    35.45

6. 16
    × 8.5
    ——
    136

7. 4.9
    × 5.7
    ——
    27.93

8. 0.92
    × 7
    ——
    6.44

9. 0.8
    × 0.6
    ——
    0.48

10. 36.2
    × 1.4
    ——
    50.68

11. 49
    × 7.3
    ——
    357.7

12. 46.8
    × 4.3
    ——
    201.24

Copy and Do

13. 6.3 × 4.8
    30.24

14. 16 × 0.15
    2.4

15. 1.6 × 9
    14.4

16. 37 × 8.4
    310.8

17. 0.7 × 0.7
    0.49

18. 8.3 × 7.6
    63.08

19. 27 × 4.9
    132.3

20. 12 × 0.9
    10.8

21. 64 × 6.8
    435.2

22. 0.67 × 23
    15.41

23. 1.59 × 3
    4.77

24. 62.5 × 7.6
    475

Apply

25. A can of nails weighs 4.8 pounds. How much do 2.5 cans weigh?
    12 pounds

26. Find the area of a rectangle that is 5.2 centimeters long and 7.6 centimeters wide.
    39.52 square centimeters

5.2 cm

7.6 cm

EXCURSION

If a year's supply of pencils for the world, laid end to end, circled the earth 5 times, how many times would an $8\frac{3}{4}$ year's supply circle the earth? Write your answer as a decimal. ___43.75 times___

304

Practice

Remind students that the product must have the same number of decimal places as the total places in the problem. Have students complete the page independently.

Excursion

Guide students in converting $8\frac{3}{4}$ to 8.75.

Extra Credit *Probability*

Give each group of two students a deck of cards with all the face cards removed. There will be 40 cards in each deck. Ask them to shuffle and draw cards, replacing the card and shuffling after each draw until they get a 7. Have them record the number of cards drawn before they get a seven and then shuffle and start to draw again. Ask them to repeat this process 20 times. Have students record their data in a logical way. Show the groups how to find the average number of cards drawn by adding all the cards drawn and dividing by 20. List each group's average on the board. Ask students to look at the averages and predict the number of cards drawn before they get a 7 another time.

Problem Solving Using a Formula

pages 305-306

Objective

To use a formula to solve a problem

Materials

Mental Math

Ask students true or false?

1. A parallelogram always has right angles. (F)
2. The product of 20 and 400 has 3 zeros. (T)
3. The fraction 1/10 is equal to the decimal .1. (T)
4. $4\frac{1}{8} < 4\frac{2}{16}$ (F)
5. An open sentence has an unknown in it. (T)
6. The average of 2, 3 and 10 is 6. (F)
7. 24 is a multiple of only 3 and 8. (F)
8. $^{14}/_{20}$ is equivalent to $^{7}/_{10}$. (T)

PROBLEM SOLVING

Using a Formula

Mr. Glove must drive his son to college. The trip takes $5\frac{1}{2}$ hours while traveling at an average rate of 50 miles per hour. How many miles away is the college?

★ SEE

We are looking for the distance Mr. Glove must travel to his son's college.

He will drive at the rate of __50__ miles per hour.

It will take him __$5\frac{1}{2}$__ hours to make the trip.

★ PLAN

We can use the formula $D = R \times T$, where D represents the distance traveled, R represents the average rate of speed of the car and T represents the time it takes the car to travel D miles. For our situation, $R = 50$ miles per hour and $T = 5.5$ hours. $\left(5\frac{1}{2} = 5.5\right)$

★ DO

$D = R \times T$

$D = 50 \times 5.5$

$$\begin{array}{r} 5.5 \\ \times\ 50 \\ \hline 275.0 \end{array}$$

$D =$ __275__ miles

Mr. Glove will travel __275__ miles to take his son to college.

★ CHECK

We can reason that if Mr. Glove travels
50 miles in 1 hour,
he will travel 100 miles in 2 hours,
150 miles in 3 hours,

__200__ miles in 4 hours,

__250__ miles in 5 hours,

and __275__ miles in $5\frac{1}{2}$ hours.

MCP All rights reserved

305

Teaching the Lesson

Write $A = L \times W$ on the board and ask students what this **formula** means. (Area equals length times width.) Ask what the symbols A, L and W represent. (area, length, width) Write $P = S + S + S + S$ on the board and remind students this formula means the perimeter of a 4-sided figure equals the sum of the sides. Tell students the symbols in a formula represent parts of the problem. Have students tell what P and S represent. (perimeter, side) Give students some values for symbols in the area and perimeter formulas and have students solve for P and A.

Have a student read the problem aloud and tell what is to be solved. (the distance to the college) Ask what information is known. (He drives for 5-1/2 hours at a speed of 50 miles per hour.) Tell students that since 1/2 is equivalent to 5/10, we can then write 1/2 as the decimal, .5. Write **1/2 = 0.5** on the board. Ask students to tell a decimal for the 5 1/2

hours driven. (5.5) Have students complete the sentences in the SEE stage, and help them use the formula in the PLAN and DO stage. Have students check their answer.

305

Apply

Use the given formulas to solve these problems.

1. The area of a square is side times side. **A = S × S.** Find the area of a square if its side measures 15 centimeters.
225 square centimeters

2. The perimeter of a square is 4 times a side. **P = 4 × S** or **P = 4S.** Find the perimeter of a square that is 12 units long on one side.
48 units

3. The perimeter of a rectangle is two lengths plus two widths. **P = 2L + 2W.** Find the perimeter of a rectangle having a width of 7 feet and a length of 12 feet.
38 feet

4. The area of a rectangle is length times width. **A = L × W** or **A = LW.** Find the area of a rectangle which is 4 centimeters long and 2 centimeters wide.
8 square centimeters

5. Use the formula in Exercise 3. Suppose the perimeter of a rectangle is 24 feet and the width is 4 feet. Explain how to find the length.
Answers will vary.

6. What if the width of a rectangle were given in feet and the length were given in inches? How would you find the perimeter?
See Solution Notes.

7. What if you doubled the length of a rectangle like the one in Exercise 4? How would this affect the area?
See Solution Notes.

8. What if you doubled both the length and the width of a rectangle? How would this affect the area?
See Solution Notes.

9. Write a formula for the number of days in **W** weeks.
Answers will vary.

10. Write a formula for the number of months in **Y** years.
Answers will vary.

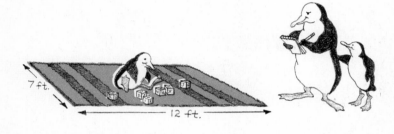

306

Extra Credit *Measurement*

Ask the students to draw a map of an imaginary park from the set of specifications you will give them. They should select the map scale that they think would work the best for the park, and write it at the bottom of the page. Have them draw the map, following the specifications, using colored pencils to make the map attractive. Sample specifications might include:

1. The park is ten miles long and five miles wide. Outline its perimeter.
2. The entrance is on the south side, two miles from the eastern edge.
3. A river runs north and south through the park, two miles west of the entrance.
4. There is a camping area two miles from both the southern and western boundaries.
5. A bridle path begins at the entrance, and runs to the northwest corner of the park.
6. There is a flagpole to the east of the entrance.

Solution Notes

Remind students that all answers must name the unit that is used, for example miles, feet, meters, etc. It may be helpful to have students write the unit word throughout the problem as they work it.

1. Remind students that area is measured in square units.
   $A = S \times S$
   $A = 15 \text{ cm} \times 15 \text{ cm}$
   $A = 225$ square centimeters
2. $P = 4 \times S$
   $P = 4 \times 12$ units
   $P = 48$ units
3. Remind students that $2 \times L$ is the same as 2L.
   $P = 2W + 2L$
   $P = 2(7 \text{ ft}) + 2(12 \text{ ft})$
   $P = 14 \text{ ft} + 24 \text{ ft}$
   $P = 38$ feet
4. $A = LW$
   $A = 4 \text{ cm} (2 \text{ cm})$
   $A = 8$ square centimeters

Higher-Order Thinking Skills

5. [Analysis] Answers should be similar to: "Subtract double the width from the perimeter and take half the difference."
6. [Synthesis] First express both dimensions in the same unit, either feet or inches; then use the formula for finding perimeter.
7. [Synthesis] It would double the area.
8. [Synthesis] It would make the area 4 times as large.
9. [Synthesis] Answers should be similar to the formula D = 7W.
10. [Synthesis] Answers should be similar to the formula M = 12Y.

Calculators and Percent

pages 307-308

Objective

To use a calculator to find a percent of a number

Materials

calculators

Mental Math

Have students name the mixed number:

1. 1¾ (3¼)
2. ¹⁶⁄₁₂ (1⅓)
3. ⁹⁄₂ (4½)
4. ¹⁰⁄₃ (3⅓)
5. ⁸⁄₃ (2⅔)
6. ¾ + ½ (1¼)
7. ¼ + 2¼ (2½)
8. 1⅓ + 2½ (3⅚)

Skill Review

Review the operations of addition, subtraction, multiplication and division on the calculator by having students work the following codes:
26 × 3 ÷ 2 = (39)
(47 + 3) ÷ 2 = (25)
784 × 16 = (12,544)
45 ÷ 3 × 20 = (300)
$17.64 × 50 = ($882).

Calculators and Percents

Valerie grew roses this year for the first time. She is buying a crystal vase to display them in her living room. What is the total amount of the bill for the vase?

We need to find the total amount of the bill.

The price of the vase is __$25__.

The rate of sales tax is __5__%.
We must find the amount of sales tax and add it to the cost of the vase.

To find the sales tax, we find 5% of __$25__.
Complete this code:

25 ⊠ 5 ⊠ (1.25)

✔ When a percent of a number is found on a calculator, the ⊜ key is not used. The number on the display is in dollars and cents.

The sales tax is __$1.25__.

To find the total amount, we add __$25__ and __$1.25__.

25 ⊕ 1.25 ⊜ (26.25)

The total cost of the vase is __$26.25__.
Complete the code to find the total cost of a crystal bell that costs $9.

9 ⊠ 5 ⊠ ⊕ 9 ⊜ (9.45)

✔ Note that on a calculator, you can find the total cost in one code.

The cost of the bell is __$9.45__.

307

Teaching the Lesson

Introducing the Problem   Have a student read the problem aloud and tell what is to be found. (the total amount of the bill for the vase) Ask students what information is given in the picture. (The bill shows the vase costs $25 and sales tax is 5%.) Have students complete the sentences and the computer codes using the percent key, to solve the problem.

Developing the Skill   Have students locate the percent key on their calculators. Tell students they will use this key to find a part of a money amount. Tell students we want to find 6 percent of $15 to find the sales tax on a $15 item. Tell students we will then add the sales tax to the $15 to find the total cost of the item. Have students clear their calculators and enter the following code: 15 × 6 % =. Ask a student to write on the board what is shown on the calculator screen. (0.9) Tell students that if we are just finding a percentage of a number, we do not use the = key. Tell stu-

dents 0.9 means 90¢ since we are figuring an amount of money. Tell students we now add $15 and 90¢ to find the total cost and write **15 + .9 =** on the board. Have students tell the total cost. ($15.90) Tell students that all of this can be done in 1 code. Write **15 × 6% + 15 =** on the board. Remind students that here we do use the = key because we are finding more than just the percentage of a number. Remind students that the dollar amount is entered first. Have students write and solve calculator codes to find: 10% of $20 ($2), 5% of $80 ($4) and 16% of $3,025 ($484). Have students then find the total cost of each item if the percentage found is an additional tax. ($22, $84, $3,509)

307

Practice

Find the percent of each number.

1. 8% of $150 = ___$12___
2. 10% of $256 = ___$25.60___
3. 25% of $364 = ___$91___
4. 50% of $750 = ___$375___

Complete the table.

5.

Amount spent	$855	$760	$2,578	$3,745
Percent of sales tax	6%	5%	7%	4%
Amount of sales tax	$51.30	$38	$180.46	$149.80
Total cost	$906.30	$798	$2,758.46	$3,894.80

Apply

Solve these problems.

6. Angelo bought a used car for $2,650. He had to pay 8% of the cost of the car as a down payment. How much did Angelo have to pay down?
$212

7. Donna earned a salary of $26,000 a year as a city planner. This year Donna also received a bonus of 5% of her salary. How much did Donna earn this year?
$27,300

EXCURSION

Valerie has a short-cut way to find the total amount of a bill, when the tax rate is given. Valerie uses this code to find the total cost.

25 ⊠ 105 % ⌐26.25⌐

Use her method to complete the chart.

Cost	Rate of Sales Tax	Total Cost
$650	6%	$689
$495	8%	$534.60
$810	9%	$882.90
$735	12%	$823.20

308

308

Chapter Test

Item	Objective
1-8	Identify place value of digit in a decimal through thousandths (See pages 291-292)
9-16	Compare and order decimals through thousandths (See pages 293-294)
17	Add decimals through thousandths (See pages 295-296)
18-20	Add ragged decimals (See pages 299-300)
21	Subtract decimals through thousandths (See pages 297-298)
22-24	Subtract ragged decimals (See pages 299-300)
25, 30, 32	Multiply a whole number and decimal (See pages 301-302)
26-29, 31	Multiply two decimals (See pages 303-304)

---

14 CHAPTER TEST

What is the value of the 4 in each number?

1. 84.21   2. 156.43   3. 20.148   4. 47.296

   ones   tenths   hundredths   tens

What is the value of the 7 in each number?

5. 639.74   6. 387.28   7. 200.07   8. 583.957

   tenths   ones   hundredths   thousandths

Write <, = or > in the circle.

9. 3.46 ⊘< 3.64   10. 8.4 ⊜= 8.40   11. 7.3 ⊘< 7.39   12. 0.231 ⊘> 0.22

13. 7.05 ⊘< 7.5   14. 0.13 ⊘> 0.031   15. 65.1 ⊘> 65.09   16. 8.7 ⊜= 8.70

Add.

17. 5.9   18. 27.25   19. 96.5   20. 53.47
   + 3.6   + 48.397   + 48.76   + 16.7
   ‾‾‾‾   ‾‾‾‾‾‾   ‾‾‾‾‾   ‾‾‾‾‾
    9.5   75.647   145.26   70.17

Subtract.

21. 8.9   22. 37.038   23. 38.7   24. 89.25
   - 2.5   - 15.969   - 19.58   - 27.5
   ‾‾‾‾   ‾‾‾‾‾‾   ‾‾‾‾‾   ‾‾‾‾‾
    6.4   21.069   19.12   61.75

Multiply.

25.   3.2   26. 12.6   27. 13.7   28.   8.9
    × 8   × 0.4   × 0.6   × 4.7
   ‾‾‾‾   ‾‾‾‾‾   ‾‾‾‾‾   ‾‾‾‾‾
    25.6   5.04   8.22   41.83

29. 24.8   30.   1.6   31.   5.1   32. 0.09
   × 7.2   ×   7   × 0.06   ×   7
   ‾‾‾‾   ‾‾‾‾‾   ‾‾‾‾‾   ‾‾‾‾‾
   178.56   11.2   0.306   0.63

MCP All rights reserved

Circle the letter of the correct answer.

1 Round 4,500 to the nearest thousand.
  a 4,000
  ⓑ 5,000
  c NG

2 13,296
  + 8,758
  a 21,954
  ⓑ 22,054
  c 22,064
  d NG

3 9,241
  − 8,658
  ⓐ 583
  b 1,417
  c 1,583
  d NG

4 Choose the better estimate of length.
  ⓐ 2 cm
  b 2 meters

5 409 × 6
  ⓐ 2,454
  b 2,456
  c 24,054
  d NG

6 16)328
  a 2 R8
  ⓑ 20 R8
  c 28
  d NG

7 Find the perimeter. **16 feet**
  a 16 sq ft
  b 64 sq ft
  c 48 feet
  ⓓ NG

8 Find the area. 6 cm / 9 cm
  a 15 cm
  b 15 sq cm
  c 30 sq cm
  ⓓ NG

9 $\frac{3}{4}$ of 16 = ?
  a 4
  b 8
  ⓒ 12
  d NG

10 Simplify $\frac{16}{24}$.
  a $\frac{1}{2}$
  ⓑ $\frac{2}{3}$
  c $\frac{3}{4}$
  d NG

11 $\frac{2}{3}$ + $\frac{3}{5}$
  a $\frac{5}{8}$
  b $1\frac{4}{5}$
  ⓒ $1\frac{4}{15}$
  d NG

12 $\frac{5}{8}$ − $\frac{1}{8}$
  ⓐ $\frac{1}{2}$
  b $\frac{3}{4}$
  c 4
  d NG

☐ score

310

Item	Objective
1	Round numbers to nearest 1,000 (See pages 33-34)
2	Add two numbers up to 5 digits (See pages 51-52)
3	Subtract two 4-digit numbers with zero in minuend (See pages 67-68)
4	Determine appropriate metric unit of length (See pages 153-156)
5	Multiply 3-digit number by 1-digit number with trading (See pages 171-172)
6	Divide 3-digit number by 2-digit number to get 2-digit quotient with remainder (See pages 209-210)
7	Find perimeter of a polygon (See pages 229-230)
8	Find area of a rectangle (See pages 231-232)
9	Find fraction of a number by multiplying and dividing (See pages 245-246)
10	Reduce fraction to lowest terms (See pages 253-254)
11	Add two fractions with uncommon denominators (See pages 275-276)
12	Subtract two fractions with common denominators (See pages 267-268)

Alternate Cumulative Review
Choose the letter of the correct answer.

1 Round 6,421 to the nearest thousand.
  a 6,400
  b 6,500
  ⓒ NG

2 12,284
  + 9,869
  a 21,153
  b 22,053
  c 22,143
  ⓓ NG

3 6,130
  − 4,796
  ⓐ 1,334
  b 1,344
  c 1,434
  d NG

4 What unit of length would you use to measure your finger?
  ⓐ cm
  b dm
  c m
  d NG

5 306 × 8 =
  ⓐ 2,448
  b 2,488
  c 2,528
  d NG

6 14)347
  a 36
  ⓑ 24 R11
  c 23 R1
  d NG

7 Find the perimeter of an equilateral triangle with 6 cm sides.
  a 10 cm
  b 13 cm
  ⓒ 18 cm
  d NG

8 Find the area of a rectangle 5 ft by 12 ft.
  a 34 sq ft
  b 50 sq ft
  ⓒ 60 sq ft
  d NG

9 $\frac{2}{3}$ of 18 =
  a 6
  b 9
  ⓒ 12
  d NG

10 Simplify $\frac{24}{36}$
  a $\frac{1}{2}$
  b $\frac{1}{3}$
  c $\frac{3}{4}$
  ⓓ NG

11 $\frac{3}{8}$ + $\frac{2}{6}$ =
  ⓐ $\frac{17}{24}$
  b $\frac{3}{4}$
  c $\frac{19}{24}$
  d NG

12 $\frac{6}{14}$ − $\frac{4}{14}$ =
  a $\frac{1}{14}$
  ⓑ $\frac{1}{7}$
  c $\frac{3}{14}$
  d NG

310

Tallies
Bar Graphs

Objective

To make and interpret tally charts and bar graphs

Materials

*large-grid paper
graph paper

Mental Math

Have students complete the comparison 14 is to 2 as:

1. 63 is to (9)
2. 21 is to (3)
3. 84 is to (12)
4. 560 is to (80)
5. 4,900 is to (700)
6. 91 is to (13)
7. 7 is to (1)
8. 42 is to (6)

Skill Review

Have students work in pairs. Have 1 student write 2 numbers for the other student to compare. Have second student write >, < or = to show the comparison. Have students change roles for more practice. Encourage students to use fractions and decimals as well as whole numbers.

---

15 GRAPHING AND PROBABILITY

Tallies and Bar Graphs

Mr. Ryan is keeping a **tally** of the different types of books checked out of the library. Complete the **bar graph** to show which type is the most popular.

Library Book Checkout	
Mystery	ЖЖ ЖЖ |||
Biography	ЖЖ |||
Sports	ЖЖ ЖЖ	
Romance	ЖЖ ЖЖ ЖЖ	

We want to show the most popular type of book on a bar graph.

There were __13__ mysteries, __8__ biographies,

__11__ sports books and __16__ romance books checked out.

The most popular type of book checked out is the __romance__ book.

Getting Started

Use the graph above to answer these questions.

1. How many sports books were checked out?

   __11__

2. How many more mystery books than biographies were checked out?

   __5__

3. How many books were checked out altogether?

   __48__

4. How many checked-out books were not romances?

   __32__

311

Teaching the Lesson

Introducing the Problem Have students read the problem and tell what is to be done. (make a bar graph to find the most popular type of book) Ask what information is given. (tallies of mystery, biography, sports and romance books checked out) Have students complete the information sentence and the bar graph. Then have them fill in the solution sentence.

Developing the Skill Draw a tally chart on the board and have students vote for their favorite breakfast food, such as eggs, cereal or pancakes. Make a slash for each vote and cross each group of 4 with a fifth slash. Ask students to tell the total votes for each food. Display a sheet of large grid paper and make a horizontal bar graph from the tally chart information. Remind students that a tally sheet or graph must be titled so that anyone seeing the work will know what information is being tallied or graphed. Ask questions from the bar graph to have students tell such information as how many students voted, how many students prefer foods other than eggs, etc. Have students make their own vertical bar graph from the same information.

311

Practice

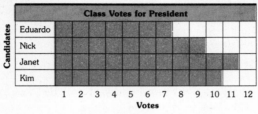

Use the tally chart to complete the bar graph and answer the questions.

Class Votes for President	
Eduardo	卌 ‖
Nick	卌 ‖‖
Janet	卌 卌 ‖
Kim	卌 卌

Class Votes for President

Candidates	1	2	3	4	5	6	7	8	9	10	11	12
Eduardo												
Nick												
Janet												
Kim												

Votes

1. Who won the election?

   Janet

2. How many more votes did Nick get than Eduardo?

   2

3. How many votes were cast altogether?

   37

4. How many votes did the boys get together?

   16

Complete the bar graphs using the information on the tally charts.

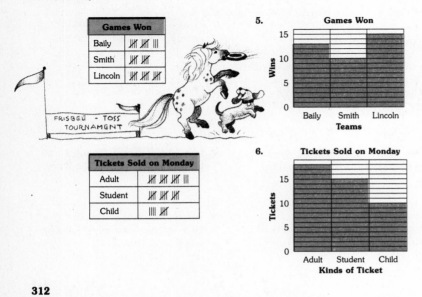

Games Won	
Baily	卌 卌 ‖‖
Smith	卌 卌
Lincoln	卌 卌 卌

5. **Games Won**

Tickets Sold on Monday	
Adult	卌 卌 卌 ‖‖
Student	卌 卌 卌
Child	‖‖‖ 卌

6. **Tickets Sold on Monday**

312

Practice

Have students complete the page independently.

Mixed Practice

1. $953 \div 86$ (11 R7)
2. 178×26 (4,628)
3. $35,789 + 2,527$ (38,316)
4. 36.8×4 (147.2)
5. $4,061 - 2,948$ (1,113)
6. $15.9 - 4.73$ (11.17)
7. $2\frac{1}{3} + 3\frac{1}{4}$ $\left(5\frac{7}{12}\right)$
8. $7.95 + 12.835 + 6.6$ (27.385)
9. $347 \div 8$ (43 R3)
10. $\frac{3}{8} + \frac{3}{4}$ $\left(1\frac{1}{8}\right)$

Correcting Common Errors

Some students may have difficulty giving the value of a bar that ends somewhere between numbers on the scale. Have them use a ruler or edge of a piece of paper to align the end of the bar with the scale and count over from the last given number to this edge.

Enrichment

Have students tally the number of families in the class who have particular kinds of cars. Tell them to make a bar graph from their information. Then have them write 5 questions for a friend to answer from their graph.

Extra Credit   *Geometry*

Give each student a circle cut from plain paper. Ask them to fold the paper twice in one direction and twice in the other. Explain that they will have divided the circle into twelve equal arcs.

Have students label each fold point with a number (1 through 12) as though the circle were a clock. Ask them to use a straightedge to connect every other point. Ask students to name the figure formed. (hexagon) Have them connect every third point and name the figure. (square) Have them connect every fourth point (triangle), and every fifth point. (dodecagon) Ask them to predict what will happen when they connect every sixth point. (They will get a straight line.)

Picture Graphs

pages 313-314

Objective

To make and interpret picture graphs

Materials

Mental Math

Have students compare the following numbers:

1. 2.5 ◯ 2.56 (<)
2. 68.02 ◯ 68 (>)
3. ⅘ ◯ ¹⁶⁄₂₀ (=)
4. 0.75 ◯ 0.8 (<)
5. 2.068 ◯ 2.06 (>)
6. 42,676 ◯ 42,776 (<)
7. 4½ ◯ ⁹⁄₂ (=)
8. 5.01 ◯ 5.10 (<)

Skill Review

Review the procedure to find an average by having students find the average of numbers such as 10, 6 and 14. (10) Be sure each average is a whole number.

Picture Graphs

A **picture graph** uses pictures to represent data on a graph. Here we can see how the construction industry changes with the seasons. How many new houses were started in June?

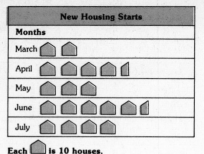

New Housing Starts	
Months	
March	🏠🏠
April	🏠🏠🏠🏠◣
May	🏠🏠🏠
June	🏠🏠🏠🏠🏠◣
July	🏠🏠🏠🏠

Each 🏠 is 10 houses.

We want to know how many houses were started in June.

Each full house on the graph means __10__ houses started.

A half of a house means __5__ houses started.

There are __5__ full houses and __1__ half house pictured for June.

(__5__ × 10) + __5__ = __55__

There were __55__ houses started in June.

Getting Started

Use the picture graph to answer these questions.

1. Which month had the fewest new houses started?

   March

2. How many houses were started in April?

   45

3. How many more houses were started in July than March?

   20

4. In which month were only 30 houses started?

   May

5. What was the monthly average of housing starts?

   38

6. Which months were above average in number of housing starts?

   April, June, July

313

Teaching the Lesson

Introducing the Problem Have students read the problem and tell what is to be found. (number of new houses started in June) Ask what a **picture graph** does. (uses pictures to show data on a graph) Ask students what data is shown in the picture graph. (number of new houses started in months of March through July) Have students complete the information sentences. Then have them solve the problem and complete the solution sentence.

Developing the Skill Tell students that the sales of baseballs in dozens by a small sporting goods store are: April, 16 ½; May, 15½; June, 14; July, 9. Tell students these sales could be shown on a tally chart or a bar graph but a picture graph is yet another way to show data on a graph. Draw a picture graph frame and write **Baseball Sales** above it. Have students name the months as you write them on the graph. Tell students that if we tried to show each individual ball sold, the graph would be huge, so

we will group the balls in dozens. Tell students we need to note that plan on the graph. Write **Each is 1 dozen balls** under the graph. Ask students the number of sales in April as you draw 16 balls and ½ ball. Tell students each ball represents 1 dozen and the ½ dozen is shown by ½ ball. Have students draw the balls to represent sales in May, June and July. Have students answer questions from the information on the graph.

313

Practice

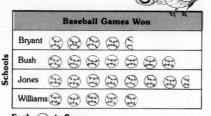

Use the picture graph to answer these questions.

1. How many games did Bush win? __14__

2. How many games did Jones win? __15__

3. Which team won exactly 10 games? __Williams__

Baseball Games Won							
Schools	Bryant	⚾	⚾	⚾	⚾	⚾	
	Bush	⚾	⚾	⚾	⚾	⚾	⚾ ⚾
	Jones	⚾	⚾	⚾	⚾	⚾	⚾ ⚾
	Williams	⚾	⚾	⚾	⚾	⚾	

Each ⚾ is 2 games.

Use the tally chart to make a picture graph. Then answer the questions.

Birds Sighted	
Bluebird	ЖЖЖ I
Robin	ЖЖ II
Cardinal	ЖЖ IIII
Blackbird	ЖЖ

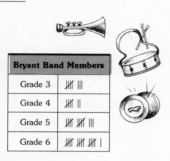

4.

Birds Sighted				
Birds	Bluebird	🐤	🐤	🐤 🐤
	Robin	🐤	🐤	🐤
	Cardinal	🐤	🐤	🐤 🐤
	Blackbird	🐤	🐤	🐤

Each 🐤 means 4 birds.

5. How many birds were sighted? __52__

6. What was the average number of each kind of bird sighted? __13__

Bryant Band Members	
Grade 3	ЖЖ III
Grade 4	ЖЖ II
Grade 5	ЖЖ ЖЖ III
Grade 6	ЖЖ ЖЖ ЖЖ I

7.

Bryant Band Members							
Grades	3	🔔	🔔	🔔	🔔		
	4	🔔	🔔	🔔			
	5	🔔	🔔	🔔	🔔	🔔	🔔
	6	🔔	🔔	🔔	🔔	🔔	🔔 🔔

Each 🔔 means 2 children.

8. How many more Grade 6 band members are there than Grade 4? __9__

9. Which grades have more than the average number of members? __5, 6__

314

Correcting Common Errors

When interpreting picture graphs, some students may ignore the key and read each picture as 1. To help them attend to the key, have them write the value of the picture as given in the key on at least one of the pictures in the graph.

Enrichment

Have students make a tally, a bar graph and a picture graph to show the student absences in your school during one week.

Practice

Remind students that each picture on a picture graph may represent more than 1. Tell students that in problems 7 and 10 they are to find an average and in number 10 they are to compare data to the average. Have students complete the page independently.

Mixed Practice

1. $3.6 \times .7$ (2.52)
2. $\$1.95 \times 18$ ($35.10)
3. $153 + 2,748 + 894$ (3,795)
4. $2.7 + 0.953 + 1.61$ (5.263)
5. $52,106 - 30,793$ (21,313)
6. $681 \div 27$ (25 R6)
7. 0.64×8 (5.12)
8. $\frac{4}{5} - \frac{3}{4}$ $\left(\frac{1}{20}\right)$
9. $9.53 - 2.7$ (6.83)
10. $\frac{3}{5} + \frac{2}{3}$ $\left(1\frac{4}{15}\right)$

Extra Credit   *Biography*

"If I had a place to stand, I could move the earth," Archimedes once boasted. This ancient genius, who lived from 287–212 B.C. in Syracuse, Sicily, made this claim after discovering the principle of the lever and pulley. One popular story tells of Archimedes bathing, and discovering that an object in water weighs exactly as much as the water whose place the object took. Archimedes was so excited about this discovery that he leaped from the tub, forgetting his clothes, and ran through the streets yelling, "Eureka!" meaning "I have found it!" Archimedes invented the Archimedean screw, used to drain Egyptian fields flooded by the Nile River. Archimedes used his inventions to help defend Syracuse from the invading Romans. He devised catapults, and even used a system of lenses and mirrors to set fire to a fleet of Roman ships. His tomb is engraved with the figure of a sphere inscribed in a cylinder, as a tribute to his work in geometry.

Line Graphs

pages 315-316

Objective

To interpret line graphs

Materials

line graph of spelling test scores from 80 to 100

Mental Math

Have students simplify:

1. $^{18}/_2$ (9)
2. $^{26}/_{21}$ ($1^5/_{21}$)
3. $5^4/_{20}$ ($5^1/_5$)
4. $^{19}/_{57}$ ($^1/_3$)
5. $^{105}/_5$ (21)
6. $7^9/_3$ (10)
7. $82^{22}/_{20}$ ($83^1/_{10}$)
8. $5^7/_{21}$ ($5^1/_3$)

Skill Review

Write on the board: **6 20 15 18 9**
Ask students to put the numbers in order from least to greatest. Repeat for more ordering of series of numbers. Include some series of fractions or decimals.

Line Graphs

A **line graph** is a good way to show changes in information. This line graph shows the high temperatures for each day for the first two weeks in May. What was the high temperature on May 6?

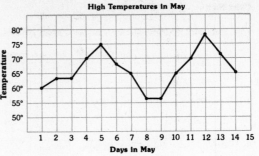

High Temperatures in May

We want to find the high temperature for May 6 on the line graph.

The ___temperature___ is shown up the side of the graph.

The ___days in May___ are shown along the bottom. To find the high temperature for May 6, we follow along the bottom of the graph until we reach the sixth day. Then we go up the vertical line until we reach the dot. The temperature is the reading in degrees directly opposite that dot. On May 6, the

high temperature was ___68°___.

Getting Started

Use the line graph above to answer the questions.

1. What was the high temperature on May 10? ___65°___

2. On what day was the high temperature exactly 72°? ___May 13___

3. On how many days was the high temperature exactly 70°? ___2___

4. What was the average high temperature for May 12, May 13 and May 14? ___72°___

315

Teaching the Lesson

Introducing the Problem Have students read the problem and tell what is to be found. (the high temperature on May 6) Ask what information is given in the **line graph.** (high temperatures in May) Discuss the graph with students and then have them complete the sentences.

Developing the Skill Tell students that a line graph is another kind of graph used to show data. Show students a line graph of spelling test scores with scores from 80 through 100 listed up the left side and the following information represented: Test 1–89, Test 2–96, Test 3–99, Test 4–80, Test 5–85, Test 6–92. Tell students the name of the line graph is Spelling Test Scores and that the numbers across the bottom are the different tests. Tell students that scores are shown up the side of the graph. Tell students that we can find a score for a particular test by going across the bottom of the graph to the number of the test, follow that line upward to the dot and go across to the left of the graph

to find the score. Have students find the score for Test 4, Test 6, etc. Ask students other questions using the line graph. Help students understand why the range of scores is from 80 through 100 rather than from 1 through 100.

Practice

Use the line graph to answer the questions.

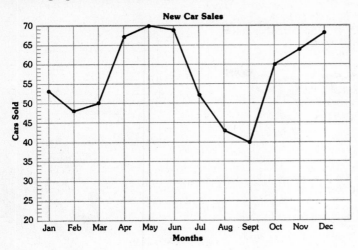

New Car Sales

1. How many cars were sold in October? _____60_____

2. How many cars were sold in February? _____48_____

3. Did sales go up or down from June through September? _____down_____

4. In what month did the sales of cars increase the most over the month before? _____October_____

5. In which month were the most cars sold? _____May_____

6. In which month were the least cars sold? _____September_____

7. How many more cars were sold in April than September? _____27_____

8. What was the average monthly car sales during the last four months? _____58 cars_____

316

Practice

Remind students to go across the bottom of the graph, up the vertical line to the dot and then look to the left of the graph to find the number. Have them complete the page independently. Students may use a ruler to help them see across from a point to the scale on the left of the graph.

Mixed Practice

1. $5,671 \times 23$ (130,433)
2. $\frac{7}{10} + \frac{2}{5} \left(1\frac{1}{10}\right)$
3. 9.4×5.7 (53.58)
4. $\frac{15}{16} - \frac{7}{8} \left(\frac{1}{16}\right)$
5. $\$25.00 - 14.37$ ($10.63)
6. $.62 \times 8$ (4.96)
7. $392 \div 6$ (65 R2)
8. $15,395 + 26,008$ (41,403)
9. $3\frac{5}{8} + 7\frac{2}{3} \left(11\frac{7}{24}\right)$
10. $895 \div 72$ (12 R31)

Extra Credit *Measurement*

Have each student use a timepiece with a second hand in this activity. Tell students to make 2 columns on their papers, one titled Guess, the other titled Check. Tell them that they are first going to guess how long it will take them to complete each of the activities you list on the board, and write it on their papers. Then, working with a partner, each student will do each activity while being timed, and write their time on their papers. Then, have students find the difference between their estimation and actual time for each activity. The exercise can be extended by adding activities to the list. Activities: 1. hop 10 times, 2. snap your fingers 20 times, 3. count backwards from 20, 4. count by 5's to 100, 5. write the numbers counting by 2's to 50, 6. tie your shoe 10 times, 7. write 40 x's on the board, 8. cut out 10 circles.

Number Pairs

pages 317-318

Objective

To locate number pairs on a grid

Materials

*line graph of spelling test
    scores
grid paper

Mental Math

Have students tell the total price if
they pay 6¢ tax on every dollar:

1. $7 ($7.42)
2. $80 ($84.80)
3. $20 ($21.20)
4. $5 ($5.30)
5. $10 ($10.60)
6. $100 ($106)
7. $2 ($2.12)
8. $9 ($9.54)

Skill Review

Have students tell which test had a
score of 96, 92, etc. as shown on the
Spelling Test Scores line graph used in
the previous lesson. Have students locate more points on the line graph as
you give other possible scores for each
test.

Number Pairs

Number pairs can be used to give
locations on a number grid. The
pair (4, 3) shows the location
of point A. What number pair
locates point D?

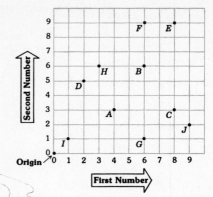

We want to know what number
pair identifies point D.

We know that point A is __4__ units

over and __3__ units up.

✔ To find a number pair, we start at the
origin on the grid. The **first** number tells
how far **over** to move. The **second** number
tells how far **up** to move.

Point D is __2__ over and __5__ up.

The number pair (__2__, __5__) locates point D.

We write: **D (2, 5).** We say: **Point D is the
number pair two, five.**

Getting Started

Write the letter identified by each number pair.

1. __G__(6, 1)   2. __J__(9, 2)   3. __H__(3, 6)   4. __B__(6, 6)

Write the number pair that identifies each letter.

5. I(__1__, __1__)   6. E(__8__, __9__)   7. C(__8__, __3__)   8. F(__6__, __9__)

317

Teaching the Lesson

Introducing the Problem Have students read the problem and tell what is to be found. (a **number pair** to locate
point D on the grid) Ask what information is known about
point A. (The number pair 4,3 shows its location.) Discuss
the grid and then have students complete the sentences to
answer the question.

Developing the Skill Display a numbered grid similar
to that on page 317 but with no points identified. Point to
the zero and tell students that the zero point in the lower
left corner of the grid is called the **origin,** meaning beginning. Locate other points as you tell students that every
point on a grid also has a name and that a number pair
names each point's location. Tell students we always start at
the origin to find the number pairs of any point on a grid.
Locate 3,6 as point **A** and tell students that to find point A's
number pair, we begin at the origin and go across the bottom of the grid to the line point A is on. Write **3 over** on

the board. Tell students we then go up line 3 to point A
and look for the line that intersects line 3 at point A. Write
6 up on the board. Write **A (3,6)** on the board and tell students we read this as **three, six.** Have students locate other
points and write and read their number pairs.

317

Practice

Write the letter identified by each number pair.

1. _L_ (5, 2)   2. _B_ (6, 4)
3. _K_ (2, 5)   4. _Z_ (1, 6)
5. _A_ (4, 6)   6. _F_ (2, 8)
7. _O_ (7, 2)   8. _C_ (1, 1)
9. _N_ (7, 4)   10. _H_ (3, 4)

```
9 |     X       D   J
8 |   F       G
7 |
6 | Z     A   M     I
5 |   K
4 |     H       B N Y
3 |
2 |         L   O     E
1 | C
  +-------------------
    0 1 2 3 4 5 6 7 8 9
```

Write the number pair that identifies each letter.

11. M(_6_ , _6_)   12. X(_4_ , _9_)   13. I(_8_ , _6_)   14. O(_7_ , _2_)
15. E(_9_ , _2_)   16. G(_6_ , _8_)   17. J(_9_ , _9_)   18. Y(_9_ , _4_)

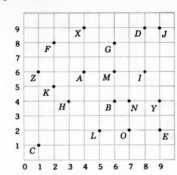

Take the numbers from the first square, and rearrange them in the second square so that each column, row and diagonal has the same sum.

15	1	11
5	9	13
7	17	3

Arrangements will vary.

318

Some students may interpret number pairs incorrectly and think of (2, 3) as 2 up and 3 over. Correct by having students read a number pair with the words "over" and "up." For example, they would read (2, 3) not as "two, three," but as "2 over, 3 up."

Enrichment

Have students locate and label 10 points on a grid. Then have them exchange papers with a friend and write the number pair for each point.

Practice

Remind students that the first number tells how far over to go and the second number tells how far up to go on the grid. Have students complete the page independently.

Excursion

New squares can be created by rotating the numbers along the outside. Have students rotate the square one more time.

Extra Credit   *Applications*

Clear a bulletin board. Use cardboard as woodwork, to outline windows and a door on it. Invite a paper hanger to your class to guide the students in measuring, cutting and hanging wall paper to cover the bulletin board, working around the door and windows. Ask the professional to take students step by step through the measurement and math processes needed for the job. Students can help write the figures and make computations on the chalkboard or on calculators. Estimation is also a part of the paper hanger's job. Ask your guest to take the class step by step through the process of estimating how much paper it would take to cover one wall of your classroom.

Graphing Number Pairs

pages 319-320

Objective

To graph number pairs to draw figures

Materials

*large-grid paper
grid paper
rulers

Mental Math

Ask students if A is 1 and Z is 26, what is the number:

1. C + C (6)
2. Z ÷ A (26)
3. ⅔ of X (16)
4. E × D (20)
5. Y ÷ E (5)
6. average of F and B (4)
7. 2.2 + Z (28.2)
8. A + B + C + D (10)

Skill Review

Review the properties of a rectangle by having students tell about its sides (Opposite sides are equal.) and angles. (4 right angles) Have students draw and label various-sized rectangles on grid paper. Remind students to use a straightedge to draw line segments.

Graphing Number Pairs

Graph the points $C(3, 6)$ and $D(7, 6)$. Use a ruler to draw $\overline{AB}$, $\overline{BD}$, $\overline{DC}$ and $\overline{CA}$. What kind of figure did you draw?

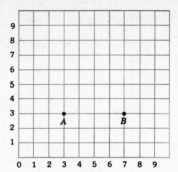

We graph and label points C and D.

Point C is __3__ units over

and __6__ units up.

Point D is __7__ units over

and __6__ units up.

We draw line segments __AB__, __BD__,

__DC__ and __CA__.

✔ Remember that $\overline{AB}$ is a line segment from point A to point B.

The figure $ABCD$ is a __rectangle__.

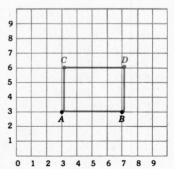

Getting Started

Graph these points. Draw line segments AB, BC and CA.

$A(2, 9)$ $B(6, 9)$
$C(4, 5)$

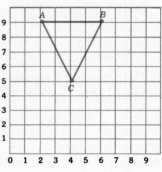

319

Teaching the Lesson

Introducing the Problem Have students read the problem and tell what is to be solved. (the kind of figure formed by points A, B, C and D) Ask students what is known. (Points A and B are located, point C is (3,6) and point D is (7,6).) Have students complete the sentences and the grids to solve the problem. Have students tell the number of units on each side of their figure to check their answer.

Developing the Skill Display a grid and have students write the numbers from 0 through 10 along the bottom and up the left side. Have a student locate A (1,1) and B (2,3), connect points A and B and tell the figure formed. (line segment AB) Tell students that **number pairs** can be used to describe other figures or pictures. Have a student locate X (3,1) and Y (4,3). Have a student draw line segment XY and name the figure formed by $\overline{AB}$ and $\overline{XY}$. (parallel lines) Continue for students to locate R (2,5), S (2,8), T (4,5) and U (4,8) and tell the figure formed. (more parallel lines)

Practice

1. Graph and label these points.

   $A(3, 6)$ $B(4, 8)$ $C(5, 6)$
   $D(4, 6)$ $F(3, 4)$
   $E(4, 5)$ $G(5, 4)$
   Use a ruler to draw $\overline{AB}$, $\overline{BC}$, $\overline{AC}$, $\overline{DE}$, $\overline{FE}$ and $\overline{EG}$.

2. Graph and label these points.

   $A(2, 2)$ $B(2, 6)$
   $C(4, 2)$ $D(4, 6)$
   $E(2, 4)$ $F(4, 4)$
   $G(7, 2)$ $H(7, 6)$
   Use a ruler to draw $\overline{AB}$, $\overline{EF}$, $\overline{DC}$ and $\overline{HG}$.

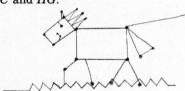

3. Graph and label these points.

   $A(1, 1)$ $J(4, 8)$ $K(5, 8)$ $D(5, 1)$
   $E(3, 4)$ $B(2, 4)$ $C(5, 4)$ $H(4, 4)$
   $G(4, 6)$ $F(3, 6)$ $L(5, 7)$ $M(4, 7)$
   Use a ruler to draw $\overline{AB}$, $\overline{BC}$, $\overline{CD}$, $\overline{EF}$, $\overline{FG}$, $\overline{HJ}$, $\overline{JK}$, $\overline{KL}$ and $\overline{LM}$.

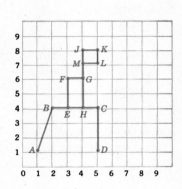

320

Practice

Have students complete the page independently.

Mixed Practice

1. $16,008 - 12,973$ (3,035)
2. 2.3×8.5 (19.55)
3. $1,278 \times 9$ (11,502)
4. $\frac{2}{5}$ of 15 (6)
5. $959 \div 72$ (13 R23)
6. $32.76 + 21.204$ (53.964)
7. Simplify: $\frac{25}{35}$ $\left(\frac{5}{7}\right)$
8. $9.5 - 6.238$ (3.262)
9. $3\frac{1}{4} + 7\frac{1}{3} + 8\frac{1}{2}$ $\left(19\frac{1}{12}\right)$
10. $\frac{8}{9} - \frac{5}{6}$ $\left(\frac{1}{18}\right)$

320

Probability

Objective

To understand probability

Materials

*paper bag
*red, yellow, green, blue crayons
*math and reading books

Mental Math

Have students multiply by 6, divide by 3 and double the numbers:

1. 6 (24)
2. 5 (20)
3. 4 (16)
4. 10 (40)
5. 12 (48)
6. 50 (200)
7. 40 (160)
8. 15 (60)

Skill Review

Review ratio by displaying 4 math books and 6 reading books. Remind students that the ratio of math books to reading books is 4 to 6 or 4/6. Ask the ratio of reading books to math books (6/4), reading books to all the books (6/10), etc.

Understanding Probability

Jean is working a **probability** experiment for her class by drawing numbers from a hat. Each number is equally likely to be drawn. What is the probability that she will pick a 3?

We want to know how likely it is that Jean will pick a 3 from the hat.

There are __5__ papers in the hat.

The papers are numbered __1__, __2__, __3__, __4__ and __5__.

Each number that is drawn is called an **outcome**. Since there are five possible outcomes in this experiment, we say the probability of Jean picking a 3, the first time, is **1 out of 5.**

The __1__ in this probability tells how many 3's there are in the hat.

The __5__ tells how many numbers there are altogether in the hat.

The probability of Jean picking a 3 is __1 out of 5__.

Getting Started

Answer these questions based on the papers in the hat.

1. What is the probability of drawing a 2?

   __1 out of 5__

2. What is the probability of drawing a 5?

   __1 out of 5__

3. What is the probability of drawing an odd number?

   __3 out of 5__

4. What is the probability of drawing a number greater than 2?

   __3 out of 5__

5. What is the probability of drawing a number divisible by 2?

   __2 out of 5__

6. What is the probability of drawing a number greater than 6?

   __0 out of 5__

321

Teaching the Lesson

Introducing the Problem Have students read the problem aloud and tell what is to be found. (the **probability** of Jean picking the number 3) Ask students what information is given in the picture. (There are 5 pieces of paper and each has a different number from 1 through 5.) Have students complete the sentences to solve the problem.

Developing the Skill Tell students that every time a baby is born there is a 1 out of 2 chance it will be a boy and a 1 out of 2 chance it will be a girl. Tell students that a boy birth or a girl birth is called the **outcome** and since there are 2 possible outcomes, we say the probability of a boy is 1 out of 2 and the probability of a girl is 1 out of 2. Show students 1 red, 1 green, 1 blue and 1 yellow crayon. Place the crayons in a paper bag and tell students the chance or probability of picking a yellow crayon from the bag is 1 out of 4, since there is only 1 yellow crayon. Tell students the probability of picking a crayon other than yellow is 3 out of 4 since 3 of the 4 crayons are not yellow. Ask students to tell the probability of picking a blue or green crayon. (2 out of 4) Ask additional probability questions.

Practice

A cube has the letters *A, B, C, D, E* and *F* printed on its faces. The outcome is the letter printed on the top of the cube after a toss.

1. What are the possible outcomes?

   A, B, C, D, E, F

2. How many different outcomes are possible?

   6

3. What is the probability of tossing an A?

   1 out of 6

4. What is the probability of tossing a C?

   1 out of 6

5. What is the probability of tossing a D?

   1 out of 6

6. What is the probability of tossing an X?

   0 out of 6

A spinner is divided into three equal parts numbered 1, 2 and 3. If the spinner lands on a line it doesn't count.

7. What are the possible outcomes?

   1, 2, 3

8. How many different outcomes are possible?

   3

9. What is the probability of spinning a 1?

   1 out of 3

10. What is the probability of spinning a 2 or 3?

    2 out of 3

11. What is the probability of spinning a 5?

    0 out of 3

12. What is the probability of spinning an odd number?

    2 out of 3

322

Students may forget to count all the possibilities for a particular outcome and also always write a probability as "1 out of . . . " When working a problem, have students first count to find all the possible outcomes and write the number; then they can count to find all the outcomes that satisfy the given condition, and write this number. They then should use only the two numbers that have been written to state the probability.

Enrichment

Have students write as many examples of probability in daily living as they can. If possible, they should give the probability of the outcome. For example, the probability of wearing brown shoes is 2 out of 3 if you have 3 pairs and 2 are brown.

Practice

Help students answer the first 2 or 3 questions about the cube and the spinner and then have them complete the page independently.

Extra Credit *Probability*

Give each pair of students a page from a newspaper. Ask them to find a section with 100 words. Have them count the number of letters in each word and record it. Have one student count the word lengths while the other makes a tally.

Then have them list the frequency with which words of differing lengths appeared in the 100 word section. Have them express the probability of finding a two letter word. ($^{10}/_{100}$ in this case)

More Probability

pages 323-324

Objective

To list all possible outcomes

Materials

*6 crayons of different colors
*5 objects

Mental Math

Have students give a ratio for:

1. months to weeks in a year. ($^{12}/_{52}$)
2. weeks to months in a year. ($^{52}/_{12}$)
3. 2's in 10 to 2's in 12. ($^{5}/_{6}$)
4. days in May to days in 1 year. ($^{31}/_{365}$)
5. consonants to vowels. ($^{21}/_{5}$)
6. vowels to consonants. ($^{5}/_{21}$)

Skill Review

Discuss the probability of students' parents saying yes to a movie request. Help students see that yes is 1 out of 3 possible outcomes, with the other 2 being no and maybe. Have students discuss how the probability of a no answer may be 3 out of 3 if the movie is not for children, time does not allow, etc. Discuss other situations of probability in daily living.

Listing Outcomes

Tim is playing a game in which he spins the wheel twice each turn. His move on the board is determined by the two numbers the arrow points to. List all the possible outcomes Tim could spin in one turn.

We want a list of all the possible number combinations Tim could spin.

We know the numbers on the wheel are $\underline{1}$, $\underline{2}$, $\underline{3}$, $\underline{4}$, $\underline{5}$, $\underline{6}$, $\underline{7}$ and $\underline{8}$.

Tim could spin:

1, 1	2, 1	3, 1	4, 1	5, 1	6, 1	7, 1	8, 1
1, 2	2, 2	3, 2	4, 2	5, 2	6, 2	7, 2	8, 2
1, 3	2, 3	3, 3	4, 3	5, 3	6, 3	7, 3	8, 3
1, 4	2, 4	3, 4	4, 4	5, 4	6, 4	7, 4	8, 4
1, 5	2, 5	3, 5	4, 5	5, 5	6, 5	7, 5	8, 5
1, 6	2, 6	3, 6	4, 6	5, 6	6, 6	7, 6	8, 6
1, 7	2, 7	3, 7	4, 7	5, 7	6, 7	7, 7	8, 7
1, 8	2, 8	3, 8	4, 8	5, 8	6, 8	7, 8	8, 8

There are $\underline{64}$ different possible outcomes.

Getting Started

Tracy tosses a nickel and a dime in the air to see if she will get heads or tails. List all the possible outcomes she could get.

H, H   T, H   H, T   T, T

323

Teaching the Lesson

Introducing the Problem Have students read the problem and tell what is to be done. (list all possible outcomes of Tim's spins in 1 turn) Ask what information is known from the problem and the picture. (Tim spins 2 times for 1 turn and the possible numbers are 1 through 6 for each spin.) Together with students, complete the sentence and list to find and count the possible outcomes. Have students check to be sure their solution is complete. Note with students that the systematic recording of the outcomes allows for easier checking.

Developing the Skill Display 4 crayons. Tell students we want to find all the possible outcomes or ways of combining 2 crayons at a time if we replace those 2 each time. Have a student pick 2 crayons and record the outcome on the board. Have the student replace the crayons and continue to combine and record until all outcomes have been found. (6) Add 2 more crayons and have students find all the outcomes. (15) Remind students that a systematic way of finding and recording the outcome prevents error and saves time.

1. Each of the letters in the word COURAGE are written on a piece of paper and put into a box. List all the possible outcomes, if two papers are pulled out at a time. (All papers are returned to the box after each turn.)

C, O	C, G	O, A	U, A	R, G	G, E
C, U	C, E	O, G	U, G	R, E	
C, R	O, U	O, E	U, E	A, G	
C, A	O, R	U, R	R, A	A, E	

2. List all the possible partner combinations for a chess game if Team A plays Team B.

A, D	A, G	B, F	C, E
A, E	B, D	B, G	C, F
A, F	B, E	C, D	C, G

Team A	Team B
Alexi	Danielle
Bruce	Erica
Carl	Flo
	Gerry

3. Complete these diagrams to show all possible combinations of candidates for president and vice-president of the safety squad. The people running for the jobs are Greg, Lee, Kristine, Lin and Heather.

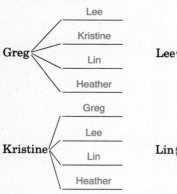

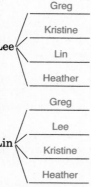

Greg
- Lee
- Kristine
- Lin
- Heather

Kristine
- Greg
- Lee
- Lin
- Heather

Lee
- Greg
- Kristine
- Lin
- Heather

Lin
- Greg
- Lee
- Kristine
- Heather

Heather
- Greg
- Lee
- Kristine
- Lin

Correcting Common Errors

Some students may miss some of the possible outcomes because they are not listing them in an organized way. Have students work with partners with 5 slips of paper numbered 1 through 5. They should pretend that they are selecting two slips from a box. Have them write all the possible pairs of numbers starting with all the numbers that could be paired with 1, then with 2, then 3, then 4, and finally 5.

Enrichment

Tell students that different kinds of ice cream sundaes can be made with 3 flavors of ice cream, 2 kinds of nuts, 4 syrup toppings and 2 kinds of sprinkles. Using one of each ingredient for every sundae, have students list all possible outcomes.

Practice

Remind students to make a systematic recording of their outcomes. Have students complete the page independently.

Mixed Practice

1. 576×25 (14,400)
2. $\frac{7}{8} + \frac{1}{2} \left(1\frac{3}{8}\right)$
3. $\$150.00 - 127.31$ ($22.69)
4. $62,975 + 398$ (63,373)
5. $7 - 3.56$ (3.44)
6. $698 \div 27$ (25 R23)
7. $0.75 + 2.3 + 4.871$ (7.921)
8. $\frac{5}{3} - \frac{5}{6} \left(\frac{5}{6}\right)$
9. 0.8×0.9 (.72)
10. $\$7.52 \div 8$ ($.94)

Extra Credit   *Numeration*

Have students write 3 word problems that include four different mathematical operations each. Have them exchange with a partner to work each other's problems. Extend the activity by having students write a word problem that involves the use of only one operation, used four times.

Problem Solving Review

pages 325-326

Objective

To select and use a strategy to solve a problem

Materials

Mental Math

Ask students if A × B = C, what is the unknown when:

1. A = 20, B = 46 (C = 920)
2. B = 2, C = 18 (A = 9)
3. A = 0.3, B = 0.10 (C = 0.03)
4. C = 45, A = 15 (B = 3)
5. A = B − 2, B = 6 (C = 24)
6. A = 6, B = 2A (C = 72)
7. C = 40, B = 5 (A = 8)
8. A = B (C = any product of a number times itself)

Review

The four-step plan can help us to be better problem solvers. A review of this plan can remind us of ways to use it.

★ **SEE**

We decide what we are looking for. We state all the facts we know that will help us to solve the problem.

★ **PLAN**

We think about the important facts and choose a plan to solve the problem. Some of the things we have learned to do to help us solve problems are:

Guessing and checking
Making an organized list
Acting it out
Looking for a pattern
Making a table
Drawing a picture or diagram
Making a model
Making a tally
Making a graph
Restating the problem
Selecting notations
Writing an open sentence
Using a formula

★ **DO**

We carry out the plan and reach a solution to the problem.

★ **CHECK**

We check the problem for careless errors.
We see if the solution makes sense.
We look for another way to work the problem.

325

Teaching the Lesson

This lesson provides practice in choosing a strategy to solve a problem. Although a strategy is suggested for each practice problem, it is important that students be free to choose their own, since the purpose of problem solving lessons is to teach the use of strategies. Helping students choose a strategy to use changes the emphasis to the solution rather than the strategy itself. It is important to let students consider a problem for a while before receiving assistance.

Have students describe the picture. Read the paragraph beside the picture to the students. Have a student read about the SEE stage aloud. Have a student read the 2 introductory sentences about the PLAN stage and then have individual students read the types of problem solving strategies. Help students give examples of problems they have learned to solve using each strategy. Have a student read the DO and CHECK stages aloud.

Apply

Use one or more of the things you have learned to do to help solve each problem.

1. Angela, Barbara, Cheryl, Dorothy and Eve are in a foot race. At the finish line, Angela is 20 meters behind Barbara. Barbara is 50 meters ahead of Cheryl. Cheryl is 10 meters behind Eve. Dorothy is 30 meters ahead of Angela. In what order did they finish?
   Dorothy, Barbara, Angela, Eve, Cheryl

2. Ling spent half of the money in his pocket for his school lunch. On the way home, he spent half of his remaining money at the music store. Ling then had $1.15 in his pocket. How much money did Ling have in his pocket before lunch?
   $4.60

3. Forgetful Fred was cleaning out his desk and discovered a library book that was 12 days overdue. Write an open sentence that will help find Fred's library debt, if the fine is 8 cents for each of the 12 days.
   96c

4. A school bus starts its morning route and picks up one passenger at the first stop, two passengers at the second stop, three at the third and so on. After the twelfth stop, how many passengers are on the bus?
   78 passengers

5. Suppose you know that it is more likely to rain tomorrow than it is to snow. What do you know is true about the probability of rain and the probability of snow?
   See Solution Notes.

6. A spinner with different colors is used in a game. Suppose the probability of spinning the color red is 3 out of 5. If you spin 20 times, how many times can you expect to spin red?
   See Solution Notes.

7. Fred and Freida are drawing a map of their town on a grid. Fred says, "The school is at (5,3)." Freida says, "No, the school is at (3,5)." What is alike about these number pairs and what is different?
   See Solution Notes.

8. Barney and Betty buy old bicycles and fix them up to sell. They bought a bicycle for $40, used $20 worth of parts to repair it, and then sold it for $80. Rewrite this problem so that they make a profit of $40.
   Answers will vary.

326

Extra Credit *Applications*

Tell students their class is in charge of planning and shopping for the 4th-grade picnic this year. Tell students to plan for lunch, activities with prizes and any supplies necessary, for a group of 55 students. Tell students they must stay within a budget of $200.00. Let students decide the best way to approach the problem, (probably by committee) and list all things they need to buy. Have them use actual store advertisements so they come up with accurate pricing. When they have finished their plans, have them discuss any major problems they were faced with and why a budget is sometimes necessary.

Solution Notes

1. Draw a picture of a number line in units of 10 with A, B, C, D and E representing the runners.
2. Work backwards for $1.15. Double it to $2.30 and double again to $4.60.
3. Write an open sentence of F = 8 × D for fine = 8 cents times the number of days.
4. Look for a pattern or create a table:

Stop	Pickup	Total
1	1	1
2	2	3
3	3	6
.	.	.
.	.	.
12	12	78

Another strategy which students may use is a formula of N(N + 1) ÷ 2.

Higher-Order Thinking Skills

5. [Comprehension] The probability of rain is greater than the probability of snow.
6. [Synthesis] Three out of 5 is the same as 12 out of 20. The student can expect to spin red 12 times although it is not assured that this will happen.
7. [Analysis] The numbers in both pairs are the same, but their order is different and will locate different points on a grid.
8. [Synthesis] Sample answers include paying $20 for the bicycle, or charging $60 for the bicycle, or any other combination of changes that make $40 the difference between costs and sale price.

Calculator Review

Objective

To review calculator use

Materials

calculators

Mental Math

Dictate the following problems:

1. ⅔ yd = (24) in.
2. 6.4 − 3.9 = (2.5)
3. ⅔ of $15 = ($10)
4. ⁹⁄₁₀ of 1 meter (90 cm)
5. 18 miles × 4 miles = (72 sq mi)
6. 10 centuries = (365,000) days
7. 7 ⅚ yd − 2 ⅓ yd = (5 ½ yd)
8. 6 ¾ ft = (81) in.

Skill Review

Remind students to clear their screens before each new problem is entered. Write the following review problems on the board for students to solve on their calculators: **18 + 142 + 64** (224), **539 × 24** (12,936), **36,540 ÷ 812** (45), **$48.20 × 64** ($3,084.80), **2,268 ÷ 54** (42), **5% of $120** ($6), **426 × 6 − 283** (2,273), **⁷⁄₉ of 8,136** (6,328), **84 × 2 × 16** (2,688)

Calculator Review

You have used a calculator to work with the basic operations involving whole numbers, money, fractions and percents. Now let's use your calculator to have some fun.

Activity 1: A Magic Square

Fill in the squares with 7, 14, 21, 28, 42, 49, 56 and 63 so each row, column, and diagonal has a sum of 105.

56	7	42
21	35	49
28	63	14

Arrangements may vary.

Activity 2: Palindromes

A number that is read the same forward or backward is called a **palindrome.** Reverse the digits and add. Continue to reverse and add each new sum until it becomes a palindrome.

Pick a number.	215
Reverse the digit.	512
Add.	727 A palindrome

Find the palindrome for each number.

1. 623   949
2. 807   6,666
3. 285   6,996
4. 518   4,664
5. 374   44,044

Activity 3: Patterns

Find the product of the first three problems. Use the pattern to guess the fourth product. Check your guess with your calculator.

1. 99 × 2 = 198
   99 × 3 = 297
   99 × 4 = 396
   99 × 7 = 693

2. 999 × 2 = 1,998
   999 × 3 = 2,997
   999 × 4 = 3,996
   999 × 7 = 6,993

3. 9,999 × 2 = 19,998
   9,999 × 3 = 29,997
   9,999 × 4 = 39,996
   9,999 × 7 = 69,993

MCP All rights reserved

327

Teaching the Lesson

Have a student read the paragraph aloud. Tell students that 1 of the activities in this lesson is about **palindromes.** Write **206** on the board and tell students to watch what happens when 206 is written backwards and then added to 206. Write **206 + 602** vertically on the board and add for a sum of **808.** Have a student read the sum forward and then backward. Ask what is interesting about this sum. (It is the same forward and backward.) Tell students this is called a palindrome because it reads the same from either end. Write **459** on the board. Have a student write the number backwards under 459 and add. (1413) Ask students if this sum is a palindrome. (no) Have a student write 1413 backward and add to 1413. (4554) Ask if this is a palindrome. (yes) Tell students there were 2 reversals and additions necessary to reach this palindrome. Write **376** on the board and have students find its palindrome and tell how many reversals and additions were needed. (15851, 3) Have students find palindromes for 862 (1441), 526 (2662), 214 (626) and

372 (5115). Students may ask if every number has a palindrome. If not, give more numbers until they see that all numbers do have palindromes but some require many reversals and additions. Finding palindromes is a good exercise for the student who makes careless errors from working too fast.

Students may need help in getting started on the activities but, again, it is beneficial to allow them to consider each for awhile. Help students see a pattern in some of the activities.

Activity 4: Cross-Number Puzzle

Across
1. 56 × 28
3. 12 × 12
5. 494 + 379
6. 1,680 ÷ 112
7. 16,808 − 9,783
8. 25 × 225
10. 64 × 16 × 56
13. 26 × 26 − 294
14. 4,739 + 4,769

Down
1. 39 + 16 + 48
2. 4,225 ÷ 65
3. 12,360 ÷ 12
4. 7,653 − 2,968
6. 219 × 75
8. 9,546 − 3,877
9. 2,938 + 2,640
11. 33,495 ÷ 35 ÷ 29
12. 16,575 ÷ 39

Activity 5: A Game of Markout for 2 People

Use the numbers 12, 15, 21, 36, 48, 64 and 52. Pick two numbers each turn and find their product, using the calculator. Then find your answer on the chart, and mark out the number with your initial. The first to mark out a horizontal or vertical path across the chart wins.

540	441	1,092	180	1,488
1,344	432	576	780	624
3,072	2,704	2,304	252	1,008
768	756	315	1,872	225
1,296	2,496	960	144	720

Activity 6: Mystery Numbers

Find each pair of mystery numbers.

1. Their sum is 70 and their difference is 16.

   43, 27

2. Their sum is 30 and their product is 216.

   18, 12

3. Their difference is 50 and their quotient is 3.

   75, 25

4. Their product is 45 and their quotient is 5.

   15, 3

328

Solution Notes

1. Students should see that repeated subtractions from 105 will lead them to the number combinations.
2. Remind students they may need to reverse and add several times to reach a palindrome.
3. Students should not need any assistance in finding the pattern here.
4. Have students complete the puzzle.
5. Group students in pairs for this activity. You may want to have students copy this chart before beginning the game. This will allow for reworking it once they see that a systematic plan reduces unnecessary calculations. Have students find the least number of calculations necessary. Students may want to create a similar chart for continued play.
6. Students may need help to start this activity. In the first problem tell students that a logical place to begin to find 2 numbers whose sum is 70 is to divide 70 by 2 and work backward.

Extra Credit *Logic*

Remind students about the concept of a **syllogism.** Write the following on the board as an example:

Mary is shorter than Heidi.
Sue is shorter than Mary.
Therefore, Sue is shorter than Heidi.

Put this syllogism on the board for students to complete, and discuss whether or not it is true:

All dogs are terriers.
Spot is a dog.
(Spot is a terrier. Not true because the first statement is false.)

Tell students to write 3 syllogisms with the last step missing, that may be true or false. Have them exchange with a partner to complete and identify each as true or false.

Chapter Test

Item	Objective
1-3	Find the average of a set of numbers (See pages 211-212)
4	Interpret a picture graph (See pages 313-314)
5	Interpret a line graph (See pages 315-316)
6-9	Locate and name number pairs on a grid (See pages 317-318)
10-11	Find probability of taking objects from set of objects (See pages 321-322)

Find the average of each set of numbers.

1. 18, 26 __22__   2. 16, 47, 51 __38__   3. 23, 35, 45, 15, 32 __30__

Use the graphs to answer questions 8 and 9.

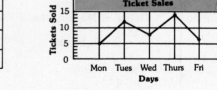

4. How many games did Fremont win? __8__

5. How many tickets were sold on Thursday? __14__

Use the graph to do exercise 6 through 9.

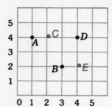

6. Give the letter for (3, 2). __B__

7. Give the number pair for point A. __(1, 4)__

8. Graph and label point C(2, 4).

9. Graph and label point E(4, 2).

Ten lettered pieces of paper are placed in a box. One letter is drawn.

10. What is the probability the letter will be a B?

__4 out of 10__

11. What is the probability the letter will be a C?

__0 out of 10__

329

329

Circle the letter of the correct answer.

1 16,245
   + 45,682

a 51,827
b 61,927
c 62,927
d NG

2 7,046
   − 4,895

a 2,151
b 3,851
c 11,941
d NG

3 36 × 28

a 360
b 908
c 1,308
d NG

4 8)109

a 1 R35
b 13 R5
c 135
d NG

5 Find the perimeter.
38 / 35 \ 38
30 \ 46 / 30

a 215 units
b 217 units
c 225 units
d NG

6 Find the area.
8 in.
6 in.

a 48 in.
b 28 sq in.
c 48 sq in.
d NG

7 $\frac{2}{3}$ of 24 = n
   $n = ?$

a 8
b 12
c 16
d NG

8 $2\frac{2}{3}$
   $+ 3\frac{1}{2}$

a $5\frac{1}{6}$
b $5\frac{3}{5}$
c $6\frac{1}{6}$
d NG

9 $\frac{15}{8}$
   $- \frac{3}{4}$

a $\frac{14}{4}$
b $\frac{1}{8}$
c $1\frac{1}{8}$
d NG

10 7.9 + 9.8

a 1.77
b 17.7
c 177
d NG

11 15.9
   − 8.73

a 7.17
b 7.23
c 7.27
d NG

12 5.9
   × 0.3

a 1.77
b 17.7
c 177
d NG

13 Find the average.
16, 24, 23

a 6
b 20
c 23
d NG

☐ score

330

Cumulative Review

page 330

Item	Objective
1	Add two 5-digit numbers (See pages 51-52)
2	Subtract two 4-digit numbers with zero in minuend (See pages 67-68)
3	Multiply two 2-digit numbers with trading (See pages 179-180)
4	Divide 3-digit number by 1-digit number to get 2-digit quotient with remainder (See pages 197-198)
5	Find perimeter of a non-regular hexagon (See pages 229-230)
6	Find area of a rectangle (See pages 231-232)
7	Find fraction of a number by multiplying and dividing (See pages 245-246)
8	Add two mixed numbers (See pages 277-278)
9	Subtract fractions with uncommon denominators (See pages 275-276)
10	Adding decimals through thousandths (See pages 295-296)
11	Subtract ragged decimals (See pages 297-300)
12	Multiply two decimals (See pages 303-304)
13	Find averages (See pages 211-212)

Alternate Cumulative Review

Choose the letter of the correct answer.

1 27,124
   + 65,983

a 92,107
b 93,007
c 93,107
d NG

2 4,027
   − 2,783

a 1,234
b 1,344
c 2,744
d NG

3 26
   ×45

a 1,070
b 1,170
c 1,180
d NG

4 6)215

a 30 R5
b 35 R5
c 36 R1
d NG

5 Find the perimeter of a rectangle 69 in. by 17 in.

a 172 in.
b 178 in.
c 182 in.
d NG

6 Find the area of a rectangle 7 m by 5 m.

a 35 sq m
b 40 sq m
c 42 sq m
d NG

7 $\frac{2}{5}$ of 75 =

a 15
b 20
c 30
d NG

8 $\frac{14}{9}$
   $- \frac{2}{3}$

a $8\frac{1}{15}$
b $8\frac{4}{5}$
c $9\frac{1}{15}$
d NG

9 $3\frac{2}{5}$
   $+ 5\frac{2}{3}$

a $\frac{8}{9}$
b 1
c $1\frac{1}{9}$

10 6.3 + 4.7 =

a 10
b 11
c 110
d NG

11 17.6 − 9.56 =

a 8.04
b 8.14
c 8.16
d NG

12 6.4
   ×0.5

a 3.20
b 32.0
c 320
d NG

13 Find the average of: 19, 25, 32, 16

a 20
b 22
c 24
d NG

330

Alternate Chapter Test

for page 19

Item	Objective
1-4	Compare and order numbers less than 100 (See pages 1-2)
5-11	Compute basic addition facts (See pages 3-4)
12-16	Add three, four or five 1-digit addends (See pages 5-6)
17-22	Understand grouping and zero properties of addition (See pages 7-8)
23-30	Compute basic subtraction facts (See pages 9-10)
31-34	Find missing addend (See pages 13-14)

Compare these numbers.

1. $9 \; \boxed{<} \; 25$ 2. $36 \; \boxed{<} \; 63$ 3. $47 \; \boxed{<} \; 49$ 4. $88 \; \boxed{>} \; 82$

Add.

5. $6 + 4 = \underline{\;10\;}$ 6. $8 + 8 = \underline{\;16\;}$ 7. $7 + 6 = \underline{\;13\;}$ 8. $6 + 9 = \underline{\;15\;}$

9.
$$\begin{array}{r} 2 \\ + 8 \\ \hline 10 \end{array}$$

10.
$$\begin{array}{r} 7 \\ + 5 \\ \hline 12 \end{array}$$

11.
$$\begin{array}{r} 9 \\ + 8 \\ \hline 17 \end{array}$$

12.
$$\begin{array}{r} 4 \\ 5 \\ + 8 \\ \hline 17 \end{array}$$

13.
$$\begin{array}{r} 3 \\ 7 \\ + 6 \\ \hline 16 \end{array}$$

14.
$$\begin{array}{r} 6 \\ 2 \\ 8 \\ + 7 \\ \hline 23 \end{array}$$

15.
$$\begin{array}{r} 7 \\ 1 \\ 6 \\ + 1 \\ \hline 15 \end{array}$$

16.
$$\begin{array}{r} 2 \\ 6 \\ 3 \\ 4 \\ + 1 \\ \hline 16 \end{array}$$

17. $(6 + 3) + 2 = \underline{\;11\;}$ 18. $3 + (7 + 1) = \underline{\;11\;}$ 19. $(8 + 1) + 4 = \underline{\;13\;}$

20. $4 + (0 + 7) = \underline{\;11\;}$ 21. $(2 + 5) + 8 = \underline{\;15\;}$ 22. $9 + (4 + 7) = \underline{\;20\;}$

Subtract.

23. $8 - 5 = \underline{\;3\;}$ 24. $16 - 7 = \underline{\;9\;}$ 25. $12 - 7 = \underline{\;5\;}$ 26. $13 - 5 = \underline{\;8\;}$

27.
$$\begin{array}{r} 11 \\ - 6 \\ \hline 5 \end{array}$$

28.
$$\begin{array}{r} 14 \\ - 7 \\ \hline 7 \end{array}$$

29.
$$\begin{array}{r} 9 \\ - 5 \\ \hline 4 \end{array}$$

30.
$$\begin{array}{r} 6 \\ - 6 \\ \hline 0 \end{array}$$

Write the missing addends.

31. $4 + n = 7$
$n = \underline{\;3\;}$

32. $n + 6 = 14$
$n = \underline{\;8\;}$

33.
$$\begin{array}{r} 9 \\ + n \\ \hline 12 \end{array}$$
$n = \underline{\;3\;}$

34.
$$\begin{array}{r} 6 \\ + n \\ \hline 12 \end{array}$$
$n = \underline{\;6\;}$

331

Write the place value of the red digits.

1. 4,6<u>2</u>1   2. 11,<u>6</u>24   3. <u>7</u>,243   4. <u>1</u>27,640

    <u>tens</u>   <u>hundreds</u>   <u>thousands</u>   <u>hundred thousands</u>

5. 64,32<u>7</u>   6. <u>4</u>,681,000   7. 1<u>6</u>1,121   8. 34,69<u>1</u>,200

    <u>ones</u>   <u>millions</u>   <u>ten thousand</u>   <u>thousands</u>

Write the numbers.

9. two hundred ninety-five   10. seven thousand, four hundred twenty-nine

    <u>295</u>   <u>7,429</u>

11. six million, four hundred thousand, fifty-one

    <u>6,400,051</u>

Write the missing words.

12. 27,653 twenty-seven <u>thousand</u>, six <u>hundred</u> fifty-three

13. 34,006,126 thirty four <u>million</u>, six <u>thousand</u>, one <u>hundred</u>, twenty-six

Compare these numbers.

14. 7,826 ⊖ 3,271   15. 4,362 ⊘ 4,623   16. 6,327 ⊘ 6,273

Round to the nearest ten.

17. 4,565 <u>4,570</u>   18. 6,212 <u>6,210</u>   19. 9,287 <u>9,290</u>

Round to the nearest hundred or dollar.

20. 657 <u>700</u>   21. 387 <u>400</u>   22. $5.27 <u>$5</u>

23. 4,582 <u>4,600</u>   24. $6.72 <u>$7</u>   25. 349 <u>300</u>

332

Alternate Chapter Test

for page 41

Item	Objective
1, 3, 5	Identify place value of digit in number less than 10,000 (See pages 27-28)
2	Identify place value of digit in number less than 1,000,000 (See pages 35-36)
4, 6, 7, 8	Identify place value in 7-, 8- and 9-digit numbers (See pages 37-38)
9-13	Read and write numbers less than 1 billion (See pages 21-22, 27-28, 35-38)
14-16	Compare and order numbers less than 10,000 (See pages 29-30)
17-19	Round numbers to nearest 10 (See pages 31-34)
20-25	Round numbers to nearest 100 or dollar (See pages 31-34)

Alternate Chapter Test

for page 59

Item	Objective
1-4	Add two 2-digit numbers (See pages 43-44)
5-8	Add two 3-digit numbers (See pages 45-46)
9-16	Add two 4-digit numbers (See pages 47-48)
17-24	Estimate sums up to 4 digits (See pages 49-50)
25-28	Add more than two numbers (See pages 53-54)

Add.

1. 47
  + 32
  ‾‾‾‾
   79

2. 61
  + 17
  ‾‾‾‾
   78

3. 93
  + 28
  ‾‾‾‾
  121

4. 27
  + 84
  ‾‾‾‾
  111

5. 246
  + 221
  ‾‾‾‾‾
   467

6. 302
  + 459
  ‾‾‾‾‾
   761

7. 364
  + 252
  ‾‾‾‾‾
   616

8. 974
  + 668
  ‾‾‾‾‾
  1,642

9. 1,324
  + 3,361
  ‾‾‾‾‾‾
   4,685

10. 5,536
  + 3,528
  ‾‾‾‾‾‾
   9,064

11. 8,317
  + 3,835
  ‾‾‾‾‾‾
  12,152

12. 4,635
  + 7,977
  ‾‾‾‾‾‾
  12,612

13. 6,452
  + 8,693
  ‾‾‾‾‾‾
  15,145

14. 7,958
  + 4,655
  ‾‾‾‾‾‾
  12,613

15. 2,976
  + 5,493
  ‾‾‾‾‾‾
   8,469

16. 9,498
  + 3,784
  ‾‾‾‾‾‾
  13,282

Estimate the sum after rounding to the nearest hundred or dollar.

17. 683
  + 326
  ‾‾‾‾‾
  1,000

18. 179
  + 420
  ‾‾‾‾‾
   600

19. 6,877
  + 2,416
  ‾‾‾‾‾‾
   9,300

20. $23.50
  + 19.64
  ‾‾‾‾‾‾
    $44

Estimate the sum after rounding to the nearest thousand.

21. 6,426
  + 2,397
  ‾‾‾‾‾‾
   8,000

22. 4,956
  + 2,631
  ‾‾‾‾‾‾
   8,000

23. 2,751
  + 3,284
  ‾‾‾‾‾‾
   6,000

24. 6,421
  + 3,791
  ‾‾‾‾‾‾
  10,000

Add.

25. 736
   254
  + 323
  ‾‾‾‾‾
  1,313

26. 6,436
   1,527
  + 3,375
  ‾‾‾‾‾‾
  11,338

27. 13,276
   3,184
  + 27,722
  ‾‾‾‾‾‾‾
  44,182

28. $45.67
   8.23
  + 36.51
  ‾‾‾‾‾‾
  $90.41

333

Subtract and check.

1.   75
    − 62
    ____
      13

2.   47
    − 32
    ____
      15

3.   94
    − 37
    ____
      57

4.   67
    − 48
    ____
      19

5.   659
    − 324
    ____
     335

6.   763
    − 256
    ____
     507

7.   927
    − 565
    ____
     362

8.   715
    − 287
    ____
     428

9.   702
    − 431
    ____
     271

10.   503
    − 218
    ____
     285

11. $6.04
    − 3.48
    ____
    $2.56

12. $7.00
    − 4.95
    ____
    $2.05

13. 8,483
    − 3,674
    _____
    4,809

14. $96.43
    − 48.27
    _____
    $48.16

15. 27,466
    − 19,598
    _____
     7,868

16. $654.18
    − 247.73
    _____
    $406.45

17. 8,005
    − 4,679
    _____
    3,326

18. $60.00
    − 15.49
    _____
    $44.51

19. 50,003
    − 36,357
    _____
    13,646

20. $700.36
     257.38
    _____
    $442.98

Estimate each difference.

Round to the nearest thousand.

21. 6,421
    − 2,716
    _____
    3,000

22. 7,056
    − 2,194
    _____
    5,000

23. 4,809
    − 3,674
    _____
    1,000

24. 9,362
    − 3,212
    _____
    6,000

Round to the nearest hundred.

25. 697
    − 356
    ____
     300

26. 387
    − 126
    ____
     300

27. 428
    − 287
    ____
     100

28. 949
    − 655
    ____
     200

Alternate Chapter Test

for page 79

Item	Objective
1-4	Subtract two 2-digit numbers, check subtraction with addition (See pages 61-62, 73-74)
5-8	Subtract two 3-digit numbers, check subtraction with addition (See pages 63-64, 73-74)
9-12	Subtract when middle digit in minuend is zero, check subtraction with addition (See pages 65-66, 73-74)
13-20	Subtract two 4- or 5-digit numbers, check subtraction with addition (See pages 67-68, 73-74)
21-28	Estimate differences of 3- and 4-digit subtractions (See pages 71-72)

Alternate Chapter Test

for page 101

Item	Objective
1-35	Recall multiplication facts through 9 (See pages 83-94)
36-41	Find value of expression with parentheses (See pages 89-90)
42-47	Find missing factors (See pages 95-96)

Multiply.

1. $\begin{array}{r} 8 \\ \times 6 \\ \hline 48 \end{array}$	2. $\begin{array}{r} 9 \\ \times 0 \\ \hline 0 \end{array}$	3. $\begin{array}{r} 8 \\ \times 3 \\ \hline 24 \end{array}$	4. $\begin{array}{r} 7 \\ \times 7 \\ \hline 49 \end{array}$	5. $\begin{array}{r} 0 \\ \times 8 \\ \hline 0 \end{array}$	6. $\begin{array}{r} 8 \\ \times 9 \\ \hline 72 \end{array}$	7. $\begin{array}{r} 9 \\ \times 3 \\ \hline 27 \end{array}$
8. $\begin{array}{r} 7 \\ \times 8 \\ \hline 56 \end{array}$	9. $\begin{array}{r} 6 \\ \times 2 \\ \hline 12 \end{array}$	10. $\begin{array}{r} 6 \\ \times 6 \\ \hline 36 \end{array}$	11. $\begin{array}{r} 2 \\ \times 6 \\ \hline 12 \end{array}$	12. $\begin{array}{r} 7 \\ \times 9 \\ \hline 63 \end{array}$	13. $\begin{array}{r} 6 \\ \times 7 \\ \hline 42 \end{array}$	14. $\begin{array}{r} 2 \\ \times 7 \\ \hline 14 \end{array}$
15. $\begin{array}{r} 1 \\ \times 6 \\ \hline 6 \end{array}$	16. $\begin{array}{r} 6 \\ \times 5 \\ \hline 30 \end{array}$	17. $\begin{array}{r} 5 \\ \times 7 \\ \hline 35 \end{array}$	18. $\begin{array}{r} 7 \\ \times 6 \\ \hline 42 \end{array}$	19. $\begin{array}{r} 6 \\ \times 9 \\ \hline 54 \end{array}$	20. $\begin{array}{r} 5 \\ \times 9 \\ \hline 45 \end{array}$	21. $\begin{array}{r} 2 \\ \times 8 \\ \hline 16 \end{array}$
22. $\begin{array}{r} 6 \\ \times 3 \\ \hline 18 \end{array}$	23. $\begin{array}{r} 5 \\ \times 5 \\ \hline 25 \end{array}$	24. $\begin{array}{r} 9 \\ \times 6 \\ \hline 54 \end{array}$	25. $\begin{array}{r} 4 \\ \times 7 \\ \hline 28 \end{array}$	26. $\begin{array}{r} 8 \\ \times 4 \\ \hline 32 \end{array}$	27. $\begin{array}{r} 4 \\ \times 5 \\ \hline 20 \end{array}$	28. $\begin{array}{r} 0 \\ \times 0 \\ \hline 0 \end{array}$
29. $\begin{array}{r} 3 \\ \times 5 \\ \hline 15 \end{array}$	30. $\begin{array}{r} 4 \\ \times 9 \\ \hline 36 \end{array}$	31. $\begin{array}{r} 7 \\ \times 4 \\ \hline 28 \end{array}$	32. $\begin{array}{r} 9 \\ \times 5 \\ \hline 45 \end{array}$	33. $\begin{array}{r} 3 \\ \times 6 \\ \hline 18 \end{array}$	34. $\begin{array}{r} 8 \\ \times 8 \\ \hline 64 \end{array}$	35. $\begin{array}{r} 4 \\ \times 8 \\ \hline 32 \end{array}$

Solve for n.

36. $(6 \times 4) + 3 = n$ $n = \underline{27}$

37. $(8 - 5) \times 3 = n$ $n = \underline{9}$

38. $5 \times (3 + 4) = n$ $n = \underline{35}$

39. $30 + (6 \times 5) = n$ $n = \underline{60}$

40. $45 - (9 \times 5) = n$ $n = \underline{0}$

41. $(6 \times 6) - 26 = n$ $n = \underline{10}$

42. $6 \times n = 42$ $n = \underline{7}$

43. $n \times 9 = 45$ $n = \underline{5}$

44. $n \times 4 = 36$ $n = \underline{9}$

45. $n \times 7 = 49$ $n = \underline{7}$

46. $n \times 3 = 24$ $n = \underline{8}$

47. $8 \times n = 0$ $n = \underline{0}$

Write the first nine multiples of each of these numbers.

1. 2   0 , 2 , 4 , 6 , 8 , 10 , 12 , 14 , 16

2. 6   0 , 6 , 12 , 18 , 24 , 30 , 36 , 42 , 48

3. 8   0 , 8 , 16 , 24 , 32 , 40 , 48 , 56 , 64

Skip-count by 9.

4. 27, 36 , 45 , 54 , 63 , 72 , 81 , 90 , 99

Skip-count by 7.

5. 21, 28 , 35 , 42 , 49 , 56 , 63 , 70 , 77

Solve for n.

6. $9 \times (0 \times 2) = n$ 7. $8 \times 6 = n$ 8. $(6 \times 3) + (9 \times 2) = n$

$n = $ 0 $n = $ 48 $n = $ 36

9. $5 \times (7 \times 2) = n$ 10. $(7 \times 0) + (7 \times 0) = n$ 11. $11 \times 7 = n$

$n = $ 70 $n = $ 0 $n = $ 77

Multiply.

12.   43
    × 2
    ───
     86

13.   13
    × 3
    ───
     39

14.   22
    × 4
    ───
     88

15.   11
    × 6
    ───
     66

16.   27
    × 3
    ───
     81

17.   24
    × 4
    ───
     96

18.   13
    × 5
    ───
     65

19.   37
    × 2
    ───
     74

20. $0.46
    ×   5
    ──────
    $2.30

21. $0.63
    ×   7
    ──────
    $4.41

22. $0.84
    ×   8
    ──────
    $6.72

23. $0.49
    ×   6
    ──────
    $2.94

336

Alternate Chapter Test
for page 119

Item	Objective
1-3, 5	Recall basic multiplication facts (See page 104)
6-11	Solve for n while following rule of order (See pages 89-90)
12-15	Multiply 2-digit number by 1-digit number without trading (See pages 107-108)
4, 16-19	Multiply 2-digit number by 1-digit number with one regrouping (See pages 109-110)
20-23	Multiply 2-digit number by 1-digit number with two regroupings (See pages 111-112)

Alternate Chapter Test

for page 141

Item	Objective
1-18	Master division facts using 2-9 as divisor (See pages 121-128)
3, 8	Master special division properties of 1 and 0 (See pages 129-130)
19-22	Solve basic division problems with remainders (See pages 131-132)
23-26	Divide 2-digit number by 1-digit quotient without remainder (See pages 133-134)
27-34	Divide 2-digit number by 1-digit number to get 2-digit quotient with remainder (See pages 135-136)

Divide.

1. $6)\overline{48}$ — 8

2. $7)\overline{28}$ — 4

3. $6)\overline{30}$ — 5

4. $3)\overline{12}$ — 4

5. $2)\overline{18}$ — 9

6. $5)\overline{20}$ — 4

7. $2)\overline{18}$ — 9

8. $9)\overline{45}$ — 5

9. $6)\overline{0}$ — 0

10. $7)\overline{49}$ — 7

11. $72 \div 9 = \underline{8}$

12. $35 \div 5 = \underline{7}$

13. $8 \div 8 = \underline{1}$

14. $40 \div 8 = \underline{5}$

15. $6 \div 6 = \underline{1}$

16. $63 \div 7 = \underline{9}$

17. $72 \div 8 = \underline{9}$

18. $64 \div 8 = \underline{8}$

Divide. Show your work.

19. $6)\overline{50}$ — 8 R2

20. $9)\overline{46}$ — 5 R1

21. $8)\overline{34}$ — 4 R2

22. $7)\overline{37}$ — 5 R2

23. $6)\overline{66}$ — 11

24. $3)\overline{96}$ — 32

25. $4)\overline{96}$ — 24

26. $2)\overline{36}$ — 18

27. $5)\overline{77}$ — 15 R2

28. $4)\overline{98}$ — 24 R2

29. $3)\overline{97}$ — 32 R1

30. $2)\overline{73}$ — 36 R1

31. $6)\overline{62}$ — 10 R2

32. $7)\overline{86}$ — 12 R2

33. $4)\overline{67}$ — 16 R3

34. $8)\overline{86}$ — 10 R6

337

337

Write the times. Include AM or PM.

1. 25 minutes before nine in the morning

8:35 AM

2. 2 hours and 25 minutes after 7:25 PM

9:50 PM

Circle the best estimate for each measurement.

3.

(6 in.)   6 ft   6 mi

4.

7 in.   7 ft   (7 yd)

5.

(1 pt)   1 qt   1 gal

6.

10 in.   (10 yd)   10 mi

7.

(36 in.)   36 ft   36 yd

8.

2 mL   (2 L)   2 kL

9.

(100 g)   100 kg

10.

1 cm   (1 m)   1 km

11.

(50 g)   50 kg

12.

2 kg   (2 T)

338

Alternate Chapter Test

for page 163

Item	Objective
1	Use AM and PM notation; find time by moving minute hand backward (See pages 145-146)
2	Use AM and PM notation; find time by moving hour and minute hands forward (See pages 145-146)
3-4	Determine appropriate customary unit of length (See pages 147-148)
5-6	Determine appropriate customary unit of capacity (See pages 149-150)
7-8	Determine appropriate customary unit of weight (See pages 151-152)
9-10	Determine appropriate metric unit of length (See pages 153-156)
11	Determine appropriate metric unit of capacity (See pages 157-158)
12	Determine appropriate metric unit of weight (See pages 159-160)

338

Alternate Chapter Test

for page 189

Item	Objective
1, 3, 4	Multiply 3-digit number by 1-digit number with trading (See pages 167-168)
2	Multiply 3-digit amount of money by 1-digit number (See pages 169-170)
5-8 22-23	Multiply 4-digit number or money by 1-digit number with or without trading ones, tens, hundreds (See pages 173-174)
9-12	Multiply two 2-digit numbers without trading (See pages 177-178)
13-16	Multiply two 2-digit numbers trading ones (See pages 179-180)
17-20 25, 26, 28	Multiply 3-digit number or money by 2-digit number (See pages 181-182)
21, 24, 27	Multiply money by 2-digit numbers (See pages 181-182)

Multiply. Use estimation to check your answers.

1.
$$\begin{array}{r} 214 \\ \times\ \ 3 \\ \hline 642 \end{array}$$

2.
$$\begin{array}{r} \$5.16 \\ \times\ \ \ \ 5 \\ \hline \$25.80 \end{array}$$

3.
$$\begin{array}{r} 747 \\ \times\ \ 6 \\ \hline 4{,}482 \end{array}$$

4.
$$\begin{array}{r} 637 \\ \times\ \ 5 \\ \hline 3{,}185 \end{array}$$

5.
$$\begin{array}{r} 6{,}307 \\ \times\ \ \ \ \ 4 \\ \hline 25{,}228 \end{array}$$

6.
$$\begin{array}{r} 3{,}652 \\ \times\ \ \ \ \ 8 \\ \hline 29{,}216 \end{array}$$

7.
$$\begin{array}{r} \$27.32 \\ \times\ \ \ \ \ \ 3 \\ \hline \$81.96 \end{array}$$

8.
$$\begin{array}{r} 8{,}456 \\ \times\ \ \ \ \ 5 \\ \hline 42{,}280 \end{array}$$

9.
$$\begin{array}{r} 12 \\ \times 24 \\ \hline 288 \end{array}$$

10.
$$\begin{array}{r} 62 \\ \times 13 \\ \hline 806 \end{array}$$

11.
$$\begin{array}{r} 42 \\ \times 22 \\ \hline 924 \end{array}$$

12.
$$\begin{array}{r} 64 \\ \times 43 \\ \hline 2{,}752 \end{array}$$

13.
$$\begin{array}{r} 83 \\ \times 45 \\ \hline 3{,}735 \end{array}$$

14.
$$\begin{array}{r} 67 \\ \times 76 \\ \hline 5{,}092 \end{array}$$

15.
$$\begin{array}{r} 93 \\ \times 68 \\ \hline 6{,}324 \end{array}$$

16.
$$\begin{array}{r} 37 \\ \times 75 \\ \hline 2{,}775 \end{array}$$

17.
$$\begin{array}{r} 527 \\ \times\ \ 48 \\ \hline 25{,}296 \end{array}$$

18.
$$\begin{array}{r} \$6.27 \\ \times\ \ \ \ 56 \\ \hline \$351.12 \end{array}$$

19.
$$\begin{array}{r} 548 \\ \times\ \ 36 \\ \hline 19{,}728 \end{array}$$

20.
$$\begin{array}{r} 458 \\ \times\ \ 77 \\ \hline 35{,}266 \end{array}$$

21.
$$\begin{array}{r} 752 \\ \times\ \ 63 \\ \hline 47{,}476 \end{array}$$

22.
$$\begin{array}{r} \$6.84 \\ \times\ \ \ \ 39 \\ \hline \$266.76 \end{array}$$

23.
$$\begin{array}{r} 498 \\ \times\ \ 75 \\ \hline 37{,}350 \end{array}$$

24.
$$\begin{array}{r} 596 \\ \times\ \ 87 \\ \hline 51{,}852 \end{array}$$

25.
$$\begin{array}{r} 397 \\ \times\ \ 54 \\ \hline 21{,}438 \end{array}$$

26.
$$\begin{array}{r} 429 \\ \times\ \ 36 \\ \hline 15{,}444 \end{array}$$

27.
$$\begin{array}{r} \$9.89 \\ \times\ \ \ \ 47 \\ \hline \$464.83 \end{array}$$

28.
$$\begin{array}{r} 434 \\ \times\ \ 38 \\ \hline 16{,}492 \end{array}$$

339

Divide. Show your work.

1. $\overset{122}{4\overline{)488}}$ 2. $\overset{222}{3\overline{)666}}$ 3. $\overset{111\ R2}{5\overline{)557}}$ 4. $\overset{131\ R1}{7\overline{)918}}$

5. $\overset{234\ R1}{3\overline{)703}}$ 6. $\overset{400}{2\overline{)800}}$ 7. $\overset{207\ R1}{3\overline{)622}}$ 8. $\overset{208\ R3}{4\overline{)835}}$

9. $\overset{73}{5\overline{)365}}$ 10. $\overset{43}{8\overline{)344}}$ 11. $\overset{72\ R1}{7\overline{)505}}$ 12. $\overset{93}{7\overline{)651}}$

13. $\overset{\$2.06}{3\overline{)\$6.18}}$ 14. $\overset{\$0.82}{9\overline{)\$7.38}}$ 15. $\overset{\$2.74}{3\overline{)\$8.22}}$ 16. $\overset{\$0.61}{6\overline{)\$3.66}}$

17. $\overset{8}{22\overline{)176}}$ 18. $\overset{6\ R8}{42\overline{)260}}$ 19. $\overset{7\ R12}{32\overline{)236}}$ 20. $\overset{4\ R3}{64\overline{)259}}$

21. $\overset{12}{24\overline{)288}}$ 22. $\overset{\$0.21}{35\overline{)\$7.35}}$ 23. $\overset{20\ R13}{47\overline{)953}}$ 24. $\overset{15}{37\overline{)555}}$

Find the average of each set of numbers.

25. 6, 4, 8, 3, 4

$\underline{\quad 5 \quad}$

26. 39, 42, 65, 26

$\underline{\quad 43 \quad}$

27. 74, 68, 116, 92, 45

$\underline{\quad 79 \quad}$

28. 262, 434, 204

$\underline{\quad 300 \quad}$

340

Alternate Chapter Test

for page 217

Item	Objective
1-2	Divide 3-digit number by 1-digit number to get 3-digit quotient without remainder (See pages 191-192)
3-4	Divide 3-digit number by 1-digit number to get 3-digit quotient with remainder (See pages 193-194)
5-8	Divide 3-digit number by 1-digit number to get 3-digit quotient with zero (See pages 195-196)
9-12	Divide 3-digit number by 1-digit number to get 2-digit quotient with or without remainder (See pages 197-198)
13-16	Divide money by 1-digit number (See pages 199-200)
17-20	Divide 3-digit number by 2-digit number to get 1-digit quotient with or without remainder (See pages 201-203)
21-24	Divide 3-digit number by 2-digit number to get 2-digit quotient with or without remainder (See pages 207-209)
25-28	Find the average of a set of numbers (See pages 211-212)

Alternate Chapter Test

for page 239

Item	Objective
1	Identify points, line segments, lines, intersecting lines and parallel lines (See pages 219-220)
2-3	Identify rays, angles and right angles (See pages 221-222)
4-6	Identify polygons by number of sides (See pages 223-224)
7-8	Find perimeter of a polygon (See pages 229-230)
9-10	Find area of rectangles and squares (See pages 231-232)
11-12	Find volume of a box (See pages 235-236)

Write the name for each figure.

1.  X ● ─────── ● Y

2. X ● ─────── ● Z

3. (angle O, M, N)

Ray XY or $\overrightarrow{XY}$ Line XZ or $\overleftrightarrow{XZ}$ Angle MNO or ∠MNO

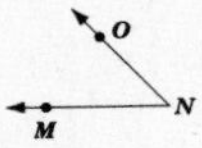

Name each kind of polygon.

4.   5. 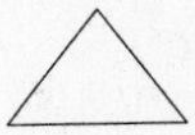   6.

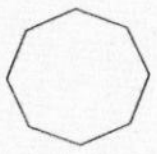

Square   Triangle   Octagon

Find the perimeter.

7.  11 cm   8. 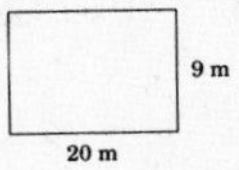 9 m, 20 m

44 cm   58 cm

Find the area.

9. 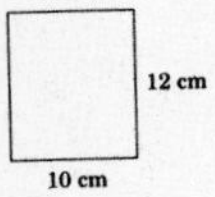 12 cm, 10 cm   10. 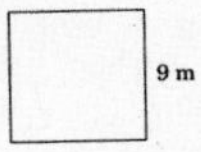 9 m

120 sq cm   81 sq m

Find the volume.

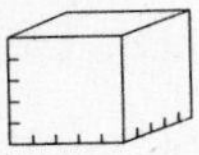

11. 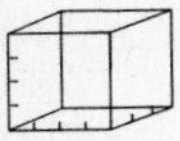   12.

48 cubic units   125 cubic units

341

Write the fraction.

1. What part of the figure is red?
$\frac{3}{4}$

2. What part of the coins are pennies?
$\frac{1}{2}$

Find the part.

3. $\frac{1}{3}$ of 45 = 15

4. $\frac{2}{5}$ of 30 = 12

5. $\frac{3}{4}$ of the price of \$56
$42

Write the missing numerators.

6. $\frac{2}{3} = \frac{6}{9}$

7. $\frac{2}{3} = \frac{8}{12}$

8. $\frac{3}{4} = \frac{6}{8}$

9. $\frac{4}{8} = \frac{12}{24}$

Write < or > in the circle.

10. $\frac{3}{4} \bigcirc > \frac{3}{8}$

11. $\frac{5}{8} \bigcirc > \frac{1}{2}$

12. $\frac{4}{12} \bigcirc < \frac{1}{2}$

13. $\frac{4}{5} \bigcirc > \frac{3}{4}$

Simplify each fraction.

14. $\frac{20}{24} = \frac{5}{6}$

15. $\frac{14}{18} = \frac{7}{9}$

16. $\frac{9}{15} = \frac{3}{5}$

17. $\frac{12}{18} = \frac{2}{3}$

Write each fraction as a mixed number.

18. $\frac{8}{6} = 1\frac{1}{3}$

19. $\frac{9}{2} = 4\frac{1}{2}$

20. $\frac{17}{6} = 2\frac{5}{6}$

21. $\frac{18}{12} = 1\frac{1}{2}$

Measure the figures to the nearest quarter inch.

22. 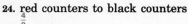 $2\frac{1}{4}$

23. $1\frac{3}{4}$

Write the comparisons as fractions.

24. red counters to black counters
$\frac{4}{8}$

25. black counters to all counters
$\frac{8}{12}$

342

Alternate Chapter Test

for page 263

Item	Objective
1-2	Write a fraction for part of whole or part of set (See pages 241-242)
3	Find unit fraction of a number (See pages 243-244)
4-5	Find a fraction of number by multiplying and dividing (See pages 245-246)
6-9	Find equivalent fractions by multiplying (See pages 249-250)
10-13	Compare and order fractions (See pages 251-252)
14-17	Reduce a fraction to lowest terms (See pages 253-254)
18-21	Write an improper fraction as mixed number (See pages 255-256)
22-23	Write mixed numbers for lengths measured to nearest ¼ inch (See pages 257-258)
24-25	Use ratios to compare two quantities (See pages 259-260)

Alternate Chapter Test

for page 285

Item	Objective
1, 3, 6-8	Add fractions with common denominators (See pages 265-266)
2, 4, 5	Subtract fractions with common denominators (See pages 267-268)
10, 14	Subtract fractions when one denominator is a multiple of other denominator (See pages 273-274)
9, 11-13 15-16, 21-22	Add or subtract two fractions with uncommon denominators (See pages 275-276)
17-20, 23	Add two mixed numbers (See pages 277-278)
24	Subtract two mixed numbers (See pages 273-274)

Add or subtract. Simplify your answers.

1. $\dfrac{3}{7} + \dfrac{2}{7} = \dfrac{5}{7}$

2. $\dfrac{6}{8} - \dfrac{3}{8} = \dfrac{3}{8}$

3. $\dfrac{7}{9} - \dfrac{3}{9} = \dfrac{4}{9}$

4. $\dfrac{7}{10} - \dfrac{3}{10} = \dfrac{2}{5}$

5. $\dfrac{10}{11} - \dfrac{7}{11} = \dfrac{3}{11}$

6. $\dfrac{3}{8} + \dfrac{3}{8} = \dfrac{3}{4}$

7. $\dfrac{5}{6} - \dfrac{2}{6} = \dfrac{1}{2}$

8. $\dfrac{3}{6} - \dfrac{1}{6} = \dfrac{1}{3}$

9. $\dfrac{1}{6} + \dfrac{1}{4} = \dfrac{5}{12}$

10. $\dfrac{5}{12} - \dfrac{1}{4} = \dfrac{1}{6}$

11. $\dfrac{10}{14} + \dfrac{2}{7} = \dfrac{3}{7}$

12. $\dfrac{6}{8} + \dfrac{8}{12} = 1\dfrac{5}{12}$

13. $\dfrac{9}{12} - \dfrac{3}{8} = \dfrac{3}{8}$

14. $\dfrac{6}{8} - \dfrac{3}{16} = \dfrac{9}{16}$

15. $\dfrac{3}{4} + \dfrac{5}{6} = 1\dfrac{7}{12}$

16. $\dfrac{11}{9} - \dfrac{2}{6} = \dfrac{8}{9}$

17. $4\dfrac{4}{9} + 1\dfrac{6}{18} = 5\dfrac{7}{9}$

18. $6\dfrac{4}{6} + 3\dfrac{1}{4} = 9\dfrac{11}{12}$

19. $24\dfrac{2}{8} + 16\dfrac{3}{4} = 41$

20. $3\dfrac{3}{5} + 8\dfrac{1}{3} = 11\dfrac{14}{15}$

21. $\dfrac{2}{3} + \dfrac{1}{5} = \dfrac{13}{15}$

22. $\dfrac{3}{4} - \dfrac{1}{7} = \dfrac{17}{28}$

23. $2\dfrac{1}{6} + 3\dfrac{3}{4} = 5\dfrac{11}{12}$

24. $8\dfrac{1}{2} - 3\dfrac{1}{6} = 5\dfrac{1}{3}$

What is the value of the 4 in each number?

1. 27.43   2. 621.84   3. 164.32   4. 297.364

<u> tenths </u>   <u> hundredths </u>   <u> ones </u>   <u> thousandths </u>

What is the value of the 7 in each number?

5. 937.63   6. 432.71   7. 562.437   8. 263.07

<u> ones </u>   <u> tenths </u>   <u> thousandths </u>   <u> hundredths </u>

Write <, = or > in the circle.

9. 2.97 $>$ 2.79   10. 6.19 $>$ 6.1   11. 0.246 $<$ 2.46   12. 9.04 $<$ 9.40

13. 6.7 $>$ 6.07   14. 0.15 $>$ 0.051   15. 8.20 $=$ 8.2   16. 72.08 $<$ 72.1

Add.

17.   5.4
  + 2.7
    8.1

18.   17.36
 + 36.255
   53.615

19.   47.3
 + 84.91
  132.21

20.   86.24
 + 19.8
  106.04

Subtract.

21.   6.4
  − 3.1
    3.3

22.   26.147
 − 15.369
   10.778

23.   47.6
 − 28.29
  19.31

24.   67.13
 − 15.4
  51.73

Multiply.

25.   4.5
  ×   4
    18

26.   23.4
 × 0.3
  7.02

27.   15.3
 × 0.7
  10.71

28.   7.6
 × 3.4
  25.84

29. 32.6
 × 4.3
 140.18

30.   8.9
 × 4
 35.6

31.   6.1
 × 0.05
 0.305

32. 0.07
 ×   8
 0.56

344

Alternate Chapter Test

for page 309

Item	Objective
1-8	Identify place value of digit in a decimal through thousandths (See pages 291-292)
9-16	Compare and order decimals through thousandths (See pages 293-294)
17	Add decimals through thousandths (See pages 295-296)
18-20	Add ragged decimals (See pages 299-300)
21	Subtract decimals through thousandths (See pages 297-298)
22-24	Subtract ragged decimals (See pages 299-300)
25, 30, 32	Multiply a whole number and decimal (See pages 301-302)
26-29, 31	Multiply two decimals (See pages 303-304)

344

Alternate Chapter Test

for page 329

Item	Objective
1-3	Find the average of a set of numbers (See pages 211-212)
4	Interpret a picture graph (See pages 313-314)
5	Interpret a line graph (See pages 315-316)
6-9	Locate and name number pairs on a grid (See pages 317-318)
10-11	Find probability of taking objects from set of objects (See pages 321-322)

Find the average of each set of numbers.

1. 32, 46 _39_   2. 24, 27, 72 _41_   3. 15, 21, 42, 62 _35_

Use the graphs to answer questions 4 and 5.

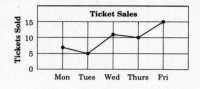

4. How many games did the Rockets win? _10_

5. How many tickets were sold on Tuesday? _5_

Use the graph to do exercise 6 through 9.

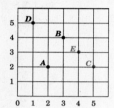

6. Give the letter for (2,2). _A_

7. Give the number for point B. _(3,4)_

8. Graph and label point C (5,2).

9. Graph and label point E (4,3).

Ten lettered pieces of paper are placed in a box. One letter is drawn.

10. What is the probability the letter will be an A?
    3 out of 10

11. What is the probability the letter will be an O?
    1 out of 10

Glossary

Addend A number that is added to another number.

Angle The figure made by two straight lines that meet at one endpoint, or vertex.

Area The measure of a surface surrounded by a boundary.
The shaded part of the square is its area.

Average The number obtained by adding two or more quantities and dividing by the number of quantities added.
The average of 2, 5 and 11 is 6;
$2 + 5 + 11 = 18; 18 \div 3 = 6$

Bar graph A representation of numerical facts using lengths of bars to show information.

Calculator code A set of numbers and symbols representing the order for pressing keys on a calculator keyboard.

Centimeter (cm) A metric unit of length.
100 centimeters = 1 meter

Common denominator Denominator that is a multiple of two or more denominators.

Congruent figures Figures of exactly the same size and shape.

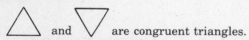
and are congruent triangles.

Congruent sides Line segments of equal length.

Cup (c) A customary unit of liquid measure.
1 cup = 8 ounces

Customary units Standard measures of length, weight, volume and capacity.
Inches, miles, pounds, cubic feet and ounces are examples of customary units.

Decimal A fractional part that uses place value and a decimal point to show tenths, hundredths and so on.
0.6 is the decimal equivalent of the fraction $\frac{3}{5}$.

Decimeter (dm) A metric unit of length.
10 decimeters = 1 meter

Denominator The number below the line in a fraction.
In $\frac{3}{5}$, 5 is the denominator.

Diagram A drawing, chart or figure used to illustrate an idea.

Difference The answer in a subtraction problem.

Digit Any one of the ten number symbols: 0, 1, 2, 3, 4, 5, 6, 7, 8 and 9.

Dividend The number that is being divided in a division problem.
In $42 \div 7 = 6$, 42 is the dividend.

Divisor The number that is being divided into the dividend.
In $42 \div 7 = 6$, 7 is the divisor.

Edge A segment that is the side of a face on a solid figure.

Entry A number or symbol recorded on a calculator.

Endpoint A point at the end of a segment or ray.

Equivalent fractions Fractions that name the same number.

Face A plane figure making up part of a solid figure.

Fact family The set of four related facts having the same three numbers in their equations.

Factor A number to be multiplied.
In $2 \times 3 = 6$, both 2 and 3 are factors.

Foot (ft) A customary unit of length.
1 foot = 12 inches

Formula A general rule expressed using symbols.

Fraction A number that names a part of a whole.
$\frac{1}{2}$ is a fraction.

Gallon (gal) A customary unit of liquid measure.
1 gallon = 4 quarts or 8 pints

Gram (g) A basic metric unit of weight.
1000 grams = 1 kilogram

Graphing Drawing a picture of relationships among numbers and quantities.

Grouping property of addition When the grouping of 3 or more addends is changed, the sum remains the same.
$(2 + 5) + 1 = 2 + (5 + 1)$

Grouping property of multiplication   When the grouping of 3 or more factors is changed, the product remains the same.

$$(5 \times 3) \times 2 = 5 \times (3 \times 2)$$

Inch (in.)   A customary unit of length.

12 inches = 1 foot

Intersect   To meet and cross over at a point.

Line AB intersects
line CD at point P.

Kilogram (kg)   A metric unit of weight.

1 kilogram = 1,000 grams

Kilometer (km)   A metric unit of length.

1 kilometer = 1,000 meters

Least common multiple (LCM)   The smallest number that is a common multiple of two or more numbers.

The LCM of 4 and 6 is 12.

Line   A set of points whose straight path extends indefinitely in opposite directions.

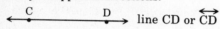
line CD or $\overleftrightarrow{CD}$

Line graph   A representation of numerical facts using points and lines on a grid to show information.

Liter (L)   A basic metric unit of liquid measure.

1 liter = 1,000 milliliters

Meter (m)   A basic metric unit of length.

1 meter = 1,000 millimeters

Metric system   A decimal system of weights and measures whose basic units are the meter, liter and gram.

Mile (mi)   A customary unit of length.

1 mile = 5, 280 feet

Milliliter (mL)   A metric unit of liquid measure.

1 milliliter = $\frac{1}{1000}$ liter

Minuend   A number or quantity from which another is subtracted.

In 18 − 5 = 13, 18 is the minuend.

Mixed number   A fractional number greater than 1 that is written as a whole number and a fraction.

$5\frac{2}{3}$ is a mixed number.

Multiple   The product of a particular number and another number.

8 is a multiple of 2.

Multiplication-addition property   The property that allows the first factor to be multiplied separately with each addend whose sum represents the second factor.

$$5 \times (4 + 6) = (5 \times 4) + (5 \times 6)$$
$$5 \times 10 \quad = 20 \quad + 30$$
$$50 \quad = 50$$

Multiples of 10   The product of any factor multiplied by 10.

Number pair   Two numbers that define one point on a grid; the first number names the distance across, and the second names the distance up.

Number sentence   Numbers and symbols that express an equation or inequality.

3 + 4 = 7; 10 > 2

Numerator   The number above the line in a fraction.

In $\frac{3}{5}$, 3 is the numerator.

One property of multiplication   Any factor multiplied by one equals the original factor.

Order property of addition   The order of the addends does not change the sum.

5 + 7 = 7 + 5

Order property of multiplication   The order of the factors does not change the product.

3 × 4 = 4 × 3

Ounce (oz)   A customary unit of weight.

16 ounces = 1 pound

Outcome   The single result in a probability experiment.

Palindrome   A number which reads the same backward and forward.

838 and 61,416 are palindromes.

Parallel lines   Lines in the same plane that do not meet.

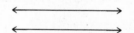

Parallelogram   A quadrilateral having two pairs of opposite, congruent, parallel sides.

Percent   A word meaning hundredths.

37 percent is written 37%

and means 0.37 or $\frac{37}{100}$.

Perimeter   The distance around a shape that is the sum of the lengths of all of its sides.

Picture graph   A representation of numerical facts using pictures.

Pint (pt)   A customary unit of liquid measure.

1 pint = 16 ounces or 2 cups

Plane figure   A shape that appears on a flat surface.

A circle, square, and triangle are plane figures.

Polygon   A many-sided plane figure.

Pound (lb)   A customary unit of weight.

1 pound = 16 ounces

Probability   A number which tells how likely it is that a certain event will happen.

Product   The answer to a multiplication problem.
   In 4 × 5 = 20, 20 is the product.

Quadrilateral   A four-sided plane figure.

Quart (qt)   A customary unit of liquid measure.
   1 quart = 32 ounces   4 quarts = 1 gallon

Quotient   The answer to a division problem.
   In 7)$\overset{9}{63}$, 9 is the quotient.

Ratio   A comparison of two quantities.
   The ratio of 3 to 4 can be written $\frac{3}{4}$.

Ray   A part of a line having one endpoint.
   E •————————→ F   ray EF or $\overrightarrow{EF}$

Rectangle   A four-sided plane figure with two pairs of opposite congruent sides and four right angles.

Remainder   The number left over in a division problem.

Right angle   An angle with the same shape as the vertex of a square; 90 degrees.

   In this square, [90°] all angles are right angles.

Rounding   Estimating a number's value by raising or lowering any of its place values.

Rule of order   The order for working operations in a mathematical sentence.
   Operations within parentheses should be worked before multiplication.
   Work all multiplications left to right.
   Work all additions and subtractions left to right.

Segment   A part of a line having two endpoints.
   A————————B   segment AB or $\overline{AB}$

Similar figures   Plane figures that have the same shape.

Simplify   To rename a fraction as an equivalent fraction or mixed number whose numerator and denominator cannot be divided by any common factor other than 1.

simplest form	simplest form
$\frac{12}{26} = \frac{1}{3}$	$\frac{36}{10} = 3\frac{6}{10} = 3\frac{3}{5}$

   Also, to rename an improper fraction as an equivalent whole number.
   simplest form
   $\frac{35}{7} = 5$

Skip-counting   Naming the multiples of a number one after another.

Solid figure   A figure that is in more than one plane.

Square   A plane figure with four congruent sides and four right angles.

Subtrahend   The number that is subtracted from the minuend.
   In 18 − 5 = 13, 5 is the subtrahend.

Sum   The answer to an addition problem.
   In 8 + 9 = 17, 17 is the sum.

Tally   Marks used to count by fives. ⱧⱧ ⱧⱧ |||

Terms   The numerator and the denominator of a fraction.

Ton (T)   A customary unit of weight.
   1 ton = 2,000 pounds

Total   The answer to an addition problem, the same as the sum.

Triangle   A three-sided plane figure.

Unit cost   The cost of a single base unit of measurement.

Unit fraction   A fraction whose numerator is 1.
   $\frac{1}{5}$ is a unit fraction.

Vertex (pl. vertices)   The point at which two sides of an angle, two sides of a plane figure, or three or more sides of a solid figure meet.

Volume   The number of cubic units needed to fill a solid figure.
   The volume of this cube is 8 cubic units.

Whole numbers   Those numbers used in counting, including zero.

Yard (yd)   A customary unit of length.
   1 yard = 3 feet or 36 inches

Zero properties of subtraction   When zero is subtracted from a minuend, the difference is the same as the minuend.
   8 − 0 = 8
   Subtracting a number from itself leaves zero.
   9 − 9 = 0

Zero property of addition   When one addend is zero, the sum is the other addend.
   3 + 0 = 3

Zero property of multiplication   When one factor is zero, the product is zero.
   4 × 0 = 0

Index

A

Addends, 3–8, 13–14

Addition
  calculator codes, 57–58
  column, 5–8, 53–54
  decimals, 295–296, 299–300
  estimating, 49–50
  facts, 3–4, 17
  fractions, 265–266, 269–272,
    275–280
  mixed numbers, 277–280
  money, 53–54
  multi-digit numbers, 45–52
  properties, 7–8
  two-digit numbers, 43–44

Angles, 221–222

Area, 231–232

Averages, 211–212

C

Calculators
  bank accounts, 99–100
  codes, 57–58
  comparison shopping, 215–216
  data from an ad, 77–78
  decimal key, 187–188
  division key, 139–140
  fractions, 283–284
  keys, 57–58
  money, 57–58, 77–78, 187–188
  multiplication key, 117–118
  percents, 307–308
  review, 327–328
  unit cost, 215–216

Centimeters, 153–156

Common factors, 253

Common multiples, 275–276

Comparing
  decimals, 293–294
  fractions, 251–252
  whole numbers, 1–2, 29–30

Congruent
  figures, 225–226
  sides, 227–228

Counting
  money, 25–26
  skip, 103–104
  tallies, 311–312

Cubic units, 235–236

Customary measurement
  capacity, 149–150
  length, 147–148
  liquid, 149–150
  weight, 151–152

D

Decimals
  adding, 295–296, 299–300
  comparing, 293–294
  multiplying, 301–304
  place value, 287–292
  subtracting, 297–300

Decimeters, 155–156

Diagrams, 137–138

Division
  averages, 211–212
  calculators, 139–140, 215–216
  checking, 205–206
  divisors,
    one-digit, 121–136, 139–140,
      191–200
    two-digit, 201–210
  facts, 121–130
  larger dividends, 197–198
  money, 199–200, 215–216
  multiples of ten, 201–204
  ones, 129–130
  partial dividends, 209–210
  remainders, 131–132, 135–136,
    193–198, 203–210
  three-digit quotients, 191–196
  trial quotients, 207–208
  two-digit quotients, 133–136,
    197–200, 203–204, 209–210
  zeros in, 129–130, 195–196

E

Estimating
  differences, 71–72
  products, 183–184
  quotients, 207–208
  sums, 49–50

F

Fact families, 11

Feet, 147–148

Four-step plan, 15–16

Fractions
  adding, 265–266, 269–272,
    275–280
  calculator, 283–284
  common denominator, 265–268,
  common multiples, 275–276
  comparing, 251–252
  equivalent, 247–250
  finding a fraction of a number,
    243–246
  in measurement, 257–258
  least common denominator,
    275–276
  simplest terms, 253–254

  mixed numbers, 255–256,
    277–280
  naming, 241–242
  non-unit, 245–256
  ratios, 259–260
  simplifying, 253–256, 265–272,
    277–280
  subtracting, 267–270, 273–276
  uncommon denominators,
    271–280
  unit, 243–244

G

Gallons, 149–150

Geometry
  angles, 221–222
  congruent figures, 225–226
  congruent sides, 225–226
  lines, 219–220
  line segments, 219–220
  plane figures, 223–232
  similar figures, 225–226
  solid figures, 233–236

Grams, 159–160

Graphs
  bar, 311–312
  line, 315–316
  number pairs, 317–320
  picture, 313–314

Grouping, property of addition, 7

Grouping property of
  multiplication, 105

I

Inches, 147–148

K

Kilograms, 159–160

Kilometers, 155–156

L

Lines, 219–220

Liters, 157–158

M

Measuring
  area, 231–232
  capacity, 149–150, 157–158
  customary units, 147–152
  length, 147–148, 153–156
  metric units, 153–160, 229–232
  perimeter, 229–230
  time, 143–146
  volume, 235–236
  weight, 151–152, 159–160

Meters, 155–156

Metric measurement
    capacity, 157–158
    length, 153–156, 229–230
    liquid, 157–158
    weight, 159–160
    volume, 235–236

Miles, 147–148

Milliliters, 157–158

Missing addends, 13–14

Missing factors, 95–96

Money
    adding, 53–54, 57–58, 77–78
    calculators, 77–78, 187–188,
        215–216
    counting, 25–26
    dividing, 199–200, 215–216
    multiplying, 113–114, 169–170
    place value, 23–24
    subtracting, 63–74, 77–78

Multiplication
    calculators, 117–118, 187–188
    decimals, 301–304
    estimation, 183–184
    facts, 81–86, 91–94
    money, 113–114, 169–174,
        181–184, 187–188
    multiples, 103–104
    multiples of ten, 175–176
    of multi-digit numbers,
        167–168, 171–174, 181–182
    of two-digit numbers, 107–112,
        175–180
    powers of 10, 165–166
    properties, 87–88, 105–106

N

Number pairs, 317–320

Numbers
    comparing, 1–2, 29–30
    rounding, 31–34

O

One property of multiplication,
    87

Order of operations, 89–90

Order property of addition, 7

Order property of multiplication,
    87

Ounces, 151–152

P

Parallelograms, 227–228

Percents, 307–308

Perimeter, 229–230

Pints, 149–150

Place Value
    decimals, 287–292
    hundreds, tens, ones, 21–22
    millions, 37–38
    money, 23–24
    ten thousands, hundred
        thousands, 35–36
    thousands, 27–28

Plane figures, 223–232

Points, 219–222

Pounds, 151–152

Powers of ten, 165–166

Probability, 321–324

Problem Solving
    acting it out, 75–76
    drawing a picture or diagram,
        137–138
    guessing and checking, 39–40
    looking for a pattern, 97–98
    making a graph, 213–214
    making a model, 161–162
    making a systematic listing or
        table, 55–56, 115–116
    making a tally, 185–186
    properties, 7, 11
    restating the problem, 237–238
    review, 325–326
    selecting suitable notation,
        261–262
    using a formula, 305–306
    using a 4-step plan, 15–16
    writing an open sentence,
        281–282

Q

Quarts, 149–150

R

Rays, 221–222

Rounding
    money, 31–32, 71–72
    whole numbers, 31–34, 49–50
        71–72

Rule of order, 89–90

S

Segments, 219–220

Similar figures, 225–226

Skip-counting, 103–104

Solid figures, 233–236

Subtraction
    calculators, 77–78, 99–100
    checking, 11–12, 73–74
    decimals, 297–300

estimating, 71–72
facts, 9–12, 18
fractions,
    common denominators,
        267–270
    uncommon denominators,
        273–276
minuends with zeros, 65–66,
    69–70
money, 63–74, 77–78
multi-digit numbers, 67–70
three-digit numbers, 63–66
two-digit numbers, 61–62

T

Time, 143–146

Tons, 151–152

V

Volume, 235–236

Y

Yards, 147–148

Z

Zero
    calculators, 57
    in addition, 7–8
    in division, 129–130, 195–196
    in multiplication, 87–88
    in subtraction, 11–12, 65–66,
        69–70